The Evolutionary
History of Life

Compiled by
Janelle Pounds

Scribbles

Year of Publication 2018

ISBN : 9789352979271

Book Published by

Scribbles

(An Imprint of Alpha Editions)

email - alphaedis@gmail.com

Produced by: PediaPress GmbH
Limburg an der Lahn
Germany
http://pediapress.com/

Contents

History of the Earth

History of Earth

The **history of Earth** concerns the development of planet Earth from its formation to the present day. Nearly all branches of natural science have contributed to understanding of the main events of Earth's past, characterized by constant geological change and biological evolution.

The geological time scale (GTS), as defined by international convention,[1] depicts the large spans of time from the beginning of the Earth to the present, and its divisions chronicle some definitive events of Earth history. (In the graphic: Ga means "billion years ago"; Ma, "million years ago".) Earth formed around 4.54 billion years ago, approximately one-third the age of the universe, by accretion from the solar nebula. Volcanic outgassing probably created the primordial atmosphere and then the ocean, but the early atmosphere contained almost no oxygen. Much of the Earth was molten because of frequent collisions with other bodies which led to extreme volcanism. While Earth was in its earliest stage (Early Earth), a giant impact collision with a planet-sized body named Theia is thought to have formed the Moon. Over time, the Earth cooled, causing the formation of a solid crust, and allowing liquid water on the surface.

The Hadean eon represents the time before a reliable (fossil) record of life; it began with the formation of the planet and ended 4.0 billion years ago. The following Archean and Proterozoic eons produced the beginnings of life on Earth and its earliest evolution. The succeeding eon is the Phanerozoic, divided into three eras: the Palaeozoic, an era of arthropods, fishes, and the first life on land; the Mesozoic, which spanned the rise, reign, and climactic extinction of the non-avian dinosaurs; and the Cenozoic, which saw the rise of mammals. Recognizable humans emerged at most 2 million years ago, a vanishingly small period on the geological scale.

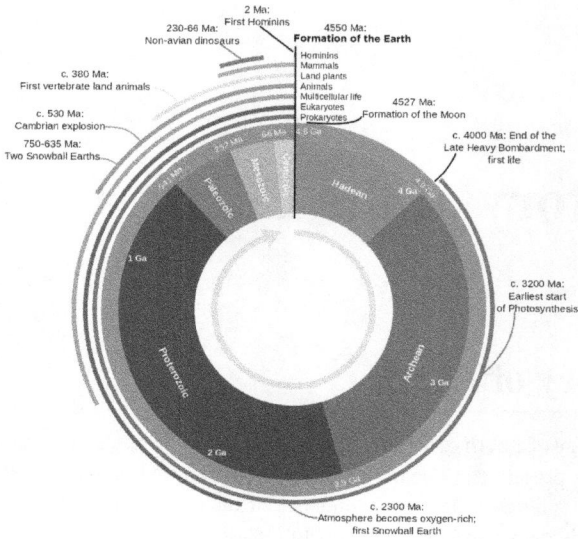

The earliest undisputed evidence of life on Earth dates at least from 3.5 billion years ago, during the Eoarchean Era, after a geological crust started to solidify following the earlier molten Hadean Eon. There are microbial mat fossils such as stromatolites found in 3.48 billion-year-old sandstone discovered in Western Australia. Other early physical evidence of a biogenic substance is graphite in 3.7 billion-year-old metasedimentary rocks discovered in south-western Greenland as well as "remains of biotic life" found in 4.1 billion-year-old rocks in Western Australia.[2] According to one of the researchers, "If life arose relatively quickly on Earth ... then it could be common in the universe."

Photosynthetic organisms appeared between 3.2 and 2.4 billion years ago and began enriching the atmosphere with oxygen. Life remained mostly small and microscopic until about 580 million years ago, when complex multicellular life arose, developed over time, and culminated in the Cambrian Explosion about 541 million years ago. This sudden diversification of life forms produced most of the major phyla known today, and divided the Proterozoic Eon from the Cambrian Period of the Paleozoic Era. It is estimated that 99 percent of all species that ever lived on Earth, over five billion, have gone extinct. Estimates on the number of Earth's current species range from 10 million to 14 million, of which about 1.2 million are documented, but over 86 percent have not been described. However, it was recently claimed that 1 trillion species currently live on Earth, with only one-thousandth of one percent described.

The Earth's crust has constantly changed since its formation, as has life has since its first appearance. Species continue to evolve, taking on new forms, splitting into daughter species, or going extinct in the face of ever-changing physical environments. The process of plate tectonics continues to shape the Earth's continents and oceans and the life they harbor. Human activity is now a dominant force affecting global change, harming the biosphere, the Earth's surface, hydrosphere, and atmosphere with the loss of wild lands, over-exploitation of the oceans, production of greenhouse gases, degradation of the ozone layer, and general degradation of soil, air, and water quality.

Eons

In geochronology, time is generally measured in mya (megayears or million years), each unit representing the period of approximately 1,000,000 years in the past. The history of Earth is divided into four great eons, starting 4,540 mya with the formation of the planet. Each eon saw the most significant changes in Earth's composition, climate and life. Each eon is subsequently divided into eras, which in turn are divided into periods, which are further divided into epochs.

Eon	Time (mya)	Description
Hadean	4,540–4,000	The Earth is formed out of debris around the solar protoplanetary disk. There is no life. Temperatures are extremely hot, with frequent volcanic activity and hellish environments. The atmosphere is nebular. Possible early oceans or bodies of liquid water. The moon is formed around this time, probably due to a protoplanet's collision into Earth.
Archean	4,000–2,500	Prokaryote life, the first form of life, emerges at the very beginning of this eon, in a process known as abiogenesis. The continents of Ur, Vaalbara and Kenorland may have been formed around this time. The atmosphere is composed of volcanic and greenhouse gases.
Protero-zoic	2,500–541	Eukaryotes, a more complex form of life, emerge, including some forms of multicellular organisms. Bacteria begin producing oxygen, shaping the third and current of Earth's atmospheres. Plants, later animals and possibly earlier forms of fungi form around this time. The early and late phases of this eon may have undergone "Snowball Earth" periods, in which all of the planet suffered below-zero temperatures. The early continents of Columbia, Rodinia and Pannotia may have formed around this time, in that order.
Phanero-zoic	541–present	Complex life, including vertebrates, begin to dominate the Earth's ocean in a process known as the Cambrian explosion. Pangaea forms and later dissolves into Laurasia and Gondwana. Gradually, life expands to land and all familiar forms of plants, animals and fungi begin appearing, including annelids, insects and reptiles. Several mass extinctions occur, among which birds, the descendants of dinosaurs, and more recently mammals emerge. Modern animals—including humans—evolve at the most recent phases of this eon.

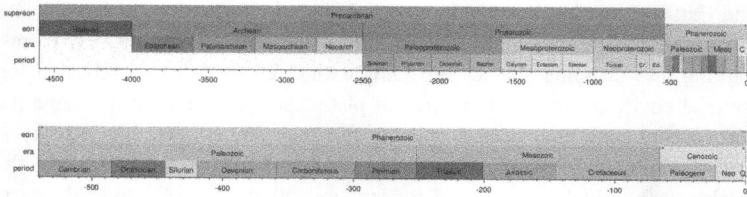

Geologic time scale

The history of the Earth can be organized chronologically according to the geologic time scale, which is split into intervals based on stratigraphic analysis. The following four timelines show the geologic time scale. The first shows the entire time from the formation of the Earth to the present, but this gives little space for the most recent eon. Therefore, the second timeline shows an expanded view of the most recent eon. In a similar way, the most recent era is expanded in the third timeline, and the most recent period is expanded in the fourth timeline.

Millions of Years

Solar System formation

The standard model for the formation of the Solar System (including the Earth) is the solar nebula hypothesis. In this model, the Solar System formed from a large, rotating cloud of interstellar dust and gas called the solar nebula. It was composed of hydrogen and helium created shortly after the Big Bang 13.8 Ga (billion years ago) and heavier elements ejected by supernovae. About 4.5 Ga, the nebula began a contraction that may have been triggered by the shock wave from a nearby supernova. A shock wave would have also made the nebula rotate. As the cloud began to accelerate, its angular momentum, gravity, and inertia flattened it into a protoplanetary disk perpendicular to its axis of rotation. Small perturbations due to collisions and the angular momentum of other large debris created the means by which kilometer-sized protoplanets began to form, orbiting the nebular center.

The center of the nebula, not having much angular momentum, collapsed rapidly, the compression heating it until nuclear fusion of hydrogen into helium began. After more contraction, a T Tauri star ignited and evolved into

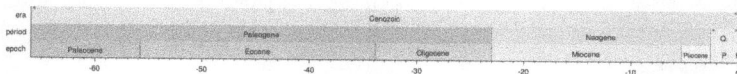

period		Quaternary			
epoch		Pleistocene			Holocene
age	Gelasian	Calabrian		Middle	Late
	-2		-1		0

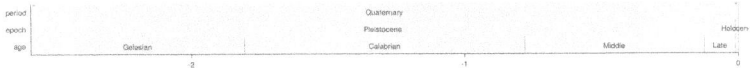

Figure 1: *An artist's rendering of a protoplanetary disk*

the Sun. Meanwhile, in the outer part of the nebula gravity caused matter to condense around density perturbations and dust particles, and the rest of the protoplanetary disk began separating into rings. In a process known as runaway accretion, successively larger fragments of dust and debris clumped together to form planets. Earth formed in this manner about 4.54 billion years ago (with an uncertainty of 1%) and was largely completed within 10–20 million years. The solar wind of the newly formed T Tauri star cleared out most of the material in the disk that had not already condensed into larger bodies. The same process is expected to produce accretion disks around virtually all newly forming stars in the universe, some of which yield planets.

The proto-Earth grew by accretion until its interior was hot enough to melt the heavy, siderophile metals. Having higher densities than the silicates, these metals sank. This so-called *iron catastrophe* resulted in the separation of a primitive mantle and a (metallic) core only 10 million years after the Earth began to form, producing the layered structure of Earth and setting up the formation of Earth's magnetic field.[3] J. A. Jacobs was the first to suggest that the inner core—a solid center distinct from the liquid outer core—is freezing and growing out of the liquid outer core due to the gradual cooling of Earth's interior (about 100 degrees Celsius per billion years).

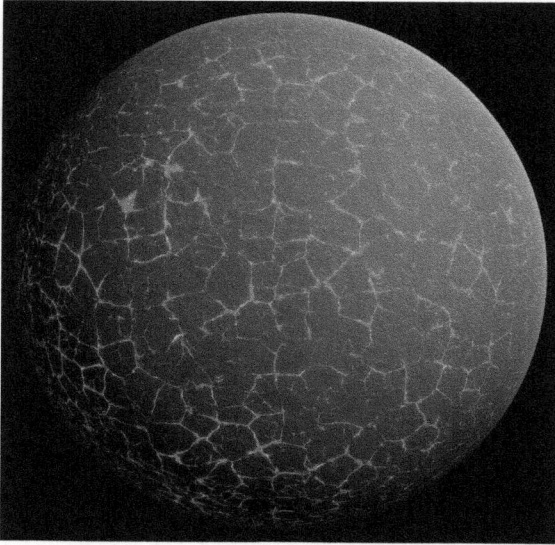

Figure 2: *Artist's conception of Hadean Eon Earth, when
it was much hotter and inhospitable to all forms of life.*

Hadean and Archean Eons

The first eon in Earth's history, the *Hadean*, begins with the Earth's formation
and is followed by the *Archean* eon at 3.8 Ga.[145] The oldest rocks found on
Earth date to about 4.0 Ga, and the oldest detrital zircon crystals in rocks to
about 4.4 Ga, soon after the formation of the Earth's crust and the Earth itself.
The giant impact hypothesis for the Moon's formation states that shortly after
formation of an initial crust, the proto-Earth was impacted by a smaller pro-
toplanet, which ejected part of the mantle and crust into space and created the
Moon.

From crater counts on other celestial bodies, it is inferred that a period of intense meteorite impacts, called the *Late Heavy Bombardment*, began about
4.1 Ga, and concluded around 3.8 Ga, at the end of the Hadean. In addition, volcanism was severe due to the large heat flow and geothermal gradient.
Nevertheless, detrital zircon crystals dated to 4.4 Ga show evidence of having undergone contact with liquid water, suggesting that the Earth already had
oceans or seas at that time.

By the beginning of the Archean, the Earth had cooled significantly. Present
life forms could not have survived at Earth's surface, because the Archean atmosphere lacked oxygen hence had no ozone layer to block ultraviolet light.

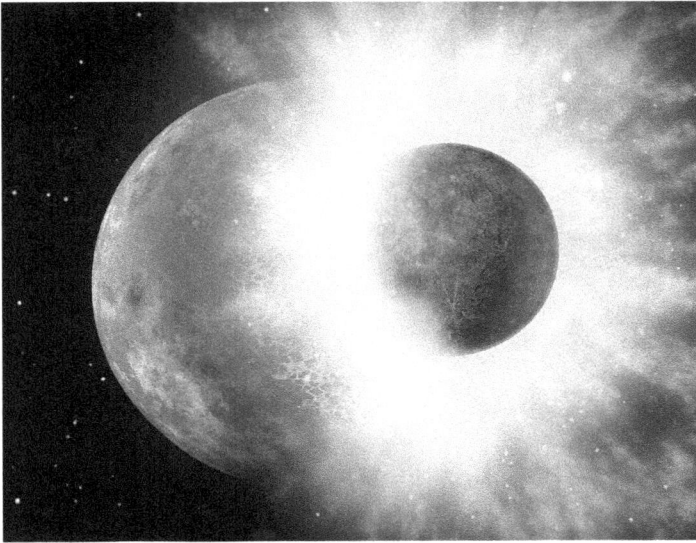

Figure 3: *Artist's impression of the enormous collision that probably formed the Moon*

Nevertheless, it is believed that primordial life began to evolve by the early
Archean, with candidate fossils dated to around 3.5 Ga. Some scientists even
speculate that life could have begun during the early Hadean, as far back as
4.4 Ga, surviving the possible Late Heavy Bombardment period in hydrothermal vents below the Earth's surface.

Formation of the Moon

Earth's only natural satellite, the Moon, is larger relative to its planet than
any other satellite in the solar system.[4] but Pluto is defined as a dwarf
planet.</ref> During the Apollo program, rocks from the Moon's surface were
brought to Earth. Radiometric dating of these rocks shows that the Moon is
4.53 ± 0.01 billion years old, formed at least 30 million years after the solar
system. New evidence suggests the Moon formed even later, 4.48 ± 0.02 Ga,
or 70–110 million years after the start of the Solar System.

Theories for the formation of the Moon must explain its late formation as well
as the following facts. First, the Moon has a low density (3.3 times that of
water, compared to 5.5 for the earth) and a small metallic core. Second, there
is virtually no water or other volatiles on the moon. Third, the Earth and Moon

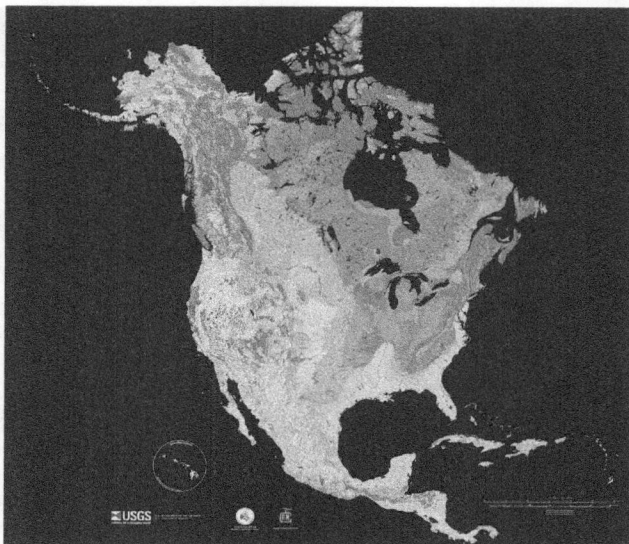

Figure 4: *Geologic map of North America, color-coded by age. The reds and pinks indicate rock from the Archean.*

have the same oxygen isotopic signature (relative abundance of the oxygen isotopes). Of the theories proposed to account for these phenomena, one is widely accepted: The *giant impact hypothesis* proposes that the Moon originated after a body the size of Mars (sometimes named Theia) struck the proto-Earth a glancing blow.[256]

The collision released about 100 million times more energy than the more recent Chicxulub impact that is believed to have caused the extinction of the dinosaurs. It was enough to vaporize some of the Earth's outer layers and melt both bodies.[256] A portion of the mantle material was ejected into orbit around the Earth. The giant impact hypothesis predicts that the Moon was depleted of metallic material, explaining its abnormal composition. The ejecta in orbit around the Earth could have condensed into a single body within a couple of weeks. Under the influence of its own gravity, the ejected material became a more spherical body: the Moon.

First continents

Mantle convection, the process that drives plate tectonics, is a result of heat flow from the Earth's interior to the Earth's surface.[2] It involves the creation of

rigid tectonic plates at mid-oceanic ridges. These plates are destroyed by subduction into the mantle at subduction zones. During the early Archean (about
3.0 Ga) the mantle was much hotter than today, probably around 1,600 °C
(2,910 °F),[82] so convection in the mantle was faster. Although a process similar to present-day plate tectonics did occur, this would have gone faster too.
It is likely that during the Hadean and Archean, subduction zones were more
common, and therefore tectonic plates were smaller.[258]

The initial crust, formed when the Earth's surface first solidified, totally disappeared from a combination of this fast Hadean plate tectonics and the intense
impacts of the Late Heavy Bombardment. However, it is thought that it was
basaltic in composition, like today's oceanic crust, because little crustal differentiation had yet taken place.[258] The first larger pieces of continental crust,
which is a product of differentiation of lighter elements during partial melting
in the lower crust, appeared at the end of the Hadean, about 4.0 Ga. What
is left of these first small continents are called cratons. These pieces of late
Hadean and early Archean crust form the cores around which today's continents grew.

The oldest rocks on Earth are found in the North American craton of Canada.
They are tonalites from about 4.0 Ga. They show traces of metamorphism
by high temperature, but also sedimentary grains that have been rounded by
erosion during transport by water, showing that rivers and seas existed then.
Cratons consist primarily of two alternating types of terranes. The first are
so-called greenstone belts, consisting of low-grade metamorphosed sedimentary rocks. These "greenstones" are similar to the sediments today found in
oceanic trenches, above subduction zones. For this reason, greenstones are
sometimes seen as evidence for subduction during the Archean. The second
type is a complex of felsic magmatic rocks. These rocks are mostly tonalite,
trondhjemite or granodiorite, types of rock similar in composition to granite
(hence such terranes are called TTG-terranes). TTG-complexes are seen as the
relics of the first continental crust, formed by partial melting in basalt.[Chapter 5]

Oceans and atmosphere

Earth is often described as having had three atmospheres. The first atmosphere, captured from the solar nebula, was composed of light (atmophile)
elements from the solar nebula, mostly hydrogen and helium. A combination
of the solar wind and Earth's heat would have driven off this atmosphere, as
a result of which the atmosphere is now depleted of these elements compared
to cosmic abundances. After the impact which created the moon, the molten
Earth released volatile gases; and later more gases were released by volcanoes,
completing a second atmosphere rich in greenhouse gases but poor in oxygen.

Stages

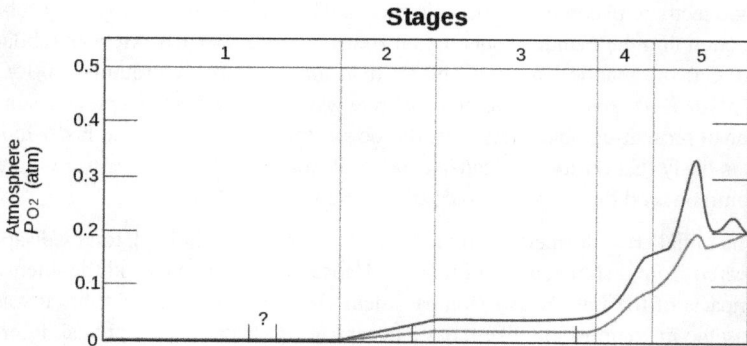

Figure 5: *Graph showing range of estimated partial pressure of atmospheric oxygen through geologic time*

:256 Finally, the third atmosphere, rich in oxygen, emerged when bacteria began to produce oxygen about 2.8 Ga.:[83-84,116-117]

In early models for the formation of the atmosphere and ocean, the second atmosphere was formed by outgassing of volatiles from the Earth's interior. Now it is considered likely that many of the volatiles were delivered during accretion by a process known as *impact degassing* in which incoming bodies vaporize on impact. The ocean and atmosphere would, therefore, have started to form even as the Earth formed. The new atmosphere probably contained water vapor, carbon dioxide, nitrogen, and smaller amounts of other gases.

Planetesimals at a distance of 1 astronomical unit (AU), the distance of the Earth from the Sun, probably did not contribute any water to the Earth because the solar nebula was too hot for ice to form and the hydration of rocks by water vapor would have taken too long. The water must have been supplied by meteorites from the outer asteroid belt and some large planetary embryos from beyond 2.5 AU. Comets may also have contributed. Though most comets are today in orbits farther away from the Sun than Neptune, computer simulations show that they were originally far more common in the inner parts of the solar system.:[130-132]

As the Earth cooled, clouds formed. Rain created the oceans. Recent evidence suggests the oceans may have begun forming as early as 4.4 Ga. By the start of the Archean eon, they already covered much of the Earth. This early formation has been difficult to explain because of a problem known as the faint young Sun paradox. Stars are known to get brighter as they age, and at the time of its formation the Sun would have been emitting only 70% of its current power. Thus, the Sun has become 30% brighter in the last 4.5 billion years.[5] Many models indicate that the Earth would have been covered in ice. A likely solution is that there was enough carbon dioxide and methane to produce a greenhouse effect. The carbon dioxide would have been produced by volcanoes and the methane by early microbes. Another greenhouse gas, ammonia, would have been ejected by volcanos but quickly destroyed by ultraviolet radiation.[83]

Origin of life

Life timeline

```
θ —
  =
500 —
1000 =
1500 =
2000 =
2500 =
3000 =
3500 =
4000 =
4500 —
```

Axis scale: million years

Also see: *Human timeline* and *Nature timeline*

One of the reasons for interest in the early atmosphere and ocean is that they form the conditions under which life first arose. There are many models, but little consensus, on how life emerged from non-living chemicals; chemical systems created in the laboratory fall well short of the minimum complexity for a living organism.

The first step in the emergence of life may have been chemical reactions that produced many of the simpler organic compounds, including nucleobases and amino acids, that are the building blocks of life. An experiment in 1953 by Stanley Miller and Harold Urey showed that such molecules could form in an atmosphere of water, methane, ammonia and hydrogen with the aid of sparks to mimic the effect of lightning. Although atmospheric composition was probably different from that used by Miller and Urey, later experiments with more realistic compositions also managed to synthesize organic molecules. Computer simulations show that extraterrestrial organic molecules could have formed in the protoplanetary disk before the formation of the Earth.

Additional complexity could have been reached from at least three possible starting points: self-replication, an organism's ability to produce offspring that are similar to itself; metabolism, its ability to feed and repair itself; and external cell membranes, which allow food to enter and waste products to leave, but exclude unwanted substances.

Replication first: RNA world

Even the simplest members of the three modern domains of life use DNA to record their "recipes" and a complex array of RNA and protein molecules to "read" these instructions and use them for growth, maintenance, and self-replication.

The discovery that a kind of RNA molecule called a ribozyme can catalyze both its own replication and the construction of proteins led to the hypothesis that earlier life-forms were based entirely on RNA. They could have formed an RNA world in which there were individuals but no species, as mutations and horizontal gene transfers would have meant that the offspring in each generation were quite likely to have different genomes from those that their parents started with. RNA would later have been replaced by DNA, which is more stable and therefore can build longer genomes, expanding the range of capabilities a single organism can have. Ribozymes remain as the main components of ribosomes, the "protein factories" of modern cells.

Although short, self-replicating RNA molecules have been artificially produced in laboratories, doubts have been raised about whether natural non-biological

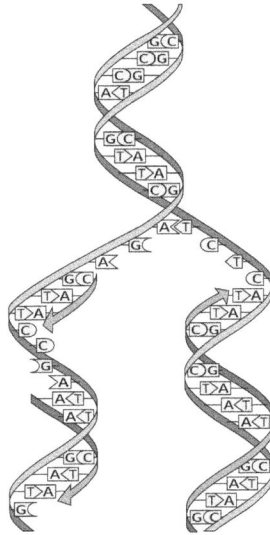

Figure 6: *The replicator in virtually all known life is de-oxyribonucleic acid. DNA is far more complex than the original replicator and its replication systems are highly elaborate.*

synthesis of RNA is possible. The earliest ribozymes may have been formed of simpler nucleic acids such as PNA, TNA or GNA, which would have been replaced later by RNA. Other pre-RNA replicators have been posited, including crystals[:150] and even quantum systems.

In 2003 it was proposed that porous metal sulfide precipitates would assist RNA synthesis at about 100 °C (212 °F) and at ocean-bottom pressures near hydrothermal vents. In this hypothesis, the proto-cells would be confined in the pores of the metal substrate until the later development of lipid membranes.

Metabolism first: iron–sulfur world

Another long-standing hypothesis is that the first life was composed of protein molecules. Amino acids, the building blocks of proteins, are easily synthesized in plausible prebiotic conditions, as are small peptides (polymers of amino acids) that make good catalysts.[:295–297] A series of experiments starting in 1997 showed that amino acids and peptides could form in the presence of carbon monoxide and hydrogen sulfide with iron sulfide and nickel sulfide as catalysts. Most of the steps in their assembly required temperatures of about 100 °C (212 °F) and moderate pressures, although one stage required 250 °C (482 °F) and a pressure equivalent to that found under 7 kilometers (4.3 mi) of rock. Hence, self-sustaining synthesis of proteins could have occurred near hydrothermal vents.

A difficulty with the metabolism-first scenario is finding a way for organisms to evolve. Without the ability to replicate as individuals, aggregates of molecules would have "compositional genomes" (counts of molecular species in the aggregate) as the target of natural selection. However, a recent model shows that such a system is unable to evolve in response to natural selection.

Membranes first: Lipid world

It has been suggested that double-walled "bubbles" of lipids like those that form the external membranes of cells may have been an essential first step. Experiments that simulated the conditions of the early Earth have reported the formation of lipids, and these can spontaneously form liposomes, double-walled "bubbles", and then reproduce themselves. Although they are not intrinsically information-carriers as nucleic acids are, they would be subject to natural selection for longevity and reproduction. Nucleic acids such as RNA might then have formed more easily within the liposomes than they would have outside.

The clay theory

Some clays, notably montmorillonite, have properties that make them plausible accelerators for the emergence of an RNA world: they grow by self-replication of their crystalline pattern, are subject to an analog of natural selection (as the clay "species" that grows fastest in a particular environment rapidly becomes dominant), and can catalyze the formation of RNA molecules. Although this idea has not become the scientific consensus, it still has active supporters.:150–158

Research in 2003 reported that montmorillonite could also accelerate the conversion of fatty acids into "bubbles", and that the bubbles could encapsulate RNA attached to the clay. Bubbles can then grow by absorbing additional lipids and dividing. The formation of the earliest cells may have been aided by similar processes.

A similar hypothesis presents self-replicating iron-rich clays as the progenitors of nucleotides, lipids and amino acids.

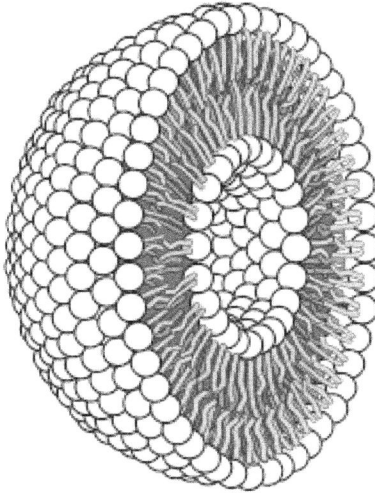

Figure 7: *Cross-section through a liposome*

Last universal ancestor

It is believed that of this multiplicity of protocells, only one line survived. Current phylogenetic evidence suggests that the last universal ancestor (LUA) lived during the early Archean eon, perhaps 3.5 Ga or earlier. This LUA cell is the ancestor of all life on Earth today. It was probably a prokaryote, possessing a cell membrane and probably ribosomes, but lacking a nucleus or membrane-bound organelles such as mitochondria or chloroplasts. Like modern cells, it used DNA as its genetic code, RNA for information transfer and protein synthesis, and enzymes to catalyze reactions. Some scientists believe that instead of a single organism being the last universal common ancestor, there were populations of organisms exchanging genes by lateral gene transfer.

Proterozoic Eon

The Proterozoic eon lasted from 2.5 Ga to 542 Ma (million years) ago.[130] In this time span, cratons grew into continents with modern sizes. The change to an oxygen-rich atmosphere was a crucial development. Life developed from prokaryotes into eukaryotes and multicellular forms. The Proterozoic saw a couple of severe ice ages called snowball Earths. After the last Snowball Earth about 600 Ma, the evolution of life on Earth accelerated. About

Figure 8: *Lithified stromatolites on the shores of Lake Thetis, Western Australia. Archean stromatolites are the first direct fossil traces of life on Earth.*

580 Ma, the Ediacaran biota formed the prelude for the Cambrian Explosion.Wikipedia:Citation needed

Oxygen revolution

The earliest cells absorbed energy and food from the surrounding environment. They used fermentation, the breakdown of more complex compounds into less complex compounds with less energy, and used the energy so liberated to grow and reproduce. Fermentation can only occur in an *anaerobic* (oxygen-free) environment. The evolution of photosynthesis made it possible for cells to derive energy from the Sun.[:377]

Most of the life that covers the surface of the Earth depends directly or indirectly on photosynthesis. The most common form, oxygenic photosynthesis, turns carbon dioxide, water, and sunlight into food. It captures the energy of sunlight in energy-rich molecules such as ATP, which then provide the energy to make sugars. To supply the electrons in the circuit, hydrogen is stripped from water, leaving oxygen as a waste product. Some organisms, including purple bacteria and green sulfur bacteria, use an anoxygenic form of photosynthesis that uses alternatives to hydrogen stripped from water as electron donors; examples are hydrogen sulfide, sulfur and iron. Such extremophile

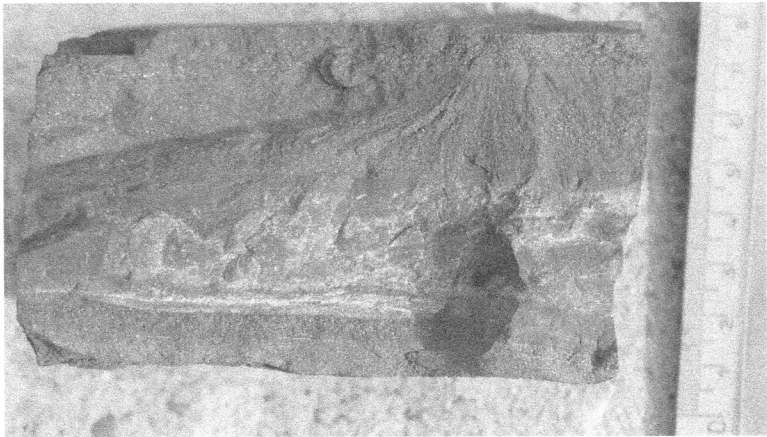

Figure 9: *A banded iron formation from the 3.15 Ga Moories Group, Bar-berton Greenstone Belt, South Africa. Red layers represent the times when oxygen was available; gray layers were formed in anoxic circumstances.*

organisms are restricted to otherwise inhospitable environments such as hot springs and hydrothermal vents.[379–382]

The simpler anoxygenic form arose about 3.8 Ga, not long after the appearance of life. The timing of oxygenic photosynthesis is more controversial; it had certainly appeared by about 2.4 Ga, but some researchers put it back as far as 3.2 Ga. The latter "probably increased global productivity by at least two or three orders of magnitude". Among the oldest remnants of oxygen-producing lifeforms are fossil stromatolites.

At first, the released oxygen was bound up with limestone, iron, and other minerals. The oxidized iron appears as red layers in geological strata called banded iron formations that formed in abundance during the Siderian period (between 2500 Ma and 2300 Ma).[133] When most of the exposed readily reacting minerals were oxidized, oxygen finally began to accumulate in the atmosphere. Though each cell only produced a minute amount of oxygen, the combined metabolism of many cells over a vast time transformed Earth's atmosphere to its current state. This was Earth's third atmosphere.[50–51:83–84,116–117]

Some oxygen was stimulated by solar ultraviolet radiation to form ozone, which collected in a layer near the upper part of the atmosphere. The ozone layer absorbed, and still absorbs, a significant amount of the ultraviolet radiation that once had passed through the atmosphere. It allowed cells to colonize the surface of the ocean and eventually the land: without the ozone layer, ultraviolet radiation bombarding land and sea would have caused unsustainable levels of mutation in exposed cells.[219-220]

Photosynthesis had another major impact. Oxygen was toxic; much life on Earth probably died out as its levels rose in what is known as the *oxygen catastrophe*. Resistant forms survived and thrived, and some developed the ability to use oxygen to increase their metabolism and obtain more energy from the same food.

Snowball Earth

The natural evolution of the Sun made it progressively more luminous during the Archean and Proterozoic eons; the Sun's luminosity increases 6% every billion years.[165] As a result, the Earth began to receive more heat from the Sun in the Proterozoic eon. However, the Earth did not get warmer. Instead, the geological record suggests it cooled dramatically during the early Proterozoic. Glacial deposits found in South Africa date back to 2.2 Ga, at which time, based on paleomagnetic evidence, they must have been located near the equator. Thus, this glaciation, known as the Huronian glaciation, may have been global. Some scientists suggest this was so severe that the Earth was frozen over from the poles to the equator, a hypothesis called Snowball Earth.

The Huronian ice age might have been caused by the increased oxygen concentration in the atmosphere, which caused the decrease of methane (CH_4) in the atmosphere. Methane is a strong greenhouse gas, but with oxygen it reacts to form CO_2, a less effective greenhouse gas.[172] When free oxygen became available in the atmosphere, the concentration of methane could have decreased dramatically, enough to counter the effect of the increasing heat flow from the Sun.

However, the term Snowball Earth is more commonly used to describe later extreme ice ages during the Cryogenian period. There were four periods, each lasting about 10 million years, between 750 and 580 million years ago, when the earth is thought to have been covered with ice apart from the highest mountains, and average temperatures were about –50 °C (–58 °F). The snowball may have been partly due to the location of the supercontinent Rodinia straddling the Equator. Carbon dioxide combines with rain to weather rocks to form carbonic acid, which is then washed out to sea, thus extracting the greenhouse gas from the atmosphere. When the continents are near the poles, the advance of ice covers the rocks, slowing the reduction in carbon dioxide, but in the Cryogienian the weathering of Rodinia was able to continue unchecked until the ice advanced to the tropics. The process may have finally been reversed by the emission of carbon dioxide from volcanoes or the destabilization

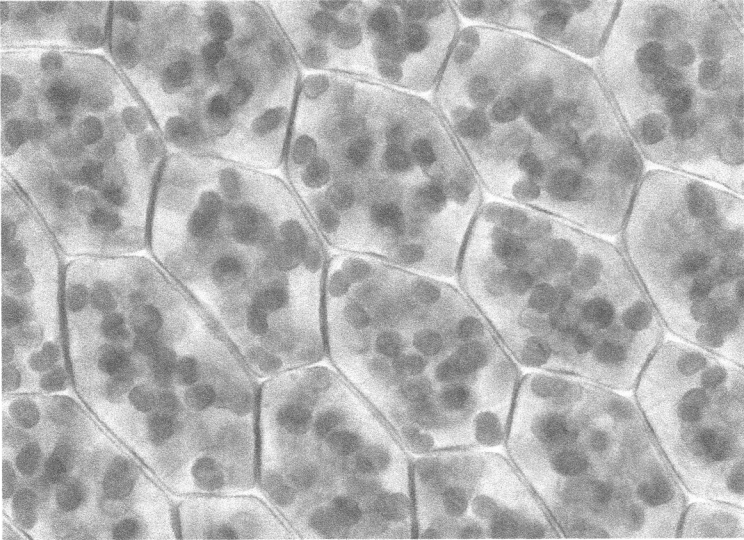

Figure 10: *Chloroplasts in the cells of a moss*

of methane gas hydrates. According to the alternative Slushball Earth theory, even at the height of the ice ages there was still open water at the Equator.

Emergence of eukaryotes

Modern taxonomy classifies life into three domains. The time of their origin is uncertain. The Bacteria domain probably first split off from the other forms of life (sometimes called Neomura), but this supposition is controversial. Soon after this, by 2 Ga, the Neomura split into the Archaea and the Eukarya. Eukaryotic cells (Eukarya) are larger and more complex than prokaryotic cells (Bacteria and Archaea), and the origin of that complexity is only now becoming known. Wikipedia:Citation needed

Around this time, the first proto-mitochondrion was formed. A bacterial cell related to today's *Rickettsia*, which had evolved to metabolize oxygen, entered a larger prokaryotic cell, which lacked that capability. Perhaps the large cell attempted to digest the smaller one but failed (possibly due to the evolution of prey defenses). The smaller cell may have tried to parasitize the larger one. In any case, the smaller cell survived inside the larger cell. Using oxygen, it metabolized the larger cell's waste products and derived more energy. Part of this excess energy was returned to the host. The smaller cell replicated inside the larger one. Soon, a stable symbiosis developed between the large cell and the smaller cells inside it. Over time, the host cell acquired some genes from the smaller cells, and the two kinds became dependent on each other: the larger cell could not survive without the energy produced by the smaller ones, and these, in turn, could not survive without the raw materials provided by the larger cell. The whole cell is now considered a single organism, and the smaller cells are classified as organelles called mitochondria.

A similar event occurred with photosynthetic cyanobacteria entering large heterotrophic cells and becoming chloroplasts.[:60–61:536–539] Probably as a result of these changes, a line of cells capable of photosynthesis split off from the other eukaryotes more than 1 billion years ago. There were probably several such inclusion events. Besides the well-established endosymbiotic theory of the cellular origin of mitochondria and chloroplasts, there are theories that cells led to peroxisomes, spirochetes led to cilia and flagella, and that perhaps a DNA virus led to the cell nucleus, though none of them are widely accepted.

Archaeans, bacteria, and eukaryotes continued to diversify and to become more complex and better adapted to their environments. Each domain repeatedly split into multiple lineages, although little is known about the history of the archaea and bacteria. Around 1.1 Ga, the supercontinent Rodinia was assembling. The plant, animal, and fungi lines had split, though they still existed as solitary cells. Some of these lived in colonies, and gradually a division of labor began to take place; for instance, cells on the periphery might have started to assume different roles from those in the interior. Although the division between a colony with specialized cells and a multicellular organism is not always clear, around 1 billion years ago, the first multicellular plants emerged, probably green algae. Possibly by around 900 Ma[:488] true multicellularity had also evolved in animals.Wikipedia:Citation needed

At first, it probably resembled today's sponges, which have totipotent cells that allow a disrupted organism to reassemble itself.[:483–487] As the division of labor was completed in all lines of multicellular organisms, cells became more specialized and more dependent on each other; isolated cells would die.Wikipedia:Citation needed

Supercontinents in the Proterozoic

Reconstructions of tectonic plate movement in the past 250 million years (the Cenozoic and Mesozoic eras) can be made reliably using fitting of continental margins, ocean floor magnetic anomalies and paleomagnetic poles. No ocean crust dates back further than that, so earlier reconstructions are more difficult. Paleomagnetic poles are supplemented by geologic evidence such as orogenic belts, which mark the edges of ancient plates, and past distributions of flora and fauna. The further back in time, the scarcer and harder to interpret the data get and the more uncertain the reconstructions.[:370]

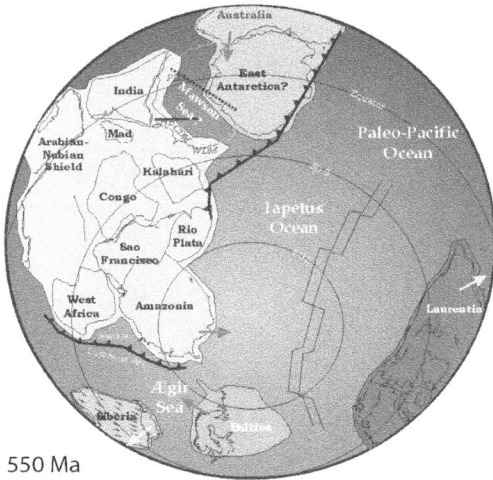

550 Ma

Figure 11: *A reconstruction of Pannotia (550 Ma).*

Throughout the history of the Earth, there have been times when continents collided and formed a supercontinent, which later broke up into new continents. About 1000 to 830 Ma, most continental mass was united in the supercontinent Rodinia.[370] Rodinia may have been preceded by Early-Middle Proterozoic continents called Nuna and Columbia.[374]

After the break-up of Rodinia about 800 Ma, the continents may have formed another short-lived supercontinent around 550 Ma. The hypothetical supercontinent is sometimes referred to as Pannotia or Vendia.[321–322] The evidence for it is a phase of continental collision known as the Pan-African orogeny, which joined the continental masses of current-day Africa, South America, Antarctica and Australia. The existence of Pannotia depends on the timing of the rifting between Gondwana (which included most of the landmass now in the Southern Hemisphere, as well as the Arabian Peninsula and the Indian subcontinent) and Laurentia (roughly equivalent to current-day North America).[374] It is at least certain that by the end of the Proterozoic eon, most of the continental mass lay united in a position around the south pole.

Figure 12: *A 580 million year old fossil of Spriggina floundensi, an animal from the Ediacaran period. Such life forms could have been ancestors to the many new forms that originated in the Cambrian Explosion.*

Late Proterozoic climate and life

The end of the Proterozoic saw at least two Snowball Earths, so severe that the surface of the oceans may have been completely frozen. This happened about 716.5 and 635 Ma, in the Cryogenian period. The intensity and mechanism of both glaciations are still under investigation and harder to explain than the early Proterozoic Snowball Earth. Most paleoclimatologists think the cold episodes were linked to the formation of the supercontinent Rodinia. Because Rodinia was centered on the equator, rates of chemical weathering increased and carbon dioxide (CO_2) was taken from the atmosphere. Because CO_2 is an important greenhouse gas, climates cooled globally.Wikipedia:Citation needed In the same way, during the Snowball Earths most of the continental surface was covered with permafrost, which decreased chemical weathering again, leading to the end of the glaciations. An alternative hypothesis is that enough carbon dioxide escaped through volcanic outgassing that the resulting greenhouse effect raised global temperatures. Increased volcanic activity resulted from the break-up of Rodinia at about the same time.Wikipedia:Citation needed

The Cryogenian period was followed by the Ediacaran period, which was characterized by a rapid development of new multicellular lifeforms. Whether

there is a connection between the end of the severe ice ages and the increase
in diversity of life is not clear, but it does not seem coincidental. The new
forms of life, called Ediacara biota, were larger and more diverse than ever.
Though the taxonomy of most Ediacaran life forms is unclear, some were ancestors of groups of modern life. Important developments were the origin of
muscular and neural cells. None of the Ediacaran fossils had hard body parts
like skeletons. These first appear after the boundary between the Proterozoic
and Phanerozoic eons or Ediacaran and Cambrian periods.Wikipedia:Citation
needed

Phanerozoic Eon

The Phanerozoic is the current eon on Earth, which started approximately 542
million years ago. It consists of three eras: The Paleozoic, Mesozoic, and
Cenozoic, and is the time when multi-cellular life greatly diversified into almost
all the organisms known today.

The Paleozoic ("old life") era was the first and longest era of the Phanerozoic eon, lasting from 542 to 251 Ma. During the Paleozoic, many modern groups of life came into existence. Life colonized the land, first plants,
then animals. Two major extinctions occurred. The continents formed at
the break-up of Pannotia and Rodinia at the end of the Proterozoic slowly
moved together again, forming the supercontinent Pangaea in the late Paleo-
zoic.Wikipedia:Citation needed

The Mesozoic ("middle life") era lasted from 251 Ma to 66 Ma. It is subdivided into the Triassic, Jurassic, and Cretaceous periods. The era began with the Permian–Triassic extinction event, the most severe extinction event in the fossil record; 95% of the species on Earth died out. It
ended with the Cretaceous–Paleogene extinction event that wiped out the di-
nosaurs.Wikipedia:Citation needed.

The Cenozoic ("new life") era began at 66 Ma, and is subdivided into the Paleogene, Neogene, and Quaternary periods. These three periods are further
split into seven sub-divisions, with the Paleogene composed of The Paleocene,
Eocene, and Oligocene, the Neocene divided into the Miocene, Pliocene,
and the Quaternary composed of the Pleistocene, and Holocene. Mammals,
birds, amphibians, crocodilians, turtles, and lepidosaurs survived the Cretaceous–Paleogene extinction event that killed off the non-avian dinosaurs and
many other forms of life, and this is the era during which they diversified into
their modern forms.Wikipedia:Citation needed

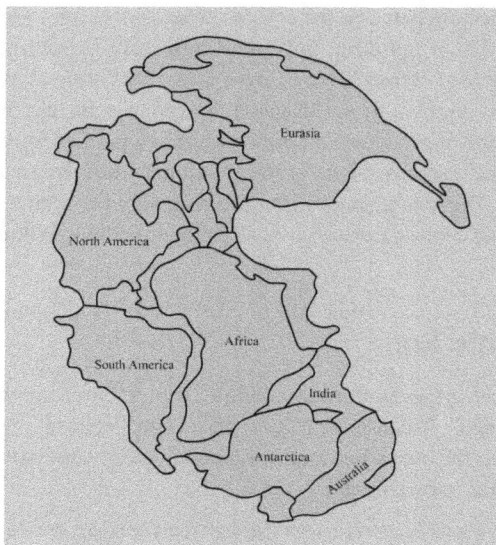

Figure 13: *Pangaea was a supercontinent that existed from about 300 to 180 Ma. The outlines of the modern continents and other landmasses are indicated on this map.*

Tectonics, paleogeography and climate

At the end of the Proterozoic, the supercontinent Pannotia had broken apart into the smaller continents Laurentia, Baltica, Siberia and Gondwana. During periods when continents move apart, more oceanic crust is formed by volcanic activity. Because young volcanic crust is relatively hotter and less dense than old oceanic crust, the ocean floors rise during such periods. This causes the sea level to rise. Therefore, in the first half of the Paleozoic, large areas of the continents were below sea level.Wikipedia:Citation needed

Early Paleozoic climates were warmer than today, but the end of the Ordovician saw a short ice age during which glaciers covered the south pole, where the huge continent Gondwana was situated. Traces of glaciation from this period are only found on former Gondwana. During the Late Ordovician ice age, a few mass extinctions took place, in which many brachiopods, trilobites, Bryozoa and corals disappeared. These marine species could probably not contend with the decreasing temperature of the sea water.

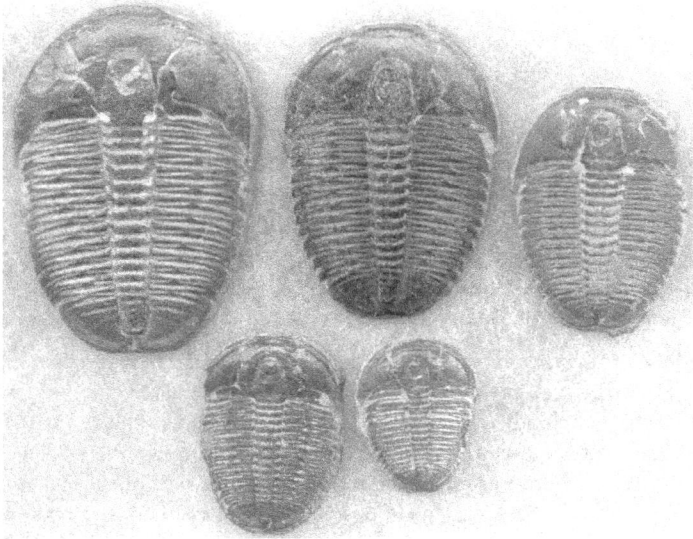

Figure 14: *Trilobites first appeared during the Cambrian period and were among the most widespread and diverse groups of Paleozoic organisms.*

The continents Laurentia and Baltica collided between 450 and 400 Ma, during the Caledonian Orogeny, to form Laurussia (also known as Euramerica). Traces of the mountain belt this collision caused can be found in Scandinavia, Scotland, and the northern Appalachians. In the Devonian period (416–359 Ma) Gondwana and Siberia began to move towards Laurussia. The collision of Siberia with Laurussia caused the Uralian Orogeny, the collision of Gondwana with Laurussia is called the Variscan or Hercynian Orogeny in Europe or the Alleghenian Orogeny in North America. The latter phase took place during the Carboniferous period (359–299 Ma) and resulted in the formation of the last supercontinent, Pangaea.

By 180 Ma, Pangaea broke up into Laurasia and Gondwana.Wikipedia:Citation needed

Cambrian explosion

The rate of the evolution of life as recorded by fossils accelerated in the Cambrian period (542–488 Ma). The sudden emergence of many new species, phyla, and forms in this period is called the Cambrian Explosion. The biological fomenting in the Cambrian Explosion was unpreceded before and since that time.:229 Whereas the Ediacaran life forms appear yet primitive and not easy to put in any modern group, at the end of the Cambrian most modern phyla were already present. The development of hard body parts such as shells, skeletons or exoskeletons in animals like molluscs, echinoderms, crinoids and arthropods (a well-known group of arthropods from the lower Paleozoic are the trilobites) made the preservation and fossilization of such life forms easier than those of their Proterozoic ancestors. For this reason, much more is known about life in and after the Cambrian than about that of older periods. Some of these Cambrian groups appear complex but are seemingly quite different from modern life; examples are *Anomalocaris* and *Haikouichthys*. More recently, however, these seem to have found a place in modern classification. Wikipedia:Citation needed

During the Cambrian, the first vertebrate animals, among them the first fishes, had appeared.[357] A creature that could have been the ancestor of the fishes, or was probably closely related to it, was *Pikaia*. It had a primitive notochord, a structure that could have developed into a vertebral column later. The first fishes with jaws (Gnathostomata) appeared during the next geological period, the Ordovician. The colonisation of new niches resulted in massive body sizes. In this way, fishes with increasing sizes evolved during the early Paleozoic, such as the titanic placoderm *Dunkleosteus*, which could grow 7 meters (23 ft) long.Wikipedia:Citation needed

The diversity of life forms did not increase greatly because of a series of mass extinctions that define widespread biostratigraphic units called *biomeres*. After each extinction pulse, the continental shelf regions were repopulated by similar life forms that may have been evolving slowly elsewhere. By the late Cambrian, the trilobites had reached their greatest diversity and dominated nearly all fossil assemblages.[34]

Colonization of land

Oxygen accumulation from photosynthesis resulted in the formation of an ozone layer that absorbed much of the Sun's ultraviolet radiation, meaning unicellular organisms that reached land were less likely to die, and prokaryotes began to multiply and become better adapted to survival out of the water. Prokaryote lineages had probably colonized the land as early as 2.6 Ga even before the origin of the eukaryotes. For a long time, the land remained barren of multicellular organisms. The supercontinent Pannotia formed around 600 Ma and then broke apart a short 50 million years later. Fish, the earliest vertebrates, evolved in the oceans around 530 Ma.[354] A major extinction event occurred near the end of the Cambrian period, which ended 488 Ma.

Several hundred million years ago, plants (probably resembling algae) and fungi started growing at the edges of the water, and then out of it.[138–140] The

Figure 15: *Artist's conception of Devonian flora*

oldest fossils of land fungi and plants date to 480–460 Ma, though molecular evidence suggests the fungi may have colonized the land as early as 1000 Ma and the plants 700 Ma. Initially remaining close to the water's edge, mutations and variations resulted in further colonization of this new environment. The timing of the first animals to leave the oceans is not precisely known: the oldest clear evidence is of arthropods on land around 450 Ma, perhaps thriving and becoming better adapted due to the vast food source provided by the terrestrial plants. There is also unconfirmed evidence that arthropods may have appeared on land as early as 530 Ma.

Evolution of tetrapods

At the end of the Ordovician period, 443 Ma, additional extinction events occurred, perhaps due to a concurrent ice age. Around 380 to 375 Ma, the first tetrapods evolved from fish. Fins evolved to become limbs that the first tetrapods used to lift their heads out of the water to breathe air. This would let them live in oxygen-poor water, or pursue small prey in shallow water. They may have later ventured on land for brief periods. Eventually, some of them became so well adapted to terrestrial life that they spent their adult lives on land, although they hatched in the water and returned to lay their eggs. This was the origin of the amphibians. About 365 Ma, another period of extinction occurred, perhaps as a result of global cooling. Plants evolved seeds, which

Figure 16: *Tiktaalik, a fish with limb-like fins and a predecessor of tetrapods. Reconstruction from fossils about 375 million years old.*

dramatically accelerated their spread on land, around this time (by approximately 360 Ma).

About 20 million years later (340 Ma[:293–296]), the amniotic egg evolved, which could be laid on land, giving a survival advantage to tetrapod embryos. This resulted in the divergence of amniotes from amphibians. Another 30 million years (310 Ma[:254–256]) saw the divergence of the synapsids (including mammals) from the sauropsids (including birds and reptiles). Other groups of organisms continued to evolve, and lines diverged—in fish, insects, bacteria, and so on—but less is known of the details.Wikipedia:Citation needed

After yet another, the most severe extinction of the period (251∼250 Ma), around 230 Ma, dinosaurs split off from their reptilian ancestors. The Triassic–Jurassic extinction event at 200 Ma spared many of the dinosaurs, and they soon became dominant among the vertebrates. Though some mammalian lines began to separate during this period, existing mammals were probably small animals resembling shrews.[:169]

The boundary between avian and non-avian dinosaurs is not clear, but *Archaeopteryx*, traditionally considered one of the first birds, lived around 150 Ma.

The earliest evidence for the angiosperms evolving flowers is during the Cretaceous period, some 20 million years later (132 Ma).

Extinctions

The first of five great mass extinctions was the Ordovician-Silurian extinction. Its possible cause was the intense glaciation of Gondwana, which eventually led to a snowball earth. 60% of marine invertebrates became extinct and 25% of all families.Wikipedia:Citation needed

The second mass extinction was the Late Devonian extinction, probably caused by the evolution of trees, which could have led to the depletion of greenhouse

Figure 17: *Dinosaurs were the dominant terrestrial vertebrates throughout most of the Mesozoic*

gases (like CO2) or the eutrophication of water. 70% of all species became extinct.Wikipedia:Citation needed

The third mass extinction was the Permian-Triassic, or the Great Dying, event was possibly caused by some combination of the Siberian Traps volcanic event, an asteroid impact, methane hydrate gasification, sea level fluctuations, and a major anoxic event. Either the proposed Wilkes Land crater in Antarctica or Bedout structure off the northwest coast of Australia may indicate an impact connection with the Permian-Triassic extinction. But it remains uncertain whether either these or other proposed Permian-Triassic boundary craters are either real impact craters or even contemporaneous with the Permian-Triassic extinction event. This was by far the deadliest extinction ever, with about 57% of all families and 83% of all genera killed.

The fourth mass extinction was the Triassic-Jurassic extinction event in which almost all synapsids and archosaurs became extinct, probably due to new competition from dinosaurs.Wikipedia:Citation needed

The fifth and most recent mass extinction was the K-T extinction. In 66 Ma, a 10-kilometer (6.2 mi) asteroid struck Earth just off the Yucatán Peninsula – somewhere in the south western tip of then Laurasia – where the Chicxulub crater is today. This ejected vast quantities of particulate matter and vapor into the air that occluded sunlight, inhibiting photosynthesis. 75% of all life, including the non-avian dinosaurs, became extinct, marking the end of the Cretaceous period and Mesozoic era.Wikipedia:Citation needed

Diversification of mammals

The first true mammals evolved in the shadows of dinosaurs and other large archosaurs that filled the world by the late Triassic. The first mammals were very small, and were probably nocturnal to escape predation. Mammal diversification truly began only after the Cretaceous-Paleogene extinction event. By the early Paleocene the earth recovered from the extinction, and mammalian diversity increased. Creatures like *Ambulocetus* took to the oceans to eventually evolve into whales, whereas some creatures, like primates, took to the trees. This all changed during the mid to late Eocene when the circum-Antarctic current formed between Antarctica and Australia which disrupted weather patterns on a global scale. Grassless savannas began to predominate much of the landscape, and mammals such as *Andrewsarchus* rose up to become the largest known terrestrial predatory mammal ever, and early whales like *Basilosaurus* took control of the seas. Wikipedia:Citation needed

The evolution of grass brought a remarkable change to the Earth's landscape, and the new open spaces created pushed mammals to get bigger and bigger. Grass started to expand in the Miocene, and the Miocene is where many modern- day mammals first appeared. Giant ungulates like *Paraceratherium* and *Deinotherium* evolved to rule the grasslands. The evolution of grass also brought primates down from the trees, and started human evolution. The first big cats evolved during this time as well. The Tethys Sea was closed off by the collision of Africa and Europe.

The formation of Panama was perhaps the most important geological event to occur in the last 60 million years. Atlantic and Pacific currents were closed off from each other, which caused the formation of the Gulf Stream, which made Europe warmer. The land bridge allowed the isolated creatures of South America to migrate over to North America, and vice versa. Various species migrated south, leading to the presence in South America of llamas, the spectacled bear, kinkajous and jaguars.Wikipedia:Citation needed

Three million years ago saw the start of the Pleistocene epoch, which featured dramatic climactic changes due to the ice ages. The ice ages led to the evolution of modern man in Saharan Africa and expansion. The mega-fauna that dominated fed on grasslands that, by now, had taken over much of the subtropical world. The large amounts of water held in the ice allowed for various bodies of water to shrink and sometimes disappear such as the North Sea and the Bering Strait. It is believed by many that a huge migration took place along

Beringia which is why, today, there are camels (which evolved and became extinct in North America), horses (which evolved and became extinct in North America), and Native Americans. The ending of the last ice age coincided with the expansion of man, along with a massive die out of ice age megafauna. This extinction, nicknamed "the Sixth Extinction", has been going ever since.Wikipedia:Citation needed

Human evolution

Hominin timeline

0 —
1 —
2 —
3 —
4 —
5 —
6 —
7 —
8 —
9 —
10 —

Axis scale: million years

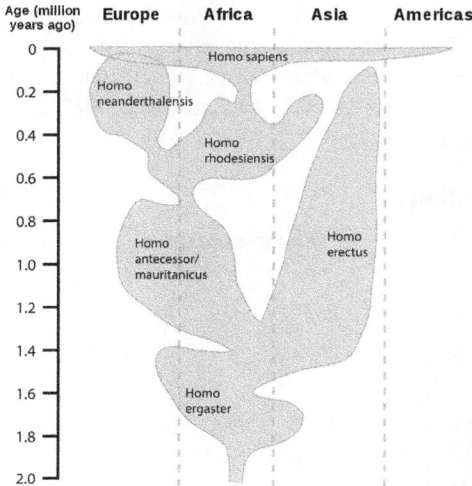

Figure 18: *A reconstruction of human history based on fossil data.*

Also see: *Life timeline* and *Nature timeline*

A small African ape living around 6 Ma was the last animal whose descendants would include both modern humans and their closest relatives, the chimpanzees.[100–101] Only two branches of its family tree have surviving descendants. Very soon after the split, for reasons that are still unclear, apes in one branch developed the ability to walk upright.[95–99] Brain size increased rapidly, and by 2 Ma, the first animals classified in the genus *Homo* had appeared.[300] Of course, the line between different species or even genera is somewhat arbitrary as organisms continuously change over generations. Around the same time, the other branch split into the ancestors of the common chimpanzee and the ancestors of the bonobo as evolution continued simultaneously in all life forms.[100–101]

The ability to control fire probably began in *Homo erectus* (or *Homo ergaster*), probably at least 790,000 years ago but perhaps as early as 1.5 Ma.[67] The use and discovery of controlled fire may even predate *Homo erectus*. Fire was possibly used by the early Lower Paleolithic (Oldowan) hominid *Homo habilis* or strong australopithecines such as *Paranthropus*.

It is more difficult to establish the origin of language; it is unclear whether *Homo erectus* could speak or if that capability had not begun until *Homo sapiens*.:67 As brain size increased, babies were born earlier, before their heads grew too large to pass through the pelvis. As a result, they exhibited more plasticity, and thus possessed an increased capacity to learn and required a longer period of dependence. Social skills became more complex, language became more sophisticated, and tools became more elaborate. This contributed to further cooperation and intellectual development.:7 Modern humans (*Homo sapiens*) are believed to have originated around 200,000 years ago or earlier in Africa; the oldest fossils date back to around 160,000 years ago.

The first humans to show signs of spirituality are the Neanderthals (usually classified as a separate species with no surviving descendants); they buried their dead, often with no sign of food or tools.:17 However, evidence of more sophisticated beliefs, such as the early Cro-Magnon cave paintings (probably with magical or religious significance):17–19 did not appear until 32,000 years ago. Cro-Magnons also left behind stone figurines such as Venus of Willendorf, probably also signifying religious belief.:17–19 By 11,000 years ago, *Homo sapiens* had reached the southern tip of South America, the last of the uninhabited continents (except for Antarctica, which remained undiscovered until 1820 AD). Tool use and communication continued to improve, and interpersonal relationships became more intricate.Wikipedia:Citation needed

Civilization

Throughout more than 90% of its history, *Homo sapiens* lived in small bands as nomadic hunter-gatherers.:8 As language became more complex, the ability to remember and communicate information resulted, according to a theory proposed by Richard Dawkins, in a new replicator: the meme. Ideas could be exchanged quickly and passed down the generations. Cultural evolution quickly outpaced biological evolution, and history proper began. Between 8500 and 7000 BC, humans in the Fertile Crescent in the Middle East began the systematic husbandry of plants and animals: agriculture. This spread to neighboring regions, and developed independently elsewhere, until most *Homo sapiens* lived sedentary lives in permanent settlements as farmers. Not all societies abandoned nomadism, especially those in isolated areas of the globe poor in domesticable plant species, such as Australia. However, among those civilizations that did adopt agriculture, the relative stability and increased productivity provided by farming allowed the population to expand.Wikipedia:Citation needed

Agriculture had a major impact; humans began to affect the environment as never before. Surplus food allowed a priestly or governing class to arise, followed by increasing division of labor. This led to Earth's first civilization at

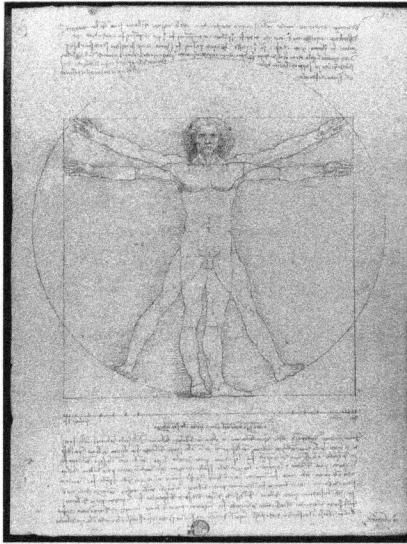

Figure 19: *Vitruvian Man by Leonardo da Vinci epitomizes the advances in art and science seen during the Renaissance.*

Sumer in the Middle East, between 4000 and 3000 BC.[15] Additional civilizations quickly arose in ancient Egypt, at the Indus River valley and in China. The invention of writing enabled complex societies to arise: record-keeping and libraries served as a storehouse of knowledge and increased the cultural transmission of information. Humans no longer had to spend all their time working for survival, enabling the first specialized occupations (e.g. craftsmen, merchants, priests, etc...). Curiosity and education drove the pursuit of knowledge and wisdom, and various disciplines, including science (in a primitive form), arose. This in turn led to the emergence of increasingly larger and more complex civilizations, such as the first empires, which at times traded with one another, or fought for territory and resources.

By around 500 BC, there were advanced civilizations in the Middle East, Iran, India, China, and Greece, at times expanding, at times entering into decline.[3] In 221 BC, China became a single polity that would grow to spread its culture throughout East Asia, and it has remained the most populous nation in the world. The fundamentals of Western civilization were largely shaped in Ancient Greece, with the world's first democratic government and major advances in philosophy, science, and mathematics, and in Ancient Rome in law, government, and engineering. The Roman Empire was Christianized by Emperor Constantine in the early 4th century and declined by the end of the 5th.

Beginning with the 7th century, Christianization of Europe began. In 610, Islam was founded and quickly became the dominant religion in Western Asia. The House of Wisdom was established in Abbasid-era Baghdad, Iraq. It is considered to have been a major intellectual center during the Islamic Golden Age, where Muslim scholars in Baghdad and Cairo flourished from the ninth to the thirteenth centuries until the Mongol sack of Baghdad in 1258 AD. In 1054 AD the Great Schism between the Roman Catholic Church and the Eastern Orthodox Church led to the prominent cultural differences between Western and Eastern Europe.Wikipedia:Citation needed

In the 14th century, the Renaissance began in Italy with advances in religion, art, and science.:317–319 At that time the Christian Church as a political entity lost much of its power. In 1492, Christopher Columbus reached the Americas, initiating great changes to the new world. European civilization began to change beginning in 1500, leading to the scientific and industrial revolutions. That continent began to exert political and cultural dominance over human societies around the world, a time known as the Colonial era (also see Age of Discovery).:295–299 In the 18th century a cultural movement known as the Age of Enlightenment further shaped the mentality of Europe and contributed to its secularization. From 1914 to 1918 and 1939 to 1945, nations around the world were embroiled in world wars. Established following World War I, the League of Nations was a first step in establishing international institutions to settle disputes peacefully. After failing to prevent World War II, mankind's bloodiest conflict, it was replaced by the United Nations. After the war, many new states were formed, declaring or being granted independence in a period of decolonization. The United States and Soviet Union became the world's dominant superpowers for a time, and they held an often-violent rivalry known as the Cold War until the dissolution of the latter. In 1992, several European nations joined in the European Union. As transportation and communication improved, the economies and political affairs of nations around the world have become increasingly intertwined. This globalization has often produced both conflict and cooperation.Wikipedia:Citation needed

Recent events

Change has continued at a rapid pace from the mid-1940s to today. Technological developments include nuclear weapons, computers, genetic engineering, and nanotechnology. Economic globalization, spurred by advances in communication and transportation technology, has influenced everyday life in many parts of the world. Cultural and institutional forms such as democracy, capitalism, and environmentalism have increased influence. Major concerns and problems such as disease, war, poverty, violent radicalism, and recently, human-caused climate change have risen as the world population increases.Wikipedia:Citation needed

Figure 20: *Astronaut Bruce McCandless II outside of the space shuttle Challenger in 1984*

In 1957, the Soviet Union launched the first artificial satellite into orbit and, soon afterward, Yuri Gagarin became the first human in space. Neil Armstrong, an American, was the first to set foot on another astronomical object, the Moon. Unmanned probes have been sent to all the known planets in the solar system, with some (such as Voyager) having left the solar system. Five space agencies, representing over fifteen countries, have worked together to build the International Space Station. Aboard it, there has been a continuous human presence in space since 2000. The World Wide Web became a part of everyday life in the 1990s, and since then has become an indispensable source of information in the developed world.Wikipedia:Citation needed

References

Further reading

<templatestyles src="Template:Refbegin/styles.css" />

- Dalrymple, G. B. (1991). *The Age of the Earth*. California: Stanford University Press. ISBN 978-0-8047-1569-0.

- Dalrymple, G. Brent (2001). "The age of the Earth in the twentieth century: a problem (mostly) solved"[6]. *Geological Society, London, Special Publications*. **190** (1): 205–221. Bibcode: 2001GSLSP.190..205D[7]. doi: 10.1144/GSL.SP.2001.190.01.14[8]. Retrieved 2012-04-13.
- Dawkins, Richard (2004). *The Ancestor's Tale: A Pilgrimage to the Dawn of Life*. Boston: Houghton Mifflin Company. ISBN 978-0-618-00583-3.
- Gradstein, F. M.; Ogg, James George; Smith, Alan Gilbert, eds. (2004). *A Geological Time Scale 2004*. Reprinted with corrections 2006. Cambridge University Press. ISBN 978-0-521-78673-7.
- Gradstein, Felix M.; Ogg, James G.; van Kranendonk, Martin (2008). On the Geological Time Scale 2008[9] (PDF) (Report). International Commission on Stratigraphy. Fig. 2. Archived from the original[10] (PDF) on 28 October 2012. Retrieved 20 April 2012.
- Levin, H. L. (2009). *The Earth through time* (9th ed.). Saunders College Publishing. ISBN 978-0-470-38774-0.
- Lunine, J. I. (1999). *Earth: evolution of a habitable world*. United Kingdom: Cambridge University Press. ISBN 978-0-521-64423-5.
- McNeill, Willam H. (1999) [1967]. *A World History* (4th ed.). New York: Oxford University Press. ISBN 978-0-19-511615-1.
- Melosh, H. J.; Vickery, A. M. & Tonks, W. B. (1993). *Impacts and the early environment and evolution of the terrestrial planets*, in Levy, H.J. & Lunine, J.I. (eds.): *Protostars and Planets III*, University of Arizona Press, Tucson, pp. 1339–1370.
- Stanley, Steven M. (2005). *Earth system history* (2nd ed.). New York: Freeman. ISBN 978-0-7167-3907-4.
- Stern, T. W.; Bleeker, W. (1998). "Age of the world's oldest rocks refined using Canada's SHRIMP: The Acasta Gneiss Complex, Northwest Territories, Canada". *Geoscience Canada*. **25**: 27–31.
- Wetherill, G. W. (1991). "Occurrence of Earth-Like Bodies in Planetary Systems". *Science*. **253** (5019): 535–538. Bibcode: 1991Sci...253..535W[11]. doi: 10.1126/science.253.5019.535[12]. PMID 17745185[13].

External links

<indicator name="spoken-icon"> </indicator>

- Davies, Paul. " Quantum leap of life[14]". *The Guardian*. 2005 December 20. – discusses speculation on the role of quantum systems in the origin of life

- Evolution timeline[15] (uses Shockwave). Animated story of life shows everything from the big bang to the formation of the earth and the development of bacteria and other organisms to the ascent of man.
- 25 biggest turning points in earth History[16] BBC
- Evolution of the Earth[17]. Timeline of the most important events in the evolution of the Earth.
- The Earth's Origins[18] on *In Our Time* at the BBC
- Ageing the Earth[19], BBC Radio 4 discussion with Richard Corfield, Hazel Rymer & Henry Gee (*In Our Time*, Nov. 20, 2003)

Earliest evidence for life on Earth

Earliest known life forms

The **earliest known life forms** on Earth are putative fossilized microorganisms found in hydrothermal vent precipitates. The earliest time that life forms first appeared on Earth is unknown. They may have lived earlier than 3.77 billion years ago, possibly as early as 4.28 billion years ago, not long after the oceans formed 4.41 billion years ago, and not long after the formation of the Earth 4.54 billion years ago. The earliest *direct* evidence of life on Earth are fossils of microorganisms permineralized in 3.465-billion-year-old Australian Apex chert rocks.

Overview

A life form, or lifeform, is an entity or being that is living. Currently, Earth remains the only place in the universe known to harbor life forms.

More than 99% of all species of life forms, amounting to over five billion species, that ever lived on Earth are estimated to be extinct.

Some estimates on the number of Earth's current species of life forms range from 10 million to 14 million, of which about 1.2 million have been documented and over 86 percent have not yet been described. However, a May 2016 scientific report estimates that 1 trillion species are currently on Earth, with only one-thousandth of one percent described. The total number of DNA base pairs on Earth is estimated at 5.0×10^{37} with a weight of 50 billion tonnes. In comparison, the total mass of the biosphere has been estimated to be as much as 4 trillion tons of carbon. In July 2016, scientists reported identifying a set of 355 genes from the Last Universal Common Ancestor (LUCA) of all organisms living on Earth.

Figure 21: *Evidence of possibly the oldest forms of life on Earth has been found in hydrothermal vent precipitates.*

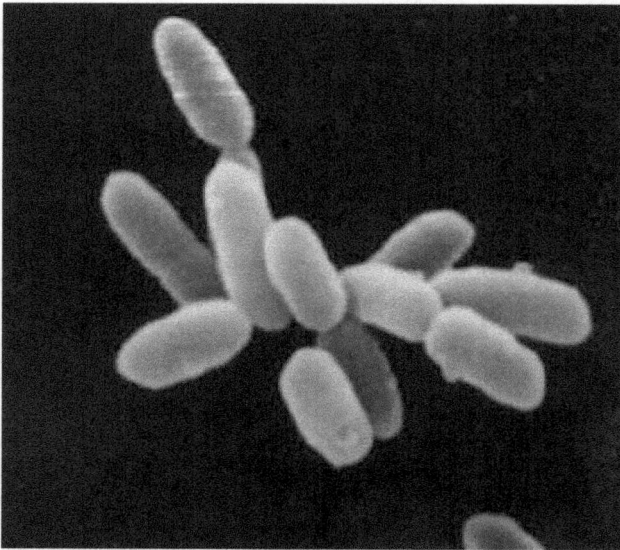

Figure 22: *Archaea, prokaroytic microbes, were first found in extreme environments, such as hydrothermal vents.*

The Earth's biosphere includes soil, hydrothermal vents, rock up to 19 km (12 mi) or deeper underground, the deepest parts of the ocean, and at least 64 km (40 mi) high into the atmosphere. Under certain test conditions, life forms have been observed to thrive in the near-weightlessness of space and to survive in the vacuum of outer space. Life forms appear to thrive in the Mariana Trench, the deepest spot in the Earth's oceans, reaching a depth of 11,034 m (36,201 ft; 6.856 mi). Other researchers reported related studies that life forms thrive inside rocks up to 580 m (1,900 ft; 0.36 mi) below the sea floor under 2,590 m (8,500 ft; 1.61 mi) of ocean off the coast of the north-western United States, as well as 2,400 m (7,900 ft; 1.5 mi) beneath the seabed off Japan. In August 2014, scientists confirmed the existence of life forms living 800 m (2,600 ft; 0.50 mi) below the ice of Antarctica.

According to one researcher, "You can find microbes everywhere — they're extremely adaptable to conditions, and survive wherever they are."

Earliest life forms

Fossil evidence informs most studies of the origin of life. The age of the Earth is about 4.54 billion years; the earliest undisputed evidence of life on Earth dates from at least 3.5 billion years ago. There is evidence that life began much earlier.

In 2017, fossilized microorganisms, or microfossils, were announced to have been discovered in hydrothermal vent precipitates in the Nuvvuagittuq Belt of Quebec, Canada that may be as old as 4.28 billion years old, the oldest record of life on Earth, suggesting "an almost instantaneous emergence of life" (in a geological time-scale sense), after ocean formation 4.41 billion years ago, and not long after the formation of the Earth 4.54 billion years ago.

"Remains of life" have been found in 4.1 billion-year-old rocks in Western Australia.

Evidence of biogenic graphite, and possibly stromatolites, was discovered in 3.7 billion-year-old metasedimentary rocks in southwestern Greenland.

In May 2017, evidence of life on land may have been found in 3.48 billion-year-old geyserite which is often found around hot springs and geysers, and other related mineral deposits, uncovered in the Pilbara Craton of Western Australia. This complements the November 2013 publication that microbial mat fossils had been found in 3.48 billion-year-old sandstone in Western Australia.

In November 2017, a study by the University of Edinburgh suggested that life on Earth may have originated from biological particles carried by streams of space dust.

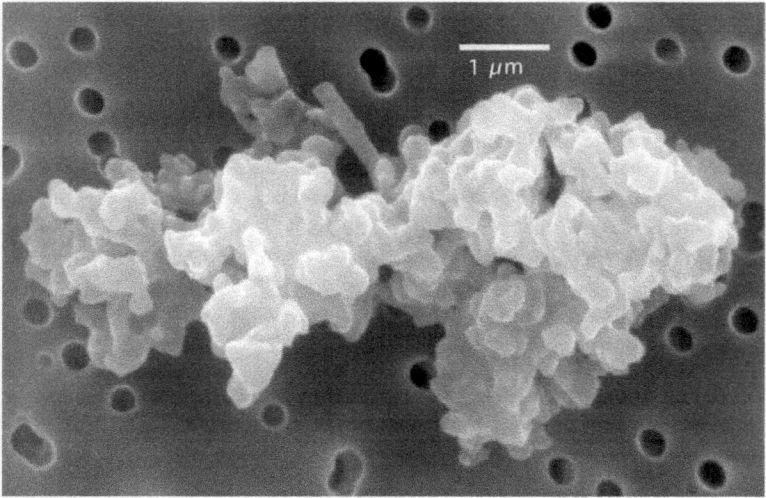

Figure 23: *Studies suggest that life on Earth may have come
from biological matter carried by space dust or meteorites.*

A December 2017 report stated that 3.465-billion-year-old Australian Apex
chert rocks once contained microorganisms, the earliest *direct* evidence of life
on Earth.

In January 2018, a study found that 4.5 billion-year-old meteorites found on
Earth contained liquid water along with prebiotic complex organic substances
that may be ingredients for life.

According to biologist Stephen Blair Hedges, "If life arose relatively quickly
on Earth ... then it could be common in the universe."

Gallery

Figure 24: *Stromatolites are made by microbes moving upward to avoid being smothered by sediment.*

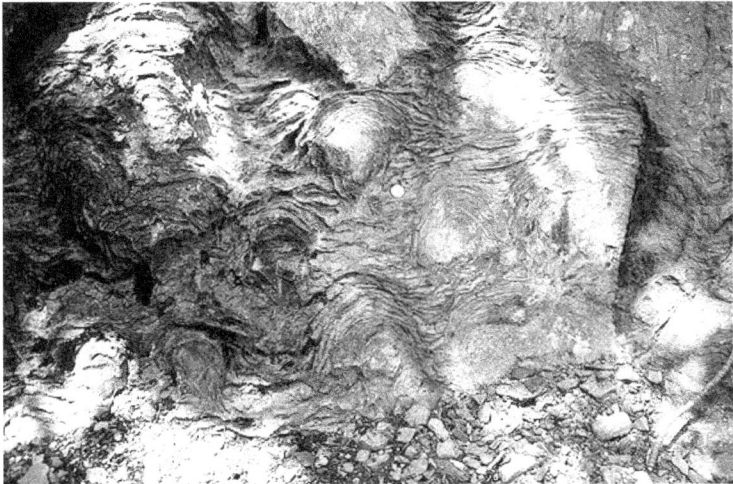

Figure 25: *Stromatolites left behind by cyanobacteria are one of the oldest fossils of life on Earth.*

Figure 26: *The cyanobacterial-algal mat, salty lake on the White Sea seaside.*

Figure 27: *Wrinkled Kinneyia-type sedimentary structures formed beneath cohesive microbial mats in peritidal zones.*

Figure 28: *Kinneyia-like structure in the Grimsby Formation (Silurian) exposed in Niagara Gorge, NY.*

External links

- Biota[20] (Taxonomicon)
- Life[21] (Systema Naturae 2000)
- Vitae[22] (BioLib)
- Wikispecies – a free directory of life
- Google Images: Earliest known life forms[23]

Origins of life on Earth

Evidence of common descent

Part of a series on
Evolutionary biology
•   Evolutionary biology portal
•   Category
•   *Book*
•   Related topics
• v
• t
• e[24]

Evidence of common descent of living organisms has been discovered by scientists researching in a variety of disciplines over many decades, demonstrating that all life on Earth comes from a single ancestor. This forms an important part of the evidence on which evolutionary theory rests, demonstrates that evolution does occur, and illustrates the processes that created Earth's biodiversity. It supports the modern evolutionary synthesis—the current scientific theory that explains how and why life changes over time. Evolutionary biologists document evidence of common descent, all the way back to the last universal common ancestor, by developing testable predictions, testing hypotheses, and constructing theories that illustrate and describe its causes.

Comparison of the DNA genetic sequences of organisms has revealed that organisms that are phylogenetically close have a higher degree of DNA sequence similarity than organisms that are phylogenetically distant. Genetic fragments such as pseudogenes, regions of DNA that are orthologous to a gene in a related organism, but are no longer active and appear to be undergoing a steady process of degeneration from cumulative mutations support common descent alongside the universal biochemical organization and molecular variance patterns found in all organisms. Additional genetic information conclusively supports the relatedness of life and has allowed scientists (since the discovery of DNA) to develop phylogenetic trees: a construction of organisms evolutionary relatedness. It has also led to the development of molecular clock techniques to date taxon divergence times and to calibrate these with the fossil record.

Fossils are important for estimating when various lineages developed in geologic time. As fossilization is an uncommon occurrence, usually requiring hard body parts and death near a site where sediments are being deposited, the fossil record only provides sparse and intermittent information about the evolution of life. Evidence of organisms prior to the development of hard body parts such as shells, bones and teeth is especially scarce, but exists in the form of ancient microfossils, as well as impressions of various soft-bodied organisms. The comparative study of the anatomy of groups of animals shows structural features that are fundamentally similar (homologous), demonstrating phylogenetic and ancestral relationships with other organisms, most especially when compared with fossils of ancient extinct organisms. Vestigial structures and comparisons in embryonic development are largely a contributing factor in anatomical resemblance in concordance with common descent. Since metabolic processes do not leave fossils, research into the evolution of the basic cellular processes is done largely by comparison of existing organisms' physiology and biochemistry. Many lineages diverged at different stages of development, so it is possible to determine when certain metabolic processes appeared by comparing the traits of the descendants of a common ancestor.

Evidence from animal coloration was gathered by some of Darwin's contemporaries; camouflage, mimicry, and warning coloration are all readily explained by natural selection. Special cases like the seasonal changes in the plumage of the ptarmigan, camouflaging it against snow in winter and against brown moorland in summer provide compelling evidence that selection is at work. Further evidence comes from the field of biogeography because evolution with common descent provides the best and most thorough explanation for a variety of facts concerning the geographical distribution of plants and animals across the world. This is especially obvious in the field of insular biogeography. Combined with the well-established geological theory of plate tectonics, common descent provides a way to combine facts about the current distribution of species with evidence from the fossil record to provide a logically consistent explanation of how the distribution of living organisms has changed over time.

The development and spread of antibiotic resistant bacteria provides evidence that evolution due to natural selection is an ongoing process in the natural world. Natural selection is ubiquitous in all research pertaining to evolution, taking note of the fact that all of the following examples in each section of the article document the process. Alongside this are observed instances of the separation of populations of species into sets of new species (speciation). Speciation has been observed in the lab and in nature. Multiple forms of such have been described and documented as examples for individual modes of speciation. Furthermore, evidence of common descent extends from direct laboratory experimentation with the selective breeding of organisms—historically and currently—and other controlled experiments involving many of the topics in the article. This article summarizes the varying disciplines that provide the evidence for evolution and the common descent of all life on Earth, accompanied by numerous and specialized examples, indicating a compelling consilience of evidence.

Evidence from comparative physiology and biochemistry

Genetics

One of the strongest evidences for common descent comes from gene sequences. Comparative sequence analysis examines the relationship between the DNA sequences of different species, producing several lines of evidence that confirm Darwin's original hypothesis of common descent. If the hypothesis of common descent is true, then species that share a common ancestor inherited that ancestor's DNA sequence, as well as mutations unique to that ancestor. More closely related species have a greater fraction of identical sequence and shared substitutions compared to more distantly related species.

The simplest and most powerful evidence is provided by phylogenetic reconstruction. Such reconstructions, especially when done using slowly evolving protein sequences, are often quite robust and can be used to reconstruct a great deal of the evolutionary history of modern organisms (and even in some instances of the evolutionary history of extinct organisms, such as the recovered gene sequences of mammoths or Neanderthals). These reconstructed phylogenies recapitulate the relationships established through morphological and biochemical studies. The most detailed reconstructions have been performed on the basis of the mitochondrial genomes shared by all eukaryotic organisms, which are short and easy to sequence; the broadest reconstructions have been performed either using the sequences of a few very ancient proteins or by using ribosomal RNA sequence.Wikipedia:Citation needed

Figure 29: *Figure 1a: While on board HMS Beagle, Charles Darwin collected numerous specimens, many new to science, which supported his later theory of evolution by natural selection.*

Phylogenetic relationships extend to a wide variety of nonfunctional sequence elements, including repeats, transposons, pseudogenes, and mutations in protein-coding sequences that do not change the amino-acid sequence. While a minority of these elements might later be found to harbor function, in aggregate they demonstrate that identity must be the product of common descent rather than common function.

Universal biochemical organisation and molecular variance patterns

All known extant (surviving) organisms are based on the same biochemical processes: genetic information encoded as nucleic acid (DNA, or RNA for many viruses), transcribed into RNA, then translated into proteins (that is, polymers of amino acids) by highly conserved ribosomes. Perhaps most tellingly, the Genetic Code (the "translation table" between DNA and amino acids) is the same for almost every organism, meaning that a piece of DNA in a bacterium codes for the same amino acid as in a human cell. ATP is used as energy currency by all extant life. A deeper understanding of developmental biology shows that common morphology is, in fact, the product of shared genetic elements. For example, although camera-like eyes are believed to have evolved independently on many separate occasions, they share a common set of light-sensing proteins (opsins), suggesting a common point of origin for all sighted creatures.[25] Another example is the familiar vertebrate body plan, whose structure is controlled by the homeobox (Hox) family of genes.

DNA sequencing

Comparison of DNA sequences allows organisms to be grouped by sequence similarity, and the resulting phylogenetic trees are typically congruent with traditional taxonomy, and are often used to strengthen or correct taxonomic classifications. Sequence comparison is considered a measure robust enough to correct erroneous assumptions in the phylogenetic tree in instances where other evidence is scarce. For example, neutral human DNA sequences are approximately 1.2% divergent (based on substitutions) from those of their nearest genetic relative, the chimpanzee, 1.6% from gorillas, and 6.6% from baboons. Genetic sequence evidence thus allows inference and quantification of genetic relatedness between humans and other apes.[26,27] The sequence of the 16S ribosomal RNA gene, a vital gene encoding a part of the ribosome, was used to find the broad phylogenetic relationships between all extant life. The analysis by Carl Woese resulted in the three-domain system, arguing for two major splits in the early evolution of life. The first split led to modern Bacteria and the subsequent split led to modern Archaea and Eukaryotes.

Some DNA sequences are shared by very different organisms. It has been predicted by the theory of evolution that the differences in such DNA sequences between two organisms should roughly resemble both the biological difference between them according to their anatomy and the time that had passed since these two organisms have separated in the course of evolution, as seen in fossil evidence. The rate of accumulating such changes should be low for some sequences, namely those that code for critical RNA or proteins, and high for others that code for less critical RNA or proteins; but for every specific sequence, the rate of change should be roughly constant over time. These results have been experimentally confirmed. Two examples are DNA sequences coding for rRNA, which is highly conserved, and DNA sequences coding for fibrinopeptides (amino acid chains that are discarded during the formation of fibrin), which are highly non-conserved.

Proteins

Proteomic evidence also supports the universal ancestry of life. Vital proteins, such as the ribosome, DNA polymerase, and RNA polymerase, are found in everything from the most primitive bacteria to the most complex mammals. The core part of the protein is conserved across all lineages of life, serving similar functions. Higher organisms have evolved additional protein subunits, largely affecting the regulation and protein-protein interaction of the core. Other overarching similarities between all lineages of extant organisms, such as DNA, RNA, amino acids, and the lipid bilayer, give support to the theory of common descent. Phylogenetic analyses of protein sequences from various organisms produce similar trees of relationship between all organisms.[28] The chirality of DNA, RNA, and amino acids is conserved across all known life. As there is no functional advantage to right- or left-handed molecular chirality, the simplest hypothesis is that the choice was made randomly by early organisms and passed on to all extant life through common descent. Further evidence for reconstructing ancestral lineages comes from junk DNA such as pseudogenes, "dead" genes that steadily accumulate mutations.

Pseudogenes

Pseudogenes, also known as noncoding DNA, are extra DNA in a genome that do not get transcribed into RNA to synthesize proteins. Some of this noncoding DNA has known functions, but much of it has no known function and is called "Junk DNA". This is an example of a vestige since replicating these genes uses energy, making it a waste in many cases. A pseudogene can be produced when a coding gene accumulates mutations that prevent it from being transcribed, making it non-functional. But since it is not transcribed, it may disappear without affecting fitness, unless it has provided some beneficial function as non-coding DNA. Non-functional pseudogenes may be passed on to later species, thereby labeling the later species as descended from the earlier species.

Other mechanisms

A large body of molecular evidence supports a variety of mechanisms for large evolutionary changes, including: genome and gene duplication, which facilitates rapid evolution by providing substantial quantities of genetic material under weak or no selective constraints; horizontal gene transfer, the process of transferring genetic material to another cell that is not an organism's offspring, allowing for species to acquire beneficial genes from each other; and recombination, capable of reassorting large numbers of different alleles and of establishing reproductive isolation. The endosymbiotic theory explains the origin of mitochondria and plastids (including chloroplasts), which are organelles of eukaryotic cells, as the incorporation of an ancient prokaryotic cell into ancient eukaryotic cell. Rather than evolving eukaryotic organelles slowly, this theory offers a mechanism for a sudden evolutionary leap by incorporating the genetic material and biochemical composition of a separate species. Evidence supporting this mechanism has been found in the protist *Hatena*: as a predator it engulfs a green algal cell, which subsequently behaves as an endosymbiont, nourishing *Hatena*, which in turn loses its feeding apparatus and behaves as an autotroph.

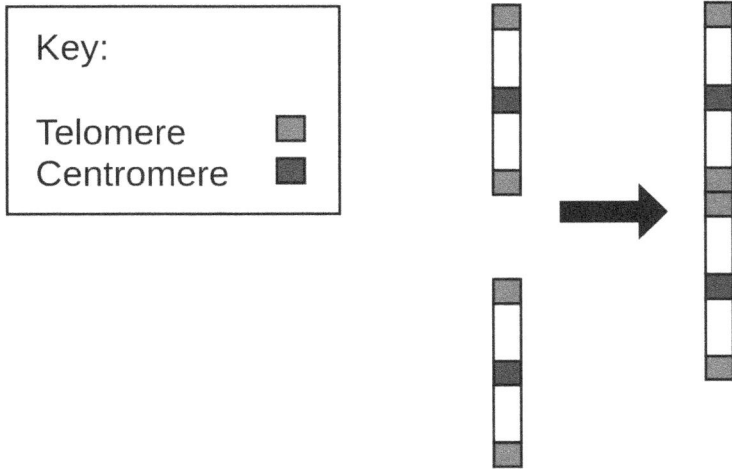

Figure 30: *Figure 1b: Fusion of ancestral chromosomes left distinctive remnants of telomeres, and a vestigial centromere*

Since metabolic processes do not leave fossils, research into the evolution of the basic cellular processes is done largely by comparison of existing organisms. Many lineages diverged when new metabolic processes appeared, and it is theoretically possible to determine when certain metabolic processes appeared by comparing the traits of the descendants of a common ancestor or by detecting their physical manifestations. As an example, the appearance of oxygen in the earth's atmosphere is linked to the evolution of photosynthesis.

Specific examples from comparative physiology and biochemistry

Chromosome 2 in humans

Evidence for the evolution of *Homo sapiens* from a common ancestor with chimpanzees is found in the number of chromosomes in humans as compared to all other members of Hominidae. All hominidae have 24 pairs of chromosomes, except humans, who have only 23 pairs. Human chromosome 2 is a result of an end-to-end fusion of two ancestral chromosomes.[29,30]

The evidence for this includes:

- The correspondence of chromosome 2 to two ape chromosomes. The closest human relative, the common chimpanzee, has near-identical DNA sequences to human chromosome 2, but they are found in two separate chromosomes. The same is true of the more distant gorilla and orangutan.[31]

- The presence of a vestigial centromere. Normally a chromosome has just one centromere, but in chromosome 2 there are remnants of a second centromere.
- The presence of vestigial telomeres. These are normally found only at the ends of a chromosome, but in chromosome 2 there are additional telomere sequences in the middle.

Chromosome 2 thus presents strong evidence in favour of the common descent of humans and other apes. According to J. W. Ijdo, "We conclude that the locus cloned in cosmids c8.1 and c29B is the relic of an ancient telomere-telomere fusion and marks the point at which two ancestral ape chromosomes fused to give rise to human chromosome 2."

Cytochrome c and b

A classic example of biochemical evidence for evolution is the variance of the ubiquitous (i.e. all living organisms have it, because it performs very basic life functions) protein Cytochrome c in living cells. The variance of cytochrome c of different organisms is measured in the number of differing amino acids, each differing amino acid being a result of a base pair substitution, a mutation. If each differing amino acid is assumed the result of one base pair substitution, it can be calculated how long ago the two species diverged by multiplying the number of base pair substitutions by the estimated time it takes for a substituted base pair of the cytochrome c gene to be successfully passed on. For example, if the average time it takes for a base pair of the cytochrome c gene to mutate is N years, the number of amino acids making up the cytochrome c protein in monkeys differ by one from that of humans, this leads to the conclusion that the two species diverged N years ago.

The primary structure of cytochrome c consists of a chain of about 100 amino acids. Many higher order organisms possess a chain of 104 amino acids.[32]

The cytochrome c molecule has been extensively studied for the glimpse it gives into evolutionary biology. Both chicken and turkeys have identical sequence homology (amino acid for amino acid), as do pigs, cows and sheep. Both humans and chimpanzees share the identical molecule, while rhesus monkeys share all but one of the amino acids: the 66th amino acid is isoleucine in the former and threonine in the latter.

What makes these homologous similarities particularly suggestive of common ancestry in the case of cytochrome c, in addition to the fact that the phylogenies derived from them match other phylogenies very well, is the high degree of functional redundancy of the cytochrome c molecule. The different existing configurations of amino acids do not significantly affect the functionality of the protein, which indicates that the base pair substitutions are not part of a directed design, but the result of random mutations that aren't subject to selection.

In addition, Cytochrome b is commonly used as a region of mitochondrial DNA to determine phylogenetic relationships between organisms due to its sequence variability. It is considered most useful in determining relationships within families and genera. Comparative studies involving cytochrome b have resulted in new classification schemes and have been used to assign newly described species to a genus, as well as deepen the understanding of evolutionary relationships.

Endogenous retroviruses

Endogenous retroviruses (or ERVs) are remnant sequences in the genome left from ancient viral infections in an organism. The retroviruses (or virogenes) are always passed on to the next generation of that organism that received the infection. This leaves the virogene left in the genome. Because this event is rare and random, finding identical chromosomal positions of a virogene in two different species suggests common ancestry. Cats (Felidae) present a notable instance of virogene sequences demonstrating common descent. The standard phylogenetic tree for Felidae have smaller cats (*Felis chaus*, *Felis silvestris*, *Felis nigripes*, and *Felis catus*) diverging from larger cats such as the subfamily Pantherinae and other carnivores. The fact that small cats have an ERV where the larger cats do not suggests that the gene was inserted into the ancestor of the small cats after the larger cats had diverged. Another example of this is with humans and chimps. Humans contain numerous ERVs that comprise a considerable percentage of the genome. Sources vary, but 1% to 8% has been proposed. Humans and chimps share seven different occurrences of virogenes, while all primates share similar retroviruses congruent with phylogeny.

Recent African origin of modern humans

Mathematical models of evolution, pioneered by the likes of Sewall Wright, Ronald Fisher and J. B. S. Haldane and extended via diffusion theory by Motoo Kimura, allow predictions about the genetic structure of evolving populations. Direct examination of the genetic structure of modern populations via DNA sequencing has allowed verification of many of these predictions. For example, the Out of Africa theory of human origins, which states that modern humans developed in Africa and a small sub-population migrated out (undergoing a population bottleneck), implies that modern populations should show the signatures of this migration pattern. Specifically, post-bottleneck populations (Europeans and Asians) should show lower overall genetic diversity and

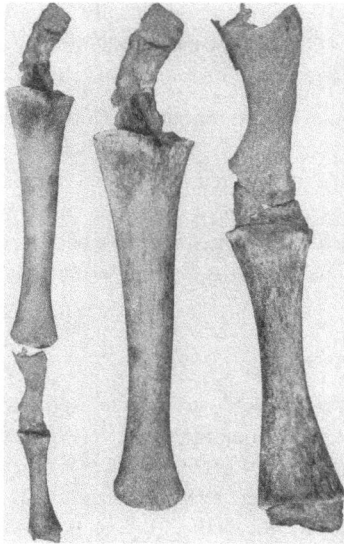

Figure 31: *Figure 2a: In July 1919, a humpback whale was caught by a ship operating out of Vancouver that had legs 4 ft 2 in (1.27 m) long. This image shows the hindlegs of another humpback whale reported in 1921 by the American Museum of Natural History.*

a more uniform distribution of allele frequencies compared to the African population. Both of these predictions are borne out by actual data from a number of studies.

Evidence from comparative anatomy

Comparative study of the anatomy of groups of animals or plants reveals that certain structural features are basically similar. For example, the basic structure of all flowers consists of sepals, petals, stigma, style and ovary; yet the size, colour, number of parts and specific structure are different for each individual species. The neural anatomy of fossilized remains may also be compared using advanced imaging techniques.

Atavisms

Once thought of as a refutation to evolutionary theory, atavisms are "now seen as potent evidence of how much genetic potential is retained...after a particular structure has disappeared from a species". "Atavisms are the reappearance of a lost character typical of remote ancestors and not seen in the parents or recent ancestors..." and are an "[indication] of the developmental plasticity that exists within embryos..." Atavisms occur because genes for previously existing phenotypical features are often preserved in DNA, even though the genes are not expressed in some or most of the organisms possessing them.[33] Numerous examples have documented the occurrence of atavisms alongside experimental research triggering their formation. Due to the complexity and interrelatedness of the factors involved in the development of atavisms, both biologists and medical professionals find it "difficult, if not impossible, to distinguish [them] from malformations."

Some examples of atavisms found in the scientific literature include:

- Hind limbs in whales. (see figure 2a)
- Reappearance of limbs in limbless vertebrates.
- Back pair of flippers on a bottlenose dolphin.
- Extra toes of the modern horse.
- Human tails (not pseudo-tails) and extra nipples in humans.
- Re-evolution of sexuality from parthenogenesis in orbitid mites.
- Teeth in chickens.
- Dewclaws in dogs.
- Reappearance of wings on wingless stick insects and earwigs.
- Atavistic muscles in several birds and mammals such as the beagle and the jerboa.
- Extra toes in guinea pigs.

Evolutionary developmental biology and embryonic development

Evolutionary developmental biology is the biological field that compares the developmental process of different organisms to determine ancestral relationships between species. A large variety of organism's genomes contain a small fraction of genes that control the organisms development. Hox genes are an example of these types of nearly universal genes in organisms pointing to an origin of common ancestry. Embryological evidence comes from the development of organisms at the embryological level with the comparison of different organisms embryos similarity. Remains of ancestral traits often appear and disappear in different stages of the embryological development process.

Some examples include:

- Hair growth and loss (lanugo) during human development.
- Development and degeneration of a yolk sac.

- Terrestrial frogs and salamanders passing through the larval stage within the egg—with features of typically aquatic larvae—but hatch ready for life on land;
- The appearance of gill-like structures (pharyngeal arch) in vertebrate embryo development. Note that in fish, the arches continue to develop as branchial arches while in humans, for example, they give rise to a variety of structures within the head and neck.

Homologous structures and divergent (adaptive) evolution

If widely separated groups of organisms are originated from a common ancestry, they are expected to have certain basic features in common. The degree of resemblance between two organisms should indicate how closely related they are in evolution:

- Groups with little in common are assumed to have diverged from a common ancestor much earlier in geological history than groups with a lot in common;
- In deciding how closely related two animals are, a comparative anatomist looks for structures that are fundamentally similar, even though they may serve different functions in the adult. Such structures are described as homologous and suggest a common origin.
- In cases where the similar structures serve different functions in adults, it may be necessary to trace their origin and embryonic development. A similar developmental origin suggests they are the same structure, and thus likely derived from a common ancestor.

When a group of organisms share a homologous structure that is specialized to perform a variety of functions to adapt different environmental conditions and modes of life, it is called adaptive radiation. The gradual spreading of organisms with adaptive radiation is known as divergent evolution.

Nested hierarchies and classification

Taxonomy is based on the fact that all organisms are related to each other in nested hierarchies based on shared characteristics. Most existing species can be organized rather easily in a nested hierarchical classification. This is evident from the Linnaean classification scheme. Based on shared derived characters, closely related organisms can be placed in one group (such as a genus), several genera can be grouped together into one family, several families can be grouped together into an order, etc.[34] The existence of these nested hierarchies was recognized by many biologists before Darwin, but he showed that his theory of evolution with its branching pattern of common descent could explain them. Darwin described how common descent could provide a logical basis for classification:

❝ **❞**

All the foregoing rules and aids and difficulties in classification are explained, if I do not greatly deceive myself, on the view that the natural system is founded on descent with modification; that the characters which naturalists consider as showing true affinity between any two or more species, are those which have been inherited from a common parent, and, in so far, all true classification is genealogical; that community of descent is the hidden bond which naturalists have been unconsciously seeking, ...

—Charles Darwin, *On the Origin of Species*, page 577

Evolutionary trees

An evolutionary tree (of Amniota, for example, the last common ancestor of mammals and reptiles, and all its descendants) illustrates the initial conditions causing evolutionary patterns of similarity (e.g., all Amniotes produce an egg that possesses the amnios) and the patterns of divergence amongst lineages (e.g., mammals and reptiles branching from the common ancestry in Amniota). Evolutionary trees provide conceptual models of evolving systems once thought limited in the domain of making predictions out of the theory. However, the method of phylogenetic bracketing is used to infer predictions with far greater probability than raw speculation. For example, paleontologists use this technique to make predictions about nonpreservable traits in fossil organisms, such as feathered dinosaurs, and molecular biologists use the technique to posit predictions about RNA metabolism and protein functions. Thus evolutionary trees are evolutionary hypotheses that refer to specific facts, such as the characteristics of organisms (e.g., scales, feathers, fur), providing evidence for the patterns of descent, and a causal explanation for modification (i.e., natural selection or neutral drift) in any given lineage (e.g., Amniota). Evolutionary biologists test evolutionary theory using phylogenetic systematic methods that measure how much the hypothesis (a particular branching pattern in an evolutionary tree) increases the likelihood of the evidence (the distribution of characters among lineages). The severity of tests for a theory increases if the predictions "are the least probable of being observed if the causal event did not occur." "Testability is a measure of how much the hypothesis increases the likelihood of the evidence."

Vestigial structures

Evidence for common descent comes from the existence of vestigial structures. These rudimentary structures are often homologous to structures that correspond in related or ancestral species. A wide range of structures exist such as mutated and non-functioning genes, parts of a flower, muscles, organs, and even behaviors. This variety can be found across many different groups of

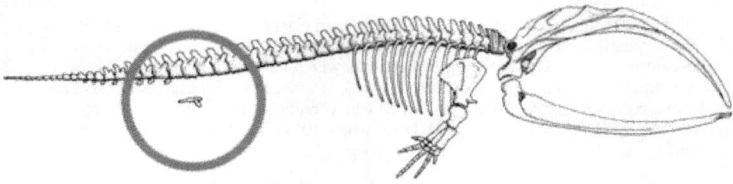

Figure 32: *Figure 2b: Skeleton of a baleen whale with the hind limb and pelvic bone structure circled in red. This bone structure stays internal during the entire life of the species.*

species. In many cases they are degenerated or underdeveloped. The existence of vestigial organs can be explained in terms of changes in the environment or modes of life of the species. Those organs are typically functional in the ancestral species but are now either semi-functional, nonfunctional, or re-purposed.

Scientific literature concerning vestigial structures abounds. One study complied 64 examples of vestigial structures found in the literature across a wide range of disciplines within the 21st century. The following non-exhaustive list summarizes Senter et al. alongside various other examples:

- The presence of remnant mitochondria (mitosomes) that have lost the ability to synthesize ATP in *Entamoeba histolytica, Trachipleistophora hominis, Cryptosporidium parvum, Blastocystis hominis,* and *Giardia intestinalis.*
- Remnant chloroplast organelles (leucoplasts) in non-photosynthetic algae species (*Plasmodium falciparum, Toxoplasma gondii, Aspasia longa, Anthophysa vegetans, Ciliophrys infusionum, Pteridomonas danica, Paraphysomonas, Spumella* and *Epifagus americana.*
- Missing stamens (unvascularized staminodes) on *Gilliesia* and *Gethyum* flowers.
- Non-functioning androecium in female flowers and non-functioning gynoecium in male flowers of the cactus species *Consolea spinosissima.*
- Remnant stamens on female flowers of *Fragaria virginiana*; all species in the genus *Schiedea*; and on *Penstemon centranthifolius, P. rostriflorus, P. ellipticus,* and *P. palmeri.*
- Vestigial anthers on *Nemophila menziesii.*

- Reduced hindlimbs and pelvic girdle embedded in the muscles of extant whales (see figure 2b). Occasionally, the genes that code for longer extremities cause a modern whale to develop legs. On October 28, 2006, a four-finned bottlenose dolphin was caught and studied due to its extra set of hind limbs. These legged Cetacea display an example of an atavism predicted from their common ancestry.

- Nonfunctional hind wings in *Carabus solieri* and other beetles.
- Remnant eyes (and eye structures) in animals that have lost sight such as blind cavefish (e.g. *Astyanax mexicanus*), mole rats, snakes, spiders, salamanders, shrimp, crayfish, and beetles.
- Vestigial eye in the extant *Rhineura floridana* and remnant jugal in the extinct *Rhineura hatchery* (reclassified as *Protorhineura hatcherii*).
- Functionless wings in flightless birds such as ostriches, kiwis, cassowaries, and emus.
- The presence of the plica semilunaris in the human eye—a vestigial remnant of the nictitating membrane.
- Harderian gland in primates.
- Reduced hind limbs and pelvic girdle structures in legless lizards, skinks, amphisbaenians, and some snakes.
- Reduced and missing olfactory apparatus in whales that still possess vestigial olfactory receptor subgenomes.
- Vestigial teeth in narwhal.
- Rudimentary digits of *Ateles geoffroyi*, *Colobus guereza*, and *Perodicticus potto*.
- Vestigial dental primordia in the embryonic tooth pattern in mice.
- Reduced or absent vomeronasal organ in humans and Old World monkeys.
- Presence of non-functional sinus hair muscles in humans used in whisker movement.
- Degenerating palmaris longus muscle in humans.
- Teleost fish, anthropoid primates (Simians), guinea pigs, some bat species, and some Passeriformes have lost the ability to synthesize vitamin C (ascorbic acid), yet still possess the genes involved. This inability is due to mutations of the L-gulono-γ-lactone oxidase (*GLO*) gene— and in primates, teleost fish, and guinea pigs it is irreversible.
- Remnant abdominal segments in cirripedes (barnacles).
- Non-mammalian vertebrate embryos depend on nutrients from the yolk sac. Humans and other mammal genomes contain broken, nonfunctioning genes that code for the production of yolk. alongside the presence of an empty yolk sac with the embryo.
- Dolphin embryonic limb buds.
- Leaf formation in some cacti species.
- Presence of a vestigial endosymbiont *Lepidodinium viride* within the dinoflagellate *Gymnodinium chlorophorum*.
- The species *Dolabrifera dolabrifera* has an ink gland but is "incapable of producing ink or its associated anti-predator proteins".

Specific examples from comparative anatomy

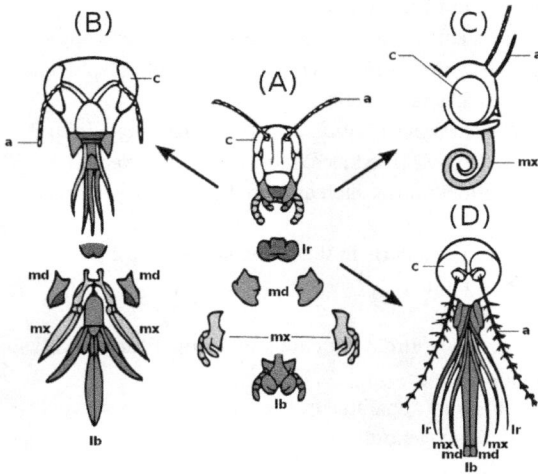

Figure 2c: Adaptation of insect mouthparts: a, antennae; c, compound eye; lb, labrium; lr, labrum; md, mandibles; mx, maxillae.

(A) Primitive state — biting and chewing: *e.g.* grasshopper. Strong mandibles and maxillae for manipulating food.

(B) Ticking and biting: *e.g.* honey bee. Labium long to lap up nectar; mandibles chew pollen and mould wax.

(C) Sucking: *e.g.* butterfly. Labrum reduced; mandibles lost; maxillae long forming sucking tube.

(D) Piercing and sucking, *e.g..* female mosquito. Labrum and maxillae form tube; mandibles form piercing stylets; labrum grooved to hold other parts.

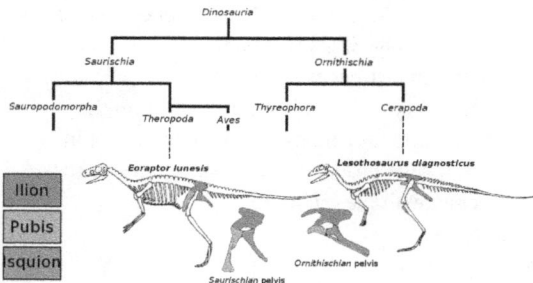

Figure 2d: Illustration of the *Eoraptor lunensis* pelvis of the order Saurischia and the *Lesothosaurus diagnosticus* pelvis of the order Ornithischia in the clade Dinosauria. The parts of the pelvis show modification over time. The clado-gram is shown to illustrate the distance of divergence between the two species.

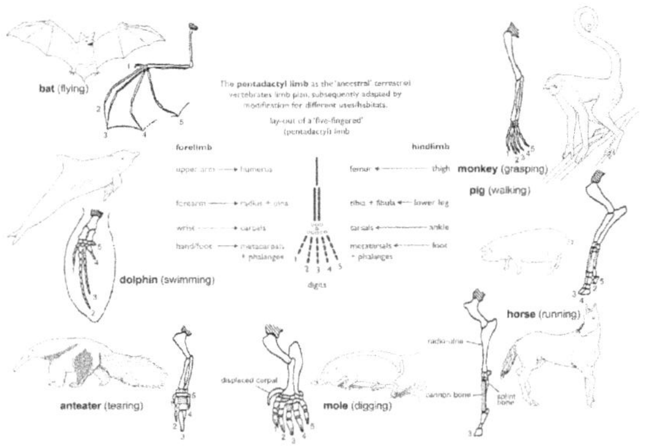

Figure 2e: The principle of homology illustrated by the adaptive radiation of the forelimb of mammals. All conform to the basic pentadactyl pattern but are modified for different usages. The third metacarpal is shaded throughout; the shoulder is crossed-hatched.

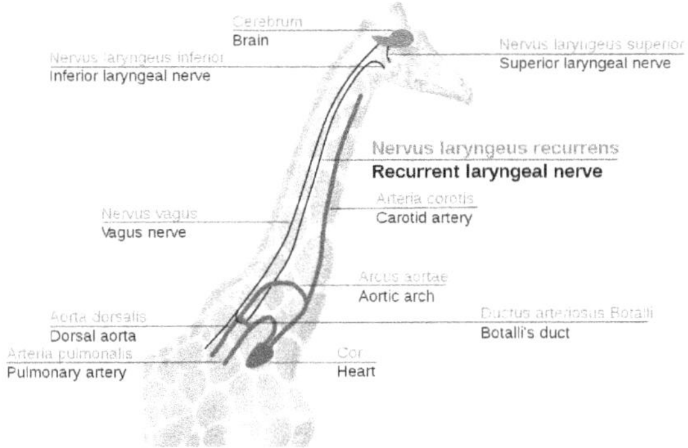

Figure 2f: The path of the recurrent laryngeal nerve in giraffes. The laryngeal nerve is compensated for by subsequent tinkering from natural selection.

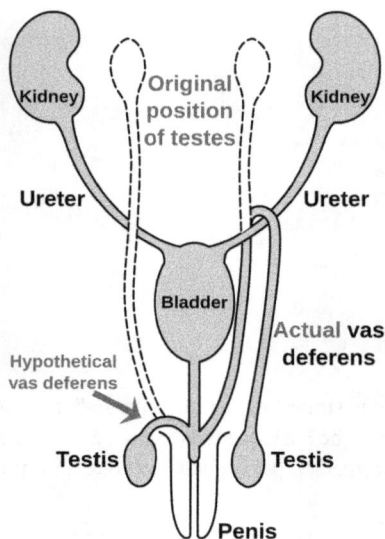

Figure 2g: Route of the vas deferens from the testis to the penis.

Insect mouthparts and appendages

Many different species of insects have mouthparts derived from the same embryonic structures, indicating that the mouthparts are modifications of a common ancestor's original features. These include a labrum (upper lip), a pair of mandibles, a hypopharynx (floor of mouth), a pair of maxillae, and a labium. (Fig. 2c) Evolution has caused enlargement and modification of these structures in some species, while it has caused the reduction and loss of them in other species. The modifications enable the insects to exploit a variety of food materials.

Insect mouthparts and antennae are considered homologues of insect legs. Parallel developments are seen in some arachnids: The anterior pair of legs may be modified as analogues of antennae, particularly in whip scorpions, which walk on six legs. These developments provide support for the theory that complex modifications often arise by duplication of components, with the duplicates modified in different directions.

Pelvic structure of dinosaurs

Similar to the pentadactyl limb in mammals, the earliest dinosaurs split into two distinct orders—the *saurischia* and *ornithischia*. They are classified as one or the other in accordance with what the fossils demonstrate. Figure 2d, shows that early *saurischians* resembled early *ornithischians*. The pattern of the pelvis in all species of dinosaurs is an example of homologous structures. Each order of dinosaur has slightly differing pelvis bones providing evidence of common descent. Additionally, modern birds show a similarity to ancient *saurischian* pelvic structures indicating the evolution of birds from dinosaurs. This can also be seen in Figure 5c as the Aves branch off the Theropoda suborder.

Pentadactyl limb

The pattern of limb bones called pentadactyl limb is an example of homologous structures (Fig. 2e). It is found in all classes of tetrapods (*i.e.* from amphibians to mammals). It can even be traced back to the fins of certain fossil fishes from which the first amphibians evolved such as tiktaalik. The limb has a single proximal bone (humerus), two distal bones (radius and ulna), a series of carpals (wrist bones), followed by five series of metacarpals (palm bones) and phalanges (digits). Throughout the tetrapods, the fundamental structures of pentadactyl limbs are the same, indicating that they originated from a common ancestor. But in the course of evolution, these fundamental structures have been modified. They have become superficially different and unrelated structures to serve different functions in adaptation to different environments and modes of life. This phenomenon is shown in the forelimbs of mammals. For example:

- In monkeys, the forelimbs are much elongated, forming a grasping hand used for climbing and swinging among trees.
- Pigs have lost their first digit, while the second and fifth digits are reduced. The remaining two digits are longer and stouter than the rest and bear a hoof for supporting the body.
- In horses, the forelimbs are highly adapted for strength and support. Fast and long-distance running is possible due to the extensive elongation of the third digit that bears a hoof.
- The mole has a pair of short, spade-like forelimbs for burrowing.
- Anteaters use their enlarged third digit for tearing into ant and termite nests.
- In cetaceans, the forelimbs become flippers for steering and maintaining equilibrium during swimming.
- In bats, the forelimbs have become highly modified and evolved into functioning wings. Four digits have become elongated, while the hook-like first digit remains free and is used to hang upside down.

Recurrent laryngeal nerve in giraffes

The recurrent laryngeal nerve is a fourth branch of the vagus nerve, which is a cranial nerve. In mammals, its path is unusually long. As a part of the vagus nerve, it comes from the brain, passes through the neck down to heart, rounds the dorsal aorta and returns up to the larynx, again through the neck. (Fig. 2f)

This path is suboptimal even for humans, but for giraffes it becomes even more suboptimal. Due to the lengths of their necks, the recurrent laryngeal nerve may be up to 4 m (13 ft) long, despite its optimal route being a distance of just several inches.

The indirect route of this nerve is the result of evolution of mammals from fish, which had no neck and had a relatively short nerve that innervated one gill slit and passed near the gill arch. Since then, the gill it innervated has become the larynx and the gill arch has become the dorsal aorta in mammals.

Route of the vas deferens

Similar to the laryngeal nerve in giraffes, the vas deferens is part of the male anatomy of many vertebrates; it transports sperm from the epididymis in anticipation of ejaculation. In humans, the vas deferens routes up from the testicle, looping over the ureter, and back down to the urethra and penis. It has been suggested that this is due to the descent of the testicles during the course of human evolution—likely associated with temperature. As the testicles descended, the vas deferens lengthened to accommodate the accidental "hook" over the ureter.

Evidence from paleontology

Figure 3a: An insect trapped in amber

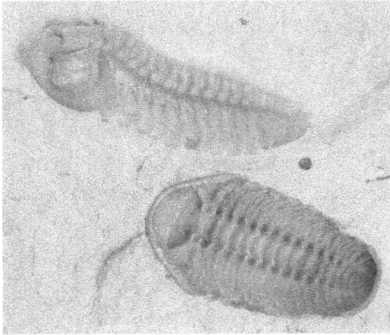

Figure 3b: Fossil trilobite, *Kainops invius*, from the early Devonian. Trilobites were hard-shelled arthropods, related to living horseshoe crabs and spiders, that first appeared in significant numbers around 540 mya, dying out 250 mya.

Figure 3c: Skull of *Cynognathus*, a eucynodont, one of a grouping of therapsids ("mammal-like reptiles") that is ancestral to all modern mammals.

Figure 3d: Charles Darwin collected fossils in South America, and found fragments of armor he thought were like giant versions of the scutes on the modern armadillos living nearby. The anatomist Richard Owen showed him that the fragments were from the gigantic extinct *Glyptodon*, related to the armadillos. This was one of the patterns of distribution that helped Darwin to develop his theory.

When organisms die, they often decompose rapidly or are consumed by scavengers, leaving no permanent evidences of their existence. However, occasionally, some organisms are preserved. The remains or traces of organisms from a past geologic age embedded in rocks by natural processes are called fossils. They are extremely important for understanding the evolutionary history of life on Earth, as they provide direct evidence of evolution and detailed information on the ancestry of organisms. Paleontology is the study of past life based on fossil records and their relations to different geologic time periods.

For fossilization to take place, the traces and remains of organisms must be quickly buried so that weathering and decomposition do not occur. Skeletal structures or other hard parts of the organisms are the most commonly occurring form of fossilized remains. There are also some trace "fossils" showing moulds, cast or imprints of some previous organisms.

As an animal dies, the organic materials gradually decay, such that the bones become porous. If the animal is subsequently buried in mud, mineral salts infiltrate into the bones and gradually fill up the pores. The bones harden into stones and are preserved as fossils. This process is known as petrification. If dead animals are covered by wind-blown sand, and if the sand is subsequently turned into mud by heavy rain or floods, the same process of mineral infiltration may occur. Apart from petrification, the dead bodies of organisms may be well preserved in ice, in hardened resin of coniferous trees (figure 3a), in tar, or in anaerobic, acidic peat. Fossilization can sometimes be a trace, an impression of a form. Examples include leaves and footprints, the fossils of which are made in layers that then harden.

Fossil record

It is possible to decipher how a particular group of organisms evolved by arranging its fossil record in a chronological sequence. Such a sequence can be determined because fossils are mainly found in sedimentary rock. Sedimentary rock is formed by layers of silt or mud on top of each other; thus, the resulting rock contains a series of horizontal layers, or strata. Each layer contains fossils typical for a specific time period when they formed. The lowest strata contain the oldest rock and the earliest fossils, while the highest strata contain the youngest rock and more recent fossils.

A succession of animals and plants can also be seen from fossil discoveries. By studying the number and complexity of different fossils at different stratigraphic levels, it has been shown that older fossil-bearing rocks contain fewer types of fossilized organisms, and they all have a simpler structure, whereas younger rocks contain a greater variety of fossils, often with increasingly complex structures.

For many years, geologists could only roughly estimate the ages of various strata and the fossils found. They did so, for instance, by estimating the time for the formation of sedimentary rock layer by layer. Today, by measuring the proportions of radioactive and stable elements in a given rock, the ages of fossils can be more precisely dated by scientists. This technique is known as radiometric dating.

Throughout the fossil record, many species that appear at an early stratigraphic level disappear at a later level. This is interpreted in evolutionary terms as indicating the times when species originated and became extinct. Geographical regions and climatic conditions have varied throughout Earth's history. Since organisms are adapted to particular environments, the constantly changing conditions favoured species that adapted to new environments through the mechanism of natural selection.

Extent of the fossil record

Despite the relative rarity of suitable conditions for fossilization, an estimated 250,000 fossil species have been named. The number of individual fossils this represents varies greatly from species to species, but many millions of fossils have been recovered: for instance, more than three million fossils from the last ice age have been recovered from the La Brea Tar Pits in Los Angeles. Many more fossils are still in the ground, in various geological formations known to contain a high fossil density, allowing estimates of the total fossil content of the formation to be made. An example of this occurs in South Africa's Beaufort Formation (part of the Karoo Supergroup, which covers most of South Africa), which is rich in vertebrate fossils, including therapsids (reptile-mammal transitional forms). It has been estimated that this formation contains 800 billion vertebrate fossils. Palentologists have documented numerous transitional forms and have constructed "an astonishingly comprehensive record of the key transitions in animal evolution". Conducting a survey of the paleontological literature, one would find that there is "abundant evidence for how all the major groups of animals are related, much of it in the form of excellent transitional fossils".

Limitations

The fossil record is an important source for scientists when tracing the evolutionary history of organisms. However, because of limitations inherent in the record, there are not fine scales of intermediate forms between related groups of species. This lack of continuous fossils in the record is a major limitation in tracing the descent of biological groups. When transitional fossils are found that show intermediate forms in what had previously been a gap in knowledge, they are often popularly referred to as "missing links".

There is a gap of about 100 million years between the beginning of the Cambrian period and the end of the Ordovician period. The early Cambrian period was the period from which numerous fossils of sponges, cnidarians (*e.g.*, jellyfish), echinoderms (*e.g.*, eocrinoids), molluscs (*e.g.*, snails) and arthropods (*e.g.*, trilobites) are found. The first animal that possessed the typical features of vertebrates, the *Arandaspis*, was dated to have existed in the later Ordovician period. Thus few, if any, fossils of an intermediate type between invertebrates and vertebrates have been found, although likely candidates include the Burgess Shale animal, *Pikaia gracilens*,[35] and its Maotianshan shales relatives, *Myllokunmingia*, *Yunnanozoon*, *Haikouella lanceolata*, and *Haikouichthys*.

Some of the reasons for the incompleteness of fossil records are:

- In general, the probability that an organism becomes fossilized is very low;

- Some species or groups are less likely to become fossils because they are soft-bodied;
- Some species or groups are less likely to become fossils because they live (and die) in conditions that are not favourable for fossilization;
- Many fossils have been destroyed through erosion and tectonic movements;
- Most fossils are fragmentary;
- Some evolutionary change occurs in populations at the limits of a species' ecological range, and as these populations are likely small, the probability of fossilization is lower (see punctuated equilibrium);
- Similarly, when environmental conditions change, the population of a species is likely to be greatly reduced, such that any evolutionary change induced by these new conditions is less likely to be fossilized;
- Most fossils convey information about external form, but little about how the organism functioned;
- Using present-day biodiversity as a guide, this suggests that the fossils unearthed represent only a small fraction of the large number of species of organisms that lived in the past.

Specific examples from paleontology

Evolution of the horse

Due to an almost-complete fossil record found in North American sedimentary deposits from the early Eocene to the present, the horse provides one of the best examples of evolutionary history (phylogeny).

This evolutionary sequence starts with a small animal called *Hyracotherium* (commonly referred to as *Eohippus*), which lived in North America about 54 million years ago then spread across to Europe and Asia. Fossil remains of *Hyracotherium* show it to have differed from the modern horse in three important respects: it was a small animal (the size of a fox), lightly built and adapted for running; the limbs were short and slender, and the feet elongated so that the digits were almost vertical, with four digits in the forelimbs and three digits in the hindlimbs; and the incisors were small, the molars having low crowns with rounded cusps covered in enamel.

The probable course of development of horses from *Hyracotherium* to *Equus* (the modern horse) involved at least 12 genera and several hundred species. The major trends seen in the development of the horse to changing environmental conditions may be summarized as follows:

- Increase in size (from 0.4 m to 1.5 m — from 15 in to 60 in);
- Lengthening of limbs and feet;
- Reduction of lateral digits;

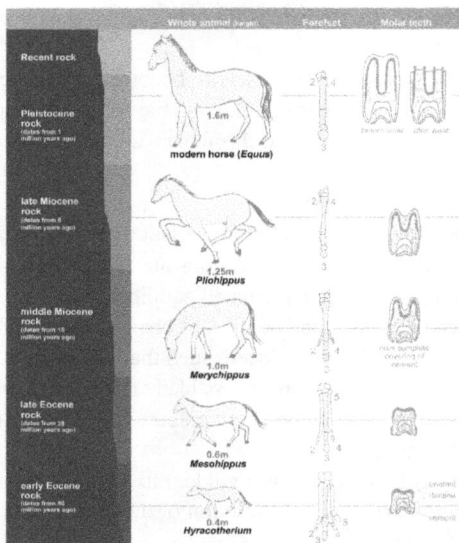

Figure 33: *Figure 3e: Evolution of the horse showing reconstruction of the fossil species obtained from successive rock strata. The foot diagrams are all front views of the left forefoot. The third metacarpal is shaded throughout. The teeth are shown in longitudinal section.*

- Increase in length and thickness of the third digit;
- Increase in width of incisors;
- Replacement of premolars by molars; and
- Increases in tooth length, crown height of molars.

Fossilized plants found in different strata show that the marshy, wooded country in which *Hyracotherium* lived became gradually drier. Survival now depended on the head being in an elevated position for gaining a good view of the surrounding countryside, and on a high turn of speed for escape from predators, hence the increase in size and the replacement of the splayed-out foot by the hoofed foot. The drier, harder ground would make the original splayed-out foot unnecessary for support. The changes in the teeth can be explained by assuming that the diet changed from soft vegetation to grass. A dominant genus from each geological period has been selected (see figure 3e) to show the slow alteration of the horse lineage from its ancestral to its modern form.[36]

Transition from fish to amphibians

Prior to 2004, paleontologists had found fossils of amphibians with necks, ears, and four legs, in rock no older than 365 million years old. In rocks more than 385 million years old they could only find fish, without these amphibian characteristics. Evolutionary theory predicted that since amphibians evolved from fish, an intermediate form should be found in rock dated between 365 and 385 million years ago. Such an intermediate form should have many fish-like characteristics, conserved from 385 million years ago or more, but also have many amphibian characteristics as well. In 2004, an expedition to islands in the Canadian arctic searching specifically for this fossil form in rocks that were 375 million years old discovered fossils of Tiktaalik. Some years later, however, scientists in Poland found evidence of fossilised tetrapod tracks predating *Tiktaalik*.

Evidence from biogeography

Data about the presence or absence of species on various continents and islands (biogeography) can provide evidence of common descent and shed light on patterns of speciation.

Continental distribution

All organisms are adapted to their environment to a greater or lesser extent. If the abiotic and biotic factors within a habitat are capable of supporting a particular species in one geographic area, then one might assume that the same species would be found in a similar habitat in a similar geographic area, e.g. in Africa and South America. This is not the case. Plant and animal species are discontinuously distributed throughout the world:

- Africa has Old World monkeys, apes, elephants, leopards, giraffes, and hornbills.
- South America has New World monkeys, cougars, jaguars, sloths, llamas, and toucans.
- Deserts in North and South America have native cacti, but deserts in Africa, Asia, and Australia have succulent (apart from *Rhipsalis baccifera*) which are native euphorbs that resemble cacti but are very different.

Even greater differences can be found if Australia is taken into consideration, though it occupies the same latitude as much of South America and Africa. Marsupials like kangaroos, bandicoots, and quolls make up about half of Australia's indigenous mammal species. By contrast, marsupials are today totally absent from Africa and form a smaller portion of the mammalian fauna of South America, where opossums, shrew opossums, and the monito del monte occur. The only living representatives of primitive egg-laying mammals (monotremes) are the echidnas and the platypus. The short-beaked echidna (*Tachyglossus aculeatus*) and its subspecies populate Australia, Tasmania, New Guinea, and Kangaroo Island while the long-beaked echidna (*Zaglossus bruijni*) lives only in New Guinea. The platypus lives in the waters of eastern Australia. They have been introduced to Tasmania, King Island, and Kangaroo Island. These Monotremes are totally absent in the rest of the world. On the other hand, Australia is missing many groups of placental mammals that are common on other continents (carnivorans, artiodactyls, shrews, squirrels, lagomorphs), although it does have indigenous bats and murine rodents; many other placentals, such as rabbits and foxes, have been introduced there by humans.

Other animal distribution examples include bears, located on all continents excluding Africa, Australia and Antarctica, and the polar bear solely in the Arctic Circle and adjacent land masses. Penguins are found only around the South Pole despite similar weather conditions at the North Pole. Families of sirenians are distributed around the earth's waters, where manatees are located in western Africa waters, northern South American waters, and West Indian waters only while the related family, the dugongs, are located only in Oceanic waters north of Australia, and the coasts surrounding the Indian Ocean. The now extinct Steller's sea cow resided in the Bering Sea.

The same kinds of fossils are found from areas known to be adjacent to one another in the past but that, through the process of continental drift, are now in widely divergent geographic locations. For example, fossils of the same types of ancient amphibians, arthropods and ferns are found in South America, Africa, India, Australia and Antarctica, which can be dated to the Paleozoic Era, when these regions were united as a single landmass called Gondwana.[37] Sometimes the descendants of these organisms can be identified and show unmistakable similarity to each other, even though they now inhabit very different regions and climates.

Island biogeography

Types of species found on islands

Evidence from island biogeography has played an important and historic role in the development of evolutionary biology. For purposes of biogeography, islands are divided into two classes. Continental islands are islands like Great Britain, and Japan that have at one time or another been part of a continent. Oceanic islands, like the Hawaiian islands, the Galápagos Islands and St. Helena, on the other hand are islands that have formed in the ocean and never

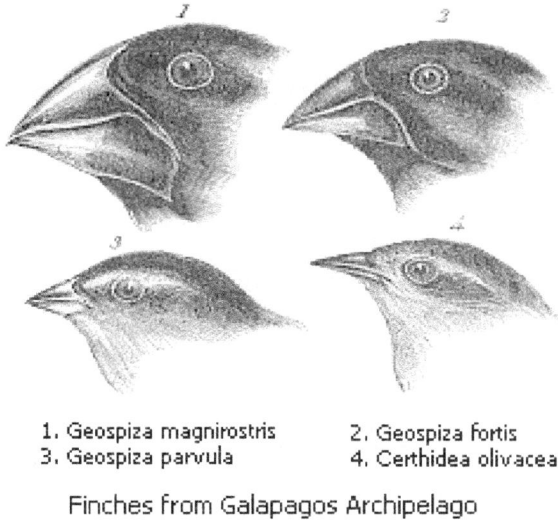

1. Geospiza magnirostris   2. Geospiza fortis
3. Geospiza parvula   4. Certhidea olivacea

Finches from Galapagos Archipelago

Figure 34: *Figure 4a: Four of the 13 finch species found on the Galápagos Archipelago, have evolved by an adaptive radiation that diversified their beak shapes to adapt them to different food sources.*

been part of any continent. Oceanic islands have distributions of native plants and animals that are unbalanced in ways that make them distinct from the biotas found on continents or continental islands. Oceanic islands do not have native terrestrial mammals (they do sometimes have bats and seals), amphibians, or fresh water fish. In some cases they have terrestrial reptiles (such as the iguanas and giant tortoises of the Galápagos Islands) but often (such as in Hawaii) they do not. This is despite the fact that when species such as rats, goats, pigs, cats, mice, and cane toads, are introduced to such islands by humans they often thrive. Starting with Charles Darwin, many scientists have conducted experiments and made observations that have shown that the types of animals and plants found, and not found, on such islands are consistent with the theory that these islands were colonized accidentally by plants and animals that were able to reach them. Such accidental colonization could occur by air, such as plant seeds carried by migratory birds, or bats and insects being blown out over the sea by the wind, or by floating from a continent or other island by sea (for example, by some kinds of plant seeds like coconuts that can survive immersion in salt water), and reptiles that can survive for extended periods on rafts of vegetation carried to sea by storms.

Endemism

Many of the species found on remote islands are endemic to a particular island or group of islands, meaning they are found nowhere else on earth. Examples of species endemic to islands include many flightless birds of New Zealand, lemurs of Madagascar, the Komodo dragon of Komodo, the dragon's blood tree of Socotra, Tuatara of New Zealand, and others. However, many such endemic species are related to species found on other nearby islands or continents; the relationship of the animals found on the Galápagos Islands to those found in South America is a well-known example. All of these facts, the types of plants and animals found on oceanic islands, the large number of endemic species found on oceanic islands, and the relationship of such species to those living on the nearest continents, are most easily explained if the islands were colonized by species from nearby continents that evolved into the endemic species now found there.

Other types of endemism do not have to include, in the strict sense, islands. Islands can mean isolated lakes or remote and isolated areas. Examples of these would include the highlands of Ethiopia, Lake Baikal, fynbos of South Africa, forests of New Caledonia, and others. Examples of endemic organisms living in isolated areas include the kagu of New Caledonia, cloud rats of the Luzon tropical pine forests of the Philippines, the boojum tree (*Fouquieria columnaris*) of the Baja California peninsula, the Baikal seal and the omul of Lake Baikal.

Adaptive radiations

Oceanic islands are frequently inhabited by clusters of closely related species that fill a variety of ecological niches, often niches that are filled by very different species on continents. Such clusters, like the finches of the Galápagos, Hawaiian honeycreepers, members of the sunflower family on the Juan Fernandez Archipelago and wood weevils on St. Helena are called adaptive radiations because they are best explained by a single species colonizing an island (or group of islands) and then diversifying to fill available ecological niches. Such radiations can be spectacular; 800 species of the fruit fly family *Drosophila*, nearly half the world's total, are endemic to the Hawaiian islands. Another illustrative example from Hawaii is the silversword alliance, which is a group of thirty species found only on those islands. Members range from the silverswords that flower spectacularly on high volcanic slopes to trees, shrubs, vines and mats that occur at various elevations from mountain top to sea level, and in Hawaiian habitats that vary from deserts to rainforests. Their closest relatives outside Hawaii, based on molecular studies, are tarweeds found on the west coast of North America. These tarweeds have sticky seeds that facilitate distribution by migrant birds. Additionally, nearly all of the species

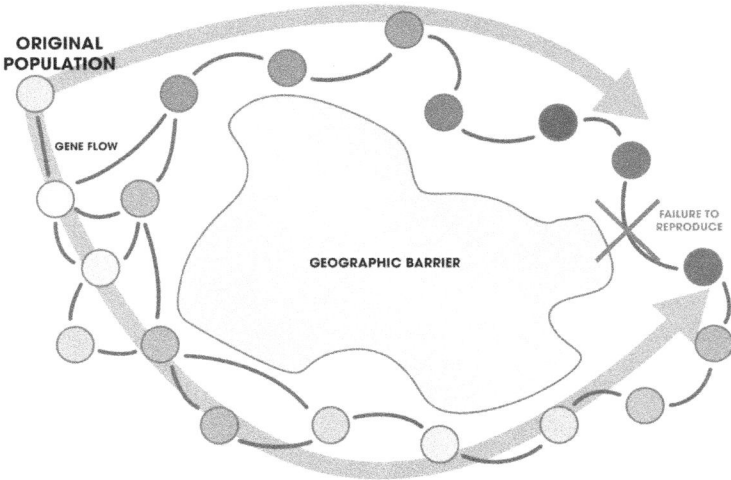

Figure 35: *In a ring species, gene flow occurs between neighboring populations, but at the ends of the "ring", the populations cannot interbreed.*

on the island can be crossed and the hybrids are often fertile, and they have been hybridized experimentally with two of the west coast tarweed species as well. Continental islands have less distinct biota, but those that have been long separated from any continent also have endemic species and adaptive radiations, such as the 75 lemur species of Madagascar, and the eleven extinct moa species of New Zealand.

Ring species

A ring species is a connected series of populations, each of which can interbreed with its neighbors, with at least two "end" populations which are too distantly related to interbreed, though with the potential for gene flow between all the populations. Ring species represent speciation and have been cited as evidence of evolution. They illustrate what happens over time as populations genetically diverge, specifically because they represent, in living populations, what normally happens over time between long deceased ancestor populations and living populations, in which the intermediates have become extinct. Richard Dawkins says that ring species "are only showing us in the spatial dimension something that must always happen in the time dimension".

Specific examples from biogeography

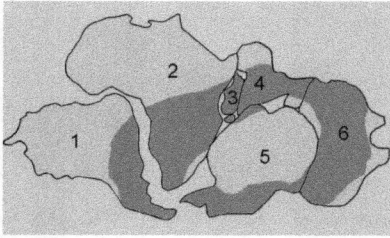

Figure 4c: Current distribution of *Glossopteris* placed on a Permian map showing the connection of the continents. (1, South America; 2, Africa; 3, Madagascar; 4, India; 5, Antarctica; and 6, Australia). Note that the map is a rough approximation of which leaves out additional land masses such as the Eurasian and North American plates.

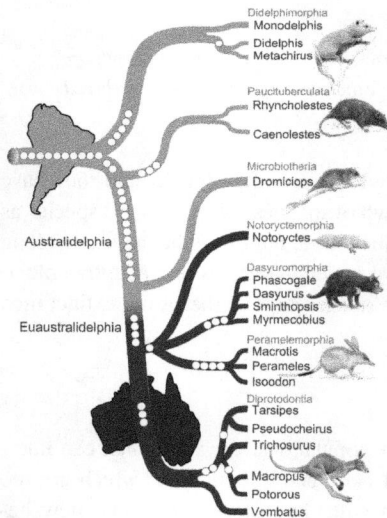

Figure 4d: A simplified phylogenetic tree of marsupials showing which groups reside on each continent.

Figure 4e: A dymaxion map of the biogeographic distribution of Camelidae species. Light blue indicates the Tertiary distribution, dark blue indicates the present-day distributions, and green indicates the introduced (feral) distributions. The yellow dot is the origin of the family Camelidae and the black arrows are the historic migration routes that explain the present day distribution.

Distribution of *Glossopteris*

The combination of continental drift and evolution can sometimes be used to predict what will be found in the fossil record. *Glossopteris* is an extinct species of seed fern plants from the Permian. *Glossopteris* appears in the fossil record around the beginning of the Permian on the ancient continent of Gondwana.[38] Continental drift explains the current biogeography of the tree. Present day *Glossopteris* fossils are found in Permian strata in southeast South America, southeast Africa, all of Madagascar, northern India, all of Australia, all of New Zealand, and scattered on the southern and northern edges of Antarctica. During the Permian, these continents were connected as Gondwana (see figure 4c) in agreement with magnetic striping, other fossil distributions, and glacial scratches pointing away from the temperate climate of the South Pole during the Permian.

Metatherian distribution

The history of metatherians (the clade containing marsupials and their extinct, primitive ancestors) provides an example of how evolutionary theory and the movement of continents can be combined to make predictions concerning fossil stratigraphy and distribution. The oldest metatherian fossils are found in present-day China. Metatherians spread westward into modern North America (still attached to Eurasia) and then to South America, which was connected to North America until around 65 mya. Marsupials reached Australia via Antarctica about 50 mya, shortly after Australia had split off suggesting a single dispersion event of just one species. Evolutionary theory suggests that the Australian marsupials descended from the older ones found in the Americas. Geologic evidence suggests that between 30 and 40 million years ago South America and Australia were still part of the Southern Hemisphere super continent of Gondwana and that they were connected by land that is now part of Antarctica. Therefore, when combining the models, scientists could predict that marsupials migrated from what is now South America, through Antarctica, and then to present-day Australia between 40 and 30 million years ago. A first marsupial fossil of the extinct family Polydolopidae was found on Seymour Island on the Antarctic Peninsula in 1982. Further fossils have subsequently been found, including members of the marsupial orders Didelphimorphia (opossum) and Microbiotheria, as well as ungulates and a member of the enigmatic extinct order Gondwanatheria, possibly *Sudamerica ameghi-noi*.[39]

Migration, isolation, and distribution of the camel

The history of the camel provides an example of how fossil evidence can be used to reconstruct migration and subsequent evolution. The fossil record indicates that the evolution of camelids started in North America (see figure 4e), from which, six million years ago, they migrated across the Bering Strait into Asia and then to Africa, and 3.5 million years ago through the Isthmus of Panama into South America. Once isolated, they evolved along their own lines, giving rise to the Bactrian camel and dromedary in Asia and Africa and the llama and its relatives in South America. Camelids then became extinct in North America at the end of the last ice age.

Evidence from selection

Examples for the evidence for evolution often stem from direct observation of natural selection in the field and the laboratory. This section is unique in that it provides a narrower context concerning the process of selection. All of the examples provided prior to this have described the evidence that evolution has occurred, but has not provided the major underlying mechanism: natural selection. This section explicitly provides evidence that natural selection occurs, has been replicated artificially, and can be replicated in laboratory experiments.

Scientists have observed and documented a multitude of events where natural selection is in action. The most well known examples are antibiotic resistance in the medical field along with better-known laboratory experiments documenting evolution's occurrence. Natural selection is tantamount to common descent in that long-term occurrence and selection pressures can lead to the diversity of life on earth as found today. All adaptations—documented and undocumented changes concerned—are caused by natural selection (and a few other minor processes). It is well established that, "...natural selection is a ubiquitous part of speciation...", and is the primary driver of speciation; therefore, the following examples of natural selection *and* speciation will often interdepend or correspond with one another. The examples below are only a small fraction of the actual experiments and observations.

Figure 36: *Figure 5a: The Chihuahua mix and Great Dane illustrate the range of sizes among dog breeds.*

Artificial selection and experimental evolution

Artificial selection demonstrates the diversity that can exist among organisms that share a relatively recent common ancestor. In artificial selection, one species is bred selectively at each generation, allowing only those organisms that exhibit desired characteristics to reproduce. These characteristics become increasingly well developed in successive generations. Artificial selection was successful long before science discovered the genetic basis. Examples of artificial selection include dog breeding, genetically modified food, flower breeding, and the cultivation of foods such as wild cabbage, and others.

Experimental evolution uses controlled experiments to test hypotheses and theories of evolution. In one early example, William Dallinger set up an experiment shortly before 1880, subjecting microbes to heat with the aim of forcing adaptive changes. His experiment ran for around seven years, and his published results were acclaimed, but he did not resume the experiment after the apparatus failed.

A large-scale example of experimental evolution is Richard Lenski's multi-generation experiment with *Escherichia coli*. Lenski observed that some strains of *E. coli* evolved a complex new ability, the ability to metabolize citrate, after tens of thousands of generations. The evolutionary biologist Jerry

Coyne commented as a critique of creationism, saying, "the thing I like most is it says you can get these complex traits evolving by a combination of unlikely events. That's just what creationists say can't happen." In addition to the metabolic changes, the different bacterial populations were found to have diverged in respect to both morphology (the overall size of the cell) and fitness (of which was measured in competition with the ancestors). The *E. coli* long-term evolution experiment that began in 1988 is still in progress, and has shown adaptations including the evolution of a strain of *E. coli* that was able to grow on citric acid in the growth media—a trait absent in all other known forms of *E. coli*, including the initial strain.

Invertebrates

Historical lead tolerance in *Daphnia*

A study of species of *Daphnia* and lead pollution in the 20th century predicted that an increase in lead pollution would lead to strong selection of lead tolerance. Researchers were able to use "resurrection ecology", hatching decades-old *Daphnia* eggs from the time when lakes were heavily polluted with lead. The hatchlings in the study were compared to current-day *Daphnia*, and demonstrated "dramatic fitness differences between old and modern phenotypes when confronted with a widespread historical environmental stressor". Essentially, the modern-day *Daphnia* were unable to resist or tolerate high levels of lead (this is due to the huge reduction of lead pollution in 21st century lakes). The old hatchlings, however, were able to tolerate high lead pollution. The authors concluded that "by employing the techniques of resurrection ecology, we were able to show clear phenotypic change over decades...".

Peppered moths

A classic example was the phenotypic change, light-to-dark color adaptation, in the peppered moth, due to pollution from the Industrial Revolution in England.

Microbes

Antimicrobial resistance

The development and spread of antibiotic-resistant bacteria is evidence for the process of evolution of species. Thus the appearance of vancomycin-resistant *Staphylococcus aureus*, and the danger it poses to hospital patients, is a direct result of evolution through natural selection. The rise of *Shigella* strains resistant to the synthetic antibiotic class of sulfonamides also demonstrates the generation of new information as an evolutionary process. Similarly, the appearance of DDT resistance in various forms of *Anopheles* mosquitoes, and the appearance of myxomatosis resistance in breeding rabbit populations in

Australia, are both evidence of the existence of evolution in situations of evolutionary selection pressure in species in which generations occur rapidly.

All classes of microbes develop resistance: including fungi (antifungal resistance), viruses (antiviral resistance), protozoa (antiprotozoal resistance), and bacteria (antibiotic resistance). This is to be expected when considering that all life exhibits universal genetic code and is therefore subject to the process of evolution through its various mechanisms.

Nylon-eating bacteria

Another example of organisms adapting to human-caused conditions are Nylon-eating bacteria: a strain of *Flavobacterium* that are capable of digesting certain byproducts of nylon 6 manufacturing. There is scientific consensus that the capacity to synthesize nylonase most probably developed as a single-step mutation that survived because it improved the fitness of the bacteria possessing the mutation. This is seen as a good example of evolution through mutation and natural selection that has been observed as it occurs and could not have come about until the production of nylon by humans.[40,41,42]

Plants and fungi

Monkeyflower radiation

Both subspecies *Mimulus aurantiacus puniceus* (red-flowered) and *Mimulus aurantiacus australis* (yellow-flowered) of monkeyflowers are isolated due to the preferences of their hummingbird and hawkmoth pollinators. The radiation of *M. aurantiacus* subspecies are mostly yellow colored; however, both *M. a.* ssp. *puniceus* and *M. a.* ssp. *flemingii* are red. Phylogenetic analysis suggests two independent origins of red-colored flowers that arose due to *cis*-regulatory mutations in the gene *MaMyb2* that is present in all *M. aurantiacus* subspecies. Further research suggested that two independent mutations did not take place, but one *MaMyb2* allele was transferred via introgressive hybridization. This study presents an example of the overlap of research in various disciplines. Gene isolation and *cis*-regulatory functions; phylogenetic analysis; geographic location and pollinator preference; and species hybridization and speciation are just some of the areas in which data can be obtained to document the occurrence of evolution.

Radiotrophic fungi

Like the codfish, human-caused pollution can come in different forms. Radiotrophic fungi is a perfect example of natural selection taking place after a chemical accident. Radiotrophic fungi appears to use the pigment melanin to convert gamma radiation into chemical energy for growth[43] and were first discovered in 2007 as black molds growing inside and around the Chernobyl Nuclear Power Plant. Research at the Albert Einstein College of Medicine showed that three melanin-containing fungi, *Cladosporium sphaerospermum*, *Wangiella dermatitidis*, and *Cryptococcus neoformans*, increased in biomass and accumulated acetate faster in an environment in which the radiation level was 500 times higher than in the normal environment.

Vertebrates

Guppies

While studying guppies (*Poecilia reticulata*) in Trinidad, biologist John Endler detected selection at work on the fish populations. To rule out alternative possibilities, Endler set up a highly controlled experiment to mimic the natural habitat by constructing ten ponds within a laboratory greenhouse at Princeton University. Each pond contained gravel to exactly match that of the natural ponds. After capturing a random sample of guppies from ponds in Trinidad, he raised and mixed them to create similar genetically diverse populations and measured each fish (spot length, spot height, spot area, relative spot length, relative spot height, total patch area, and standard body lengths). For the experiment he added *Crenicichla alta* (*P. reticulata*'s main predator) in four of the ponds, *Rivulus hartii* (a non-predator fish) in four of the ponds, and left the remaining two ponds empty with only the guppies. After 10 generations, comparisons were made between each pond's guppy populations and measurements were taken again. Endler found that the populations had evolved dramatically different color patterns in the control and non-predator pools and drab color patterns in the predator pool. Predation pressure had caused a selection against standing out from background gravel.

In parallel, during this experiment, Endler conducted a field experiment in Trinidad where he caught guppies from ponds where they had predators and relocated them to ponds upstream where the predators did not live. After 15 generations, Endler found that the relocated guppies had evolved dramatic and colorful patterns. Essentially, both experiments showed convergence due to similar selection pressures (i.e. predator selection against contrasting color patterns and sexual selection for contrasting color patterns).

In a later study by David Reznick, the field population was examined 11 years later after Endler relocated the guppies to high streams. The study found that

Figure 37: *Figure 5b: Endler's Trinadadian guppies (Poecilia reticulata)*

the populations has evolved in a number of different ways: bright color patterns, late maturation, larger sizes, smaller litter sizes, and larger offspring within litters. Further studies of *P. reticulata* and their predators in the streams of Trinidad have indicated that varying modes of selection through predation have not only changed the guppies color patterns, sizes, and behaviors, but their life histories and life history patterns.

Humans

Natural selection is observed in contemporary human populations, with recent findings demonstrating that the population at risk of the severe debilitating disease kuru has significant over-representation of an immune variant of the prion protein gene G127V versus non-immune alleles. Scientists postulate one of the reasons for the rapid selection of this genetic variant is the lethality of the disease in non-immune persons. Other reported evolutionary trends in other populations include a lengthening of the reproductive period, reduction in cholesterol levels, blood glucose and blood pressure.

A well known example of selection occurring in human populations is lactose tolerance. Lactose intolerance is the inability to metabolize lactose, because of a lack of the required enzyme lactase in the digestive system. The normal mammalian condition is for the young of a species to experience reduced lactase production at the end of the weaning period (a species-specific length of time). In humans, in non-dairy consuming societies, lactase production usually drops about 90% during the first four years of life, although the exact drop over time varies widely.[44] Lactase activity persistence in adults is associated with two polymorphisms: C/T 13910 and G/A 22018 located in the *MCM6* gene.

This gene difference eliminates the shutdown in lactase production, making it possible for members of these populations to continue consumption of raw milk and other fresh and fermented dairy products throughout their lives without difficulty. This appears to be an evolutionarily recent (around 10,000 years ago [and 7,500 years ago in Europe]) adaptation to dairy consumption, and has occurred independently in both northern Europe and east Africa in populations with a historically pastoral lifestyle.

Italian wall lizards

In 1971, ten adult specimens of *Podarcis sicula* (the Italian wall lizard) were transported from the Croatian island of Pod Kopište to the island Pod Mrčaru (about 3.5 km to the east). Both islands lie in the Adriatic Sea near Lastovo, where the lizards founded a new bottlenecked population. The two islands have similar size, elevation, microclimate, and a general absence of terrestrial predators and the *P. sicula* expanded for decades without human interference, even out-competing the (now locally extinct) *Podarcis melisellensis* population.

In the 1990s, scientists returned to Pod Mrčaru and found that the lizards there differed greatly from those on Kopište. While mitochondrial DNA analyses have verified that *P. sicula* currently on Mrčaru are genetically very similar to the Kopište source population, the new Mrčaru population of *P. sicula* had a larger average size, shorter hind limbs, lower maximal sprint speed and altered response to simulated predatory attacks compared to the original Kopište population. These changes were attributed to "relaxed predation intensity" and greater protection from vegetation on Mrčaru.

In 2008, further analysis revealed that the Mrčaru population of *P. sicula* have significantly different head morphology (longer, wider, and taller heads) and increased bite force compared to the original Kopište population. This change in head shape corresponded with a shift in diet: Kopište *P. sicula* are primarily insectivorous, but those on Mrčaru eat substantially more plant matter. The changes in foraging style may have contributed to a greater population density and decreased territorial behavior of the Mrčaru population.

Another difference found between the two populations was the discovery, in the Mrčaru lizards, of cecal valves, which slow down food passage and provide fermenting chambers, allowing commensal microorganisms to convert cellulose to nutrients digestible by the lizards. Additionally, the researchers discovered that nematodes were common in the guts of Mrčaru lizards, but absent from Kopište *P. sicula*, which do not have cecal valves. The cecal valves, which occur in less than 1 percent of all known species of scaled reptiles, have been described as an "adaptive novelty, a brand new feature not present in the ancestral population and newly evolved in these lizards".

PAH resistance in killifish

A similar study was also done regarding the polycyclic aromatic hydrocarbons (PAHs) that pollute the waters of the Elizabeth River in Portsmouth, Virginia. This chemical is a product of creosote, a type of tar. The Atlantic killifish (*Fundulus heteroclitus*) has evolved a resistance to PAHs involving the AHR gene (the same gene involved in the tomcods). This particular study focused on the resistance to "acute toxicity and cardiac teratogenesis" caused by PAHs. that mutated within the tomcods in the Hudson River.

PCB resistance in codfish

An example involving the direct observation of gene modification due to selection pressures is the resistance to PCBs in codfish. After General Electric dumped polychlorinated biphenyls (PCBs) in the Hudson River from 1947 through 1976, tomcods (*Microgadus tomcod*) living in the river were found to have evolved an increased resistance to the compound's toxic effects. The tolerance to the toxins is due to a change in the coding section of specific gene. Genetic samples were taken from the cods from 8 different rivers in the New England region: the St. Lawrence River, Miramichi River, Margaree River, Squamscott River, Niantic River, the Shinnecock Basic, the Hudson River, and the Hackensack River. Genetic analysis found that in the population of tomcods in the four southernmost rivers, the gene AHR2 (aryl hydrocarbon receptor 2) was present as an allele with a difference of two amino acid deletions. This deletion conferred a resistance to PCB in the fish species and was found in 99% of Hudson River tomcods, 92% in the Hackensack River, 6% in the Niantic River, and 5% in Shinnecock Bay. This pattern along the sampled bodies of waters infers a direct correlation of selective pressures leading to the evolution of PCB resistance in Atlantic tomcod fish.

Urban wildlife

Urban wildlife is a broad and easily observable case of human-caused selection pressure on wildlife. With the growth in human habitats, different animals have adapted to survive within these urban environments. These types of environments can exert selection pressures on organisms, often leading to new adaptations. For example, the weed *Crepis sancta*, found in France, has two types of seed, heavy and fluffy. The heavy ones land nearby to the parent plant, whereas fluffy seeds float further away on the wind. In urban environments, seeds that float far often land on infertile concrete. Within about 5–12 generations, the weed evolves to produce significantly heavier seeds than its rural relatives. Other examples of urban wildlife are rock pigeons and species of crows adapting to city environments around the world; African penguins in

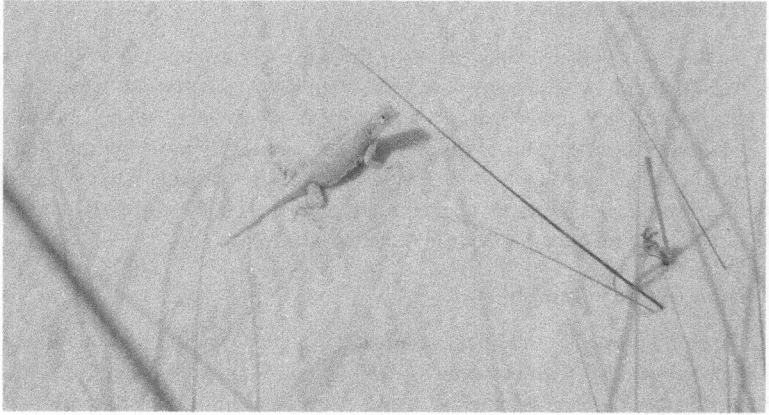

Figure 38: *Figure 5c: H. m. ruthveni, a White
Sands ecotonal variant of Holbrookia maculata*

Simon's Town; baboons in South Africa; and a variety of insects living in human habitations. Studies have been conducted and have found striking changes to animals' (more specifically mammals') behavior and physical brain size due to their interactions with human-created environments.

White Sands lizards

Animals that exhibit ecotonal variations allow for research concerning the mechanisms that maintain population differentiation. A wealth of information about natural selection, genotypic, and phenotypic variation; adaptation and ecomorphology; and social signaling has been acquired from the studies of three species of lizards located in the White Sands desert of New Mexico. *Holbrookia maculata, Aspidoscelis inornata,* and *Sceloporus undulatus* exhibit ecotonal populations that match both the dark soils and the white sands in the region. Research conducted on these species has found significant phenotypic and genotypic differences between the dark and light populations due to strong selection pressures. For example, *H. maculata* exhibits the strongest phenotypic difference (matches best with the substrate) of the light colored population coinciding with the least amount of gene flow between the populations and the highest genetic differences when compared to the other two lizard species.

New Mexico's White Sands are a recent geologic formation (approximately 6000 years old to possibly 2000 years old). This recent origin of these gypsum sand dunes suggests that species exhibiting lighter-colored variations have evolved in a relatively short time frame. The three lizard species previously mentioned have been found to display variable social signal coloration in co-existence with their ecotonal variants. Not only have the three species convergently evolved their lighter variants due to the selection pressures from the environment, they've also evolved ecomorphological differences: morphology, behavior (in is case, escape behavior), and performance (in this case, sprint speed) collectively. Roches' work found surprising results in the escape behavior of *H. maculata* and *S. undulatus*. When dark morphs were placed on white sands, their startle response was significantly diminished. This result could be due to varying factors relating to sand temperature or visual acuity; however, regardless of the cause, "...failure of mismatched lizards to sprint could be maladaptive when faced with a predator".

Evidence from speciation

Speciation is the evolutionary process by which new biological species arise. Biologists research species using different theoretical frameworks for what constitutes a species (see species problem and species complex) and there exists debate with regard to delineation. Nevertheless, much of the current research suggests that, "...speciation is a process of emerging genealogical distinctness, rather than a discontinuity affecting all genes simultaneously" and, in allopatry (the most common form of speciation), "reproductive isolation is a byproduct of evolutionary change in isolated populations, and thus can be considered an evolutionary accident". Speciation occurs as the result of the latter (allopatry); however, a variety of differing agents have been documented and are often defined and classified in various forms (e.g. peripatric, parapatric, sympatric, polyploidization, hybridization, etc.). Instances of speciation have been observed in both nature and the laboratory. A.-B Florin and A. Ödeen note that, "strong laboratory evidence for allopatric speciation is lacking..."; however, contrary to laboratory studies (focused specifically on models of allopatric speciation), "speciation most definitely occurs; [and] the vast amount of evidence from nature makes it unreasonable to argue otherwise". Coyne and Orr compiled a list of 19 laboratory experiments on *Drosophila* presenting examples of allopatric speciation by divergent selection concluding that, "reproductive isolation in allopatry can evolve as a byproduct of divergent selection".

Research documenting speciation is abundant. Biologists have documented numerous examples of speciation in nature—with evolution having produced far more species than any observer would consider necessary. For example, there are well over 350,000 described species of beetles.[45] Examples of speciation come from the observations of island biogeography and the process of adaptive radiation, both explained previously. Evidence of common descent can also be found through paleontological studies of speciation within geologic strata. The examples described below represent different modes of speciation and provide strong evidence for common descent. It is important to acknowledge that not all speciation research directly observes divergence from "start-to-finish". This is by virtue of research delimitation and definition ambiguity, and occasionally leads research towards historical reconstructions. In light of this, examples abound, and the following are by no means exhaustive—comprising only a small fraction of the instances observed. Once again, take note of the established fact that, "...natural selection is a ubiquitous part of speciation...", and is the primary driver of speciation, so; hereinafter, examples of speciation will often interdepend and correspond with selection.

Fossils

Limitations exist within the fossil record when considering the concept of what constitutes a species. Paleontologists largely rely on a different framework: the morphological species concept. Due to the absence of information such as reproductive behavior or genetic material in fossils, paleontologists distinguish species by their phenotypic differences. Extensive investigation of the fossil record has led to numerous theories concerning speciation (in the context of paleontology) with many of the studies suggesting that stasis, punctuation, and lineage branching are common. In 1995, D. H. Erwin, et al. published a major work—*New Approaches to Speciation in the Fossil Record*—which compiled 58 studies of fossil speciation (between 1972 and 1995) finding most of the examples suggesting stasis (involving anagenesis or punctuation) and 16 studies suggesting speciation. Despite stasis appearing to be the predominate conclusion at first glance, this particular meta-study investigated deeper, concluding that, "...no single pattern appears dominate..." with "...the preponderance of studies illustrating *both* stasis and gradualism in the history of a single lineage". Many of the studies conducted utilize seafloor sediments that can provide a significant amount of data concerning planktonic microfossils. The succession of fossils in stratigraphy can be used to determine evolutionary trends among fossil organisms. In addition, incidences of speciation can be interpreted from the data and numerous studies have been conducted documenting both morphological evolution and speciation.

Globorotalia

Extensive research on the planktonic foraminifer *Globorotalia truncatulinoides* has provided insight into paleobiogeographical and paleoenvironmental studies alongside the relationship between the environment and evolution. In an extensive study of the paleobiogeography of *G. truncatulinoides*, researchers found evidence that suggested the formation of a new species (via the sympatric speciation framework). Cores taken of the sediment containing the three

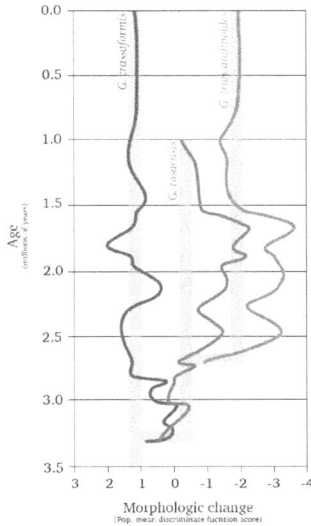

Figure 39: *Figure 6a: Morphologic change of Globorotalia crassaformis, G. tosaensis, and G. truncatulinoides over 3.5 Ma. Superimposed is a phylogenetic tree of the group. Adapted from Lazarus et al. (1995).*

species *G. crassaformis*, *G. tosaensis*, and *G. truncatulinoides* found that before 2.7 Ma, only *G. crassaformis* and *G. tosaensis* existed. A speciation event occurred at that time, whereby intermediate forms existed for quite some time. Eventually *G. tosaensis* disappears from the record (suggesting extinction) but exists as an intermediate between the extant *G. crassaformis* and *G. truncatulinoides*. This record of the fossils also matched the already existing phylogeny constructed by morphological characters of the three species. See figure 6a.

Radiolaria

In a large study of five species of radiolarians (*Calocycletta caepa*, *Pterocanium prismatium*, *Pseudoculous vema*, *Eucyrtidium calvertense*, and *Eucyrtidium matuyamai*), the researchers documented considerable evolutionary change in each lineage. Alongside this, trends with the closely related species *E. calvertense* and *E. matuyamai* showed that about 1.9 Mya *E. calvertense* invaded a new region of the Pacific, becoming isolated from the main population. The stratigraphy of this species clearly shows that this isolated population evolved into *E. Matuyamai*. It then reinvaded the region of the still-existing and static *E. calvertense* population whereby a sudden decrease in body size occurred. Eventually the invader *E. matuyamai* disappeared from the stratum (presumably due to extinction) coinciding with a desistance of size reduction of the *E. calvertense* population. From that point on, the change in size leveled to a constant. The authors suggest competition-induced character displacement.

Rhizosolenia

Researchers conducted measurements on 5,000 *Rhizosolenia* (a planktonic diatom) specimens from eight sedimentary cores in the Pacific Ocean. The core samples spanned two million years and were chronologized using sedimentary magnetic field reversal measurements. All the core samples yielded a similar pattern of divergence: with a single lineage (*R. bergonii*) occurring before 3.1 Mya and two morphologically distinct lineages (daughter species: *R. praebergonii*) appearing after. The parameters used to measure the samples were consistent throughout each core. An additional study of the daughter species *R. praebergonii* found that, after the divergence, it invaded the Indian Ocean.

Turborotalia

A recent study was conducted involving the planktonic foraminifer Turborotalia. The authors extracted "51 stratigraphically ordered samples from a site within the oceanographically stable tropical North Pacific gyre". Two hundred individual species were examined using ten specific morphological traits (size, compression index, chamber aspect ratio, chamber inflation, aperture aspect ratio, test height, test expansion, umbilical angle, coiling direction, and the number of chambers in the final whorl). Utilizing multivariate statistical clustering methods, the study found that the species continued to evolve nondirectionally within the Eocene from 45 Ma to about 36 Ma. However, from 36 Ma to approximately 34 Ma, the stratigraphic layers showed two distinct clusters with significantly defining characteristics distinguishing one another from a single species. The authors concluded that speciation must have occurred and that the two new species were ancestral to the prior species. Just as in most of evolutionary biology, this example represents the interdisciplinary nature of the field and the necessary collection of data from various fields (e.g. oceanography, paleontology) and the integration of mathematical analysis (e.g. biometry).

Vertebrates

There exists evidence for vertebrate speciation despite limitations imposed by the fossil record. Studies have been conducted documenting similar patterns seen in marine invertebrates. For example, extensive research documenting rates of morphological change, evolutionary trends, and speciation patterns in small mammals has significantly contributed to the scientific literature; once more, demonstrating that evolution (and speciation) occurred in the past and lends support common ancestry.

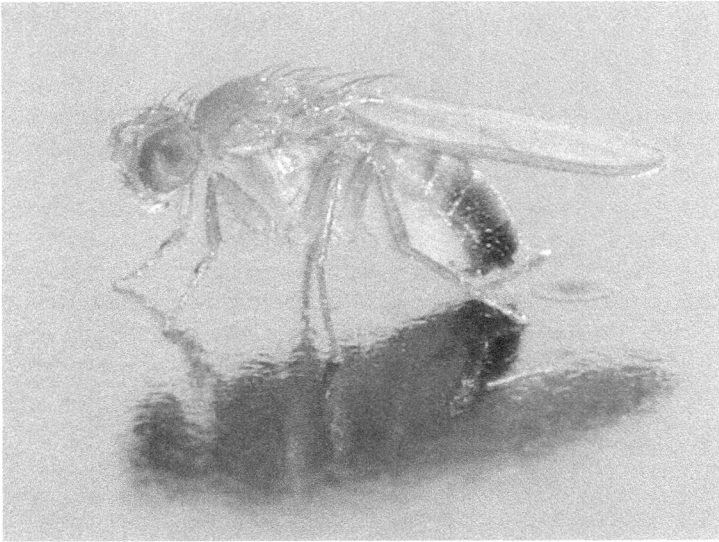

Figure 40: *Figure 6b:* A common fruit fly (Drosophila melanogaster)

A study of four mammalian genera: *Hyopsodus, Pelycodus, Haplomylus* (three from the Eocene), and *Plesiadapis* (from the Paleocene) found that—through a large number of stratigraphic layers and specimen sampling—each group exhibited, "gradual phyletic evolution, overall size increase, iterative evolution of small species, and character divergence following the origin of each new lineage". The authors of this study concluded that speciation was discernible. In another study concerning morphological trends and rates of evolution found that the European arvicolid rodent radiated into 52 distinct lineages over a time frame of 5 million years while documenting examples of phyletic gradualism, punctuation, and stasis.

Invertebrates

Drosophila melanogaster

William R. Rice and George W. Salt found experimental evidence of sympatric speciation in the common fruit fly. They collected a population of *Drosophila melanogaster* from Davis, California and placed the pupae into a habitat maze. Newborn flies had to investigate the maze to find food. The flies had three choices to take in finding food. Light and dark (phototaxis), up and down (geotaxis), and the scent of acetaldehyde and the scent of ethanol (chemotaxis) were the three options. This eventually divided the flies into 42 spatio-temporal habitats.

They then cultured two strains that chose opposite habitats. One of the strains emerged early, immediately flying upward in the dark attracted to the acetaldehyde. The other strain emerged late and immediately flew downward, attracted to light and ethanol. Pupae from the two strains were then placed together in the maze and allowed to mate at the food site. They then were collected. A selective penalty was imposed on the female flies that switched habitats. This entailed that none of their gametes would pass on to the next generation. After 25 generations of this mating test, it showed reproductive isolation between the two strains. They repeated the experiment again without creating the penalty against habitat switching and the result was the same; reproductive isolation was produced.

Gall wasps

A study of the gall-forming wasp species *Belonocnema treatae* found that populations inhabiting different host plants (*Quercus geminata* and *Q. Virginiana*) exhibited different body size and gall morphology alongside a strong expression of sexual isolation. The study hypothesized that *B. treatae* populations inhabiting different host plants would show evidence of divergent selection promoting speciation. The researchers sampled gall wasp species and oak tree localities, measured body size (right hand tibia of each wasp), and counted gall chamber numbers. In addition to measurements, they conducted mating assays and statistical analyses. Genetic analysis was also conducted on two mtDNA sites (416 base pairs from cytochrome C and 593 base pairs from cytochrome oxidase) to "control for the confounding effects of time since divergence among allopatric populations".

In an additional study, the researchers studied two gall wasp species *B. treatae* and *Disholcaspis quercusvirens* and found strong morphological and behavioral variation among host-associated populations. This study further confounded prerequisites to speciation.

Hawthorn fly

One example of evolution at work is the case of the hawthorn fly, *Rhagoletis pomonella*, also known as the apple maggot fly, which appears to be undergoing sympatric speciation. Different populations of hawthorn fly feed on different fruits. A distinct population emerged in North America in the 19th century some time after apples, a non-native species, were introduced. This apple-feeding population normally feeds only on apples and not on the historically preferred fruit of hawthorns. The current hawthorn feeding population does not normally feed on apples. Some evidence, such as the fact that six out of thirteen allozyme loci are different, that hawthorn flies mature later in the season and take longer to mature than apple flies; and that there is little evidence of interbreeding (researchers have documented a 4–6% hybridization rate) suggests that speciation is occurring.

London Underground mosquito

The London Underground mosquito is a species of mosquito in the genus *Culex* found in the London Underground. It evolved from the overground species *Culex pipiens*. This mosquito, although first discovered in the London Underground system, has been found in underground systems around the world. It is suggested that it may have adapted to human-made underground systems since the last century from local above-ground *Culex pipiens*, although more recent evidence suggests that it is a southern mosquito variety related to *Culex pipiens* that has adapted to the warm underground spaces of northern cities.

The two species have very different behaviours, are extremely difficult to mate, and with different allele frequency, consistent with genetic drift during a founder event. More specifically, this mosquito, *Culex pipiens molestus*, breeds all-year round, is cold intolerant, and bites rats, mice, and humans, in contrast to the above ground species *Culex pipiens* that is cold tolerant, hibernates in the winter, and bites only birds. When the two varieties were cross-bred the eggs were infertile suggesting reproductive isolation.

The genetic data indicates that the *molestus* form in the London Underground mosquito appears to have a common ancestry, rather than the population at each station being related to the nearest aboveground population (i.e. the *pipiens* form). Byrne and Nichols' working hypothesis was that adaptation to the underground environment had occurred locally in London only once. These widely separated populations are distinguished by very minor genetic differences, which suggest that the molestus form developed: a single mtDNA difference shared among the underground populations of ten Russian cities; a single fixed microsatellite difference in populations spanning Europe, Japan, Australia, the middle East and Atlantic islands.

Snapping shrimp and the isthmus of Panama

Debate exists determining when the isthmus of Panama closed. Much of the evidence supports a closure approximately 2.7 to 3.5 mya using "...multiple lines of evidence and independent surveys". However, a recent study suggests an earlier, transient bridge existed 13 to 15 mya. Regardless of the timing of the isthmus closer, biologists can study the species on the Pacific and Caribbean sides in, what has been called, "one of the greatest natural experiments in evolution." Studies of snapping shrimp in the genus *Alpheus* have provided direct evidence of allopatric speciation events, and contributed to the literature concerning rates of molecular evolution. Phylogenetic reconstructions using "multilocus datasets and coalescent-based analytical methods" support the relationships of the species in the group and molecular clock techniques support the separation of 15 pairs of *Alpheus* species between 3 and 15 million years ago.

Plants

The botanist Verne Grant pioneered the field of plant speciation with his research and major publications on the topic. As stated before, many biologists rely on the biological species concept, with some modern researchers utilizing the phylogenetic species concept. Debate exists in the field concerning which framework should be applied in the research. Regardless, reproductive isolation is the primary role in the process of speciation and has been studied extensively by biologists in their respective disciplines.

Both hybridization and polyploidy have also been found to be major contributors to plant speciation. With the advent of molecular markers, "hybridization [is] considerably more frequent than previously believed". In addition to these two modes leading to speciation, pollinator preference and isolation, chromosomal rearrangements, and divergent natural selection have become critical to the speciation of plants. Furthermore, recent research suggests that sexual selection, epigenetic drivers, and the creation of incompatible allele combinations caused by balancing selection also contribute to the formation of new species. Instances of these modes have been researched in both the laboratory and in nature. Studies have also suggested that, due to "the sessile nature of plants... [it increases] the relative importance of ecological speciation..."

Hybridization between two different species sometimes leads to a distinct phenotype. This phenotype can also be fitter than the parental lineage and as such, natural selection may then favor these individuals. Eventually, if reproductive isolation is achieved, it may lead to a separate species. However, reproductive isolation between hybrids and their parents is particularly difficult to achieve and thus hybrid speciation is considered a rare event. However, hybridization resulting in reproductive isolation is considered an important means of speciation in plants, since polyploidy (having more than two copies of each chromosome) is tolerated in plants more readily than in animals.

Polyploidy is important in hybrids as it allows reproduction, with the two different sets of chromosomes each being able to pair with an identical partner during meiosis. Polyploids also have more genetic diversity, which allows them to avoid inbreeding depression in small populations. Hybridization without change in chromosome number is called homoploid hybrid speciation. It is considered very rare but has been shown in *Heliconius* butterflies and sunflowers. Polyploid speciation, which involves changes in chromosome number, is a more common phenomenon, especially in plant species.

Polyploidy is a mechanism that has caused many rapid speciation events in sympatry because offspring of, for example, tetraploid x diploid matings often result in triploid sterile progeny. Not all polyploids are reproductively isolated from their parental plants, and gene flow may still occur for example through triploid hybrid x diploid matings that produce tetraploids, or matings between meiotically unreduced gametes from diploids and gametes from tetraploids. It has been suggested that many of the existing plant and most animal species have undergone an event of polyploidization in their evolutionary history. Reproduction of successful polyploid species is sometimes asexual, by parthenogenesis or apomixis, as for unknown reasons many asexual organisms are polyploid. Rare instances of polyploid mammals are known, but most often result in prenatal death.

Researchers consider reproductive isolation as key to speciation. A major aspect of speciation research is to determine the nature of the barriers that inhibit reproduction. Botanists often consider the zoological classifications of prezygotic and postzygotic barriers as inadequate. The examples provided below give insight into the process of speciation.

Mimulus peregrinus

The creation of a new allopolyploid species of monkeyflower (*Mimulus peregrinus*) was observed on the banks of the Shortcleuch Water—a river in Leadhills, South Lanarkshire, Scotland. Parented from the cross of the two species *Mimulus guttatus* (containing 14 pairs of chromosomes) and *Mimulus luteus* (containing 30-31 pairs from a chromosome duplication), *M. peregrinus* has six copies of its chromosomes (caused by the duplication of the sterile hybrid triploid). Due to the nature of these species, they have the ability to self-fertilize. Because of its number of chromosomes it is not able to pair with *M. guttatus*, *M. luteus*, or their sterile triploid offspring. *M. peregrinus* will either die, producing no offspring, or reproduce with itself effectively leading to a new species.

Raphanobrassica

Raphanobrassica includes all intergeneric hybrids between the genera *Raphanus* (radish) and *Brassica* (cabbages, etc.). The *Raphanobrassica* is an allopolyploid cross between the radish (*Raphanus sativus*) and cabbage (*Brassica oleracea*). Plants of this parentage are now known as radicole. Two other fertile forms of *Raphanobrassica* are known. Raparadish, an allopolyploid hybrid between *Raphanus sativus* and *Brassica rapa* is grown as a fodder crop.

"Raphanofortii" is the allopolyploid hybrid between *Brassica tournefortii* and *Raphanus caudatus*. The *Raphanobrassica* is a fascinating plant, because (in spite of its hybrid nature), it is not sterile. This has led some botanists to propose that the accidental hybridization of a flower by pollen of another species in nature could be a mechanism of speciation common in higher plants.

Senecio (groundsel)

The Welsh groundsel is an allopolyploid, a plant that contains sets of chromosomes originating from two different species. Its ancestor was *Senecio × baxteri*, an infertile hybrid that can arise spontaneously when the closely related groundsel (*Senecio vulgaris*) and Oxford ragwort (*Senecio squalidus*) grow alongside each other. Sometime in the early 20th century, an accidental doubling of the number of chromosomes in an *S. × baxteri* plant led to the formation of a new fertile species.

The York groundsel (*Senecio eboracensis*) is a hybrid species of the self-incompatible *Senecio squalidus* (also known as Oxford ragwort) and the self-compatible *Senecio vulgaris* (also known as common groundsel). Like *S. vulgaris*, *S. eboracensis* is self-compatible; however, it shows little or no natural crossing with its parent species, and is therefore reproductively isolated, indicating that strong breed barriers exist between this new hybrid and its parents. It resulted from a backcrossing of the F1 hybrid of its parents to *S. vulgaris*. *S. vulgaris* is native to Britain, while *S. squalidus* was introduced from Sicily in the early 18th century; therefore, *S. eboracensis* has speciated from those two species within the last 300 years.

Other hybrids descended from the same two parents are known. Some are infertile, such as *S. x baxteri*. Other fertile hybrids are also known, including *S. vulgaris* var. *hibernicus*, now common in Britain, and the allohexaploid *S. cambrensis*, which according to molecular evidence probably originated independently at least three times in different locations. Morphological and genetic evidence support the status of *S. eboracensis* as separate from other known hybrids.

Thale cress

Kirsten Bomblies et al. from the Max Planck Institute for Developmental Biology discovered two genes in the thale cress plant, *Arabidopsis thaliana*. When both genes are inherited by an individual, it ignites a reaction in the hybrid plant that turns its own immune system against it. In the parents, the genes were not detrimental, but they evolved separately to react defectively when combined. To test this, Bomblies crossed 280 genetically different strains of *Arabidopsis* in 861 distinct ways and found that 2 percent of the resulting hybrids were necrotic. Along with allocating the same indicators, the 20 plants

Figure 41: *Figure 6c: Arabidopsis thaliana (colloqui-
ally known as thale cress, mouse-ear cress or arabidopsis)*

also shared a comparable collection of genetic activity in a group of 1,080
genes. In almost all of the cases, Bomblies discovered that only two genes were
required to cause the autoimmune response. Bomblies looked at one hybrid in
detail and found that one of the two genes belonged to the NB-LRR class, a
common group of disease resistance genes involved in recognizing new infections. When Bomblies removed the problematic gene, the hybrids developed
normally. Over successive generations, these incompatibilities could create
divisions between different plant strains, reducing their chances of successful
mating and turning distinct strains into separate species.

Tragopogon (salsify)

Tragopogon is one example where hybrid speciation has been observed.
In the early 20th century, humans introduced three species of salsify into
North America. These species, the western salsify (*Tragopogon dubius*), the
meadow salsify (*Tragopogon pratensis*), and the oyster plant (*Tragopogon
porrifolius*), are now common weeds in urban wastelands. In the 1950s,
botanists found two new species in the regions of Idaho and Washington, where
the three already known species overlapped. One new species, *Tragopogon
miscellus*, is a tetraploid hybrid of *T. dubius* and *T. pratensis*. The other new
species, *Tragopogon mirus*, is also an allopolyploid, but its ancestors were *T.*

Figure 42: *Figure 6d: Purple salsify, Tragopogon porrifolius*

dubius and *T. porrifolius*. These new species are usually referred to as "the Ownbey hybrids" after the botanist who first described them. The *T. mirus* population grows mainly by reproduction of its own members, but additional episodes of hybridization continue to add to the *T. mirus* population.

T. dubius and *T. pratensis* mated in Europe but were never able to hybridize. A study published in March 2011 found that when these two plants were introduced to North America in the 1920s, they mated and doubled the number of chromosomes in there hybrid *Tragopogon miscellus* allowing for a "reset" of its genes, which in turn, allows for greater genetic variation. Professor Doug Soltis of the University of Florida said, "We caught evolution in the act...New and diverse patterns of gene expression may allow the new species to rapidly adapt in new environments". This observable event of speciation through hybridization further advances the evidence for the common descent of organisms and the time frame in which the new species arose in its new environment. The hybridizations have been reproduced artificially in laboratories from 2004 to present day.

Vertebrates

Blackcap

The bird species, *Sylvia atricapilla*, commonly referred to as blackcaps, lives in Germany and flies southwest to Spain while a smaller group flies northwest to Great Britain during the winter. Gregor Rolshausen from the University of Freiburg found that the genetic separation of the two populations is already in progress. The differences found have arisen in about 30 generations. With DNA sequencing, the individuals can be assigned to a correct group with an 85% accuracy. Stuart Bearhop from the University of Exeter reported that birds wintering in England tend to mate only among themselves, and not usually with those wintering in the Mediterranean.[46] It is still inference to say that the populations will become two different species, but researchers expect it due to the continued genetic and geographic separation.

Mollies

The shortfin molly (*Poecilia mexicana*) is a small fish that lives in the Sulfur Caves of Mexico. Years of study on the species have found that two distinct populations of mollies—the dark interior fish and the bright surface water fish—are becoming more genetically divergent. The populations have no obvious barrier separating the two; however, it was found that the mollies are hunted by a large water bug (*Belostoma spp*). Tobler collected the bug and both types of mollies, placed them in large plastic bottles, and put them back in the cave. After a day, it was found that, in the light, the cave-adapted fish endured the most damage, with four out of every five stab-wounds from the water bugs sharp mouthparts. In the dark, the situation was the opposite. The mollies' senses can detect a predator's threat in their own habitats, but not in the other ones. Moving from one habitat to the other significantly increases the risk of dying. Tobler plans on further experiments, but believes that it is a good example of the rise of a new species.

Polar bear

A remarkable example of natural selection, geographic isolation, and speciation in progress is the relationship between the polar bear (*Ursus maritimus*) and the brown bear (*Ursus arctos*). Considered separate species throughout their ranges; however, it has been documented that they possess the capability to interbreed and produce fertile offspring. This introgressive hybridization has occurred both in the wild and in captivity and has been documented and verified with DNA testing. The oldest known fossil evidence of polar bears dates around 130,000 to 110,000 years ago; however, molecular data has revealed varying estimates of divergence time. Mitochondrial DNA analysis has given an estimate of 150,000 years ago while nuclear genome analysis has shown an approximate divergence of 603,000 years ago. Recent research using the complete genomes (rather than mtDNA or partial nuclear genomes) establishes the divergence of polar and brown bears between 479-343 thousand years ago. Despite the differences in divergence rates, molecular research suggests the sister species have undergone a highly complex process of speciation and admixture between the two.

Polar bears have acquired significant anatomical and physiological differences from the brown bear that allow it to comfortably survive in conditions that the brown bear likely could not. Notable examples include the ability to swim sixty miles or more at a time in freezing waters, fur that blends with the snow, and to stay warm in the arctic environment, an elongated neck that makes it easier to keep their heads above water while swimming, and oversized and heavy-matted webbed feet that act as paddles when swimming. It has also evolved small papillae and vacuole-like suction cups on the soles to make them less likely to slip on the ice, alongside smaller ears for a reduction of heat loss, eyelids that act like sunglasses, accommodations for their all-meat diet, a large stomach capacity to enable opportunistic feeding, and the ability to fast for up to nine months while recycling their urea. This example presents a macro-evolutionary change involving an amalgamation of several fields of evolutionary biology, e.g. adaptation through natural selection, geographic isolation, speciation, and hybridization.

Evidence from coloration

Animal coloration provided important early evidence for evolution by natural selection, at a time when little direct evidence was available. Three major functions of coloration were discovered in the second half of the 19th century, and subsequently used as evidence of selection: camouflage (protective coloration); mimicry, both Batesian and Müllerian; and aposematism. After the circumstantial evidence provided by Darwin in *On the Origin of Species*, and given the absence of mechanisms for genetic variation or heredity at that time, naturalists including Darwin's contemporaries, Henry Walter Bates and Fritz Müller sought evidence from what they could observe in the field.

Mimicry and aposematism

Bates and Müller described forms of mimicry that now carry their names, based on their observations of tropical butterflies. These highly specific patterns of coloration are readily explained by natural selection, since predators such as birds which hunt by sight will more often catch and kill insects that are less good mimics of distasteful models than those that are better mimics; but the patterns are otherwise hard to explain. Darwinists such as Alfred Russel

Figure 43: *Natural selection has driven the ptarmigan to change from snow camouflage in winter to disruptive coloration suiting moorland in summer.*

Wallace and Edward Bagnall Poulton, and in the 20th century Hugh Cott and Bernard Kettlewell, sought evidence that natural selection was taking place.

Camouflage

In 1889, Wallace noted that snow camouflage, especially plumage and pelage that changed with the seasons, suggested an obvious explanation as an adaptation for concealment. Poulton's 1890 book, *The Colours of Animals*, written during Darwinism's lowest ebb, used all the forms of coloration to argue the case for natural selection. Cott described many kinds of camouflage, mimicry and warning coloration in his 1940 book *Adaptive Coloration in Animals*, and in particular his drawings of coincident disruptive coloration in frogs convinced other biologists that these deceptive markings were products of natural selection. Kettlewell experimented on peppered moth evolution, showing that the species had adapted as pollution changed the environment; this provided compelling evidence of Darwinian evolution.

Evidence from mathematical modeling

Computer science allows the iteration of self-changing complex systems to be studied, allowing a mathematical understanding of the nature of the processes behind evolution; providing evidence for the hidden causes of known evolutionary events. The evolution of specific cellular mechanisms like spliceosomes that can turn the cell's genome into a vast workshop of billions of interchangeable parts that can create tools that create tools that create tools that create us can be studied for the first time in an exact way.

"It has taken more than five decades, but the electronic computer is now powerful enough to simulate evolution",[47] assisting bioinformatics in its attempt to solve biological problems.

Computational evolutionary biology has enabled researchers to trace the evolution of a large number of organisms by measuring changes in their DNA, rather than through physical taxonomy or physiological observations alone. It has compared entire genomes permitting the study of more complex evolutionary events, such as gene duplication, horizontal gene transfer, and the prediction of factors important in speciation. It has also helped build complex computational models of populations to predict the outcome of the system over time and track and share information on an increasingly large number of species and organisms.

Future endeavors are to reconstruct a now more complex tree of life.

Christoph Adami, a professor at the Keck Graduate Institute made this point in *Evolution of biological complexity*:

> To make a case for or against a trend in the evolution of complexity in biological evolution, complexity must be both rigorously defined and measurable. A recent information-theoretic (but intuitively evident) definition identifies genomic complexity with the amount of information a sequence stores about its environment. We investigate the evolution of genomic complexity in populations of digital organisms and monitor in detail the evolutionary transitions that increase complexity. We show that, because natural selection forces genomes to behave as a natural "Maxwell Demon", within a fixed environment, genomic complexity is forced to increase.

David J. Earl and Michael W. Deem—professors at Rice University made this point in *Evolvability is a selectable trait*:

> Not only has life evolved, but life has evolved to evolve. That is, correlations within protein structure have evolved, and mechanisms to manipulate these correlations have evolved in tandem. The rates at which the various events within the hierarchy of evolutionary moves occur are not random or arbitrary but are selected by Darwinian evolution. Sensibly, rapid or

extreme environmental change leads to selection for greater evolvability. This selection is not forbidden by causality and is strongest on the largest-scale moves within the mutational hierarchy. Many observations within evolutionary biology, heretofore considered evolutionary happenstance or accidents, are explained by selection for evolvability. For example, the vertebrate immune system shows that the variable environment of anti-gens has provided selective pressure for the use of adaptable codons and low-fidelity polymerases during somatic hypermutation. A similar driving force for biased codon usage as a result of productively high mutation rates is observed in the hemagglutinin protein of influenza A.

"Computer simulations of the evolution of linear sequences have demonstrated the importance of recombination of blocks of sequence rather than point muta-genesis alone. Repeated cycles of point mutagenesis, recombination, and se-lection should allow *in vitro* molecular evolution of complex sequences, such as proteins." Evolutionary molecular engineering, also called directed evolu-tion or *in vitro* molecular evolution involves the iterated cycle of mutation, multiplication with recombination, and selection of the fittest of individual molecules (proteins, DNA, and RNA). Natural evolution can be relived show-ing us possible paths from catalytic cycles based on proteins to based on RNA to based on DNA.[48,49]

Sources

- *Biological science*. Oxford. 2002.
- Clegg, C.J. (1998). *Genetics & evolution*. London: J. Murray. ISBN 0-7195-7552-4.
- Coyne, Jerry A. (2009). *Why Evolution is True*. New York: Oxford University Press. ISBN 978-0-19-923084-6.
- Darwin, Charles November 24, 1859. *On the Origin of Species by means of Natural Selection or the Preservation of Favoured Races in the Struggle for Life*. London: John Murray, Albemarle Street. 502 pages. Reprinted: Gramercy (May 22, 1995). ISBN 0-517-12320-7
- Dawkins, Richard (2009). *The Greatest Show on Earth: The Evidence for Evolution*. Bantam Press. ISBN 978-1-4165-9478-9.
- Endler, John A. (1986). *Natural selection in the wild*. New Jersey: Princeton University Press. ISBN 0-691-08387-8.
- Fitzhugh, Kirk (2006). "-1-'Evidence' For Evolution Versus 'Evidence' For Intelligent Design: Parallel Confusions"[50]. Natural History Museum of Los Angeles County.
- Grant, Peter R.; Grant, B. Rosemary (2014). *40 Years of Evolution: Darwin's Finches on Daphne Major Island*. Princeton University Press. p. 432. ISBN 978-0691160467.

- Hill, A.; Behrensmeyer, A. K. (1980). *Fossils in the making: vertebrate taphonomy and paleoecology*. Chicago: University of Chicago Press. ISBN 0-226-04169-7.
- Kluge, Arnold G. (1999). "The Science of Phylogenetic Systematics: Explanation, Prediction and Test"[51] (PDF). *Cladistics*. **15** (4): 429–436. doi: 10.1111/j.1096-0031.1999.tb00279.x[52].
- Laurin, Michel (2000). "Early tetrapod evolution"[53] (PDF). *Tree*. **15** (3): 118–123. doi: 10.1016/s0169-5347(99)01780-2[54].
- Mayr, Ernst (2001). *What evolution is*. New York: Basic Books. ISBN 0-465-04426-3.
- Paul, C.R.C.; Donovan, S.K. (1998). *The adequacy of the fossil record*. New York: John Wiley. ISBN 0-471-96988-5.
- Prothero, Donald (2007). *Evolution: What the Fossils Say and Why it Matters*. Columbia University Press. ISBN 978-0231139625.
- Shubin, Neil (2008). *Your Inner Fish:A Journey Into the 3.5 Billion-Year History of the Human Body*. Random House. ISBN 978-0-375-42447-2.
- Sober, Elliott (2008). *Evidence and Evolution: The logic behind the science*. Cambridge University Press. ISBN 978-0-521-87188-4.

External links

- National Academies Evolution Resources[55]
- TalkOrigins Archive – 29+ Evidences for Macroevolution: The Scientific Case for Common Descent[56]
- TalkOrigins Archive – Transitional Vertebrate Fossils FAQ[57]
- Understanding Evolution: Your one-stop source for information on evolution[58]
- National Academy Press: Teaching About Evolution and the Nature of Science[59]
- Evolution[60] — Provided by *PBS*.
- Evolution News from Genome News Network (GNN)[61]
- Evolution by Natural Selection[62] — An introduction to the logic of the theory of evolution by natural selection
- Howstuffworks.com — How Evolution Works[63]
- 15 Evolutionary Gems[64]

Common descent

Part of a series on
Evolutionary biology

- ᵞᵉ **Evolutionary biology portal**
- ◯ **Category**
- ⌐ *Book*
- **Related topics**

- v
- t
- e⁶⁵

Common descent describes how, in evolutionary biology, a group of organisms share a most recent common ancestor. There is "massive" evidence of common descent of all life on Earth from the last universal common ancestor (LUCA). In July 2016, scientists reported identifying a set of 355 genes from the LUCA, by comparing the genomes of the three domains of life, archaea, bacteria, and eukaryotes.

Common ancestry between organisms of different species arises during speciation, in which new species are established from a single ancestral population. Organisms which share a more-recent common ancestor are more closely related. The most recent common ancestor of all currently living organisms is the last universal ancestor, which lived about 3.9 billion years ago. The two earliest evidences for life on Earth are graphite found to be biogenic in 3.7 billion-year-old metasedimentary rocks discovered in western Greenland and microbial mat fossils found in 3.48 billion-year-old sandstone discovered in Western Australia. All currently living organisms on Earth share a common genetic heritage, though the suggestion of substantial horizontal gene transfer during early evolution has led to questions about the monophyly (single ancestry) of life. 6,331 groups of genes common to all living animals have been identified; these may have arisen from a single common ancestor that lived 650 million years ago in the Precambrian.

Universal common descent through an evolutionary process was first proposed by the English naturalist Charles Darwin in the concluding sentence of his 1859 book *On the Origin of Species*:

There is grandeur in this view of life, with its several powers, having been originally breathed into a few forms or into one; and that, whilst this planet has gone cycling on according to the fixed law of gravity, from so simple a beginning endless forms most beautiful and most wonderful have been, and are being, evolved.

History

In the 1740s, the French mathematician Pierre Louis Maupertuis made the first known suggestion that all organisms had a common ancestor, and had diverged through random variation and natural selection. In *Essai de cosmologie* (1750), Maupertuis noted:

May we not say that, in the fortuitous combination of the productions of Nature, since only those creatures could survive in whose organizations a certain degree of adaptation was present, there is nothing extraordinary in the fact that such adaptation is actually found in all these species which now exist? Chance, one might say, turned out a vast number of individuals; a small proportion of these were organized in such a manner that the animals' organs could satisfy their needs. A much greater number showed neither adaptation nor order; these last have all perished.... Thus the species which we see today are but a small part of all those that a blind destiny has produced.

In 1790, the philosopher Immanuel Kant wrote in *Kritik der Urteilskraft* (*Critique of Judgement*) that the similarity[66] of animal forms implies a common original type, and thus a common parent.[67]

In 1794, Charles Darwin's grandfather, Erasmus Darwin, asked:

[W]ould it be too bold to imagine, that in the great length of time, since the earth began to exist, perhaps millions of ages before the commencement of the history of mankind, would it be too bold to imagine, that all warm-blooded animals have arisen from one living filament, which the great First Cause endued with animality, with the power of acquiring new parts attended with new propensities, directed by irritations, sensations, volitions, and associations; and thus possessing the faculty of continuing to improve by its own inherent activity, and of delivering down those improvements by generation to its posterity, world without end?

Charles Darwin's views about common descent, as expressed in *On the Origin of Species*, were that it was probable that there was only one progenitor for all life forms:

*Therefore I should infer from analogy that probably all the organic be-
ings which have ever lived on this earth have descended from some one
primordial form, into which life was first breathed.*

Common descent was widely accepted amongst the scientific community after
Darwin's publication.[68]   In 1907, Vernon Kellogg commented that "practically no naturalists of position and recognized attainment doubt the theory of
descent."[69]

In 2008, biologist T. Ryan Gregory noted that:

*No reliable observation has ever been found to contradict the general no-
tion of common descent. It should come as no surprise, then, that the
scientific community at large has accepted evolutionary descent as a historical reality since Darwin's time and considers it among the most reliably established and fundamentally important facts in all of science.*[70]

Evidence

Common biochemistry

All known forms of life are based on the same fundamental biochemical
organization: genetic information encoded in DNA, transcribed into RNA,
through the effect of protein- and RNA-enzymes, then translated into proteins by (highly similar) ribosomes, with ATP, NADPH and others as energy
sources. Analysis of small sequence differences in widely shared substances
such as cytochrome c further supports universal common descent. Some 23
proteins are found in all organisms, serving as enzymes carrying out core functions like DNA replication. The fact that only one such set of enzymes exists
is convincing evidence of a single ancestry. 6,331 genes common to all living
animals have been identified; these may have arisen from a single common
ancestor that lived 650 million years ago in the Precambrian.

Common genetic code

Amino acids	nonpolar	polar	basic	acidic	Stop codon

Standard genetic code

1st base	2nd base							
	T		**C**		**A**		**G**	
T	TTT	Phenyl-alanine	TCT	Serine	TAT	Tyrosine	TGT	Cysteine
	TTC		TCC		TAC		TGC	
	TTA	Leucine	TCA		TAA	Stop	TGA	Stop
	TTG		TCG		TAG	Stop	TGG	Tryptophan
C	CTT		CCT	Proline	CAT	Histidine	CGT	Arginine
	CTC		CCC		CAC		CGC	
	CTA		CCA		CAA	Glutamine	CGA	
	CTG		CCG		CAG		CGG	
A	ATT	Isoleucine	ACT	Threonine	AAT	Asparagine	AGT	Serine
	ATC		ACC		AAC		AGC	
	ATA		ACA		AAA	Lysine	AGA	Arginine
	ATG	Methionine	ACG		AAG		AGG	
G	GTT	Valine	GCT	Alanine	GAT	Aspartic acid	GGT	Glycine
	GTC		GCC		GAC		GGC	
	GTA		GCA		GAA	Glutamic acid	GGA	
	GTG		GCG		GAG		GGG	

The genetic code (the "translation table" according to which DNA information is translated into amino acids, and hence proteins) is nearly identical for all known lifeforms, from bacteria and archaea to animals and plants. The universality of this code is generally regarded by biologists as definitive evidence in favor of universal common descent.

The way that codons (DNA triplets) are mapped to amino acids seems to be strongly optimised. Richard Egel argues that in particular the hydrophobic (non-polar) side-chains are well organised, suggesting that these enabled the earliest organisms to create peptides with water-repelling regions able to support the essential electron exchange (redox) reactions for energy transfer.

Selectively neutral similarities

Similarities which have no adaptive relevance cannot be explained by convergent evolution, and therefore they provide compelling support for universal common descent. Such evidence has come from two areas: amino acid sequences and DNA sequences. Proteins with the same three-dimensional structure need not have identical amino acid sequences; any irrelevant similarity between the sequences is evidence for common descent. In certain cases, there are several codons (DNA triplets) that code redundantly for the same amino acid. Since many species use the same codon at the same place to specify an amino acid that can be represented by more than one codon, that is evidence for their sharing a recent common ancestor. Had the amino acid sequences come from different ancestors, they would have been coded for by any of the redundant codons, and since the correct amino acids would already have been in place, natural selection would not have driven any change in the codons, however much time was available. Genetic drift could change the codons, but it would be extremely unlikely to make all the redundant codons in a whole sequence match exactly across multiple lineages. Similarly, shared nucleotide sequences, especially where these are apparently neutral such as the positioning of introns and pseudogenes, provide strong evidence of common ancestry.

Other similarities

Biologists oftenWikipedia:Manual of Style/Dates and numbers point to the universality of many aspects of cellular life as supportive evidence to the more compelling evidence listed above. These similarities include the energy carrier adenosine triphosphate (ATP), and the fact that all amino acids found in proteins are left-handed. It is, however, possible that these similarities resulted because of the laws of physics and chemistry - rather than through universal common descent - and therefore resulted in convergent evolution. In contrast, there is evidence for homology of the central subunits of Transmembrane ATPases throughout all living organisms, especially how the rotating elements are bound to the membrane. This supports the assumption of a LUCA as a cellular organism, although primordial membranes may have been semipermeable and evolved later to the membranes of modern bacteria, and on a second path to those of modern archaea also.

Phylogenetic trees

Another important piece of evidence is from detailed phylogenetic trees (i.e., "genealogic trees" of species) mapping out the proposed divisions and common ancestors of all living species. In 2010, Douglas L. Theobald published a statistical analysis of available genetic data, mapping them to phylogenetic trees, that gave "strong quantitative support, by a formal test, for the unity of life."

Traditionally, these trees have been built using morphological methods, such as appearance, embryology, etc. Recently, it has been possible to construct these trees using molecular data, based on similarities and differences between genetic and protein sequences. All these methods produce essentially similar results, even though most genetic variation has no influence over external morphology. That phylogenetic trees based on different types of information agree with each other is strong evidence of a real underlying common descent.

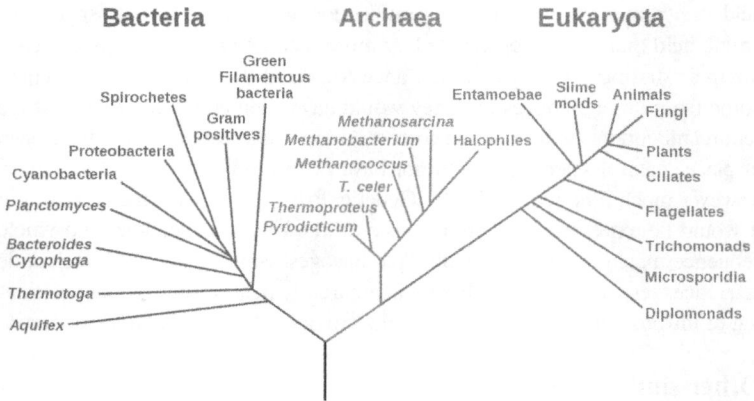

Figure 44: *A phylogenetic tree based on ribosomal RNA genes implies a single origin for all life.*

Potential objections

Gene exchange clouds phylogenetic analysis

Theobald noted that substantial horizontal gene transfer could have occurred
during early evolution. Bacteria today remain capable of gene exchange between distantly-related lineages. This weakens the basic assumption of phylogenetic analysis, that similarity of genomes implies common ancestry, because sufficient gene exchange would allow lineages to share much of their
genome whether or not they shared an ancestor (monophyly). This has led
to questions about the single ancestry of life. However, biologists consider it
very unlikely that completely unrelated proto-organisms could have exchanged
genes, as their different coding mechanisms would have resulted only in garble
rather than functioning systems. Later, however, many organisms all derived
from a single ancestor could readily have shared genes that all worked in the
same way, and it appears that they have.

Convergent evolution

If early organisms had been driven by the same environmental conditions to
evolve similar biochemistry convergently, they might independently have acquired similar genetic sequences. Theobald's "formal test" was accordingly
criticised by Takahiro Yonezawa and colleagues for not including consideration of convergence. They argued that Theobald's test was insufficient to
distinguish between the competing hypotheses. Theobald has defended his

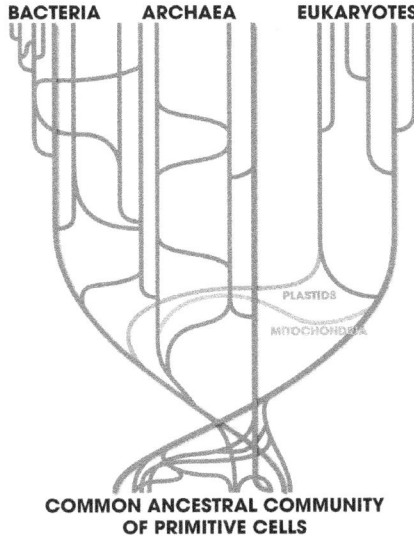

Figure 45: *2005 tree of life shows many horizontal gene transfers, implying multiple possible origins.*

method against this claim, arguing that his tests distinguish between phylogenetic structure and mere sequence similarity. Therefore, Theobald argued, his results show that "real universally conserved proteins are homologous."

Bibliography

<templatestyles src="Template:Refbegin/styles.css" />

- Crombie, A. C.; Hoskin, Michael (1970). "The Scientific Movement and the Diffusion of Scientific Ideas, 1688–1751". In Bromley, J. S. *The Rise of Great Britain and Russia, 1688–1715/25*. The New Cambridge Modern History. **6**. London: Cambridge University Press. ISBN 978-0-521-07524-4. LCCN 57014935[71]. OCLC 7588392[72].
- Darwin, Charles (1859). *On the Origin of Species by Means of Natural Selection, or the Preservation of Favoured Races in the Struggle for Life* (1st ed.). London: John Murray. LCCN 06017473[73]. OCLC 741260650[74]. The book is available from The Complete Work of Charles Darwin Online[75]. Retrieved 2015-11-23.
- Darwin, Erasmus (1818) [Originally published 1794]. *Zoonomia; or the Laws of Organic Life*. **1** (4th American ed.). Philadelphia, PA: Edward

Earle. Zoonomia; or The laws of organic life: in three parts (Volume 1) (1818)[76] on the Internet Archive Retrieved 2015-11-23.

- Harris, C. Leon (1981). *Evolution: Genesis and Revelations: With Readings from Empedocles to Wilson*. Albany, NY: State University of New York Press. ISBN 978-0-87395-487-7. LCCN 81002555[77]. OCLC 7278190[78].

- Kant, Immanuel (1987) [Originally published 1790 in Prussia as *Kritik der Urteilskraft*]. *Critique of Judgment*. Translated, with an introduction, by Werner S. Pluhar; foreword by Mary J. Gregor. Indianapolis, IN: Hackett Publishing Company. ISBN 978-0-87220-025-8. LCCN 86014852[79]. OCLC 13796153[80].

- Treasure, Geoffrey (1985). *The Making of Modern Europe, 1648-1780*. New York: Methuen. ISBN 978-0-416-72370-0. LCCN 85000255[81]. OCLC 11623262[82].

External links

- 29+ Evidences for Macroevolution: The Scientific Case for Common Descent[83] from the TalkOrigins Archive.
- The Tree of Life Web Project[84]

Homology (biology)

<indicator name="good-star"> ⊕ </indicator>

In biology, **homology** is the existence of shared ancestry between a pair of structures, or genes, in different taxa. A common example of homologous structures is the forelimbs of vertebrates, where the wings of bats, the arms of primates, the front flippers of whales and the forelegs of dogs and horses are all derived from the same ancestral tetrapod structure. Evolutionary biology explains homologous structures adapted to different purposes as the result of descent with modification from a common ancestor. Homology was explained by Charles Darwin's theory of evolution in 1859, but had been observed before this, from Aristotle onwards, and it was explicitly analysed by Pierre Belon in 1555. The term was applied to biology by the anatomist Richard Owen in 1843.

In developmental biology, organs that developed in the embryo in the same manner and from similar origins, such as from matching primordia in successive segments of the same animal, are serially homologous. Examples include the legs of a centipede, the maxillary palp and labial palp of an insect, and the spinous processes of successive vertebrae in a vertebral column. Male and

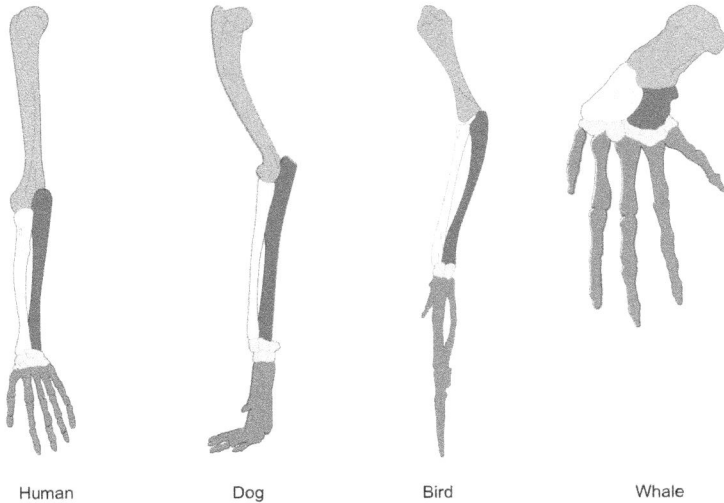

Figure 46: *The principle of homology: The biological relationships (shown by colours) of the bones in the forelimbs of vertebrates were used by Charles Darwin as an argument in favor of evolution.*

female reproductive organs are homologous if they develop from the same embryonic tissue, as do the ovaries and testicles of mammals including humans.

Sequence homology between protein or DNA sequences is similarly defined in terms of shared ancestry. Two segments of DNA can have shared ancestry because of either a speciation event (orthologs) or a duplication event (paralogs). Homology among proteins or DNA is inferred from their sequence similarity. Significant similarity is strong evidence that two sequences are related by divergent evolution from a common ancestor. Alignments of multiple sequences are used to discover the homologous regions.

Homology remains controversial in animal behaviour, but there is suggestive evidence that, for example, dominance hierarchies are homologous across the primates.

History

Homology was noticed by Aristotle (c. 350 BC), and was explicitly analysed by Pierre Belon in his 1555 *Book of Birds*, where he systematically compared the skeletons of birds and humans. The pattern of similarity was interpreted as part of the static great chain of being through the mediaeval and early modern

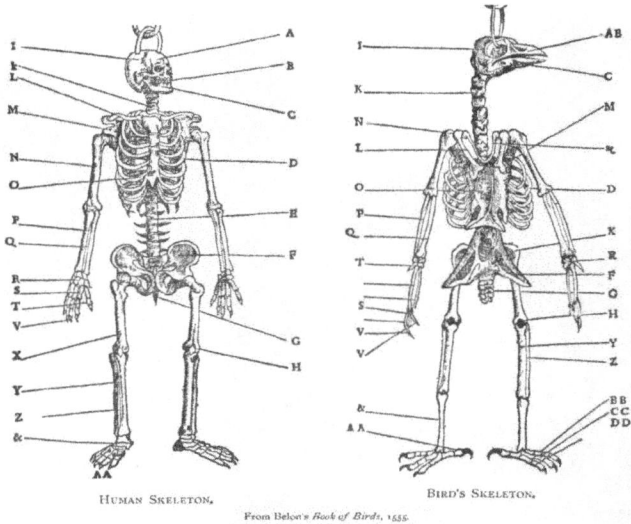

Figure 47: *Pierre Belon compared the skeletons of birds and humans in his Book of Birds (1555).*

periods: it was not then seen as implying evolutionary change. In the German *Naturphilosophie* tradition, homology was of special interest as demonstrating unity in nature. In 1790, Goethe stated his foliar theory in his essay "Metamorphosis of Plants", showing that flower part are derived from leaves. The serial homology of limbs was described late in the 18th century. The French zoologist Etienne Geoffroy Saint-Hilaire showed in 1818 in his *theorie d'analogue* ("theory of homologues") that structures were shared between fishes, reptiles, birds, and mammals. When Geoffroy went further and sought homologies between Georges Cuvier's *embranchements*, such as vertebrates and molluscs, his claims triggered the 1830 Cuvier-Geoffroy debate. Geoffroy stated the principle of connections, namely that what is important is the relative position of different structures and their connections to each other. The Estonian embryologist Karl Ernst von Baer stated what are now called von Baer's laws in 1828, noting that related animals begin their development as similar embryos and then diverge: thus, animals in the same family are more closely related and diverge later than animals which are only in the same order and have fewer homologies. von Baer's theory recognises that each taxon (such as a family) has distinctive shared features, and that embryonic development parallels the taxonomic hierarchy: not the same as recapitulation theory. The term "homology" was first used in biology by the anatomist Richard Owen in

Figure 48: *The front wings of beetles such as this five-horned rhinoceros beetle have evolved into elytra, hard wing-cases. These are homologous with the front wings of other insects.*

1843 when studying the similarities of vertebrate fins and limbs, contrasting it with the matching term "analogy" which he used to describe different structures with the same function. In 1859, Charles Darwin explained homologous structures as meaning that the organisms concerned shared a body plan from a common ancestor, and that taxa were branches of a single tree of life.

Definition

The word homology, coined in about 1656, is derived from the Greek ὁμόλογος *homologos* from ὁμός *homos* "same" and λόγος *logos* "relation".[85,86]</ref>

Biological structures or sequences in different taxa are homologous if they are derived from a common ancestor. Homology thus implies divergent evolution. For example, many insects (such as dragonflies) possess two pairs of flying wings. In beetles, the first pair of wings has evolved into a pair of hard wing covers, while in Dipteran flies the second pair of wings has evolved into small halteres used for balance.[87]

Similarly, the forelimbs of ancestral vertebrates have evolved into the front flippers of whales, the wings of birds, the running forelegs of dogs, deer, and

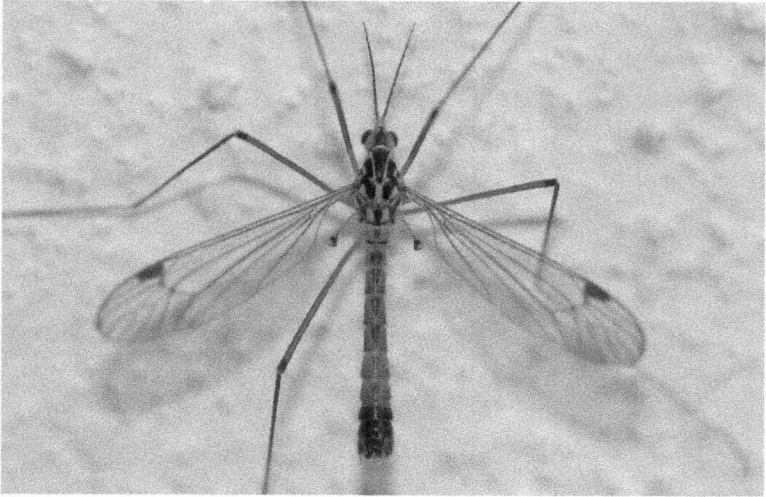

Figure 49: *The hind wings of Dipteran flies such as this crane-
fly have evolved divergently to form small club-like halteres.
These are homologous with the hind wings of other insects.*

horses, the short forelegs of frogs and lizards, and the grasping hands of primates including humans. The same major forearm bones (humerus, radius,
and ulna[88]) are found in fossils of lobe-finned fish such as *Eusthenopteron*.

Homology vs analogy

The opposite of homologous organs are analogous organs which do similar
jobs in two taxa that were not present in their last common ancestor but rather
evolved separately. For example, the wings of insects and birds evolved independently in widely separated groups, and converged functionally to support
powered flight, so they are analogous. Similarly, the wings of a sycamore
maple seed and the wings of a bird are analogous but not homologous, as they
develop from quite different structures. A structure can be homologous at one
level, but only analogous at another. Pterosaur, bird and bat wings are analogous as wings, but homologous as forelimbs because the organ served as a
forearm (not a wing) in the last common ancestor of tetrapods, and evolved in
different ways in the three groups. Thus, in the pterosaurs, the "wing" involves
both the forelimb and the hindlimb. Analogy is called homoplasy in cladistics,
and convergent or parallel evolution in evolutionary biology.[89]

Figure 50: *Sycamore maple fruits have wings analogous but not homologous to a bird's.*

In cladistics

Specialised terms are used in taxonomic research. Primary homology is a researcher's initial hypothesis based on similar structure or anatomical connections, suggesting that a character state in two or more taxa share is shared due to common ancestry. Primary homology may be conceptually broken down further: we may consider all of the states of the same character as "homologous" parts of a single, unspecified, transformation series. This has been referred to as topographical correspondence. For example, in an aligned DNA sequence matrix, all of the A, G, C, T or implied gaps at a given nucleotide site are homologous in this way. Character state identity is the hypothesis that the particular condition in two or more taxa is "the same" as far as our character coding scheme is concerned. Thus, two Adenines at the same aligned nucleotide site are hypothesized to be homologous unless that hypothesis is subsequently contradicted by other evidence. Secondary homology is implied by parsimony analysis, where a character state that arises only once on a tree is taken to be homologous.[90] As implied in this definition, many cladists consider secondary homology to be synonymous with synapomorphy, a shared derived character or trait state that distinguishes a clade from other organisms.[91]

Human	Mouse	Zebrafish	Drosophila

WT

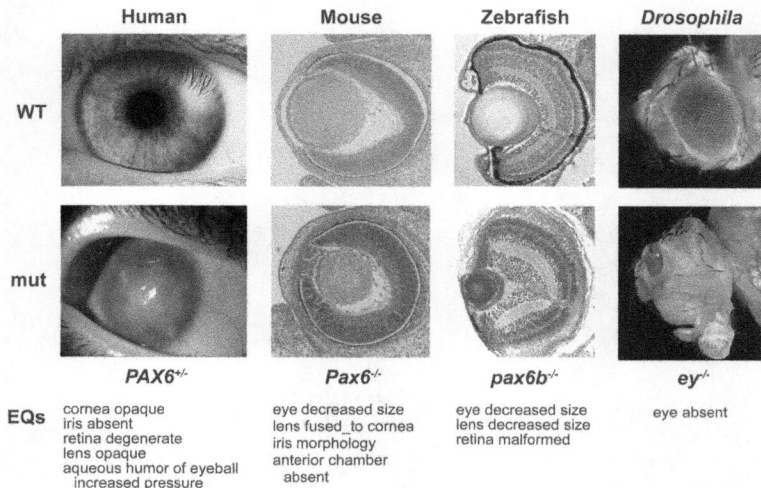

mut

PAX6⁺ᐟ⁻	Pax6⁻ᐟ⁻	pax6b⁻ᐟ⁻	ey⁻ᐟ⁻

EQs

| cornea opaque
iris absent
retina degenerate
lens opaque
aqueous humor of eyeball
increased pressure | eye decreased size
lens fused_to cornea
iris morphology
anterior chamber
absent | eye decreased size
lens decreased size
retina malformed | eye absent |

Figure 51: *pax6 alterations result in similar changes to eye morphology and function across a wide range of taxa.*

Shared ancestral character states, symplesiomorphies, represent either synapomorphies of a more inclusive group, or complementary states (often absences) that unite no natural group of organisms. For example, the presence of wings is a synapomorphy for pterygote insects, but a symplesiomorphy for holometabolous insects. Absence of wings in non-pterygote insects and other organisms is a complementary symplesiomorphy that unites no group (for example, absence of wings provides no evidence of common ancestry of silverfish, spiders and annelid worms). On the other hand, absence (or secondary loss) of wings is a synapomorphy for fleas. Patterns such as these lead many cladists to consider the concept of homology and the concept of synapomorphy to be equivalent.[92] It should be noted that some cladists follow the pre-cladistic definition of homology of Haas and Simpson,[93] and view both synapomorphies and symplesiomorphies as homologous character states

In different taxa

Homologies provide the fundamental basis for all biological classification, although some may be highly counter-intuitive. For example, deep homologies like the *pax6* genes that control the development of the eyes of vertebrates and arthropods were unexpected, as the organs are anatomically dissimilar and appeared to have evolved entirely independently.

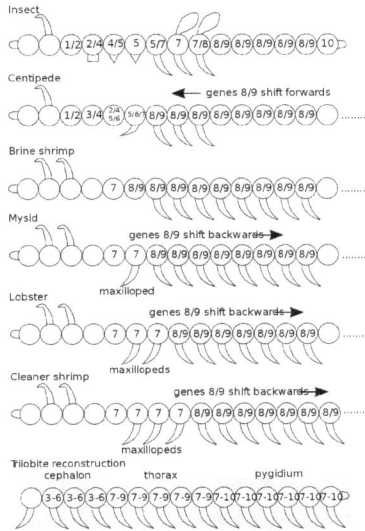

Figure 52: *Hox genes in arthropod segmentation*

In arthropods

The embryonic body segments (somites) of different arthropods taxa have diverged from a simple body plan with many similar appendages which are serially homologous, into a variety of body plans with fewer segments equipped with specialised appendages. The homologies between these have been discovered by comparing genes in evolutionary developmental biology.

Somite (body segment)	Trilobite (Trilobitomorpha)	Spider (Chelicerata)	Centipede (Myriapoda)	Insect (Hexapoda)	Shrimp (Crustacea)
1	antennae	chelicerae (jaws and fangs)	antennae	antennae	1st antennae
2	1st legs	pedipalps	-	-	2nd antennae
3	2nd legs	1st legs	mandibles	mandibles	mandibles (jaws)
4	3rd legs	2nd legs	1st maxillae	1st maxillae	1st maxillae
5	4th legs	3rd legs	2nd maxillae	2nd maxillae	2nd maxillae

6	5th legs	4th legs	collum (no legs)	1st legs	1st legs
7	6th legs	-	1st legs	2nd legs	2nd legs
8	7th legs	-	2nd legs	3rd legs	3rd legs
9	8th legs	-	3rd legs	-	4th legs
10	9th legs	-	4th legs	-	5th legs

Among insects, the stinger of the female honey bee is a modified ovipositor, homologous with ovipositors in other insects such as the Orthoptera, Hemiptera, and those Hymenoptera without stingers.

In mammals

The three small bones in the middle ear of mammals including humans, the malleus, incus, and stapes, are today used to transmit sound from the eardrum to the inner ear. The malleus and incus develop in the embryo from structures that form jaw bones (the quadrate and the articular) in lizards, and in fossils of lizard-like ancestors of mammals. Both lines of evidence show that these bones are homologous, sharing a common ancestor.

Among the many homologies in mammal reproductive systems, ovaries and testicles are homologous.

Rudimentary organs such as the human tailbone, now much reduced from their functional state, are readily understood as signs of evolution, the explanation being that they were cut down by natural selection from functioning organs when their functions were no longer needed, but make no sense at all if species are considered to be fixed. The tailbone is homologous to the tails of other primates.[94]

In plants

Leaves, stems, and roots

In many plants, defensive or storage structures are made by modifications of the development of primary leaves, stems, and roots. Leaves are variously modified from photosynthetic structures to form the insect-trapping pitchers of pitcher plants, the insect-trapping jaws of Venus flytrap, and the spines of cactuses, all homologous.

Primary organs	Defensive structures	Storage structures
Leaves	Spines	Swollen leaves (e.g. succulents)
Stems	Thorns	Tubers (e.g. potato), rhizomes (e.g. ginger), fleshy stems (e.g. cacti)
Roots	-	Root tubers (e.g. sweet potato), taproot (e.g. carrot)

Certain compound leaves of flowering plants are partially homologous both to leaves and shoots, because their development has evolved from a genetic mosaic of leaf and shoot development.

Figure 53: *One pinnate leaf of European ash*

Figure 54: *Detail of palm leaf*

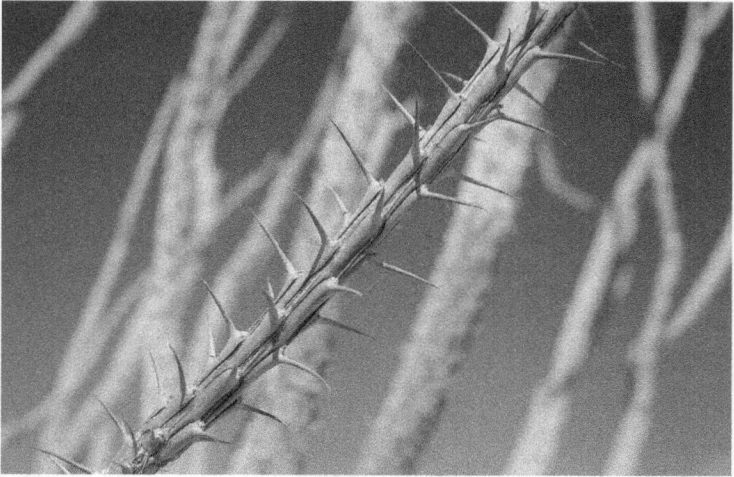

Figure 55: *Leaf petioles adapted as spines in Fouquieria splendens*

Figure 56: *The very large leaves of the banana, Musa acuminata*

Figure 57: *Succulent water storage leaf of Aloe*

Figure 58: *Insect-trapping leaf of Venus flytrap*

Figure 59: *Insect-trapping leaf of pitcher plant*

Figure 60: *Food storage leaves in an onion bulb*

Figure 61: *The ABC model of flower development. Class A genes affect sepals and petals, class B genes affect petals and stamens, class C genes affect stamens and carpels. In two specific whorls of the floral meristem, each class of organ identity genes is switched on.*

Flower parts

The four types of flower parts, namely carpels, stamens, petals, and sepals, are homologous with and derived from leaves, as Goethe correctly noted in 1790. The development of these parts through a pattern of gene expression in the growing zones (meristems) is described by the ABC model of flower development. Each of the four types of flower parts is serially repeated in concentric whorls, controlled by a small number of genes acting in various combinations. Thus, A genes working alone result in sepal formation; A and B together produce petals; B and C together create stamens; C alone produces carpels. When none of the genes are active, leaves are formed. Two more groups of genes, D to form ovules and E for the floral whorls, complete the model. The genes are evidently ancient, as old as the flowering plants themselves.

Developmental biology

Developmental biology can identify homologous structures that arose from the same tissue in embryogenesis. For example, adult snakes have no legs, but their early embryos have limb-buds for hind legs, which are soon lost as the embryos develop. The implication that the ancestors of snakes had hind legs is confirmed by fossil evidence: the Cretaceous snake *Pachyrhachis problematicus* had hind legs complete with hip bones (ilium, pubis, ischium), thigh bone (femur), leg bones (tibia, fibula) and foot bones (calcaneum, astragalus) as in tetrapods with legs today.

Figure 62: *The Cretaceous snake Pachyrhachis*
problematicus had hind legs (circled).

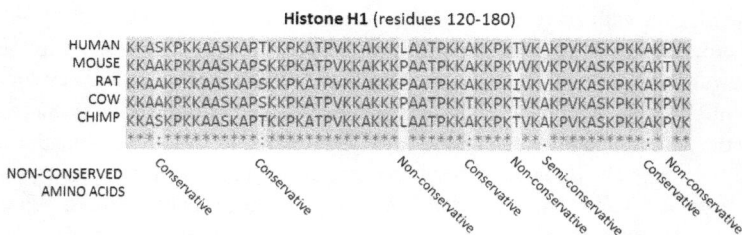

Figure 63: *A multiple sequence alignment of mammalian histone*
H1 proteins. Alignment positions conserved across all five species
analysed are highlighted in grey. Positions with conservative, semi-
conservative, and non-conservative amino acid replacements are indicated.

Sequence homology

As with anatomical structures, sequence homology between protein or DNA sequences is defined in terms of shared ancestry. Two segments of DNA can have shared ancestry because of either a speciation event (orthologs) or a duplication event (paralogs). Homology among proteins or DNA is typically in-

Figure 64: *Dominance hierarchy behaviour, as in these weeper capuchin monkeys, may be homologous across the primates.*

ferred from their sequence similarity. Significant similarity is strong evidence that two sequences are related by divergent evolution of a common ancestor. Alignments of multiple sequences are used to indicate which regions of each sequence are homologous.

Homologous sequences are orthologous if they are descended from the same ancestral sequence separated by a speciation event: when a species diverges into two separate species, the copies of a single gene in the two resulting species are said to be *orthologous*. The term "ortholog" was coined in 1970 by the molecular evolutionist Walter Fitch.

Homologous sequences are paralogous if they were created by a duplication event within the genome. For gene duplication events, if a gene in an organism is duplicated to occupy two different positions in the same genome, then the two copies are paralogous. Paralogous genes often belong to the same species. They can shape the structure of whole genomes and thus explain genome evolution to a large extent. Examples include the Homeobox (Hox) genes in animals. These genes not only underwent gene duplications within chromosomes but also whole genome duplications. As a result, Hox genes in most vertebrates are spread across multiple chromosomes: the HoxA–D clusters are the best studied.

In behaviour

It has been suggested that some behaviours might be homologous, based either on sharing across related taxa or on common origins of the behaviour in an individual's development, though this remains controversial. For example, D. W. Rajecki and Randall C. Flanery, using data on humans and on nonhuman primates, argue that patterns of behaviour in dominance hierarchies are homologous across the primates.

Further reading

- Brigandt, Ingo (2011) "Essay: Homology."[95] In: *The Embryo Project Encyclopedia*. ISSN 1940-5030[96]. http://embryo.asu.edu/handle/10776/1754[95]
- Carroll, Sean B. (2006). *Endless Forms Most Beautiful*. New York: W.W. Norton & Co. ISBN 0-297-85094-6.
- Carroll, Sean B. (2006). *The making of the fittest: DNA and the ultimate forensic record of evolution*. New York: W.W. Norton & Co. ISBN 0-393-06163-9.
- DePinna, M.C. (1991). "Concepts and tests of homology in the cladistic paradigm". *Cladistics*. **7** (4): 367–94. doi: 10.1111/j.1096-0031.1991.tb00045.x[97].
- Dewey, C.N.; Pachter, L. (April 2006). "Evolution at the nucleotide level: the problem of multiple whole-genome alignment". *Human Molecular Genetics*. **15** (Spec No 1): R51–6. doi: 10.1093/hmg/ddl056[98]. PMID 16651369[99].
- Fitch, W.M. (May 2000). "Homology a personal view on some of the problems". *Trends in Genetics*. **16** (5): 227–31. doi: 10.1016/S0168-9525(00)02005-9[100]. PMID 10782117[101].
- Gegenbaur, G. (1898). *Vergleichende Anatomie der Wirbelthiere ...* Leipzig.
- Haeckel, Ernst (1866). *Generelle Morphologie der Organismen*. Bd 1-2. Berlin.
- Larson, Edward J. (2004). *Evolution: The Remarkable History of Scientific Theory*. Modern Library. ISBN 0-679-64288-9.
- Owen, Richard (1847). *On the archetype and homologies of the vertebrate skeleton*. London.
- Mindell D.P., Meyer A. (2001). "Homology evolving"[102] (PDF). *Trends in Ecology and Evolution*. **16** (8): 434–40. doi: 10.1016/S0169-5347(01)02206-6[103]. Archived from the original[104] (PDF) on 2010-06-27.
- Kuzniar, A.; van Ham, R.C.; Pongor, S.; Leunissen, J.A. (November 2008). "The quest for orthologs: finding the corresponding gene across genomes". *Trends Genet*. **24** (11): 539–51. doi: 10.1016/j.tig.2008.08.009[105]. PMID 18819722[106].

External links

- 🌐 Media related to Homology at Wikimedia Commons

Last universal common ancestor

Part of a series on
Evolutionary biology

- 🌿 **Evolutionary biology portal**
- ⬭ **Category**
- ⌐ *Book*
- **Related topics**

- <u>v</u>
- <u>t</u>
- <u>e</u>[107]

The **last universal common ancestor** (**LUCA**), also called the **last universal ancestor** (**LUA**), **cenancestor**, or (incorrectly[108]) **progenote**, is the most recent population of organisms from which all organisms now living on Earth have a common descent. LUCA is the most recent common ancestor of all current life on Earth. LUCA is not thought to be the first living organism on Earth, but only one of many early organisms, all but one of which died out. LUCA is estimated to have lived some 3.5 to 3.8 billion years ago (sometime in the Paleoarchean era). The composition of LUCA is not directly accessible as a fossil, but can be studied by comparing the genomes of its descendants, organisms living today. By this means, a 2016 study identified a set of 355 genes inferred to have been present in LUCA. This would imply it was already a complex life form with many co-adapted features including transcription and translation mechanisms to convert information between DNA, RNA, and proteins. Some of those genes, however, could have been acquired later by horizontal gene transfer between archaea and bacteria.

The earliest evidence of life on Earth is biogenic graphite found in 3.7 billion-year-old metamorphized sedimentary rocks discovered in Western Greenland and microbial mat fossils found in 3.48 billion-year-old sandstone discovered in Western Australia. A 2015 study found potentially biogenic carbon from

4.1 billion years ago in ancient rocks in Western Australia, but such findings would indicate the existence of different conditions on Earth during that period from those generally assumed today, and point to an earlier origin of life. In 2017, there was a published description of putative fossilized microorganisms that are at least 3.77 billion, and possibly 4.28 billion years old, in ferruginous sedimentary rocks in Quebec, Canada.

Charles Darwin proposed the theory of universal common descent through an evolutionary process in his book *On the Origin of Species* in 1859, saying, "Therefore I should infer from analogy that probably all the organic beings which have ever lived on this earth have descended from some one primordial form, into which life was first breathed."

Features

By analysis of the presumed LUCA's offspring groups, the LUCA appears to have been a small, single-celled organism. It likely had a ring-shaped coil of DNA floating freely within the cell, like modern bacteria. Morphologically, it would likely not have been exceptionally distinctive among a collection of generalized, small-size, modern-day bacteria. However, Carl Woese *et al.*, who first proposed the currently-used three domain system based on an analysis of ribosomal RNA (rRNA) sequences of bacteria, archaea, and eukaryotes, stated that the LUCA would have been a "...simpler, more rudimentary entity than the individual ancestors that spawned the three [domains] (and their descendants)" regarding its genetic machinery.

While the gross anatomy of LUCA can only be reconstructed with much uncertainty, its internal mechanisms can be described in some detail, based on the properties currently shared by all independently living organisms on Earth.

The genetic code was most likely based on DNA. Some studies suggest that LUCA may have lacked DNA and been defined wholly through RNA. If DNA was present, it was composed of four nucleotides (deoxyadenosine, deoxycytidine, deoxythymidine, and deoxyguanosine), to the exclusion of other possible deoxynucleotides. The DNA was kept double-stranded by a template-dependent enzyme, DNA polymerase. The integrity of the DNA was maintained by a group of maintenance enzymes, including DNA topoisomerase, DNA ligase and other DNA repair enzymes. The DNA was also protected by DNA-binding proteins such as histones. The genetic code was composed of three-nucleotide codons, thus producing 64 different codons. Since only 20 amino acids were used, multiple codons code for the same amino acids. If the code was DNA-based, it operated as follows. The genetic code was expressed via RNA intermediates, which were single-stranded. The RNA was produced by a DNA-dependent RNA polymerase using nucleotides similar to those of

DNA, with the exception that the nucleotide thymidine in DNA was replaced by uridine in RNA.

The genetic code was expressed into proteins. These were assembled from free amino acids by translation of a messenger RNA by a mechanism composed of ribosomes, transfer RNAs, and a group of related proteins. The ribosomes were composed of two subunits, one big 50S and one small 30S. Each ribosomal subunit was composed of a core of ribosomal RNA surrounded by ribosomal proteins. Both types of RNA molecules (ribosomal and transfer RNAs) played an important role in the catalytic activity of the ribosomes. Only 20 amino acids were used, to the exclusion of countless other amino acids. Only the L-isomers of the amino acids were used. ATP was used as an energy intermediate. Several hundred enzymes made of protein catalyzed chemical reactions that extract energy from fats, sugars, and amino acids, and that synthesize fats, sugars, amino acids, and nucleic acid bases using arbitrary chemical pathways.

The cell contained a water-based cytoplasm that was surrounded and effectively enclosed by a lipid bilayer membrane. Inside the cell, the concentration of sodium was lower, and potassium was higher, than outside. This gradient was maintained by specific ion transporters (also referred to as *ion pumps*). The cell multiplied by duplicating all its contents followed by cellular division. The cell used chemiosmosis to produce energy. It also reduced CO_2 and oxidized H_2 (methanogenesis or acetogenesis) via acetyl-thioesters.

The cell probably lived in conditions found in deep sea vents caused by ocean water interacting with magma beneath the ocean floor.

Hypotheses

In 1859, Charles Darwin published *On the Origin of Species* in which he twice stated the hypothesis that there was only one progenitor for all life forms. In the summation he states, "Therefore I should infer from analogy that probably all the organic beings which have ever lived on this earth have descended from some one primordial form, into which life was first breathed."[109] The very last sentence begins with a restatement of the hypothesis: "There is grandeur in this view of life, with its several powers, having been originally breathed into a few forms or into one..."

When the LUCA was hypothesized, cladograms based on genetic distance between living cells indicated that Archaea split early from the rest of life. This was inferred from the fact that the archaeans known at that time were highly resistant to environmental extremes such as high salinity, temperature or acidity, leading some scientists to suggest that the LUCA evolved in areas like the

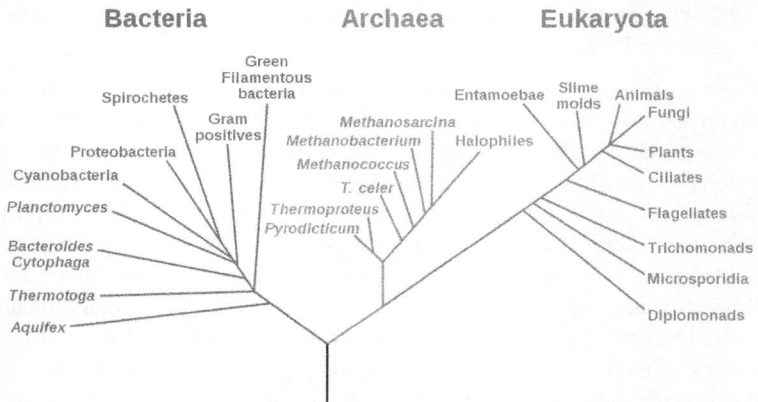

Figure 65: *A 1990 phylogenetic tree linking all major groups of living organisms to the LUCA (the black trunk at the bottom), based on ribosomal RNA sequence data.*

deep ocean vents, where such extremes prevail today. Archaea, however, were later discovered in less hostile environments, and are now believed to be more closely related to the Eukaryota than to the Bacteria, although many details are still unknown.

In 2010, based on "the vast array of molecular sequences now available from all domains of life," a formal test of universal common ancestry was published. The formal test favored the existence of a universal common ancestor over a wide class of alternative hypotheses that included horizontal gene transfer. While the formal test overwhelmingly favored the existence of a single LUCA, this does not imply that the LUCA was ever alone. Instead, it was one of several early microbes. However, given that many other nucleotides are possible besides those that are actually used in DNA and RNA today, it is almost certain that all organisms do have a single common ancestor. This is because it is extremely unlikely that organisms which descended from separate incidents where organic molecules initially came together to form cell-like structures would be able to complete a horizontal gene transfer without garbling each other's genes, converting them into noncoding segments. Further, many more amino acids are chemically possible than the twenty found in modern protein molecules. These lines of chemical evidence, taken into account for the formal statistical test by Theobald (2010), point to a single cell having been the LUCA in that, although other early microbes probably existed, only the LUCA's descendents survived beyond the Paleoarchean Era. With a common framework in the AT/GC rule and the standard twenty amino acids,

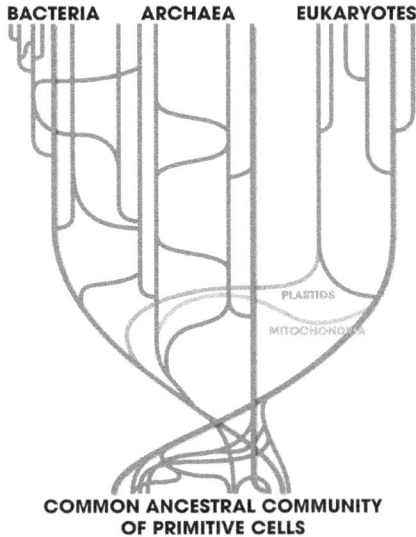

BACTERIA   ARCHAEA   EUKARYOTES

PLASTIDS

MITOCHONDRIA

**COMMON ANCESTRAL COMMUNITY
OF PRIMITIVE CELLS**

Figure 66: *2005 tree of life showing horizontal gene transfers between branches,
giving rise to an interconnected network rather than a plain hierarchy*

horizontal gene transfer would have been feasible and could have been very common later on among the progeny of that single cell.

In 1998, Carl Woese proposed (1) that no individual organism can be considered a LUCA, and (2) that the genetic heritage of all modern organisms derived through horizontal gene transfer among an ancient community of organisms. While the results described by the later papers Theobald (2010) and Saey (2010) demonstrate the existence of a single LUCA, the argument in Woese (1998) can still be applied to Ur-organismsWikipedia:Please clarify. At the beginnings of life, ancestry was not as linear as it is today because the genetic code took time to evolve. Before high fidelity replication, organisms could not be easily mapped on a phylogenetic tree. Not to be confused with the Ur-organism, however, the LUCA lived after the genetic code and at least some rudimentary early form of molecular proofreading had already evolved. It was not the very first cell, but rather, the one whose descendants survived beyond the very early stages of microbial evolution.

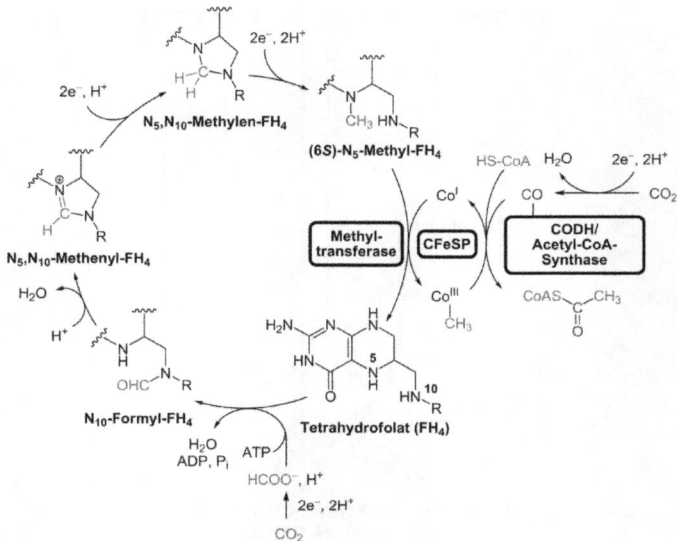

Figure 67: *The LUCA used the Wood–Ljungdahl or reductive acetyl–CoA pathway to fix carbon.*

Location of the root

The most commonly accepted location of the root of the tree of life is between a monophyletic domain Bacteria and a clade formed by Archaea and Eukaryota of what is referred to as the "traditional tree of life" based on several molecular studies.[110,111,112,113] A very small minority of studies have concluded differently, namely that the root is in the domain Bacteria, either in the phylum Firmicutes or that the phylum Chloroflexi is basal to a clade with Archaea and Eukaryotes and the rest of Bacteria as proposed by Thomas Cavalier-Smith.

Research published in 2016, by William F. Martin, by genetically analyzing 6.1 million protein coding genes from sequenced prokaryotic genomes of various phylogenetic trees, identified 355 protein clusters from amongst 286,514 protein clusters that were probably common to the LUCA. The results "depict LUCA as anaerobic, CO_2-fixing, H_2-dependent with a Wood–Ljungdahl pathway (the reductive acetyl-coenzyme A pathway), N_2-fixing and thermophilic. LUCA's biochemistry was replete with FeS clusters and radical reaction mechanisms." The cofactors also reveal "dependence upon transition metals, flavins, S-adenosyl methionine, coenzyme A, ferredoxin, molybdopterin, corrins and selenium. Its genetic code required nucleoside modifications and S-adenosylmethionine-dependent methylations."[114,115,116]

Taken as a whole, the results of the study are "quite specific".[117] They show that methanogenic clostria was a basal clade in the 355 lineages examined, and that the LUCA may therefore have inhabited an anaerobic hydrothermal vent setting in a geochemically active environment rich in H_2, CO_2, and iron. However it is not completely clear how to interpret the commonality found in this research.

These findings could mean that life on Earth originated in such hydrothermal vents, but could also be interpreted as showing that life was restricted to such locations at some later time in its early history (especially if the Late Heavy Bombardment hypothesis is correct). The identification of these genes as being present in LUCA has also been criticized, as they may simply represent subsequent horizontal gene transfers between the Archaea and the Bacteria, rather than proteins present in LUCA.

External links

- ☙ Media related to Last universal ancestor at Wikimedia Commons

Iron–sulfur world hypothesis

The **iron–sulfur world hypothesis** is a set of proposals for the origin of life and the early evolution of life advanced in a series of articles between 1988 and 1992 by Günter Wächtershäuser, a Munich patent lawyer with a degree in chemistry, who had been encouraged and supported by philosopher Karl R. Popper to publish his ideas. The hypothesis proposes that early life may have formed on the surface of iron sulfide minerals, hence the name. It was developed by retrodiction from extant biochemistry in conjunction with chemical experiments.

Origin of life

Pioneer organism

Wächtershäuser proposes that the earliest form of life, termed "pioneer organism", originated in a volcanic hydrothermal flow at high pressure and high (100 °C) temperature. It had a composite structure of a mineral base with catalytic transition metal centers (predominantly iron and nickel, but also perhaps cobalt, manganese, tungsten and zinc). The catalytic centers catalyzed autotrophic carbon fixation pathways generating small molecule (non-polymer) organic compounds from inorganic gases (e.g. carbon monoxide, carbon dioxide, hydrogen cyanide and hydrogen sulfide). These organic compounds were retained on or in the mineral base as organic ligands of the transition metal centers with a flow retention time in correspondence with their mineral bonding strength thereby defining an autocatalytic "surface metabolism". The catalytic transition metal centers became autocatalytic by being accelerated by their organic products turned ligands. The carbon fixation metabolism became autocatalytic by forming a metabolic cycle in the form of a primitive sulfur-dependent version of the reductive citric acid cycle. Accelerated catalysts expanded the metabolism and new metabolic products further accelerated the catalysts. The idea is that once such a primitive autocatalytic metabolism was established, its intrinsically synthetic chemistry began to produce ever more complex organic compounds, ever more complex pathways and ever more complex catalytic centers.

Nutrient conversions

The water gas shift reaction ($CO + H_2O \rightarrow CO_2 + H_2$) occurs in volcanic fluids with diverse catalysts or without catalysts. The combination of ferrous sulfide and hydrogen sulfide as reducing agents in conjunction with pyrite formation – $FeS + H_2S \rightarrow FeS_2 + 2H^+ + 2e^-$ (or H_2 instead of $2H^+ + 2e^-$) – has been demonstrated under mild volcanic conditions. This key result has been disputed. Nitrogen fixation has been demonstrated for the isotope $^{15}N_2$ in conjunction with pyrite formation. Ammonia forms from nitrate with FeS/H_2S as reductant. Methylmercaptan [CH_3-SH] and carbon oxysulfide [COS] form from CO_2 and FeS/H_2S, or from CO and H_2 in the presence of NiS.

Synthetic reactions

Reaction of carbon monoxide (CO), hydrogen sulfide (H_2S) and methanethiol CH_3SH in the presence of nickel sulfide and iron sulfide generates the methyl thioester of acetic acid [CH_3-CO-SCH_3] and presumably thioacetic acid (CH_3-CO-SH) as the simplest activated acetic acid analogues of acetyl-CoA. These activated acetic acid derivatives serve as starting materials for subsequent exergonic synthetic steps. They also serve for energy coupling with endergonic reactions, notably the formation of (phospho)anhydride compounds. However, Huber and Wächtershäuser reported a low 0.5% acetate yields based on input of CH_3SH (Methanethiol) (8 mM) in the presence of 350 mM CO. This is about 500 times and 3700 times the highest CH_3SH and CO concentrations respectively measured to date in a natural hydrothermal vent fluid.

Reaction of nickel hydroxide with hydrogen cyanide (HCN) (in the presence or absence of ferrous hydroxide, hydrogen sulfide or methyl mercaptan) generates nickel cyanide, which reacts with carbon monoxide (CO) to generate pairs of α-hydroxy and α-amino acids: e.g. glycolate/glycine, lactate/alanine, glycerate/serine; as well as pyruvic acid in significant quantities. Pyruvic acid is also formed at high pressure and high temperature from CO, H_2O, FeS in the presence of nonyl mercaptan. Reaction of pyruvic acid or other α-keto acids with ammonia in the presence of ferrous hydroxide or in the presence of ferrous sulfide and hydrogen sulfide generates alanine or other α-amino acids. Reaction of α-amino acids in aqueous solution with COS or with CO and H_2S generates a peptide cycle wherein dipeptides, tripeptides etc. are formed and subsequently degraded via N-terminal hydantoin moieties and N-terminal urea moieties and subsequent cleavage of the N-terminal amino acid unit.[118]

Proposed reaction mechanism for reduction of CO_2 on FeS: Ying et al. (2007) have proved that direct transformation of mackinawite (FeS) to pyrite (FeS_2) on reaction with H_2S till 300 °C is not possible without the presence of critical amount of oxidant. In the absence of any oxidant, FeS reacts with H_2S up to 300 °C to give pyrrhotite. Farid et al. have proved experimentally that mackinawite (FeS) has ability to reduce CO_2 to CO at temperature higher than 300 °C. They reported that the surface of FeS is oxidized, which on reaction with H_2S gives pyrite (FeS_2). It is expected that CO reacts with H_2O in the Drobner experiment to give H_2.

Early evolution

Early evolution is defined as beginning with the origin of life and ending with the last universal common ancestor (LUCA). According to the iron–sulfur world theory it covers a coevolution of cellular organization (cellularization), the genetic machinery and enzymatization of the metabolism.

Cellularization

Cellularization occurs in several stages. It begins with the formation of primitive lipids (e.g. fatty acids or isoprenoid acids) in the surface metabolism. These lipids accumulate on or in the mineral base. This lipophilizes the outer or inner surfaces of the mineral base, which promotes condensation reactions over hydrolytic reactions by lowering the activity of water and protons.

In the next stage lipid membranes are formed. While still anchored to the mineral base they form a semi-cell bounded partly by the mineral base and partly by the membrane. Further lipid evolution leads to self-supporting lipid membranes and closed cells. The earliest closed cells are pre-cells (*sensu* Kandler) because they allow frequent exchange of genetic material (e.g. by fusions). According to Woese, this frequent exchange of genetic material is the cause for the existence of the common stem in the tree of life and for a very rapid early evolution.

Proto-ecological systems

William Martin and Michael Russell suggest that the first cellular life forms may have evolved inside alkaline hydrothermal vents at seafloor spreading zones in the deep sea. These structures consist of microscale caverns that are coated by thin membraneous metal sulfide walls. Therefore, these structures would resolve several critical points germane to Wächtershäuser's suggestions at once:

1. the micro-caverns provide a means of concentrating newly synthesised molecules, thereby increasing the chance of forming oligomers;
2. the steep temperature gradients inside the hydrothermal vent allow for establishing "optimum zones" of partial reactions in different regions of the vent (e.g. monomer synthesis in the hotter, oligomerisation in the colder parts);
3. the flow of hydrothermal water through the structure provides a constant source of building blocks and energy (chemical disequilibrium between hydrothermal hydrogen and marine carbon dioxide);
4. the model allows for a succession of different steps of cellular evolution (prebiotic chemistry, monomer and oligomer synthesis, peptide and protein synthesis, RNA world, ribonucleoprotein assembly and DNA world) in a single structure, facilitating exchange between all developmental stages;
5. synthesis of lipids as a means of "closing" the cells against the environment is not necessary, until basically all cellular functions are developed.

This model locates the "last universal common ancestor" (LUCA) within the inorganically formed physical confines of an alkaline hydrothermal vent, rather than assuming the existence of a free-living form of LUCA. The last evolutionary step en route to bona fide free-living cells would be the synthesis of a lipid membrane that finally allows the organisms to leave the microcavern system of the vent. This postulated late acquisition of the biosynthesis of lipids as directed by genetically encoded peptides is consistent with the presence of completely different types of membrane lipids in archaea and bacteria (plus eukaryotes). The kind of vent at the foreground of their suggestion is chemically more similar to the warm (ca. 100 °C) off ridge vents such as Lost City than to the more familiar black smoker type vents (ca. 350 °C).

In an abiotic world, a thermocline of temperatures and a chemocline in concentration is associated with the pre-biotic synthesis of organic molecules, hotter in proximity to the chemically rich vent, cooler but also less chemically rich at greater distances. The migration of synthesized compounds from areas of high concentration to areas of low concentration gives a directionality that provides both source and sink in a self-organizing fashion, enabling a proto-metabolic process by which acetic acid production and its eventual oxidization can be spatially organized.

In this way many of the individual reactions that are today found in central metabolism could initially have occurred independent of any developing cell membrane. Each vent microcompartment is functionally equivalent to a single cell. Chemical communities having greater structural integrity and resilience to wildly fluctuating conditions are then selected for; their success would lead to local zones of depletion for important precursor chemicals. Progressive incorporation of these precursor components within a cell membrane would gradually increase metabolic complexity within the cell membrane, whilst leading to greater environmental simplicity in the external environment. In principle, this could lead to the development of complex catalytic sets capable of self-maintenance.

Russell adds a significant factor to these ideas, by pointing out that semi-permeable mackinawite (an iron sulfide mineral) and silicate membranes could naturally develop under these conditions and electrochemically link reactions separated in space, if not in time.[119]

Panspermia

Panspermia (from Ancient Greek πᾶν *(pan)*, meaning 'all', and σπέρμα *(sperma)*, meaning 'seed') is the hypothesis that life exists throughout the Universe, distributed by space dust, meteoroids, asteroids, comets, planetoids, and also by spacecraft carrying unintended contamination by microorganisms.[120]

Panspermia hypotheses propose (for example) that microscopic life-forms that can survive the effects of space (such as extremophiles) can become trapped in debris ejected into space after collisions between planets and small Solar System bodies that harbor life. Some organisms may travel dormant for an extended amount of time before colliding randomly with other planets or intermingling with protoplanetary disks. Under certain ideal impact circumstances (into a body of water, for example), and ideal conditions on a new planet's surfaces, it is possible that the surviving organisms could become active and begin to colonize their new environment. Panspermia studies concentrate not on how life began, but on the methods that may cause its distribution in the Universe.[121,122,123]

Pseudo-panspermia (sometimes called *"soft panspermia"* or *"molecular panspermia"*) argues that the pre-biotic organic building-blocks of life originated in space, became incorporated in the solar nebula from which planets condensed, and were further—and continuously—distributed to planetary surfaces where life then emerged (abiogenesis). From the early 1970s it started

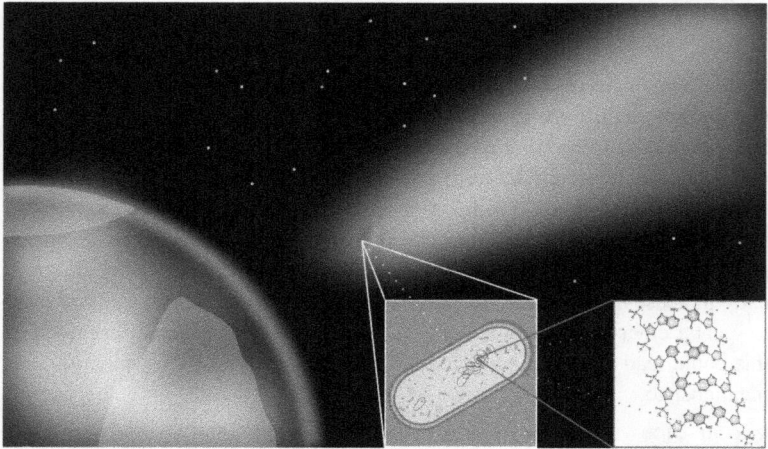

Figure 68: *Panspermia proposes that bodies such as comets transported life forms such as bacteria - complete with their DNA - through space to the Earth.*

to become evident that interstellar dust included a large component of organic molecules. Interstellar molecules are formed by chemical reactions within very sparse interstellar or circumstellar clouds of dust and gas. The dust plays a critical role in shielding the molecules from the ionizing effect of ultraviolet radiation emitted by stars.

The chemistry leading to life may have begun shortly after the Big Bang, 13.8 billion years ago, during a habitable epoch when the Universe was only 10 to 17 million years old. Though the presence of life is confirmed only on the Earth, some scientists think that extraterrestrial life is not only plausible, but probable or inevitable. Probes and instruments have started examining other planets and moons in the Solar System and in other planetary systems for evidence of having once supported simple life, and projects such as SETI attempt to detect radio transmissions from possible extra-terrestrial civilizations.

History

The first known mention of the term was in the writings of the 5th-century BC Greek philosopher Anaxagoras.[124] Panspermia began to assume a more scientific form through the proposals of Jöns Jacob Berzelius (1834), Hermann E. Richter (1865), Kelvin (1871), Hermann von Helmholtz (1879) and finally reaching the level of a detailed scientific hypothesis through the efforts of the Swedish chemist Svante Arrhenius (1903).[125]

Fred Hoyle (1915–2001) and Chandra Wickramasinghe (born 1939) were influential proponents of panspermia. In 1974 they proposed the hypothesis that some dust in interstellar space was largely organic (containing carbon), which Wickramasinghe later proved to be correct. Hoyle and Wickramasinghe further contended that life forms continue to enter the Earth's atmosphere, and may be responsible for epidemic outbreaks, new diseases, and the genetic novelty necessary for macroevolution.

In an Origins Symposium presentation on April 7, 2009, physicist Stephen Hawking stated his opinion about what humans may find when venturing into space, such as the possibility of alien life through the theory of panspermia: "Life could spread from planet to planet or from stellar system to stellar system, carried on meteors."

Three series of astrobiology experiments have been conducted outside the International Space Station between 2008 and 2015 (EXPOSE) where a wide variety of biomolecules, microorganisms, and their spores were exposed to the solar flux and vacuum of space for about 1.5 years. Some organisms survived in an inactive state for considerable lengths of time, and those samples sheltered by simulated meteorite material provide experimental evidence for the likelihood of the hypothetical scenario of lithopanspermia.

Several simulations in laboratories and in low Earth orbit suggest that ejection, entry and impact is survivable for some simple organisms. In 2015, remains of biotic material were found in 4.1 billion-year-old rocks in Western Australia, when the young Earth was about 400 million years old.[126] According to one researcher, "If life arose relatively quickly on Earth ... then it could be common in the universe."

In April 2018 a Russian team published a paper which disclosed that they found DNA on the exterior of the ISS from land and marine bacteria similar to those previously observed in superficial micro layers at the Barents and Kara seas' coastal zones. They conclude "The presence of the wild land and marine bacteria DNA on the ISS suggests their possible transfer from the stratosphere into the ionosphere with the ascending branch of the global atmospheric electrical circuit. Alternatively, the wild land and marine bacteria as well as the ISS bacteria may all have an ultimate space origin."

Proposed mechanisms

Panspermia can be said to be either interstellar (between star systems) or interplanetary (between planets in the same star system); its transport mechanisms may include comets, radiation pressure and lithopanspermia (microorganisms embedded in rocks). Interplanetary transfer of nonliving material is well documented, as evidenced by meteorites of Martian origin found on Earth. Space probes may also be a viable transport mechanism for interplanetary cross-pollination in the Solar System or even beyond. However, space agencies have implemented planetary protection procedures to reduce the risk of planetary contamination,[127,128] although, as recently discovered, some microorganisms, such as Tersicoccus phoenicis, may be resistant to procedures used in spacecraft assembly clean room facilities. In 2012, mathematician Edward Belbruno and astronomers Amaya Moro-Martín and Renu Malhotra proposed that gravitational low-energy transfer of rocks among the young planets of stars in their birth cluster is commonplace, and not rare in the general galactic stellar population.[129] Deliberate directed panspermia from space to seed Earth or sent from Earth to seed other planetary systems have also been proposed. One twist to the hypothesis by engineer Thomas Dehel (2006), proposes that plasmoid magnetic fields ejected from the magnetosphere may move the few spores lifted from the Earth's atmosphere with sufficient speed to cross interstellar space to other systems before the spores can be destroyed.

Radiopanspermia

In 1903, Svante Arrhenius published in his article *The Distribution of Life in Space*,[130] the hypothesis now called radiopanspermia, that microscopic forms of life can be propagated in space, driven by the radiation pressure from stars. Arrhenius argued that particles at a critical size below 1.5 μm would be propagated at high speed by radiation pressure of the Sun. However, because its effectiveness decreases with increasing size of the particle, this mechanism holds for very tiny particles only, such as single bacterial spores. The main criticism of radiopanspermia hypothesis came from Iosif Shklovsky and Carl Sagan, who pointed out the proofs of the lethal action of space radiations (UV and X-rays) in the cosmos. Regardless of the evidence, Wallis and Wickramasinghe argued in 2004 that the transport of individual bacteria or clumps of bacteria, is overwhelmingly more important than lithopanspermia in terms of numbers of microbes transferred, even accounting for the death rate of unprotected bacteria in transit.

Then, data gathered by the orbital experiments ERA, BIOPAN, EXOSTACK and EXPOSE, determined that isolated spores, including those of *B. subtilis*, were killed by several orders of magnitude if exposed to the full space environment for a mere few seconds, but if shielded against solar UV, the spores were capable of surviving in space for up to six years while embedded in clay or meteorite powder (artificial meteorites). Though minimal protection is required to shelter a spore against UV radiation, exposure to solar UV and cosmic ionizing radiation of unprotected DNA, break it up into its bases. Also, exposing

DNA to the ultrahigh vacuum of space alone is sufficient to cause DNA damage, so the transport of unprotected DNA or RNA during interplanetary flights powered solely by light pressure is extremely unlikely. The feasibility of other means of transport for the more massive shielded spores into the outer Solar System – for example, through gravitational capture by comets – is at this time unknown.

Based on experimental data on radiation effects and DNA stability, it has been concluded that for such long travel times, boulder-sized rocks which are greater than or equal to 1 meter in diameter are required to effectively shield resistant microorganisms, such as bacterial spores against galactic cosmic radiation. These results clearly negate the radiopanspermia hypothesis, which requires single spores accelerated by the radiation pressure of the Sun, requiring many years to travel between the planets, and support the likelihood of interplanetary transfer of microorganisms within asteroids or comets, the so-called **lithopanspermia** hypothesis.

Lithopanspermia

Lithopanspermia, the transfer of organisms in rocks from one planet to another either through interplanetary or interstellar space, remains speculative. Although there is no evidence that lithopanspermia has occurred in the Solar System, the various stages have become amenable to experimental testing.

- **Planetary ejection** — For lithopanspermia to occur, researchers have suggested that microorganisms must survive ejection from a planetary surface which involves extreme forces of acceleration and shock with associated temperature excursions. Hypothetical values of shock pressures experienced by ejected rocks are obtained with Martian meteorites, which suggest the shock pressures of approximately 5 to 55 GPa, acceleration of 3 Mm/s^2 and jerk of 6 Gm/s^3 and post-shock temperature increases of about 1 K to 1000 K. To determine the effect of acceleration during ejection on microorganisms, rifle and ultracentrifuge methods were successfully used under simulated outer space conditions.
- **Survival in transit** — The survival of microorganisms has been studied extensively using both simulated facilities and in low Earth orbit. A large number of microorganisms have been selected for exposure experiments. It is possible to separate these microorganisms into two groups, the human-borne, and the extremophiles. Studying the human-borne microorganisms is significant for human welfare and future manned missions; whilst the extremophiles are vital for studying the physiological requirements of survival in space.

- **Atmospheric entry** — An important aspect of the lithopanspermia hypothesis to test is that microbes situated on or within rocks could survive hypervelocity entry from space through Earth's atmosphere (Cockell, 2008). As with planetary ejection, this is experimentally tractable, with sounding rockets and orbital vehicles being used for microbiological experiments. *B. subtilis* spores inoculated onto granite domes were subjected to hypervelocity atmospheric transit (twice) by launch to a ~120 km altitude on an Orion two-stage rocket. The spores were shown to have survived on the sides of the rock, but they did not survive on the forward-facing surface that was subjected to a maximum temperature of 145 °C. In separate experiments, as part of the ESA STONE experiment, numerous organisms were embedded in different types or rocks and were mounted in the heat shield of six Foton re-entry capsules. During reentry, the rock samples were subjected to temperatures and pressure loads comparable to those experienced in meteorites. The exogenous arrival of photosynthetic microorganisms could have quite profound consequences for the course of biological evolution on the inoculated planet. As photosynthetic organisms must be close to the surface of a rock to obtain sufficient light energy, atmospheric transit might act as a filter against them by ablating the surface layers of the rock. Although cyanobacteria have been shown to survive the desiccating, freezing conditions of space in orbital experiments, this would be of no benefit as the STONE experiment showed that they cannot survive atmospheric entry. Thus, non-photosynthetic organisms deep within rocks have a chance to survive the exit and entry process. (See also: Impact survival.) Research presented at the European Planetary Science Congress in 2015 suggests that ejection, entry and impact is survivable for some simple organisms.

Accidental panspermia

Thomas Gold, a professor of astronomy, suggested in 1960 the hypothesis of "Cosmic Garbage", that life on Earth might have originated accidentally from a pile of waste products dumped on Earth long ago by extraterrestrial beings.[131]

Directed panspermia

Directed panspermia concerns the deliberate transport of microorganisms in space, sent to Earth to start life here, or sent from Earth to seed new planetary systems with life by introduced species of microorganisms on lifeless planets. The Nobel prize winner Francis Crick, along with Leslie Orgel proposed that life may have been purposely spread by an advanced extraterrestrial civilization, but considering an early "RNA world" Crick noted later that life may have originated on Earth.[132] It has been suggested that 'directed' panspermia was proposed in order to counteract various objections, including the argument that microbes would be inactivated by the space environment and cosmic radiation before they could make a chance encounter with Earth.

Conversely, active directed panspermia has been proposed to secure and expand life in space. This may be motivated by biotic ethics that values, and seeks to propagate, the basic patterns of our organic gene/protein life-form. The panbiotic program would seed new planetary systems nearby, and clusters of new stars in interstellar clouds. These young targets, where local life would not have formed yet, avoid any interference with local life.

For example, microbial payloads launched by solar sails at speeds up to 0.0001 c (30,000 m/s) would reach targets at 10 to 100 light-years in 0.1 million to 1 million years. Fleets of microbial capsules can be aimed at clusters of new stars in star-forming clouds, where they may land on planets or captured by asteroids and comets and later delivered to planets. Payloads may contain extremophiles for diverse environments and cyanobacteria similar to early microorganisms. Hardy multicellular organisms (rotifer cysts) may be included to induce higher evolution.

The probability of hitting the target zone can be calculated from $P(target) = \frac{A(target)}{\pi(dy)^2} = \frac{ar(target)^2v^2}{(tp)^2d^4}$ where A(target) is the cross-section of the target area, dy is the positional uncertainty at arrival; a – constant (depending on units), r(target) is the radius of the target area; v the velocity of the probe; (tp) the targeting precision (arcsec/yr); and d the distance to the target, guided by high-resolution astrometry of 1×10^{-5} arcsec/yr (all units in SIU). These calculations show that relatively near target stars(Alpha PsA, Beta Pictoris) can be seeded by milligrams of launched microbes; while seeding the Rho Ophiochus star-forming cloud requires hundreds of kilograms of dispersed capsules.

Directed panspermia to secure and expand life in space is becoming possible because of developments in solar sails, precise astrometry, extrasolar planets, extremophiles and microbial genetic engineering. After determining the composition of chosen meteorites, astroecologists performed laboratory experiments that suggest that many colonizing microorganisms and some plants could obtain many of their chemical nutrients from asteroid and cometary materials. However, the scientists noted that phosphate (PO_4) and nitrate (NO_3–N) critically limit nutrition to many terrestrial lifeforms. With such materials, and energy from long-lived stars, microscopic life planted by directed panspermia could find an immense future in the galaxy.

A number of publications since 1979 have proposed the idea that directed panspermia could be demonstrated to be the origin of all life on Earth if a distinctive 'signature' message were found, deliberately implanted into either the genome or the genetic code of the first microorganisms by our hypothetical progenitor.

In 2013 a team of physicists claimed that they had found mathematical and semiotic patterns in the genetic code which they think is evidence for such a signature. This claim has been refuted by biologist PZ Myers who said, writing in Pharyngula:

> *Unfortunately, what they've so honestly described is good old honest garbage ... Their methods failed to recognize a well-known functional association in the genetic code; they did not rule out the operation of natural law before rushing to falsely infer design ... We certainly don't need to invoke panspermia. Nothing in the genetic code requires design. and the authors haven't demonstrated otherwise.*

In a later peer-reviewed article, the authors address the operation of natural law in an extensive statistical test, and draw the same conclusion as in the previous article. In special sections they also discuss methodological concerns raised by PZ Myers and some others.

Pseudo-panspermia

Pseudo-panspermia (sometimes called soft panspermia, molecular panspermia or quasi-panspermia) proposes that the organic molecules used for life originated in space and were incorporated in the solar nebula, from which the planets condensed and were further —and continuously— distributed to planetary surfaces where life then emerged (abiogenesis). From the early 1970s it was becoming evident that interstellar dust consisted of a large component of organic molecules. The first suggestion came from Chandra Wickramasinghe, who proposed a polymeric composition based on the molecule formaldehyde (CH_2O). Interstellar molecules are formed by chemical reactions within very sparse interstellar or circumstellar clouds of dust and gas. Usually this occurs when a molecule becomes ionized, often as the result of an interaction with cosmic rays. This positively charged molecule then draws in a nearby reactant by electrostatic attraction of the neutral molecule's electrons. Molecules can also be generated by reactions between neutral atoms and molecules, although this process is generally slower. The dust plays a critical role of shielding the molecules from the ionizing effect of ultraviolet radiation emitted by stars.

A 2008 analysis of $^{12}C/^{13}C$ isotopic ratios of organic compounds found in the Murchison meteorite indicates a non-terrestrial origin for these molecules rather than terrestrial contamination. Biologically relevant molecules identified so far include uracil, an RNA nucleobase, and xanthine. These results demonstrate that many organic compounds which are components of life on

Earth were already present in the early Solar System and may have played a key role in life's origin.

In August 2009, NASA scientists identified one of the fundamental chemical building-blocks of life (the amino acid glycine) in a comet for the first time.

In August 2011, a report, based on NASA studies with meteorites found on Earth, was published suggesting building blocks of DNA (adenine, guanine and related organic molecules) may have been formed extraterrestrially in outer space. In October 2011, scientists reported that cosmic dust contains complex organic matter ("amorphous organic solids with a mixed aromatic-aliphatic structure") that could be created naturally, and rapidly, by stars. One of the scientists suggested that these complex organic compounds may have been related to the development of life on Earth and said that, "If this is the case, life on Earth may have had an easier time getting started as these organics can serve as basic ingredients for life."

In August 2012, and in a world first, astronomers at Copenhagen University reported the detection of a specific sugar molecule, glycolaldehyde, in a distant star system. The molecule was found around the protostellar binary *IRAS 16293-2422*, which is located 400 light years from Earth. Glycolaldehyde is needed to form ribonucleic acid, or RNA, which is similar in function to DNA. This finding suggests that complex organic molecules may form in stellar systems prior to the formation of planets, eventually arriving on young planets early in their formation.

In September 2012, NASA scientists reported that polycyclic aromatic hydrocarbons (PAHs), subjected to interstellar medium (ISM) conditions, are transformed, through hydrogenation, oxygenation and hydroxylation, to more complex organics – "a step along the path toward amino acids and nucleotides, the raw materials of proteins and DNA, respectively". Further, as a result of these transformations, the PAHs lose their spectroscopic signature which could be one of the reasons "for the lack of PAH detection in interstellar ice grains, particularly the outer regions of cold, dense clouds or the upper molecular layers of protoplanetary disks."

In 2013, the Atacama Large Millimeter Array (ALMA Project) confirmed that researchers have discovered an important pair of prebiotic molecules in the icy particles in interstellar space (ISM). The chemicals, found in a giant cloud of gas about 25,000 light-years from Earth in ISM, may be a precursor to a key component of DNA and the other may have a role in the formation of an important amino acid. Researchers found a molecule called cyanomethanimine, which produces adenine, one of the four nucleobases that form the "rungs" in the ladder-like structure of DNA. The other molecule, called ethanamine, is thought to play a role in forming alanine, one of the twenty amino acids in the genetic code. Previously, scientists thought such processes took place in the very tenuous gas between the stars. The new discoveries, however, suggest that the chemical formation sequences for these molecules occurred not in gas, but on the surfaces of ice grains in interstellar space. NASA ALMA scientist Anthony Remijan stated that finding these molecules in an interstellar gas cloud means that important building blocks for DNA and amino acids can 'seed' newly formed planets with the chemical precursors for life.[133]

In March 2013, a simulation experiment indicate that dipeptides (pairs of amino acids) that can be building blocks of proteins, can be created in interstellar dust.

In February 2014, NASA announced a greatly upgraded database[134] for tracking polycyclic aromatic hydrocarbons (PAHs) in the universe. According to scientists, more than 20% of the carbon in the universe may be associated with PAHs, possible starting materials for the formation of life. PAHs seem to have been formed shortly after the Big Bang, are widespread throughout the universe, and are associated with new stars and exoplanets.

In March 2015, NASA scientists reported that, for the first time, complex DNA and RNA organic compounds of life, including uracil, cytosine and thymine, have been formed in the laboratory under outer space conditions, using starting chemicals, such as pyrimidine, found in meteorites. Pyrimidine, like polycyclic aromatic hydrocarbons (PAHs), the most carbon-rich chemical found in the Universe, may have been formed in red giants or in interstellar dust and gas clouds, according to the scientists.

In May 2016, the Rosetta Mission team reported the presence of glycine, methylamine and ethylamine in the coma of 67P/Churyumov-Gerasimenko.[135] This, plus the detection of phosphorus, is consistent with the hypothesis that comets played a crucial role in the emergence of life on Earth.

Extraterrestrial life

The chemistry of life may have begun shortly after the Big Bang, 13.8 billion years ago, during a habitable epoch when the Universe was only 10–17 million years old. According to the panspermia hypothesis, microscopic life—distributed by meteoroids, asteroids and other small Solar System bodies—may exist throughout the universe. Nonetheless, Earth is the only place in the universe known by humans to harbor life. The sheer number of planets in the Milky Way galaxy, however, may make it probable that life has arisen somewhere else in the galaxy and the universe. It is generally agreed that the conditions required for the evolution of intelligent life as we know it are probably exceedingly rare in the universe, while simultaneously noting that simple single-celled microorganisms may be more likely.

The extrasolar planet results from the Kepler mission estimate 100–400 billion exoplanets, with over 3,500 as candidates or confirmed exoplanets. On 4 November 2013, astronomers reported, based on Kepler space mission data, that there could be as many as 40 billion Earth-sized planets orbiting in the habitable zones of sun-like stars and red dwarf stars within the Milky Way Galaxy. 11 billion of these estimated planets may be orbiting sun-like stars. The nearest such planet may be 12 light-years away, according to the scientists.

It is estimated that space travel over cosmic distances would take an incredibly long time to an outside observer, and with vast amounts of energy required. However, there are reasons to hypothesize that faster-than-light interstellar space travel might be feasible. This has been explored by NASA scientists since at least 1995.

Hypotheses on extraterrestrial sources of illnesses

Hoyle and Wickramasinghe have speculated that several outbreaks of illnesses on Earth are of extraterrestrial origins, including the 1918 flu pandemic, and certain outbreaks of polio and mad cow disease. For the 1918 flu pandemic they hypothesized that cometary dust brought the virus to Earth simultaneously at multiple locations—a view almost universally dismissed by experts on this pandemic. Hoyle also speculated that HIV came from outer space. After Hoyle's death, *The Lancet* published a letter to the editor from Wickramasinghe and two of his colleagues, in which they hypothesized that the virus that causes severe acute respiratory syndrome (SARS) could be extraterrestrial in origin and not originated from chickens. *The Lancet* subsequently published three responses to this letter, showing that the hypothesis was not evidence-based, and casting doubts on the quality of the experiments referenced by Wickramasinghe in his letter. A 2008 encyclopedia notes that "Like other claims linking terrestrial disease to extraterrestrial pathogens, this proposal was rejected by the greater research community."

In April 2016, Jiangwen Qu of the Department of Infectious Disease Control in China presented a statistical study suggesting that "extremes of sunspot activity to within plus or minus 1 year may precipitate influenza pandemics." He discussed possible mechanisms of epidemic initiation and early spread, including speculation on primary causation by externally derived viral variants from space via cometary dust.

Case studies

- A meteorite originating from Mars known as ALH84001 was shown in 1996 to contain microscopic structures resembling small terrestrial nanobacteria. When the discovery was announced, many immediately conjectured that these were fossils and were the first evidence of extraterrestrial life — making headlines around the world. Public interest soon started to dwindle as most experts started to agree that these structures were not indicative of life, but could instead be formed abiotically from organic molecules. However, in November 2009, a team of scientists at Johnson Space Center, including David McKay, reasserted that there was "strong evidence that life may have existed on ancient Mars", after having reexamined the meteorite and finding magnetite crystals.

- On May 11, 2001, two researchers from the University of Naples claimed to have found viable extraterrestrial bacteria inside a meteorite. Geologist Bruno D'Argenio and molecular biologist Giuseppe Geraci claim the bacteria were wedged inside the crystal structure of minerals, but were resurrected when a sample of the rock was placed in a culture medium.

- An Indian and British team of researchers led by Chandra Wickramasinghe reported on 2001 that air samples over Hyderabad, India, gathered from the stratosphere by the Indian Space Research Organisation (ISRO) on Jan 21, 2001, contained clumps of living cells. Wickramasinghe calls this "unambiguous evidence for the presence of clumps of living cells in air samples from as high as 41 km, above which no air from lower down would normally be transported". Two bacterial and one fungal species were later independently isolated from these filters which were identified as *Bacillus simplex*, *Staphylococcus pasteuri* and *Engyodontium album* respectively. Pushkar Ganesh Vaidya from the Indian Astrobiology Research Centre reported in 2009 that "the three microorganisms captured during the balloon experiment do not exhibit any distinct adaptations expected to be seen in microorganisms occupying a cometary niche".

- In 2005 an improved experiment was conducted by ISRO. On April 20, 2005, air samples were collected from the upper atmosphere at altitudes ranging from 20 km to more than 40 km. The samples were tested at two labs in India. The labs found 12 bacterial and 6 different fungal species in these samples. The fungi were *Penicillium decumbens*, *Cladosporium cladosporioides*, *Alternaria sp.* and *Tilletiopsis albescens*. Out of the 12 bacterial samples, three were identified as new species and named *Janibacter hoylei* (after Fred Hoyle), *Bacillus isronensis* (named after ISRO) and *Bacillus aryabhattai* (named after the ancient Indian mathematician, Aryabhata). These three new species showed that they were more resistant to UV radiation than similar bacteria.[136]

Some other researchers have retrieved bacteria from the stratosphere since the 1970s.[137] Atmospheric sampling by NASA in 2010 before and after hurricanes, collected 314 different types of bacteria; the study suggests that large-scale convection during tropical storms and hurricanes can then carry this material from the surface higher up into the atmosphere.

- Another proposed mechanism of spores in the stratosphere is lifting by weather and Earth magnetism up to the ionosphere into low Earth orbit, where Russian astronauts retrieved DNA from a known sterile exterior surface of the International Space Station.[138] The Russian scientists then also speculated the possibility "that common terrestrial bacteria are constantly being resupplied from space."

- On January 10, 2013, Chandra Wickramasinghe found fossil diatom frustules in what he thinks is a new kind of carbonaceous meteorite called Polonnaruwa that landed in the North Central Province of Sri Lanka on 29 December 2012. Early on, there was criticism that Wickramasinghe's report was not an examination of an actual meteorite but of some terrestrial rock passed off as a meteorite.

Wickramasinghe's team remark that they are aware that a large number of unrelated stones have been submitted for analysis, and have no knowledge regarding the nature, source or origin of the stones their critics have examined, so Wickramasinghe clarifies that he is using the stones submitted by the Medical Research Institute in Sri Lanka. In response to the criticism from other scientists, Wickramasinghe performed X-ray diffraction and isotope analyses to verify its meteoritic origin. His analysis revealed a 95% silica and 3% quartz content, and interpreted this result as a "carbonaceous meteorite of unknown type". In addition, Wickramasinghe's team remarked that the temperature at which sand must be heated by lightning to melt and form a fulgurite (1770 °C) would have vaporized and burned all carbon-rich organisms and melted and thus destroyed the delicately marked silica frustules of the diatoms, and that the oxygen isotope data confirms its meteoric origin. Wickramasinghe's team also argues that since living diatoms require nitrogen fixation to synthetize amino acids, proteins, DNA, RNA and other life-critical biomolecules, a population of extraterrestrial cyanobacteria must have been a required component of the comet (Polonnaruwa meteorite) "ecosystem".

- In 2013, Dale Warren Griffin, a microbiologist working at the United States Geological Survey noted that viruses are the most numerous entities on Earth. Griffin speculates that viruses evolved in comets and on other planets and moons may be pathogenic to humans, so he proposed to also look for viruses on moons and planets of the Solar System.

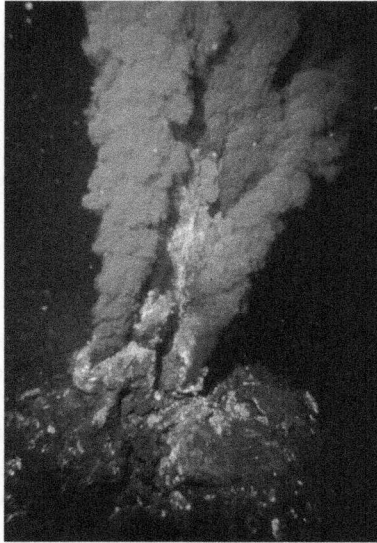

Figure 69: *Hydrothermal vents are able to support extremophile bacteria on Earth and may also support life in other parts of the cosmos.*

Hoaxes

A separate fragment of the Orgueil meteorite (kept in a sealed glass jar since its discovery) was found in 1965 to have a seed capsule embedded in it, whilst the original glassy layer on the outside remained undisturbed. Despite great initial excitement, the seed was found to be that of a European Juncaceae or Rush plant that had been glued into the fragment and camouflaged using coal dust. The outer "fusion layer" was in fact glue. Whilst the perpetrator of this hoax is unknown, it is thought that they sought to influence the 19th century debate on spontaneous generation — rather than panspermia — by demonstrating the transformation of inorganic to biological matter.

Extremophiles

Until the 1970s, life was thought to depend on its access to sunlight. Even life in the ocean depths, where sunlight cannot reach, was believed to obtain its nourishment either from consuming organic detritus rained down from the surface waters or from eating animals that did. However, in 1977, during an exploratory dive to the Galapagos Rift in the deep-sea exploration submersible *Alvin*, scientists discovered colonies of assorted creatures clustered around undersea volcanic features known as black smokers. It was soon determined that the basis for this food chain is a form of bacterium that derives its energy from oxidation of reactive chemicals, such as hydrogen or hydrogen sulfide, that bubble up from the Earth's interior. This chemosynthesis revolutionized the study of biology by revealing that terrestrial life need not be Sun-dependent; it only requires water and an energy gradient in order to exist.

It is now known that extremophiles, microorganisms with extraordinary capability to thrive in the harshest environments on Earth, can specialize to thrive in the deep-sea, ice, boiling water, acid, the water core of nuclear reactors, salt crystals, toxic waste and in a range of other extreme habitats that were previously thought to be inhospitable for life.[139] Living bacteria found in ice core samples retrieved from 3,700 metres (12,100 ft) deep at Lake Vostok in Antarctica, have provided data for extrapolations to the likelihood of microorganisms surviving frozen in extraterrestrial habitats or during interplanetary transport. Also, bacteria have been discovered living within warm rock deep in the Earth's crust.

In order to test some these organisms' potential resilience in outer space, plant seeds and spores of bacteria, fungi and ferns have been exposed to the harsh space environment. Spores are produced as part of the normal life cycle of many plants, algae, fungi and some protozoans, and some bacteria produce endospores or cysts during times of stress. These structures may be highly resilient to ultraviolet and gamma radiation, desiccation, lysozyme, temperature, starvation and chemical disinfectants, while metabolically inactive. Spores germinate when favourable conditions are restored after exposure to conditions fatal to the parent organism.

Although computer models suggest that a captured meteoroid would typically take some tens of millions of years before collision with a planet, there are documented viable Earthly bacterial spores that are 40 million years old that are very resistant to radiation, and others able to resume life after being dormant for 25 million years,[140] suggesting that lithopanspermia life-transfers are possible via meteorites exceeding 1 m in size.

The discovery of deep-sea ecosystems, along with advancements in the fields of astrobiology, observational astronomy and discovery of large varieties of extremophiles, opened up a new avenue in astrobiology by massively expanding the number of possible extraterrestrial habitats and possible transport of hardy microbial life through vast distances.

Figure 70: *EURECA facility deployment in 1992*

Research in outer space

The question of whether certain microorganisms can survive in the harsh environment of outer space has intrigued biologists since the beginning of space-flight, and opportunities were provided to expose samples to space. The first American tests were made in 1966, during the Gemini IX and XII missions, when samples of bacteriophage T1 and spores of *Penicillium roqueforti* were exposed to outer space for 16.8 h and 6.5 h, respectively. Other basic life sciences research in low Earth orbit started in 1966 with the Soviet biosatellite program Bion and the U.S. Biosatellite program. Thus, the plausibility of panspermia can be evaluated by examining life forms on Earth for their capacity to survive in space. The following experiments carried on low Earth orbit specifically tested some aspects of panspermia or lithopanspermia:

ERA

The Exobiology Radiation Assembly (ERA) was a 1992 experiment on board the European Retrievable Carrier (EURECA) on the biological effects of space radiation. EURECA was an unmanned 4.5 tonne satellite with a payload of 15 experiments. It was an astrobiology mission developed by the European Space Agency (ESA). Spores of different strains of *Bacillus subtilis* and the *Escherichia coli* plasmid pUC19 were exposed to selected conditions of space (space vacuum and/or defined wavebands and intensities of solar ultraviolet radiation). After the approximately 11-month mission, their responses were studied in terms of survival, mutagenesis in the *his* (*B. subtilis*) or *lac* locus (pUC19), induction of DNA strand breaks, efficiency of DNA repair systems, and the role of external protective agents. The data were compared with those of a simultaneously running ground control experiment:

- The survival of spores treated with the vacuum of space, however shielded against solar radiation, is substantially increased, if they are exposed in multilayers and/or in the presence of glucose as protective.
- All spores in "artificial meteorites", i.e. embedded in clays or simulated Martian soil, are killed.
- Vacuum treatment leads to an increase of mutation frequency in spores, but not in plasmid DNA.
- Extraterrestrial solar ultraviolet radiation is mutagenic, induces strand breaks in the DNA and reduces survival substantially.
- Action spectroscopy confirms results of previous space experiments of a synergistic action of space vacuum and solar UV radiation with DNA being the critical target.
- The decrease in viability of the microorganisms could be correlated with the increase in DNA damage.
- The purple membranes, amino acids and urea were not measurably affected by the dehydrating condition of open space, if sheltered from solar radiation. Plasmid DNA, however, suffered a significant amount of strand breaks under these conditions.

BIOPAN

BIOPAN is a multi-user experimental facility installed on the external surface of the Russian Foton descent capsule. Experiments developed for BIOPAN are designed to investigate the effect of the space environment on biological material after exposure between 13 and 17 days. The experiments in BIOPAN are exposed to solar and cosmic radiation, the space vacuum and weightlessness, or a selection thereof. Of the 6 missions flown so far on BIOPAN between 1992 and 2007, dozens of experiments were conducted, and some analyzed the likelihood of panspermia. Some bacteria, lichens (*Xanthoria elegans*, *Rhizocarpon geographicum* and their mycobiont cultures, the black Antarctic microfungi *Cryomyces minteri* and *Cryomyces antarcticus*), spores, and even one animal (tardigrades) were found to have survived the harsh outer space environment and cosmic radiation.

Figure 71: *EXOSTACK on the Long Duration Exposure Facility satellite.*

EXOSTACK

The German EXOSTACK experiment was deployed on 7 April 1984 on board the Long Duration Exposure Facility statellite. 30% of *Bacillus subtilis* spores survived the nearly 6 years exposure when embedded in salt crystals, whereas 80% survived in the presence of glucose, which stabilize the structure of the cellular macromolecules, especially during vacuum-induced dehydration.

If shielded against solar UV, spores of *B. subtilis* were capable of surviving in space for up to 6 years, especially if embedded in clay or meteorite powder (artificial meteorites). The data support the likelihood of interplanetary transfer of microorganisms within meteorites, the so-called lithopanspermia hypothesis.

EXPOSE

EXPOSE is a multi-user facility mounted outside the International Space Station dedicated to astrobiology experiments. There have been three EXPOSE experiments flown between 2008 and 2015: EXPOSE-E, EXPOSE-R and EXPOSE-R2.

Results from the orbital missions, especially the experiments *SEEDS* and *LiFE*, concluded that after an 18-month exposure, some seeds and lichens

Figure 72: *Location of the astrobiology EXPOSE-E and EXPOSE-R facilities on the International Space Station*

(*Stichococcus sp.* and *Acarospora sp.*, a lichenized fungal genus) may be capable to survive interplanetary travel if sheltered inside comets or rocks from cosmic radiation and UV radiation. The *LIFE*, *SPORES*, and *SEEDS* parts of the experiments provided information about the likelihood of lithopanspermia. These studies will provide experimental data to the lithopanspermia hypothesis, and they will provide basic data to planetary protection issues.

Tanpopo

The Tanpopo mission is an orbital astrobiology experiment by Japan that is currently investigating the possible interplanetary transfer of life, organic compounds, and possible terrestrial particles in low Earth orbit. The Tanpopo experiment is taking place at the Exposed Facility located on the exterior of Kibo module of the International Space Station. The mission will collect cosmic dusts and other particles for three years by using an ultra-low density silica gel called aerogel. The purpose is to assess the panspermia hypothesis and the possibility of natural interplanetary transport of life and its precursors.[141,142] Some of these aerogels will be replaced every one or two years through 2018.[143] Sample collection began in May 2015, and the first samples were be returned to Earth in mid-2016.

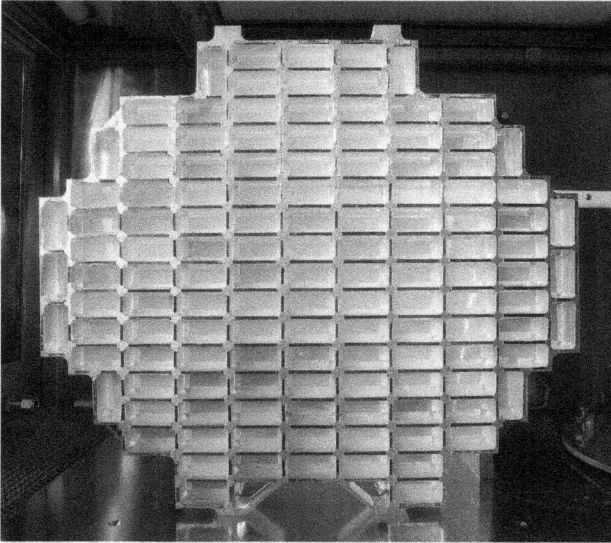

Figure 73: *Dust collector with aerogel blocks*

Criticism

Panspermia is often criticized because it does not answer the question of the origin of life but merely places it on another celestial body. It was also criticized because it was thought it could not be tested experimentally.

Wallis and Wickramasinghe argued in 2004 that the transport of individual bacteria or clumps of bacteria, is overwhelmingly more important than lithopanspermia in terms of numbers of microbes transferred, even accounting for the death rate of unprotected bacteria in transit. Then it was found that isolated spores of *B. subtilis* were killed by several orders of magnitude if exposed to the full space environment for a mere few seconds. These results clearly negate the original panspermia hypothesis, which requires single spores as space travelers accelerated by the radiation pressure of the Sun, requiring many years to travel between the planets. However, if shielded against solar UV, spores of *Bacillus subtilis* were capable of surviving in space for up to 6 years, especially if embedded in clay or meteorite powder (artificial meteorites). The data support the likelihood of interplanetary transfer of microorganisms within meteorites, the so-called **lithopanspermia** hypothesis.

Further reading

- Crick, F (1981), *Life, Its Origin and Nature*, Simon & Schuster, ISBN 0-7088-2235-5.
- Hoyle, F (1983), *The Intelligent Universe*, London: Michael Joseph, ISBN 0-7181-2298-4.

External links

シ ウ 𝓎 λ ル ↄ 呆 維 𝒲	Look up *panspermia* in Wiktionary, the free dictionary.

- A.E. Zlobin, 2013, Tunguska similar impacts and origin of life (mathematical theory of origin of life; incoming of pattern recognition algorithm due to comets)[144]
- Francis Crick's notes[145] for a lecture on directed panspermia, dated 5 November 1976.
- "Earth sows its seeds in space". *Nature News*. 23 February 2004. doi: 10.1038/news040216-20[146] (inactive 2017-01-14).
- Warmflash, D.; Weiss, B. (24 October 2005). "Did Life Come from Another World?"[147]. *Scientific American*. **293** (5): 64–71. Bibcode: 2005SciAm.293e..64W[148]. doi: 10.1038/scientificamerican1105-64[149].

Environmental and evolutionary impact of microbial mats

Microbial mat

A **microbial mat** is a multi-layered sheet of microorganisms, mainly bacteria and archaea. Microbial mats grow at interfaces between different types of material, mostly on submerged or moist surfaces, but a few survive in deserts. They colonize environments ranging in temperature from –40 °C to 120 °C. A few are found as endosymbionts of animals.

Although only a few centimetres thick at most, microbial mats create a wide range of internal chemical environments, and hence generally consist of layers of microorganisms that can feed on or at least tolerate the dominant chemicals at their level and which are usually of closely related species. In moist conditions mats are usually held together by slimy substances secreted by the microorganisms, and in many cases some of the microorganisms form tangled webs of filaments which make the mat tougher. The best known physical forms are flat mats and stubby pillars called stromatolites, but there are also spherical forms.

Microbial mats are the earliest form of life on Earth for which there is good fossil evidence, from 3,500[150] million years ago, and have been the most important members and maintainers of the planet's ecosystems. Originally they depended on hydrothermal vents for energy and chemical "food", but the development of photosynthesis allow mats to proliferate outside of these environments by utilizing a more widely available energy source, sunlight. The final and most significant stage of this liberation was the development of oxygen-producing photosynthesis, since the main chemical inputs for this are carbon dioxide and water.

Figure 74: *The cyanobacterial-algal mat, salty lake on the White Sea seaside*

As a result, microbial mats began to produce the atmosphere we know to-day, in which free oxygen is a vital component. At around the same time they may also have been the birthplace of the more complex eukaryote type of cell, of which all multicellular organisms are composed. Microbial mats were abundant on the shallow seabed until the Cambrian substrate revolution, when animals living in shallow seas increased their burrowing capabilities and thus broke up the surfaces of mats and let oxygenated water into the deeper layers, poisoning the oxygen-intolerant microorganisms that lived there. Although this revolution drove mats off soft floors of shallow seas, they still flourish in many environments where burrowing is limited or impossible, including rocky seabeds and shores, hyper-saline and brackish lagoons, and are found on the floors of the deep oceans.

Because of microbial mats' ability to use almost anything as "food", there is considerable interest in industrial uses of mats, especially for water treatment and for cleaning up pollution.

Description

Life timeline

θ —
500 —
1000 ≡
1500 ≡
2000 ≡
2500 ≡
3000 ≡
3500 ≡
4000 ≡
4500 —

Axis scale: million years

Also see: *Human timeline* and *Nature timeline*

Microbial mats have also been referred to as "algal mats" and "bacterial mats" in older scientific literature. They are a type of biofilm that is large enough to see with the naked eye and robust enough to survive moderate physical stresses.

Figure 75: *Stromatolites are formed by some microbial mats as the microbes slowly move upwards to avoid being smothered by sediment.*

These colonies of bacteria form on surfaces at many types of interface, for example between water and the sediment or rock at the bottom, between air and rock or sediment, between soil and bed-rock, etc. Such interfaces form vertical chemical gradients, i.e. vertical variations in chemical composition, which make different levels suitable for different types of bacteria and thus divide microbial mats into layers, which may be sharply defined or may merge more gradually into each other. A variety of microbes are able to transcend the limits of diffusion by using "nanowires" to shuttle electrons from their metabolic reactions up to two centimetres deep in the sediment – for example, electrons can be transferred from reactions involving hydrogen sulfide deeper within the sediment to oxygen in the water, which acts as an electron acceptor.

The best-known types of microbial mat may be flat laminated mats, which form on approximately horizontal surfaces, and stromatolites, stubby pillars built as the microbes slowly move upwards to avoid being smothered by sediment deposited on them by water. However, there are also spherical mats, some on the outside of pellets of rock or other firm material and others *inside* spheres of sediment.

Structure

A microbial mat consists of several layers, each of which is dominated by specific types of microorganism, mainly bacteria. Although the composition of individual mats varies depending on the environment, as a general rule the by-products of each group of microorganisms serve as "food" for other groups. In effect each mat forms its own food chain, with one or a few groups at the top of the food chain as their by-products are not consumed by other groups. Different types of microorganism dominate different layers based on their comparative advantage for living in that layer. In other words, they live in positions where they can out-perform other groups rather than where they would absolutely be most comfortable — ecological relationships between different groups are a combination of competition and co-operation. Since the metabolic capabilities of bacteria (what they can "eat" and what conditions they can tolerate) generally depend on their phylogeny (i.e. the most closely related groups have the most similar metabolisms), the different layers of a mat are divided both by their different metabolic contributions to the community and by their phylogenetic relationships.

In a wet environment where sunlight is the main source of energy, the uppermost layers are generally dominated by aerobic photosynthesizing cyanobacteria (blue-green bacteria whose color is caused by their having chlorophyll), while the lowest layers are generally dominated by anaerobic sulfate-reducing bacteria. Sometimes there are intermediate (oxygenated only in the daytime) layers inhabited by facultative anaerobic bacteria. For example, in hypersaline ponds near Guerrero Negro (Mexico) various kind of mats were explored. There are some mats with a middle purple layer inhabited by photosynthesizing purple bacteria.[151] Some other mats have a white layer inhabited by chemotrophic sulfide-oxidizing bacteria and beneath them an olive layer inhabited by photosynthesizing green sulfur bacteria and heterotrophic bacteria.[152] However, this layer structure is not changeless during a day: some species of cyanobacteria migrate to deeper layers at morning, and go back at evening, to avoid intensive solar light and UV radiation at mid-day.[153]

Microbial mats are generally held together and bound to their substrates by slimy extracellular polymeric substances which they secrete. In many cases some of the bacteria form filaments (threads), which tangle and thus increase the colonies' structural strength, especially if the filaments have sheaths (tough outer coverings).

This combination of slime and tangled threads attracts other microorganisms which become part of the mat community, for example protozoa, some of which feed on the mat-forming bacteria, and diatoms, which often seal the surfaces of submerged microbial mats with thin, parchment-like coverings.

Figure 76: *Wrinkled Kinneyia-type sedimentary structures formed beneath cohesive microbial mats in peritidal zones. The image shows the location, in the Burgsvik beds of Sweden, where the texture was first identified as evidence of a microbial mat.*

Marine mats may grow to a few centimeters in thickness, of which only the top few millimeters are oxygenated.

Types of environment colonized

Underwater microbial mats have been described as layers that live by exploiting and to some extent modifying local chemical gradients, i.e. variations in the chemical composition. Thinner, less complex biofilms live in many sub-aerial environments, for example on rocks, on mineral particles such as sand, and within soil. They have to survive for long periods without liquid water, often in a dormant state. Microbial mats that live in tidal zones, such as those found in the Sippewissett salt marsh, often contain a large proportion of similar microorganisms that can survive for several hours without water.

Microbial mats and less complex types of biofilm are found at temperature ranges from –40 °C to +120 °C, because variations in pressure affect the temperatures at which water remains liquid.

They even appear as endosymbionts in some animals, for example in the hindguts of some echinoids.

Figure 77: *Kinneyia-like structure in the Grimsby Formation (Silurian) exposed in Niagara Gorge, New York.*

Ecological and geological importance

Microbial mats use all of the types of metabolism and feeding strategy that have evolved on Earth—anoxygenic and oxygenic photosynthesis; anaerobic and aerobic chemotrophy (using chemicals rather than sunshine as a source of energy); organic and inorganic respiration and fermentation (i..e converting food into energy with and without using oxygen in the process); autotrophy (producing food from inorganic compounds) and heterotrophy (producing food only from organic compounds, by some combination of predation and detritivory).

Most sedimentary rocks and ore deposits have grown by a reef-like build-up rather than by "falling" out of the water, and this build-up has been at least influenced and perhaps sometimes caused by the actions of microbes. Stromatolites, bioherms (domes or columns similar internally to stromatolites) and biostromes (distinct sheets of sediment) are among such microbe-influenced build-ups. Other types of microbial mat have created wrinkled "elephant skin" textures in marine sediments, although it was many years before these textures were recognized as trace fossils of mats. Microbial mats have increased the concentration of metal in many ore deposits, and without this it would not be feasible to mine them—examples include iron (both sulfide and oxide ores), uranium, copper, silver and gold deposits.

Role in the history of life

History of life

θ —
500 —
1000 —
1500 —
2000 —
2500 —
3000 —
3500 —
4000 —
4500 —

Phanerozoic eon

Proterozoic eon

Archean eon

Hadean eon

A very brief history of life on Earth.

Axis scale is in millions of years ago.

The earliest mats

Microbial mats are among the oldest clear signs of life, as microbially induced sedimentary structures (MISS) formed 3,480[154] million years ago have been found in western Australia. At that early stage the mats' structure may already have been similar to that of modern mats that do not include photosynthesizing bacteria. It is even possible that non-photosynthesizing mats were present as early as 4,000[155] million years ago. If so, their energy source would have been hydrothermal vents (high-pressure hot springs around submerged volcanoes), and the evolutionary split between bacteria and archea may also have occurred around this time.[156]

The earliest mats were probably small, single-species biofilms of chemotrophs that relied on hydrothermal vents to supply both energy and chemical "food". Within a short time (by geological standards) the build-up of dead microorganisms would have created an ecological niche for scavenging heterotrophs, possibly methane-emitting and sulfate-reducing organisms that would have formed new layers in the mats and enriched their supply of biologically useful chemicals.

Photosynthesis

It is generally thought that photosynthesis, the biological generation of energy from light, evolved shortly after 3,000[157] million years ago (3 billion). However an isotope analysis suggests that oxygenic photosynthesis may have been widespread as early as 3,500[150] million years ago. The eminent researcher into Earth's earliest life, William Schopf, argues that, if one did not know their age, one would classify some of the fossil organisms in Australian stromatolites from 3,500[150] million years ago as cyanobacteria, which are oxygen-producing photosynthesizers. There are several different types of photosynthetic reaction, and analysis of bacterial DNA indicates that photosynthesis first arose in anoxygenic purple bacteria, while the oxygenic photosynthesis seen in cyanobacteria and much later in plants was the last to evolve.

The earliest photosynthesis may have been powered by infra-red light, using modified versions of pigments whose original function was to detect infra-red heat emissions from hydrothermal vents. The development of photosynthetic energy generation enabled the microorganisms first to colonize wider areas around vents and then to use sunlight as an energy source. The role of the hydrothermal vents was now limited to supplying reduced metals into the oceans as a whole rather than being the main supporters of life in specific locations. Heterotrophic scavengers would have accompanied the photosynthesizers in their migration out of the "hydrothermal ghetto".

The evolution of purple bacteria, which do not produce or use oxygen but can tolerate it, enabled mats to colonize areas that locally had relatively high concentrations of oxygen, which is toxic to organisms that are not adapted to it. Microbial mats would have been separated into oxidized and reduced layers, and this specialization would have increased their productivity. It may be possible to confirm this model by analyzing the isotope ratios of both carbon and sulfur in sediments laid down in shallow water.

The last major stage in the evolution of microbial mats was the appearance of cyanobacteria, photsynthesizers which both produce and use oxygen. This gave undersea mats their typical modern structure: an oxygen-rich top layer of cyanobacteria; a layer of photsynthesizing purple bacteria that could tolerate oxygen; and oxygen-free, H_2S-dominated lower layers of heterotrophic scavengers, mainly methane-emitting and sulfate-reducing organisms.

It is estimated that the appearance of oxygenic photosynthesis increased biological productivity by a factor of between 100 and 1,000. All photosynthetic reactions require a reducing agent, but the significance of oxygenic photosynthesis is that it uses water as a reducing agent, and water is much more plentiful than the geologically produced reducing agents on which photosynthesis previously depended. The resulting increases in the populations of photosynthesizing bacteria in the top layers of microbial mats would have caused corresponding population increases among the chemotrophic and heterotrophic microorganisms that inhabited the lower layers and which fed respectively on the by-products of the photosynthesizers and on the corpses and / or living bodies of the other mat organisms. These increases would have made microbial mats the planet's dominant ecosystems. From this point onwards life itself produced significantly more of the resources it needed than did geochemical processes.

Oxygenic photosynthesis in microbial mats would also have increased the free oxygen content of the Earth's atmosphere, both directly by emitting oxygen and because the mats emitted molecular hydrogen (H_2), some of which would have escaped from the Earth's atmosphere before it could re-combine with free oxygen to form more water. Microbial mats thus played a major role in the evolution of organisms which could first tolerate free oxygen and then use it as an energy source. Oxygen is toxic to organisms that are not adapted to it, but greatly increases the metabolic efficiency of oxygen-adapted organisms — for example anaerobic fermentation produces a net yield of two molecules of adenosine triphosphate, cells' internal "fuel", per molecule of glucose, while aerobic respiration produces a net yield of 36. The oxygenation of the atmosphere was a prerequisite for the evolution of the more complex eukaryote type of cell, from which all multicellular organisms are built.

Cyanobacteria have the most complete biochemical "toolkits" of all the mat-forming organisms: the photosynthesis mechanisms of both green bacteria and purple bacteria; oxygen production; and the Calvin cycle, which converts carbon dioxide and water into carbohydrates and sugars. It is likely that they acquired many of these sub-systems from existing mat organisms, by some combination of horizontal gene transfer and endosymbiosis followed by fusion. Whatever the causes, cyanobacteria are the most self-sufficient of the mat organisms and were well-adapted to strike out on their own both as floating mats and as the first of the phytoplankton, which forms the basis of most marine food chains.

Origin of eukaryotes

The time at which eukaryotes first appeared is still uncertain: there is reasonable evidence that fossils dated between 1,600[158] million years ago and 2,100[159] million years ago represent eukaryotes, but the presence of steranes in Australian shales may indicate that eukaryotes were present 2,700[160] million years ago. There is still debate about the origins of eukaryotes, and many of the theories focus on the idea that a bacterium first became an endosymbiont of an anaerobic archean and then fused with it to become one organism. If such endosymbiosis was an important factor, microbial mats would have encouraged it. There are two possible variations of this scenario:

- The boundary between the oxygenated and oxygen-free zones of a mat would have moved up when photosynthesis shut down at night and back down when photosynthesis resumed after the next sunrise. Symbiosis between independent aerobic and anaerobic organisms would have enabled both to live comfortably in the zone that was subject to oxygen "tides", and subsequent endosymbiosis would have made such partnerships more mobile.
- The initial partnership may have been between anaerobic archea that required molecular hydrogen (H_2) and heterotrophic bacteria that produced it and could live both with and without oxygen.

Life on land

Microbial mats from ~ 1,200[161] million years ago provide the first evidence of life in the terrestrial realm.

The earliest multicellular "animals"

File:Cambrian substrate revolution 02.png

Before and after the Cambrian substrate revolution

The Ediacara biota are the earliest widely accepted evidence of multicellular "animals". Most Ediacaran strata with the "elephant skin" texture characteristic of microbial mats contain fossils, and Ediacaran fossils are hardly ever found in beds that do not contain these microbial mats. Adolf Seilacher categorized the "animals" as: "mat encrusters", which were permanently attached to the mat; "mat scratchers", which grazed the surface of the mat without destroying it; "mat stickers", suspension feeders that were partially embedded in the mat; and "undermat miners", which burrowed underneath the mat and fed on decomposing mat material.

The Cambrian substrate revolution

In the Early Cambrian, however, organisms began to burrow vertically for protection or food, breaking down the microbial mats, and thus allowing water and oxygen to penetrate a considerable distance below the surface and kill the oxygen-intolerant microorganisms in the lower layers. As a result of this Cambrian substrate revolution, marine microbial mats are confined to environments in which burrowing is non-existent or negligible: very harsh environments, such as hyper-saline lagoons or brackish estuaries, which are uninhabitable for the burrowing organisms that broke up the mats; rocky "floors" which the burrowers cannot penetrate; the depths of the oceans, where burrowing activity today is at a similar level to that in the shallow coastal seas before the revolution.

Current status

Although the Cambrian substrate revolution opened up new niches for animals, it was not catastrophic for microbial mats, but it did greatly reduce their extent.

How microbial mats help paleontologists

Most fossils preserve only the hard parts of organisms, e.g. shells. The rare cases where soft-bodied fossils are preserved (the remains of soft-bodied organisms and also of the soft parts of organisms for which only hard parts such as shells are usually found) are extremely valuable because they provide information about organisms that are hardly ever fossilized and much more information than is usually available about those for which only the hard parts are usually preserved. Microbial mats help to preserve soft-bodied fossils by:

- Capturing corpses on the sticky surfaces of mats and thus preventing them from floating or drifting away.
- Physically protecting them from being eaten by scavengers and broken up by burrowing animals, and protecting fossil-bearing sediments from erosion. For example, the speed of water current required to erode sediment bound by a mat is 20–30 times as great as the speed required to erode a bare sediment.
- Preventing or reducing decay both by physically screening the remains from decay-causing bacteria and by creating chemical conditions that are hostile to decay-causing bacteria.
- Preserving tracks and burrows by protecting them from erosion. Many trace fossils date from significantly earlier than the body fossils of animals that are thought to have been capable of making them and thus improve paleontologists' estimates of when animals with these capabilities first appeared.

Industrial uses

The ability of microbial mat communities to use a vast range of "foods" has recently led to interest in industrial uses. There have been trials of microbial mats for purifying water, both for human use and in fish farming, and studies of their potential for cleaning up oil spills. As a result of the growing commercial potential, there have been applications for and grants of patents relating to the growing, installation and use of microbial mats, mainly for cleaning up pollutants and waste products.[162]

References

- Seckbach S (2010) *Microbial Mats: Modern and Ancient Microorganisms in Stratified Systems*[163] Springer, ISBN 978-90-481-3798-5.

External links

- Jürgen Schieber. "Microbial Mat Page"[164]. Retrieved 2008-07-01. – outline of microbial mats and pictures of mats in various situations and at various magnifications.

Great Oxygenation Event

The **Great Oxygenation Event**, the beginning of which is commonly known in scientific media as the **Great Oxidation Event** (**GOE**, also called the **Oxygen Catastrophe**, **Oxygen Crisis**, **Oxygen Holocaust**, **Oxygen Revolution**, or **Great Oxidation**) was the biologically induced appearance of dioxygen (O_2) in Earth's atmosphere. Geological, isotopic, and chemical evidence suggests that this major environmental change happened around 2.45 billion years ago (2.45 Ga), during the Siderian period, at the beginning of the Proterozoic eon. The causes of the event remain unclear. As of 2016[166], the geochemical and biomarker evidence for the development of oxygenic photosynthesis before the Great Oxidation Event has been mostly inconclusive.

Oceanic cyanobacteria, which evolved into coordinated (but not multicellular or even colonial) macroscopic forms more than 2.3 billion years ago (approximately 200 million years before the GOE), are believedWikipedia:Manual of Style/Words to watch#Unsupported attributions to have become the first microbes to produce oxygen by photosynthesis. Before the GOE, any free oxygen they produced was chemically captured by dissolved iron or by organic matter. The GOE started when these oxygen sinks became saturated, at which point oxygen produced by the cyanobacteria started escaping into the atmosphere.

The increased production of oxygen set Earth's original atmosphere off-balance. Free oxygen is toxic to obligate anaerobic organisms, and the rising concentrations may have destroyed most such organisms at the time.

A spike in chromium contained in ancient rock-deposits formed underwater shows the accumulation had been washed off from the continental shelves. Chromium is not easily dissolved and its release from rocks would have required the presence of a powerful acid. One such acid, sulfuric acid (H_2SO_4), might have formed through bacterial reactions with pyrite. Mats of oxygen-producing cyanobacteria can produce a thin layer, one or two millimeters thick, of oxygenated water in an otherwise anoxic environment even under thick ice; before oxygen started accumulating in the atmosphere, these organisms would already have adapted to oxygen.[167] Additionally, the free oxygen would have reacted with atmospheric methane, a greenhouse gas, greatly reducing its concentration and triggering the Huronian glaciation, possibly the longest episode of glaciation in Earth's history and called "snowball Earth".

Eventually, the evolution of aerobic organisms that consumed oxygen established an equilibrium in its availability. Free oxygen has been an important constituent of the atmosphere ever since.

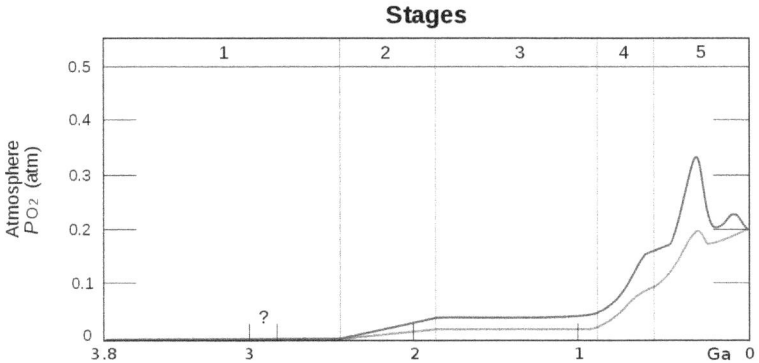

Figure 78: *O$_2$ build-up in the Earth's atmosphere. Red and green lines represent the range of the estimates while time is measured in billions of years ago (Ga). Stage 1 (3.85–2.45 Ga): Practically no O$_2$ in the atmosphere. The oceans were also largely anoxic with the possible exception of O$_2$ in the shallow oceans. Stage 2 (2.45–1.85 Ga): O$_2$ produced, rising to values of 0.02 and 0.04 atm, but absorbed in oceans and seabed rock. Stage 3 (1.85–0.85 Ga): O$_2$ starts to gas out of the oceans, but is absorbed by land surfaces. No significant change in terms of oxygen level. Stages 4 and 5 (0.85–present): O$_2$ sinks filled and the gas accumulates.*[165]

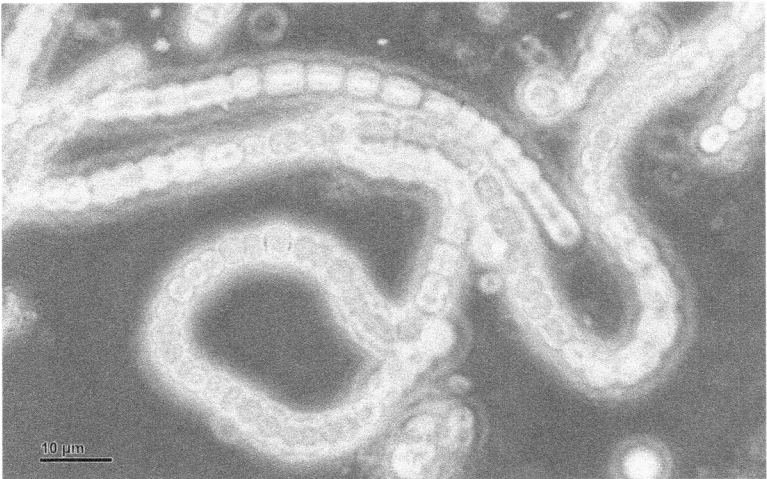

Figure 79: *Cyanobacteria: responsible for the build-up of oxygen in the Earth's atmosphere*

Timing

Life timeline

```
0 —
500
1000
1500
2000
2500
3000
3500
4000
4500
```

Axis scale: million years

Also see: *Human timeline* and *Nature timeline*

The most widely accepted chronology of the Great Oxygenation Event suggests that free oxygen was first produced by prokaryotic and then later eukaryotic organisms that carried out photosynthesis more efficiently, producing oxygen as a waste product. The first oxygen-producing organisms arose long before the GOE, perhaps as early as 3,400[168] million years ago.

Initially, the oxygen they produced would have quickly been removed from the atmosphere by the chemical weathering of reducing (oxidizable) minerals, most notably iron. This 'mass rusting' led to the deposition of iron(III) oxide in the form of banded-iron formations such as the sediments in Minnesota and Pilbara, Western Australia. The saturation of these mineral sinks, and the resulting persistence of oxygen in the atmosphere, led within 50 million years to the start of the GOE. Oxygen could have accumulated very rapidly: at today's rates of photosynthesis (much greater than those in the Precambrian without land plants), modern atmospheric O_2 levels could be produced in only 2,000 years.

Another hypothesis is that oxygen producers did not evolve until a few million years before the major rise in atmospheric oxygen concentration. This is based on a particular interpretation of a supposed oxygen indicator used in previous studies, the mass-independent fractionation of sulfur isotopes. This hypothesis would eliminate the need to explain a lag in time between the evolution of oxyphotosynthetic microbes and the rise in free oxygen.

In either case, oxygen did eventually accumulate in the atmosphere, with two major consequences.

Firstly, it oxidized atmospheric methane (a strong greenhouse gas) to carbon dioxide (a weaker one) and water. This decreased the greenhouse effect of the Earth's atmosphere, causing planetary cooling, and triggered the Huronian glaciation. Starting around 2.4 billion years ago, this lasted 300-400 million years, and may have been the longest ever snowball Earth episode.[169]

Secondly, the increased oxygen concentrations provided a new opportunity for biological diversification, as well as tremendous changes in the nature of chemical interactions between rocks, sand, clay, and other geological substrates and the Earth's air, oceans, and other surface waters. Despite the natural recycling of organic matter, life had remained energetically limited until the widespread availability of oxygen. This breakthrough in metabolic evolution greatly increased the free energy available to living organisms, with global environmental impacts. For example, mitochondria evolved after the GOE, giving organisms the energy to exploit new, more complex morphologies interacting in increasingly complex ecosystems.

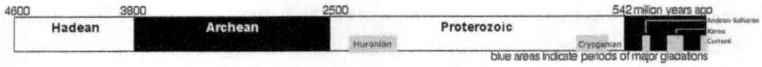

Figure 80: *Timeline of glaciations, shown in blue.*

Figure 81: *2.1 billion year old rock showing banded iron formation*

Time lag theory

There may have been a gap of up to 900 million years between the start of photosynthetic oxygen production and the geologically rapid increase in atmospheric oxygen about 2.5–2.4 billion years ago. Several hypotheses propose to explain this time lag.

Tectonic trigger

The oxygen increase had to await tectonically driven changes in the Earth, including the appearance of shelf seas, where reduced organic carbon could reach the sediments and be buried. The newly produced oxygen was first consumed in various chemical reactions in the oceans, primarily with iron. Evidence is found in older rocks that contain massive banded iron formations apparently laid down as this iron and oxygen first combined; most present-day iron ore lies in these deposits. Evidence suggests oxygen levels spiked each time smaller land masses collided to form a super-continent. Tectonic pressure thrust up mountain chains, which eroded to release nutrients into the ocean to feed photosynthetic cyanobacteria.

Nickel famine

Early chemosynthetic organisms likely produced methane, an important trap for molecular oxygen, since methane readily oxidizes to carbon dioxide (CO_2) and water in the presence of UV radiation. Modern methanogens require nickel as an enzyme cofactor. As the Earth's crust cooled and the supply of volcanic nickel dwindled, oxygen-producing algae began to out-perform methane producers, and the oxygen percentage of the atmosphere steadily increased. From 2.7 to 2.4 billion years ago, the rate of deposition of nickel declined steadily from a level 400 times today's.

Bistability

Another hypothesis posits a model of the atmosphere that exhibits bistability: two steady states of oxygen concentration. The state of stable low oxygen concentration (0.02%) experiences a high rate of methane oxidation. If some event raises oxygen levels beyond a moderate threshold, the formation of an ozone layer shields UV rays and decreases methane oxidation, raising oxygen further to a stable state of 21% or more. The Great Oxygenation Event can then be understood as a transition from the lower to the upper steady states.

Hydrogen gas

Another theory credits the appearance of cyanobacteria with suppressing hydrogen gas and increasing oxygen.

Some bacteria in the early oceans could separate water into hydrogen and oxygen. Under the Sun's rays, hydrogen molecules were incorporated into organic compounds, with oxygen as a by-product. If the hydrogen-heavy compounds were buried, it would have allowed oxygen to accumulate in the atmosphere.

However, in 2001 scientists realized that the hydrogen would instead escape into space through a process called methane photolysis, in which methane releases its hydrogen in a reaction with oxygen. This could explain why the early Earth stayed warm enough to sustain oxygen-producing lifeforms.

Late evolution of oxy-photosynthesis theory

The oxygen indicator might have been misinterpreted. During the proposed lag era in the previous theory, there was a change in sediments from mass-independently fractionated (MIF) sulfur to mass-dependently fractionated (MDF) sulfur. This was assumed to show the appearance of oxygen in the atmosphere, since oxygen would have prevented the photolysis of sulfur dioxide, which causes MIF. However, the change from MIF to MDF of sulfur isotopes may instead have been caused by an increase in glacial weathering, or the homogenization of the marine sulfur pool as a result of an increased thermal gradient during the Huronian glaciation period (which in this interpretation was not caused by oxygenation).

Role in mineral diversification

The Great Oxygenation Event triggered an explosive growth in the diversity of minerals, with many elements occurring in one or more oxidized forms near the Earth's surface. It is estimated that the GOE was directly responsible for more than 2,500 of the total of about 4,500 minerals found on Earth today. Most of these new minerals were formed as hydrated and oxidized forms due to dynamic mantle and crust processes.[170]

Million years ago. Age of Earth = 4,560

Origin of eukaryotes

It has been proposed that a local rise in oxygen levels due to cyanobacterial photosynthesis in ancient microenvironments was highly toxic to the surrounding biota, and that this selective pressure drove the evolutionary transformation of an archaeal lineage into the first eukaryotes. Oxidative stress involving production of reactive oxygen species (ROS) might have acted in synergy with other environmental stresses (such as ultraviolet radiation and/or desiccation) to drive selection in an early archaeal lineage towards eukaryosis. This archaeal ancestor may already have had DNA repair mechanisms based on DNA pairing and recombination and possibly some kind of cell fusion mechanism.[171] The detrimental effects of internal ROS (produced by endosymbiont proto-mitochondria) on the archaeal genome could have promoted the evolution of meiotic sex from these humble beginnings. Selective pressure for efficient DNA repair of oxidative DNA damages may have driven the evolution of eukaryotic sex involving such features as cell-cell fusions, cytoskeleton-mediated chromosome movements and emergence of the nuclear membrane. Thus the evolution of eukaryotic sex and eukaryogenesis were likely inseparable processes that evolved in large part to facilitate DNA repair.[172] Constant pressure of endogenous ROS has been proposed to explain the ubiquitous maintenance of meiotic sex in eukaryotes.

External links

- First breath: Earth's billion-year struggle for oxygen[173] *New Scientist*, #2746, 5 February 2010 by Nick Lane.[174]

Diversification of eukaryotes

Eukaryote

Eukaryotes
Temporal range: Orosirian - Present 1850–0Ma
Had'nArchean   Proterozoic   Pha.

Eukaryotes and some examples of their diversity – clockwise from top left: Red mason bee, *Boletus edulis*, Common chimpanzee, *Isotricha intestinalis*, Persian buttercup, and *Volvox carteri*

Scientific classification 🖉	
Domain:	**Eukaryota** (Chatton, 1925) Whittaker & Margulis, 1978
Supergroups and kingdoms	

- Archaeplastida
    Kingdom **Plantae** – Plants
- Hacrobia
- SAR (Stramenopiles + Alveolata + Rhizaria)
- Excavata
- Amoebozoa
- Opisthokonta
    Kingdom **Animalia** – Animals
    Kingdom **Fungi**

Eukaryotic organisms that cannot be classified under the kingdoms Plantae, Animalia or Fungi are sometimes grouped in the kingdom **Protista**.

Eukaryotes (/juːˈkærioʊt, <wbr />-ət/) are organisms whose cells have a nucleus enclosed within membranes, unlike Prokaryotes (Bacteria and other Archaea). Eukaryotes belong to the domain **Eukaryota** or **Eukarya**. Their name comes from the Greek εὖ (*eu*, "well" or "true") and κάρυον (*karyon*, "nut" or "kernel"). Eukaryotic cells also contain other membrane-bound organelles such as mitochondria and the Golgi apparatus, and in addition, some cells of plants and algae contain chloroplasts. Unlike unicellular archaea and bacteria, eukaryotes may also be multicellular and include organisms consisting of many cell types forming different kinds of tissue.

Eukaryotes can reproduce both asexually through mitosis and sexually through meiosis and gamete fusion. In mitosis, one cell divides to produce two genetically identical cells. In meiosis, DNA replication is followed by two rounds of cell division to produce four haploid daughter cells. These act as sex cells (gametes). Each gamete has just one set of chromosomes, each a unique mix of the corresponding pair of parental chromosomes resulting from genetic recombination during meiosis.

The domain Eukaryota appears to be monophyletic, and makes up one of the domains of life in the three-domain system. The two other domains, Bacteria and Archaea, are prokaryotes and have none of the above features. Eukaryotes represent a tiny minority of all living things. However, due to their generally much larger size, their collective worldwide biomass is estimated to be about equal to that of prokaryotes. Eukaryotes evolved approximately 1.6–2.1 billion years ago, during the Proterozoic eon.

History

In 1905 and 1910, the Russian biologist Konstantin Mereschkowski (1855–1921) argued that plastids were reduced cyanobacteria in a symbiosis with a non-photosynthetic (heterotrophic) host that was itself formed by symbiosis between an amoeba-like host and a bacterium-like cell that formed the nucleus. Plants had thus inherited photosynthesis from cyanobacteria.

Figure 82: *Konstantin Mereschkowski proposed a symbiotic origin for cells with nuclei.*

The concept of the eukaryote has been attributed to the French biologist Edouard Chatton (1883-1947). The terms prokaryote and eukaryote were more definitively reintroduced by the Canadian microbiologist Roger Stanier and the Dutch-American microbiologist C. B. van Niel in 1962. In his 1938 work *Titres et Travaux Scientifiques*, Chatton had proposed the two terms, calling the bacteria prokaryotes and organisms with nuclei in their cells eukaryotes. However he mentioned this in only one paragraph, and the idea was effectively ignored until Chatton's statement was rediscovered by Stanier and van Niel.

In 1967, Lynn Margulis provided microbiological evidence for endosymbiosis as the origin of chloroplasts and mitochondria in eukaryotic cells in her paper, *On the origin of mitosing cells*. In the 1970s, Carl Woese explored microbial phylogenetics, studying variations in 16S ribosomal RNA. This helped to uncover the origin of the eukaryotes and the symbiogenesis of two important eukaryote organelles, mitochondria and chloroplasts. In 1977, Woese and George Fox introduced a "third form of life", which they called the Archaebacteria; in 1990, Woese, Otto Kandler and Mark L. Wheeler renamed this the Archaea.

In 1979, G. W. Gould and G. J. Dring suggested that the eukaryotic cell's nucleus came from the ability of Gram-positive bacteria to form endospores.

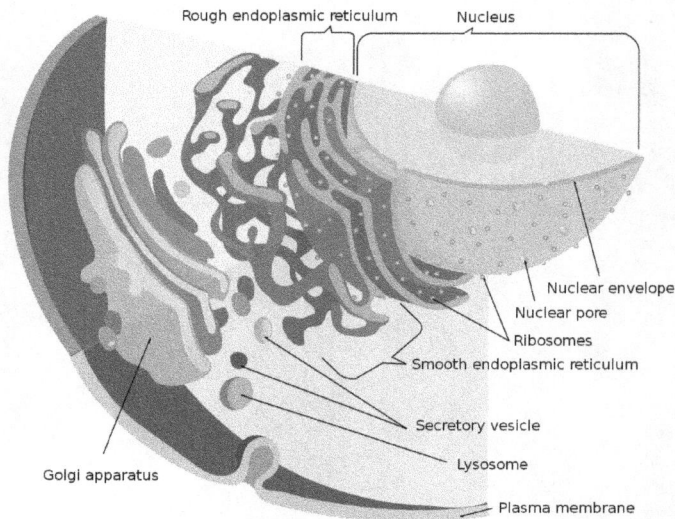

Figure 83: *The endomembrane system and its components*

In 1987 and later papers, Thomas Cavalier-Smith proposed instead that the membranes of the nucleus and endoplasmic reticulum first formed by infolding a prokaryote's plasma membrane. In the 1990s, several other biologists proposed endosymbiotic origins for the nucleus, effectively reviving Mereschkowsky's theory.

Cell features

Eukaryotic cells are typically much larger than those of prokaryotes having a volume of around 10,000 times greater than the prokaryotic cell. They have a variety of internal membrane-bound structures, called organelles, and a cytoskeleton composed of microtubules, microfilaments, and intermediate filaments, which play an important role in defining the cell's organization and shape. Eukaryotic DNA is divided into several linear bundles called chromosomes, which are separated by a microtubular spindle during nuclear division.

Internal membrane

Eukaryote cells include a variety of membrane-bound structures, collectively referred to as the endomembrane system. Simple compartments, called vesicles and vacuoles, can form by budding off other membranes. Many cells ingest food and other materials through a process of endocytosis, where the outer membrane invaginates and then pinches off to form a vesicle. It is probable that most other membrane-bound organelles are ultimately derived from such vesicles. Alternatively some products produced by the cell can leave in a vesicle through exocytosis.

The nucleus is surrounded by a double membrane (commonly referred to as a nuclear membrane or nuclear envelope), with pores that allow material to move in and out. Various tube- and sheet-like extensions of the nuclear membrane form the endoplasmic reticulum, which is involved in protein transport and maturation. It includes the rough endoplasmic reticulum where ribosomes are attached to synthesize proteins, which enter the interior space or lumen. Subsequently, they generally enter vesicles, which bud off from the smooth endoplasmic reticulum. In most eukaryotes, these protein-carrying vesicles are released and further modified in stacks of flattened vesicles (cisternae), the Golgi apparatus.

Vesicles may be specialized for various purposes. For instance, lysosomes contain digestive enzymes that break down most biomolecules in the cytoplasm. Peroxisomes are used to break down peroxide, which is otherwise toxic. Many protozoans have contractile vacuoles, which collect and expel excess water, and extrusomes, which expel material used to deflect predators or capture prey. In higher plants, most of a cell's volume is taken up by a central vacuole, which mostly contains water and primarily maintains its osmotic pressure.

Mitochondria and plastids

Mitochondria are organelles found in all but one[175]</ref> eukaryote. Mitochondria provide energy to the eukaryote cell by converting sugars into ATP. They have two surrounding membranes, each a phospholipid bi-layer; the inner of which is folded into invaginations called cristae where aerobic respiration takes place.

Mitochondria contain their own DNA, which has close structural similarities to bacterial DNA, and which encodes rRNA and tRNA genes that produce RNA which is closer in structure to bacterial RNA than to eukaryote RNA. They are now generally held to have developed from endosymbiotic prokaryotes, probably proteobacteria.

Some eukaryotes, such as the metamonads such as *Giardia* and *Trichomonas*, and the amoebozoan Pelomyxa, appear to lack mitochondria, but all have been found to contain mitochondrion-derived organelles, such as hydrogenosomes and mitosomes, and thus have lost their mitochondria secondarily. They obtain energy by enzymatic action on nutrients absorbed from the environment. The

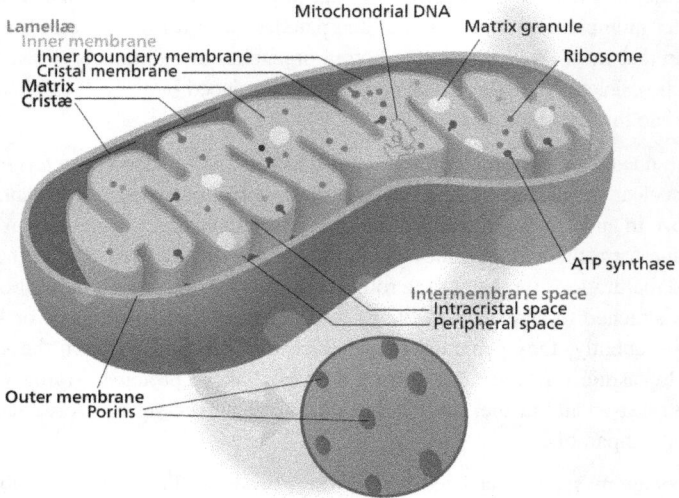

Figure 84: *Simplified structure of a mitochondrion*

metamonad *Monocercomonoides* has also acquired, by lateral gene transfer, a cytosolic sulfur mobilisation system which provides the clusters of iron and sulfur required for protein synthesis. The normal mitochondrial iron-sulfur cluster pathway has been lost secondarily.

Plants and various groups of algae also have plastids. Plastids also have their own DNA and are developed from endosymbionts, in this case cyanobacteria. They usually take the form of chloroplasts which, like cyanobacteria, contain chlorophyll and produce organic compounds (such as glucose) through photosynthesis. Others are involved in storing food. Although plastids probably had a single origin, not all plastid-containing groups are closely related. Instead, some eukaryotes have obtained them from others through secondary endosymbiosis or ingestion. The capture and sequestering of photosynthetic cells and chloroplasts occurs in many types of modern eukaryotic organisms and is known as kleptoplasty.

Endosymbiotic origins have also been proposed for the nucleus, and for eukaryotic flagella.

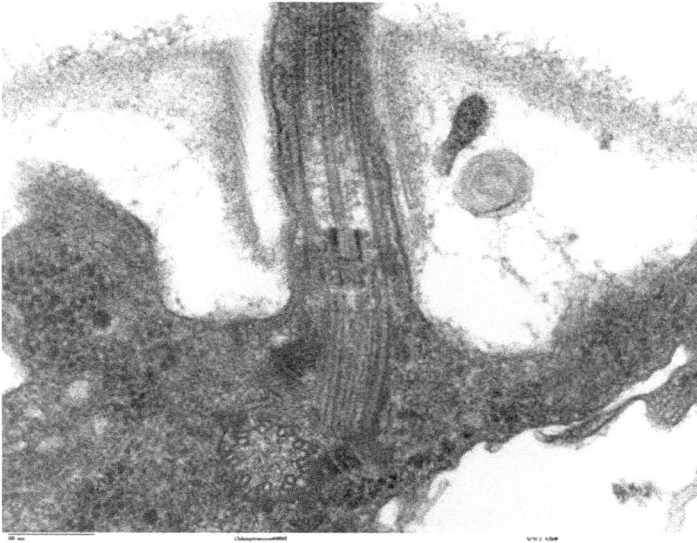

Figure 85: *Longitudinal section through the flagellum of Chlamydomonas reinhardtii*

Cytoskeletal structures

Many eukaryotes have long slender motile cytoplasmic projections, called flagella, or similar structures called cilia. Flagella and cilia are sometimes referred to as undulipodia,[176] and are variously involved in movement, feeding, and sensation. They are composed mainly of tubulin. These are entirely distinct from prokaryotic flagellae. They are supported by a bundle of microtubules arising from a centriole, characteristically arranged as nine doublets surrounding two singlets. Flagella also may have hairs, or mastigonemes, and scales connecting membranes and internal rods. Their interior is continuous with the cell's cytoplasm.

Microfilamental structures composed of actin and actin binding proteins, e.g., α-actinin, fimbrin, filamin are present in submembraneous cortical layers and bundles, as well. Motor proteins of microtubules, e.g., dynein or kinesin and actin, e.g., myosins provide dynamic character of the network.

Centrioles are often present even in cells and groups that do not have flagella, but conifers and flowering plants have neither. They generally occur in groups that give rise to various microtubular roots. These form a primary component of the cytoskeletal structure, and are often assembled over the course of several cell divisions, with one flagellum retained from the parent and the other derived from it. Centrioles produce the spindle during nuclear division.

The significance of cytoskeletal structures is underlined in the determination of shape of the cells, as well as their being essential components of migratory responses like chemotaxis and chemokinesis. Some protists have various other microtubule-supported organelles. These include the radiolaria and heliozoa, which produce axopodia used in flotation or to capture prey, and the hapto-phytes, which have a peculiar flagellum-like organelle called the haptonema.

Cell wall

The cells of plants and algae, fungi and most chromalveolates have a cell wall, a layer outside the cell membrane, providing the cell with structural support, pro-tection, and a filtering mechanism. The cell wall also prevents over-expansion when water enters the cell.

The major polysaccharides making up the primary cell wall of land plants are cellulose, hemicellulose, and pectin. The cellulose microfibrils are linked via hemicellulosic tethers to form the cellulose-hemicellulose network, which is embedded in the pectin matrix. The most common hemicellulose in the pri-mary cell wall is xyloglucan.

Differences among eukaryotic cells

There are many different types of eukaryotic cells, though animals and plants are the most familiar eukaryotes, and thus provide an excellent starting point for understanding eukaryotic structure. Fungi and many protists have some substantial differences, however.

Animal cell

All animals are eukaryotic. Animal cells are distinct from those of other eu-karyotes, most notably plants, as they lack cell walls and chloroplasts and have smaller vacuoles. Due to the lack of a cell wall, animal cells can adopt a variety of shapes. A phagocytic cell can even engulf other structures.

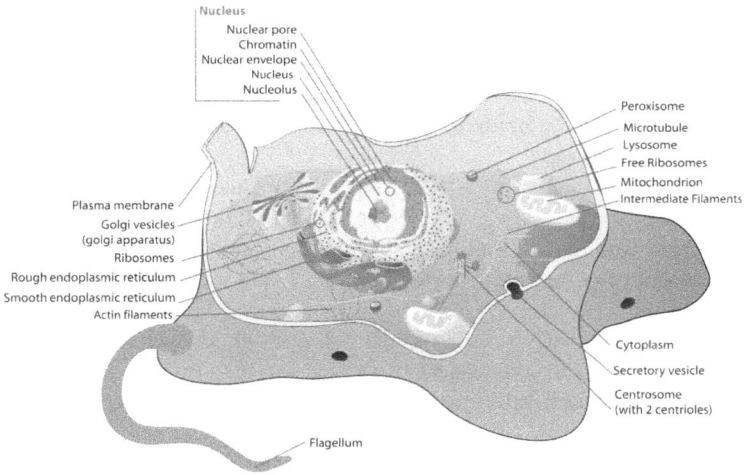

Figure 86: *Structure of a typical animal cell*

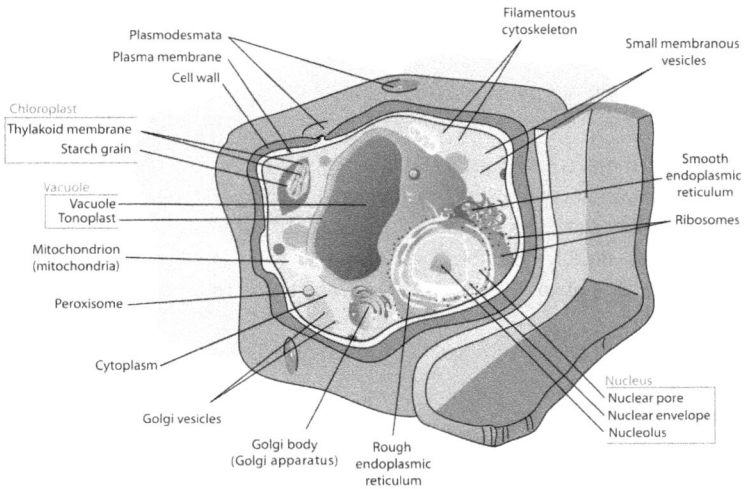

Figure 87: *Structure of a typical plant cell*

Plant cell

Plant cells are quite different from the cells of the other eukaryotic organisms. Their distinctive features are:

- A large central vacuole (enclosed by a membrane, the tonoplast), which maintains the cell's turgor and controls movement of molecules between the cytosol and sap
- A primary cell wall containing cellulose, hemicellulose and pectin, deposited by the protoplast on the outside of the cell membrane; this contrasts with the cell walls of fungi, which contain chitin, and the cell envelopes of prokaryotes, in which peptidoglycans are the main structural molecules
- The plasmodesmata, pores in the cell wall that link adjacent cells and allow plant cells to communicate with adjacent cells. Animals have a different but functionally analogous system of gap junctions between adjacent cells.
- Plastids, especially chloroplasts that contain chlorophyll, the pigment that gives plants their green color and allows them to perform photosynthesis
- Bryophytes and seedless vascular plants only have flagellae and centrioles in the sperm cells. Sperm of cycads and *Ginkgo* are large, complex cells that swim with hundreds to thousands of flagellae.
- Conifers (Pinophyta) and flowering plants (Angiospermae) lack the flagellae and centrioles that are present in animal cells.

Fungal cell

The cells of fungi are most similar to animal cells, with the following exceptions:

- A cell wall that contains chitin
- Less definition between cells; the hyphae of higher fungi have porous partitions called septa, which allow the passage of cytoplasm, organelles, and, sometimes, nuclei. Primitive fungi have few or no septa, so each organism is essentially a giant multinucleate supercell; these fungi are described as coenocytic.
- Only the most primitive fungi, chytrids, have flagella.

Other eukaryotic cells

Some groups of eukaryotes have unique organelles, such as the cyanelles (unusual chloroplasts) of the glaucophytes, the haptonema of the haptophytes, or the ejectosomes of the cryptomonads. Other structures, such as pseudopodia, are found in various eukaryote groups in different forms, such as the lobose amoebozoans or the reticulose foraminiferans.

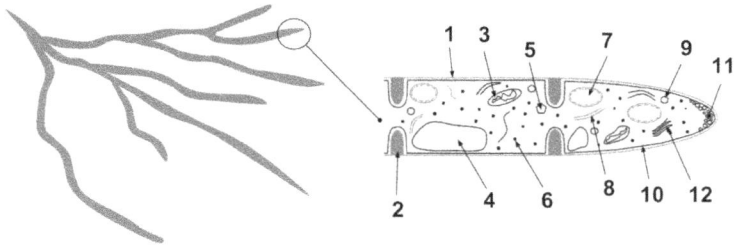

Figure 88: *Fungal Hyphae Cells*
*1- Hyphal wall 2- Septum 3- Mitochondrion 4- Vacuole 5- Ergosterol
crystal 6- Ribosome 7- Nucleus 8- Endoplasmic reticulum 9- Lipid
body 10- Plasma membrane 11- Spitzenkörper 12- Golgi apparatus*

Reproduction

Cell division generally takes place asexually by mitosis, a process that allows each daughter nucleus to receive one copy of each chromosome. Most eukaryotes also have a life cycle that involves sexual reproduction, alternating between a haploid phase, where only one copy of each chromosome is present in each cell and a diploid phase, wherein two copies of each chromosome are present in each cell. The diploid phase is formed by fusion of two haploid gametes to form a zygote, which may divide by mitosis or undergo chromosome reduction by meiosis. There is considerable variation in this pattern. Animals have no multicellular haploid phase, but each plant generation can consist of haploid and diploid multicellular phases.

Eukaryotes have a smaller surface area to volume ratio than prokaryotes, and thus have lower metabolic rates and longer generation times.

The evolution of sexual reproduction may be a primordial and fundamental characteristic of eukaryotes. Based on a phylogenetic analysis, Dacks and Roger proposed that facultative sex was present in the common ancestor of all eukaryotes. A core set of genes that function in meiosis is present in both *Trichomonas vaginalis* and *Giardia intestinalis*, two organisms previously thought to be asexual. Since these two species are descendants of lineages that diverged early from the eukaryotic evolutionary tree, it was inferred that core meiotic genes, and hence sex, were likely present in a common ancestor of all eukaryotes. Eukaryotic species once thought to be asexual, such

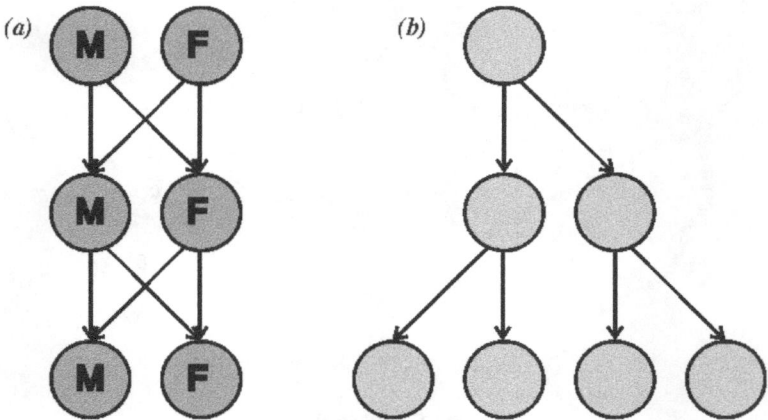

Figure 89: *This diagram illustrates the twofold cost of sex. If each individual were to contribute to the same number of offspring (two),
(a) the sexual population remains the same size each generation,
where the (b) asexual population doubles in size each generation.*

as parasitic protozoa of the genus *Leishmania*, have been shown to have a sexual cycle. Also, evidence now indicates that amoebae, previously regarded
as asexual, are anciently sexual and that the majority of present-day asexual
groups likely arose recently and independently.

Classification

In antiquity, the two lineages of animals and plants were recognized. They
were given the taxonomic rank of Kingdom by Linnaeus. Though he included
the fungi with plants with some reservations, it was later realized that they are
quite distinct and warrant a separate kingdom, the composition of which was
not entirely clear until the 1980s. The various single-cell eukaryotes were originally placed with plants or animals when they became known. In 1830, the
German biologist Georg A. Goldfuss coined the word *protozoa* to refer to organisms such as ciliates, and this group was expanded until it encompassed all
single-celled eukaryotes, and given their own kingdom, the Protista, by Ernst
Haeckel in 1866. The eukaryotes thus came to be composed of four kingdoms:

- Kingdom Protista
- Kingdom Plantae
- Kingdom Fungi
- Kingdom Animalia

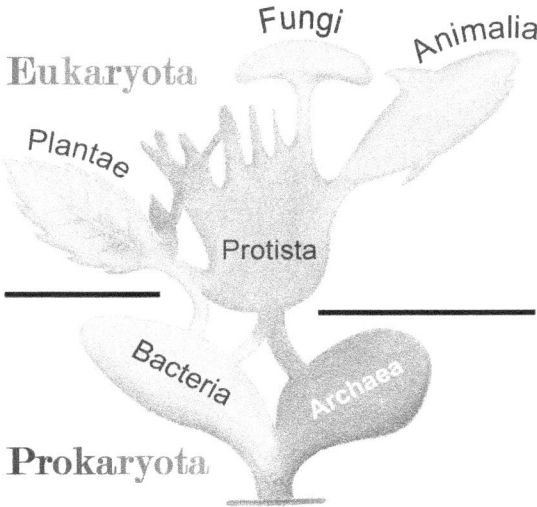

Figure 90: *Phylogenetic and symbiogenetic tree of living organisms, showing a view of the origins of eukaryotes & prokaryotes*

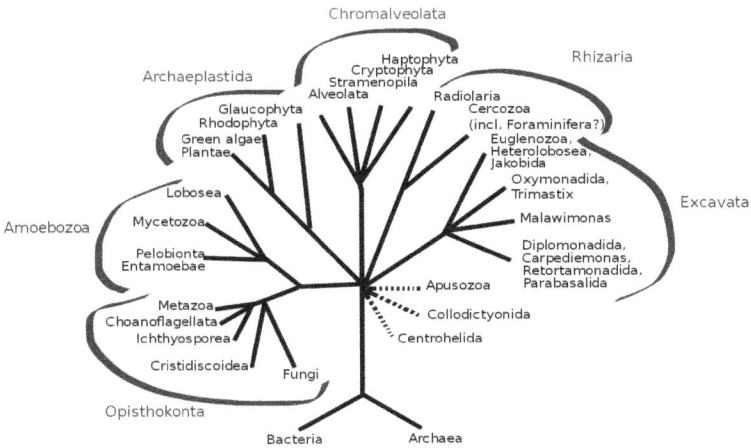

Figure 91: *One hypothesis of eukaryotic relationships. The Opisthokonta group includes both animals (Metazoa) and fungi. Plants (Plantae) are placed in Archaeplastida.*

eukaryote species

macrolepidoptera (butterflies; owlet
moths; other large moths)
6%

brachycera (house, fruit, flesh,
hover, blow, horse etc. flies)
4.6%
apocrita (wasps, bees and ants)
6.7%

weevils
3.6%
leaf and longhorn beetles
3.3%
other beetles
5%
staphyliniformia (rove beetles,
scarabs, stag beetles, etc.)
5.5%
paraneoptera (bugs, lice etc.)
3.5%

other bikonts (algae etc.)
3.4%
monocots
3.9%
rosids
4.5%
asterids
5.3%

holomycota (fungi etc.)
5.2%
deuterostomes (vertebrates,
starfish, etc.)
3.7%
lophotrochozoa (molluscs,
annelids, etc.)
6%
chelicerates (arachnids etc.)
5.4%
other pancrustacea ("crustaceans"
and "entognatha")
3.9%

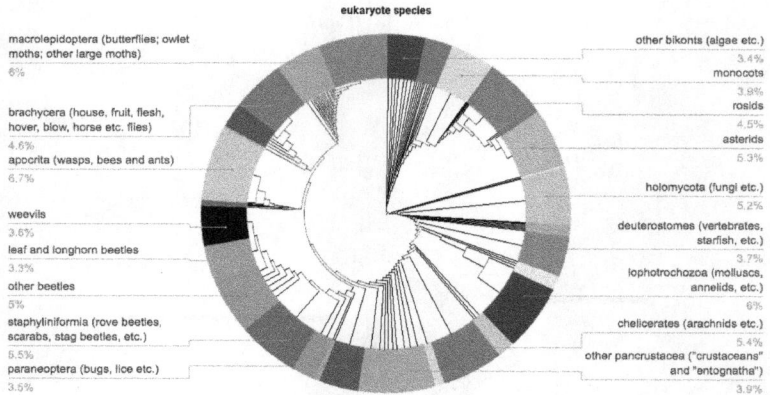

Figure 92: *A pie chart of described eukaryote species (except for Excavata),
together with a tree showing possible relationships between the groups*

The protists were understood to be "primitive forms", and thus an evolutionary grade, united by their primitive unicellular nature. The disentanglement
of the deep splits in the tree of life only really started with DNA sequencing,
leading to a system of domains rather than kingdoms as top level rank being
put forward by Carl Woese, uniting all the eukaryote kingdoms under the eukaryote domain. At the same time, work on the protist tree intensified, and
is still actively going on today. Several alternative classifications have been
forwarded, though there is no consensus in the field.

Eukaryotes are a clade usually assessed to be sister to Heimdallarchaeota in
the Asgard grouping in the Archaea. The basal groupings are the Opimoda,
Diphoda, the Discoba, and the Loukozoa. The Eukaryote root is usually assessed to be near or even in Discoba.

A classification produced in 2005 for the International Society of Protistolo-
gists, which reflected the consensus of the time, divided the eukaryotes into six
supposedly monophyletic 'supergroups'. However, in the same year (2005),
doubts were expressed as to whether some of these supergroups were monophyletic, particularly the Chromalveolata, and a review in 2006 noted the lack
of evidence for several of the supposed six supergroups. A revised classification in 2012 recognizes five supergroups.

Archaeplastida (or Primoplantae)	Land plants, green algae, red algae, and glaucophytes
SAR supergroup	Stramenopiles (brown algae, diatoms, etc.), Alveolata, and Rhizaria (Foraminifera, Radiolaria, and various other amoeboid protozoa).
Excavata	Various flagellate protozoa
Amoebozoa	Most lobose amoeboids and slime molds
Opisthokonta	Animals, fungi, choanoflagellates, etc.

There are also smaller groups of eukaryotes whose position is uncertain or seems to fall outside the major groups — in particular, Haptophyta, Cryptophyta, Centrohelida, Telonemia, Picozoa, Apusomonadida, Ancyromonadida, Breviatea, and the genus *Collodictyon*. Overall, it seems that, although progress has been made, there are still very significant uncertainties in the evolutionary history and classification of eukaryotes. As Roger & Simpson said in 2009 "with the current pace of change in our understanding of the eukaryote tree of life, we should proceed with caution."

In an article published in *Nature Microbiology* in April 2016 the authors, "reinforced once again that the life we see around us – plants, animals, humans and other so-called eukaryotes – represent a tiny percentage of the world's biodiversity." They classified eukaryote "based on the inheritance of their information systems as opposed to lipid or other cellular structures." Jillian F. Banfield of the University of California, Berkeley and fellow scientists used a super computer to generate a diagram of a new tree of life based on DNA from 3000 species including 2,072 known species and 1,011 newly reported microbial organisms, whose DNA they had gathered from diverse environments. As the capacity to sequence DNA became easier, Banfield and team were able to do metagenomic sequencing—"sequencing whole communities of organisms at once and picking out the individual groups based on their genes alone."

Phylogeny

The rRNA trees constructed during the 1980s and 1990s left most eukaryotes in an unresolved "crown" group (not technically a true crown), which was usually divided by the form of the mitochondrial cristae; see crown eukaryotes. The few groups that lack mitochondria branched separately, and so the absence was believed to be primitive; but this is now considered an artifact of long-branch attraction, and they are known to have lost them secondarily.

As of 2011[177], there is widespread agreement that the Rhizaria belong with the Stramenopiles and the Alveolata, in a clade dubbed the SAR supergroup, so that Rhizaria is not one of the main eukaryote groups; also that the Amoebozoa and Opisthokonta are each monophyletic and form a clade, often called the unikonts. Beyond this, there does not appear to be a consensus.

It has been estimated that there may be 75 distinct lineages of eukaryotes. Most of these lineages are protists.

The known eukaryote genome sizes vary from 8.2 megabases (Mb) in *Babesia bovis* to 112,000–220,050 Mb in the dinoflagellate *Prorocentrum micans*, showing that the genome of the ancestral eukaryote has undergone considerable variation during its evolution. The last common ancestor of all eukaryotes is believed to have been a phagotrophic protist with a nucleus, at least one centriole and cilium, facultatively aerobic mitochondria, sex (meiosis and syngamy), a dormant cyst with a cell wall of chitin and/or cellulose and peroxisomes. Later endosymbiosis led to the spread of plastids in some lineages.

Five supergroups

A global tree of eukaryotes from a consensus of phylogenetic evidence (in particular, phylogenomics), rare genomic signatures, and morphological characteristics is presented in Adl *et al.* 2012 and Burki 2014/2016 with the Cryptophyta and picozoa having emerged within the Archaeplastida. A similar inclusion of Glaucophyta, Cryptista (and also, unusually, Haptista) has also been made.

<templatestyles src="Template:Clade/styles.css" />

Eukaryotes

<templatestyles src="Template:Clade/styles.css" />

DiaphoretickesAmorphea

<templatestyles src="Template:Clade/styles.css" />

Ar-   <templatestyles src="Template:Clade/styles.css"
chae- />
plas-
tida

<templatestyles src="Template:Clade/styles.css" />

Red algae (Rhodophyta)

picozoa

<templatestyles src="Template:Clade/styles.css" />

<templatestyles src="Template:Clade/styles.css" />

Glaucophyta

Green plants (Viridiplantae)

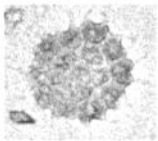

Cryptista

<templatestyles src="Template:Clade/styles.css" />

<templatestyles src="Template:Clade/styles.css" />

Haptista

Ancoracysta

<templatestyles src="Template:Clade/styles.css" />
Telonemia

SAR   <templatestyles src="Template:Clade/styles.css" />
Hal-   <templatestyles src="Template:Clade/-
varia   styles.css" />

Stramenopiles

Alveolata

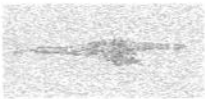

Rhizaria

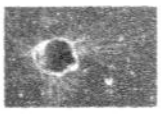

Discoba

<templatestyles src="Template:Clade/styles.css" />

Amoebozoa

<templatestyles src="Template:Clade/styles.css" />
Apusomonadida

Opisthokonta   <templatestyles src="Template:Clade/styles.css" />

Holomycota (inc. fungi)

Holozoa (inc. animals)

In some analyses, the Hacrobia group (Haptophyta + Cryptophyta) is placed next to Archaeplastida, but in other ones it is nested inside the Archaeplastida. However, several recent studies have concluded that Haptophyta and Cryptophyta do not form a monophyletic group. The former could be a sister group to the SAR group, the latter cluster with the Archaeplastida (plants in the broad sense).

The division of the eukaryotes into two primary clades, bikonts (Archaeplastida + SAR + Excavata) and unikonts (Amoebozoa + Opisthokonta), derived from an ancestral biflagellar organism and an ancestral uniflagellar organism, respectively, had been suggested earlier. A 2012 study produced a somewhat similar division, although noting that the terms "unikonts" and "bikonts" were not used in the original sense.

A highly converged and congruent set of trees appears in Derelle et al (2015), Ren et al (2016), Yang et al (2017) and Cavalier-Smith (2015) including the supplementary information, resulting in a more conservative and consolidated tree. It is combined with some results from Cavalier-Smith for the basal Opimoda. The main remaining controversies are the root, and the exact positioning of the Rhodophyta and the bikonts Rhizaria, Haptista, Cryptista, Picozoa and Telonemia, many of which may be endosymbyotic eukaryote-eukaryote hybrids. Archeaplastida developed the Chloroplasts probably by endosymbiosis of an ancestor related to a currently extant cyanobacterium, Gloeomargarita lithophora.

<templatestyles src="Template:Clade/styles.css" />

Eu-
kary-
otes
<templatestyles src="Template:Clade/styles.css" />

Diphoda   <templatestyles src="Template:Clade/styles.css" />
    Di-   <templatestyles src="Template:Clade/styles.css" />
    aphoret-   <templatestyles src="Template:Clade/styles.css" />
    ickes   Archaeplastida   <templatestyles src="Template:Clade/styles.css" />
                                                Glaucophyta

                                            <templatestyles src="Template:Clade/styles.css" />

                                                Rhodophyta

                                                Viridiplantae

                (+ Gloeomar-
                garita lithophora)
                Hacrobia   <templatestyles src="Template:Clade/styles.css" />

                                            Haptista

                                            Cryptista

            SAR   <templatestyles src="Template:Clade/styles.css" />
                Halvaria   <templatestyles src="Template:Clade/styles.css" />

                                Stramenopiles

                                Alveolata

                                Rhizaria

            Discoba

Opimoda   <templatestyles src="Template:Clade/styles.css" />
        Metamonada

            <templatestyles src="Template:Clade/styles.css" />
            Ancyromonas

            <templatestyles src="Template:Clade/styles.css" />
            Malawimonas

        Po-   <templatestyles src="Template:Clade/styles.css" />
        di-   CRuMs   Diphyllatea, Rigifilida, Mantamonas
        ata

                Amor-   <templatestyles src="Template:Clade/styles.css" />
                phea   Amoebozoa

                    Obazoa   <templatestyles src="Template:Clade/styles.css" />
                            Breviata

                            <templatestyles src="Template:Clade/styles.css" />

                                Apusomonadida

                                Opisthokonta

Cavalier-Smith's tree

Thomas Cavalier-Smith 2010, 2013, 2014, 2017 and 2018 places the eu-
karyotic tree's root between Excavata (with ventral feeding groove supported
by a microtubular root) and the grooveless Euglenozoa, and monophyletic
Chromista, correlated to a single endosymbyotic event of capturing a red-algae.
He et al specifically supports rooting eukaryotic tree between a monophyletic
Discoba (Discicristata + Jakobida) and a Amorphea-Diaphoretickes clade.

<templatestyles src="Template:Clade/styles.css" />

Eu-   <templatestyles src="Template:Clade/styles.css" />
kary-
otes

Euglenozoa

<templatestyles src="Template:Clade/styles.css" />
Percolozoa

<templatestyles src="Template:Clade/styles.css" />
Eolouka   <templatestyles src="Template:Clade/styles.css" />
              Tsukubamonas globosa

              Jakobea

Neokary-   <templatestyles src="Template:Clade/styles.css" />
ota   Corti-   <templatestyles src="Template:Clade/styles.css" />
            cata   Archaeplastida   <templatestyles src="Template:Clade/styles.css" />
                                          Glaucophytes

                                          <templatestyles src="Template:Clade/styles.css" />
                                          Rhodophytes

                                          Viridiplantae

                      Chromista (+ Rhodophyte)   <templatestyles src="Template:Clade/styles.css" />
                                                    Hacrobia

                                                    SAR

            Sco-   <templatestyles src="Template:Clade/styles.css" />
            tokary-   Malawimonadea
            ota

                      <templatestyles src="Template:Clade/styles.css" />
                      Metamonada

                      Po-   <templatestyles src="Template:Clade/styles.css" />
                      di-   Ancyromonadida
                      ata

                              <templatestyles src="Template:Clade/styles.css" />
                                  <templatestyles src="Template:Clade/styles.css" />
                              Mantamonadia

                              Diphyllatea

                      Amor-   <templatestyles src="Template:Clade/styles.css" />
                      phea   Amoebozoa

                              Obazoa   <templatestyles src="Template:Clade/styles.css"
                                       />
                                       Breviatea

                                       <templatestyles src="Template:Clade/-
                                       styles.css" />

                                       Apusomonadida

                                       Opisthokonta

            Opi-
            moda

Origin of eukaryotes

Fossils

The origin of the eukaryotic cell is a milestone in the evolution of life, since eukaryotes include all complex cells and almost all multicellular organisms. The timing of this series of events is hard to determine; Knoll (2006) suggests they developed approximately 1.6–2.1 billion years ago. Some acritarchs are known from at least 1.65 billion years ago, and the possible alga *Grypania* has been found as far back as 2.1 billion years ago. The *Geosiphon*-like fossil fungus *Diskagma* has been found in paleosols 2.2 billion years old.

Organized living structures have been found in the black shales of the Palaeo-proterozoic Francevillian B Formation in Gabon, dated at 2.1 billion years old. Eukaryotic life could have evolved at that time. Fossils that are clearly related to modern groups start appearing an estimated 1.2 billion years ago, in the form of a red alga, though recent work suggests the existence of fossilized filamentous algae in the Vindhya basin dating back perhaps to 1.6 to 1.7 billion years ago.

Biomarkers suggest that at least stem eukaryotes arose even earlier. The presence of steranes in Australian shales indicates that eukaryotes were present in these rocks dated at 2.7 billion years old, although it was suggested they could originate from samples contamination.

Whenever their origins, eukaryotes may not have become ecologically dominant until much later; a massive uptick in the zinc composition of marine sediments 800[178] million years ago has been attributed to the rise of substantial populations of eukaryotes, which preferentially consume and incorporate zinc relative to prokaryotes.

Relationship to Archaea

The nuclear DNA and genetic machinery of eukaryotes is more similar to Archaea than Bacteria, leading to a controversial suggestion that eukaryotes should be grouped with Archaea in the clade Neomura. In other respects, such as membrane composition, eukaryotes are similar to Bacteria. Three main explanations for this have been proposed:

- Eukaryotes resulted from the complete fusion of two or more cells, wherein the cytoplasm formed from a eubacterium, and the nucleus from an archaeon, from a virus, or from a pre-cell.
- Eukaryotes developed from Archaea, and acquired their eubacterial characteristics through the endosymbiosis of a proto-mitochondrion of eubacterial origin.

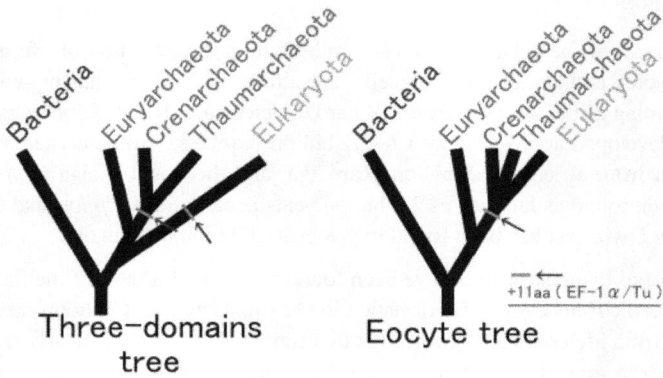

Figure 93: *The three-domains tree and the Eocyte hypothesis*

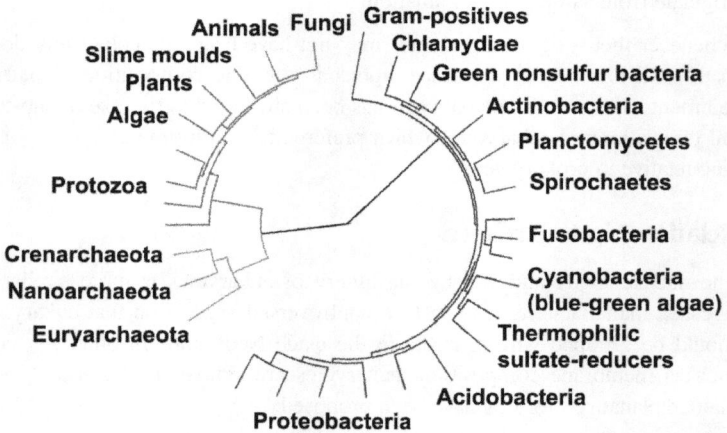

Figure 94: *Phylogenetic tree showing a possible relationship between the eukaryotes and other forms of life; eukaryotes are colored red, archaea green and bacteria blue.*

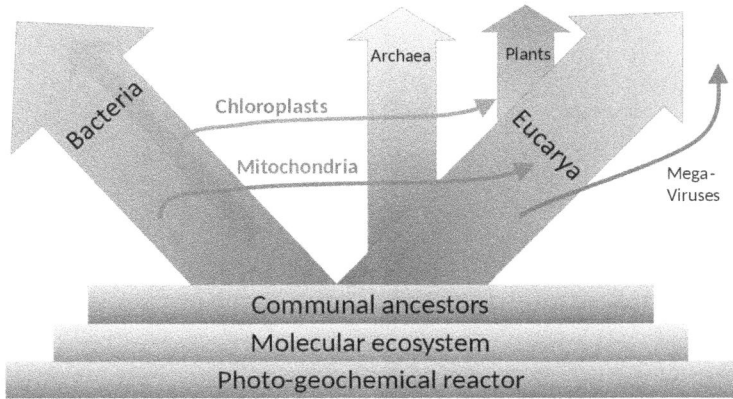

Figure 95: *Diagram of the origin of life with the Eukaryotes appearing early, not derived from Prokaryotes, as proposed by Richard Egel in 2012. This view implies that the UCA was relatively large and complex.*

- Eukaryotes and Archaea developed separately from a modified eubacterium.

Alternative proposals include:

- The chronocyte hypothesis postulates that a primitive eukaryotic cell was formed by the endosymbiosis of both archaea and bacteria by a third type of cell, termed a chronocyte.
- The universal common ancestor (UCA) of the current tree of life was a complex organism that survived a mass extinction event rather than an early stage in the evolution of life. Eukaryotes and in particular akaryotes (Bacteria and Archaea) evolved through reductive loss, so that similarities result from differential retention of original features.

Assuming no other group is involved, there are three possible phylogenies for the Bacteria, Archaea and Eukaryota in which each is monophyletic. These are labelled 1 to 3 in the table below. The eocyte hypothesis is a modification of hypothesis 2 in which the Archaea are paraphyletic. (The table and the names for the hypotheses are based on Harish and Kurland, 2017.)

Alternative hypotheses for the base of the tree of life

1 – Two empires	2 – Three domains	3 – Gupta	4 – Eocyte
<templatestyles src="Template:Clade/styles.css" /> UCA <templatestyles src="Template:Clade/styles.css" /> <templatestyles src="Template:Clade/styles.css" />	<templatestyles src="Template:Clade/styles.css" /> UCA <templatestyles src="Template:Clade/styles.css" /> <templatestyles src="Template:Clade/styles.css" />	<templatestyles src="Template:Clade/styles.css" /> UCA <templatestyles src="Template:Clade/styles.css" /> <templatestyles src="Template:Clade/styles.css" />	<templatestyles src="Template:Clade/styles.css" />
Archaea	Eukaryota	Eukaryota	UCA <templatestyles src="Template:Clade/styles.css" /> <templatestyles src="Template:Clade/styles.css" /> <templatestyles src="Template:Clade/styles.css" />
Bacteria	Archaea	Bacteria	Eukaryota
Eukaryota	Bacteria	Archaea	Archaea-Crenarchaeota
			Archaea-Euryarchaeota
			Bacteria

In recent years, most researchers have favoured either the three domains (3D) or the eocyte hypotheses. An rRNA analyses supports the eocyte scenario, apparently with the Eukaryote root in Excavata. A cladogram supporting the eocyte hypothesis, positioning eukaryotes within Archaea, based on phylogenomic analyses of the Asgard archaea, is:

<templatestyles src="Template:Clade/styles.css" />

Proteoar-   <templatestyles src="Template:Clade/styles.css" />
chaeota

     TACK   <templatestyles src="Template:Clade/styles.css" />
               Korarchaeota

               <templatestyles src="Template:Clade/styles.css" />
                 Crenarchaeota

               <templatestyles src="Template:Clade/styles.css" />
                 <templatestyles src="Template:Clade/styles.css" />
                   Aigarchaeota

                   Geoarchaeota

               <templatestyles src="Template:Clade/styles.css" />
                 Thaumarchaeota

                 Bathyarchaeota

     Asgard   <templatestyles src="Template:Clade/styles.css" />
               Lokiarchaeota

               Odinarchaeota

               Thorarchaeota

               <templatestyles src="Template:Clade/styles.css" />
                 Heimdallarchaeota

       (+α−Proteobacteria)   Eukaryota

In this scenario, the Asgard group is seen as a sister taxon of the TACK group, which comprises Crenarchaeota (formerly named eocytes), Thaumarchaeota, and others.

In 2017, there has been significant pushback against this scenario, arguing that the eukaryotes did not emerge within the Archaea. Cunha *et al.* produced analyses supporting the three domains (3D) or Woese hypothesis (2 in the table above) and rejecting the eocyte hypothesis (4 above). Harish and Kurland found strong support for the earlier two empires (2D) or Mayr hypothesis (1 in the table above), based on analyses of the coding sequences of protein domains.

They rejected the eocyte hypothesis as the least likely. A possible interpretation of their analysis is that the universal common ancestor (UCA) of the current tree of life was a complex organism that survived an evolutionary bottleneck, rather than a simpler organism arising early in the history of life.

Endomembrane system and mitochondria

The origins of the endomembrane system and mitochondria are also unclear. The **phagotrophic hypothesis** proposes that eukaryotic-type membranes lacking a cell wall originated first, with the development of endocytosis, whereas mitochondria were acquired by ingestion as endosymbionts. The **syntrophic hypothesis** proposes that the proto-eukaryote relied on the proto-mitochondrion for food, and so ultimately grew to surround it. Here the membranes originated after the engulfment of the mitochondrion, in part thanks to mitochondrial genes (the hydrogen hypothesis is one particular version).

In a study using genomes to construct supertrees, Pisani *et al.* (2007) suggest that, along with evidence that there was never a mitochondrion-less eukaryote, eukaryotes evolved from a syntrophy between an archaea closely related to Thermoplasmatales and an α-proteobacterium, likely a symbiosis driven by sulfur or hydrogen. The mitochondrion and its genome is a remnant of the α-proteobacterial endosymbiont.

Hypotheses

Different hypotheses have been proposed as to how eukaryotic cells came into existence. These hypotheses can be classified into two distinct classes – autogenous models and chimeric models.

Autogenous models

An autogenous model for the origin of eukaryotes.

Autogenous models propose that a proto-eukaryotic cell containing a nucleus existed first, and later acquired mitochondria. According to this model, a large prokaryote developed invaginations in its plasma membrane in order to obtain enough surface area to service its cytoplasmic volume. As the invaginations differentiated in function, some became separate compartments—giving rise to the endomembrane system, including the endoplasmic reticulum, golgi apparatus, nuclear membrane, and single membrane structures such as lysosomes. Mitochondria are proposed to come from the endosymbiosis of an aerobic proteobacterium, and it is assumed that all the eukaryotic lineages that did not acquire mitochondria became extinct. Chloroplasts came about from another endosymbiotic event involving cyanobacteria. Since all eukaryotes have mitochondria, but not all have chloroplasts, the serial endosymbiosis theory proposes that mitochondria came first.

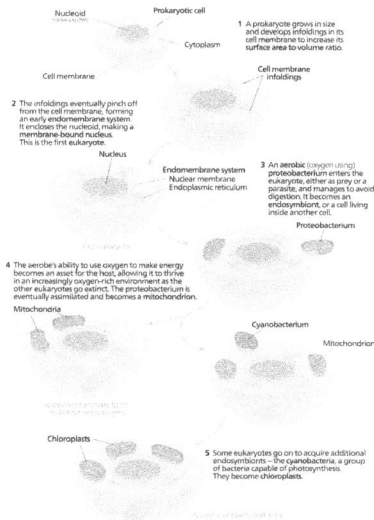

Chimeric models

Chimeric models claim that two prokaryotic cells existed initially – an archaeon and a bacterium. These cells underwent a merging process, either by a physical fusion or by endosymbiosis, thereby leading to the formation of a eukaryotic cell. Within these chimeric models, some studies further claim that mitochondria originated from a bacterial ancestor while others emphasize the role of endosymbiotic processes behind the origin of mitochondria.

Based on the process of mutualistic symbiosis, the hypotheses can be categorized as – the serial endosymbiotic theory (SET), the hydrogen hypothesis (mostly a process of symbiosis where hydrogen transfer takes place among different species), and the syntrophy hypothesis.

According to serial endosymbiotic theory (championed by Lynn Margulis), a union between a motile anaerobic bacterium (like *Spirochaeta*) and a thermoacidophilic crenarchaeon (like *Thermoplasma* which is sulfidogenic in nature) gave rise to the present day eukaryotes. This union established a motile organism capable of living in the already existing acidic and sulfurous waters. Oxygen is known to cause toxicity to organisms that lack the required metabolic machinery. Thus, the archaeon provided the bacterium with a highly beneficial reduced environment (sulfur and sulfate were reduced to sulfide). In microaerophilic conditions, oxygen was reduced to water thereby creating a mutual benefit platform. The bacterium on the other hand, contributed the necessary fermentation products and electron acceptors along with its motility feature to the archaeon thereby gaining a swimming motility for the organism. From a consortium of bacterial and archaeal DNA originated the nuclear genome of eukaryotic cells. Spirochetes gave rise to the motile features of eukaryotic cells. Endosymbiotic unifications of the ancestors of alpha-proteobacteria and cyanobacteria, led to the origin of mitochondria and plastids respectively. For example, *Thiodendron* has been known to have originated via an ectosymbiotic process based on a similar syntrophy of sulfur existing between the two types of bacteria – *Desulphobacter* and *Spirochaeta*. However, such an association based on motile symbiosis have never been observed practically. Also there is no evidence of archaeans and spirochetes adapting to intense acid-based environments.

In the hydrogen hypothesis, the symbiotic linkage of an anaerobic and autotrophic methanogenic archaeon (host) with an alpha-proteobacterium (the symbiont) gave rise to the eukaryotes. The host utilized hydrogen (H_2) and carbon dioxide (CO_2) to produce methane while the symbiont, capable of aerobic respiration, expelled H_2 and CO_2 as byproducts of anaerobic fermentation process. The host's methanogenic environment worked as a sink for H_2, which resulted in heightened bacterial fermentation. Endosymbiotic gene transfer (EGT) acted as a catalyst for the host to acquire the symbionts' carbohydrate metabolism and turn heterotrophic in nature. Subsequently, the host's methane forming capability was lost. Thus, the origins of the heterotrophic organelle (symbiont) are identical to the origins of the eukaryotic lineage. In this hypothesis, the presence of H_2 represents the selective force that forged eukaryotes out of prokaryotes.Wikipedia:Citation needed

The syntrophy hypothesis was developed in contrast to the hydrogen hypothesis and proposes the existence of two symbiotic events. According to this theory, the origin of eukaryotic cells was based on metabolic symbiosis (syntrophy) between a methanogenic archaeon and a delta-proteobacterium. This syntrophic symbiosis was initially facilitated by H_2 transfer between different species under anaerobic environments. In earlier stages, an alpha-proteobacterium became a member of this integration, and later developed into the mitochondrion. Gene transfer from a delta-proteobacterium to an archaeon led to the methanogenic archaeon developing into a nucleus. The archaeon constituted the genetic apparatus, while the delta-proteobacterium contributed towards the cytoplasmic features. This theory incorporates two selective forces at the time of nucleus evolution – (a) presence of metabolic partitioning to avoid the harmful effects of the co-existence of anabolic and catabolic cellular pathways, and (b) prevention of abnormal protein biosynthesis due to a vast spread of introns in the archaeal genes after acquiring the mitochondrion and losing methanogenesis.Wikipedia:Citation needed

References

℗ This article incorporates public domain material from the NCBI document "Science Primer"[179]. Wikipedia:Link rot

External links

> 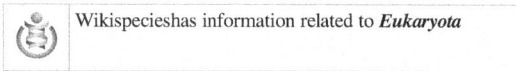 Wikispecieshas information related to *Eukaryota*

- Eukaryotes (Tree of Life web site)[180]
- *Eukaryote*[181] at the Encyclopedia of Life ✐

Plastid

The **plastid** (Greek: πλαστός; plastós: formed, molded – plural **plastids**) is a double-membrane organelle found in the cells of plants, algae, and some other eukaryotic organisms. Plastids were discovered and named by Ernst Haeckel, but A. F. W. Schimper was the first to provide a clear definition. Plastids are the site of manufacture and storage of important chemical compounds used by the cells of autotrophic eukaryotes. They often contain pigments used in photosynthesis, and the types of pigments in a plastid determine the cell's color. They have a common evolutionary origin and possess a double-stranded DNA molecule that is circular, like that of prokaryotic cells.

In plants

Those plastids that contain chlorophyll can carry out photosynthesis and are called chloroplasts. Plastids can also store products like starch and can synthesize fatty acids and terpenes, which can be used for producing energy and as raw material for the synthesis of other molecules. For example, the components of the plant cuticle and its epicuticular wax are synthesized by the epidermal cells from palmitic acid, which is synthesized in the chloroplasts of the mesophyll tissue.[182] All plastids are derived from proplastids, which are present in the meristematic regions of the plant. Proplastids and young chloroplasts commonly divide by binary fission, but more mature chloroplasts also have this capacity.

In plants, plastids may differentiate into several forms, depending upon which function they play in the cell. Undifferentiated plastids (*proplastids*) may develop into any of the following variants:

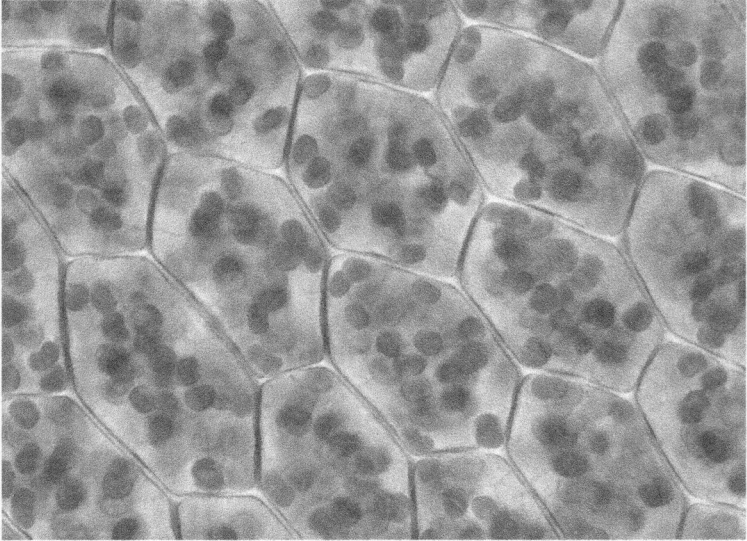

Figure 96: *Plant cells with visible chloroplasts.*

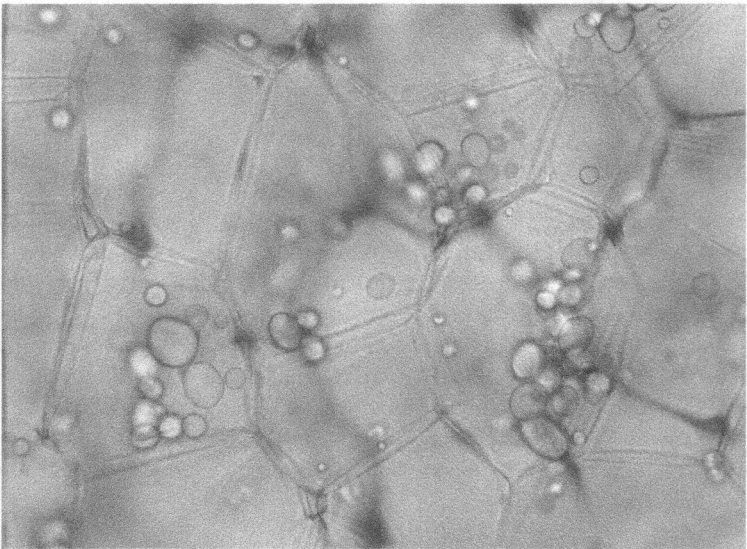

Figure 97: *Leucoplasts in plant cells.*

Plastids

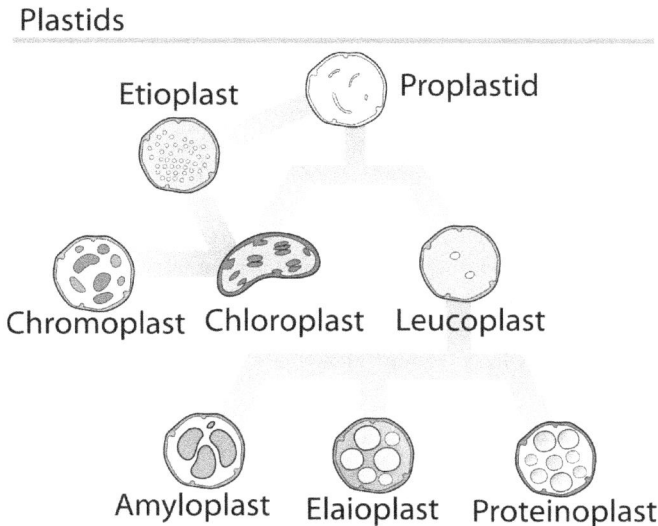

Etioplast   Proplastid

Chromoplast   Chloroplast   Leucoplast

Amyloplast   Elaioplast   Proteinoplast

- Chloroplasts green plastids: for photosynthesis; *see also etioplasts, the predecessors of chloroplasts*
- Chromoplasts coloured plastids: for pigment synthesis and storage
- Gerontoplasts: control the dismantling of the photosynthetic apparatus during plant senescence
- Leucoplasts colourless plastids: for monoterpene synthesis; *leucoplasts sometimes differentiate into more specialized plastids:*
  - Amyloplasts: for starch storage and detecting gravity (for geotropism)
  - Elaioplasts: for storing fat
  - Proteinoplasts: for storing and modifying protein
  - Tannosomes: for synthesizing and producing tannins and polyphenols

Depending on their morphology and function, plastids have the ability to differentiate, or redifferentiate, between these and other forms.

Each plastid creates multiple copies of a circular 75–250 kilobase plastome. The number of genome copies per plastid is variable, ranging from more than 1000 in rapidly dividing cells, which, in general, contain few plastids, to 100 or fewer in mature cells, where plastid divisions have given rise to a large number of plastids. The plastome contains about 100 genes encoding ribosomal and transfer ribonucleic acids (rRNAs and tRNAs) as well as proteins involved in photosynthesis and plastid gene transcription and translation. However, these proteins only represent a small fraction of the total protein set-up necessary to build and maintain the structure and function of a particular type of plastid.

Plant nuclear genes encode the vast majority of plastid proteins, and the expression of plastid genes and nuclear genes is tightly co-regulated to coordinate proper development of plastids in relation to cell differentiation.

Plastid DNA exists as large protein-DNA complexes associated with the inner envelope membrane and called 'plastid nucleoids'. Each nucleoid particle may contain more than 10 copies of the plastid DNA. The proplastid contains a single nucleoid located in the centre of the plastid. The developing plastid has many nucleoids, localized at the periphery of the plastid, bound to the inner envelope membrane. During the development of proplastids to chloroplasts, and when plastids convert from one type to another, nucleoids change in morphology, size and location within the organelle. The remodelling of nucleoids is believed to occur by modifications to the composition and abundance of nucleoid proteins.

Many plastids, particularly those responsible for photosynthesis, possess numerous internal membrane layers.

In plant cells, long thin protuberances called stromules sometimes form and extend from the main plastid body into the cytosol and interconnect several plastids. Proteins, and presumably smaller molecules, can move within stromules. Most cultured cells that are relatively large compared to other plant cells have very long and abundant stromules that extend to the cell periphery.

In 2014, evidence of possible plastid genome loss was found in *Rafflesia lagascae,* a non-photosynthetic parasitic flowering plant, and in *Polytomella,* a genus of non-photosynthetic green algae. Extensive searches for plastid genes in both *Rafflesia* and *Polytomella* yielded no results, however the conclusion that their plastomes are entirely missing is still controversial. Some scientists argue that plastid genome loss is unlikely since even non-photosynthetic plastids contain genes necessary to complete various biosynthetic pathways, such as heme biosynthesis.

In algae

In algae, the term leucoplast is used for all unpigmented plastids and their function differs from the leucoplasts of plants. Etioplasts, amyloplasts and chromoplasts are plant-specific and do not occur in algae.Wikipedia:Citation needed Plastids in algae and hornworts may also differ from plant plastids in that they contain pyrenoids.

Glaucophyte algae contain muroplasts, which are similar to chloroplasts except that they have a peptidoglycan cell wall that is similar to that of prokaryotes. Red algae contain rhodoplasts, which are red chloroplasts that allow them to photosynthesise to a depth of up to 268 m. The chloroplasts of plants differ from the rhodoplasts of red algae in their ability to synthesize starch, which is stored in the form of granules within the plastids. In red algae, floridean starch is synthesized and stored outside the plastids in the cytosol.

Inheritance

Most plants inherit the plastids from only one parent. In general, angiosperms inherit plastids from the female gamete, whereas many gymnosperms inherit plastids from the male pollen. Algae also inherit plastids from only one parent. The plastid DNA of the other parent is, thus, completely lost.

In normal intraspecific crossings (resulting in normal hybrids of one species), the inheritance of plastid DNA appears to be quite strictly 100% uniparental. In interspecific hybridisations, however, the inheritance of plastids appears to be more erratic. Although plastids inherit mainly maternally in interspecific hybridisations, there are many reports of hybrids of flowering plants that contain plastids of the father. Approximately 20% of angiosperms, including alfalfa (*Medicago sativa*), normally show biparental inheritance of plastids.

DNA damage and repair

Plastid DNA of maize seedlings is subject to increased damage as the seedlings develop. The DNA is damaged in oxidative environments created by photo-oxidative reactions and photosynthetic/respiratory electron transfer. Some DNA molecules are repaired while DNA with unrepaired damage appears to be degraded to non-functional fragments.

DNA repair proteins are encoded by the cell's nuclear genome but can be translocated to plastids where they maintain genome stability/integrity by repairing the plastid's DNA. As an example, in chloroplasts of the moss *Physcomitrella patens*, a protein employed in DNA mismatch repair (Msh1) interacts with proteins employed in recombinational repair (RecA and RecG) to maintain plastid genome stability.

Origin

Plastids are thought to have originated from endosymbiotic cyanobacteria. This symbiosis evolved around 1.5 billion years ago and enabled eukaryotes to carry out oxygenic photosynthesis. Three evolutionary lineages have since emerged in which the plastids are named differently: chloroplasts in green algae and plants, rhodoplasts in red algae and muroplasts in the glaucophytes. The plastids differ both in their pigmentation and in their ultrastructure. For example, chloroplasts in plants and green algae have lost all phycobilisomes, the light harvesting complexes found in cyanobacteria, red algae and glaucophytes, but instead contain stroma and grana thylakoids. The glaucocystophycean plastid—in contrast to chloroplasts and rhodoplasts—is still surrounded by the remains of the cyanobacterial cell wall. All these primary plastids are surrounded by two membranes.

Complex plastids start by secondary endosymbiosis (where a eukaryotic organism engulfs another eukaryotic organism that contains a primary plastid resulting in its endosymbiotic fixation), when a eukaryote engulfs a red or green alga and retains the algal plastid, which is typically surrounded by more than two membranes. In some cases these plastids may be reduced in their metabolic and/or photosynthetic capacity. Algae with complex plastids derived by secondary endosymbiosis of a red alga include the heterokonts, haptophytes, cryptomonads, and most dinoflagellates (= rhodoplasts). Those that endosymbiosed a green alga include the euglenids and chlorarachniophytes (= chloroplasts). The Apicomplexa, a phylum of obligate parasitic protozoa including the causative agents of malaria (*Plasmodium* spp.), toxoplasmosis (*Toxoplasma gondii*), and many other human or animal diseases also harbor a complex plastid (although this organelle has been lost in some apicomplexans, such as *Cryptosporidium parvum*, which causes cryptosporidiosis). The 'apicoplast' is no longer capable of photosynthesis, but is an essential organelle, and a promising target for antiparasitic drug development.

Some dinoflagellates and sea slugs, in particular of the genus *Elysia*, take up algae as food and keep the plastid of the digested alga to profit from the photosynthesis; after a while, the plastids are also digested. This process is known as kleptoplasty, from the Greek, *kleptes*, thief.

Sources

- A Novel View of Chloroplast Structure[183]: contains fluorescence images of chloroplasts and stromules as well as an easy to read chapter.
- Wycliffe P, Sitbon F, Wernersson J, Ezcurra I, Ellerström M, Rask L (October 2005). "Continuous expression in tobacco leaves of a *Brassica napus* PEND homologue blocks differentiation of plastids and development of palisade cells". *Plant J*. **44** (1): 1–15. doi: 10.1111/j.1365-313X.2005.02482.x[184]. PMID 16167891[185].
- Birky CW (2001). "The inheritance of genes in mitochondria and chloroplasts: laws, mechanisms, and models"[186]. *Annu. Rev. Genet.* **35**: 125–48. doi: 10.1146/annurev.genet.35.102401.090231[187]. PMID 11700280[188]. PDF[189]

Further reading

- Chan CX, Bhattacharya D (2010). "The origins of plastids"[190]. *Nature Education*. **3** (9): 84.
- Bhattacharya, D., ed. (1997). *Origins of Algae and their Plastids*. New York: Springer-Verlag/Wein. ISBN 3-211-83036-7.
- Gould SB, Waller RR, McFadden GI (2008). Plastid evolution. Annu Rev Plant Biol 59: 491–517.
- Keeling P (2010). The endosymbiotic origin, diversification and fate of plastids.[191] Philos Trans R Soc Lond B Biol Sci. 365(1541): 729-48

External links

- Transplastomic plants for biocontainment (biological confinement of transgenes)[192] — Co-extra research project on coexistence and traceability of GM and non-GM supply chains
- Tree of Life Eukaryotes[193]

Sexual reproduction and multicellular organisms

Evolution of sexual reproduction

The **evolution of sexual reproduction** describes how sexually reproducing animals, plants, fungi and protists evolved from a common ancestor that was a single celled eukaryotic species. There are a few species which have secondarily lost the ability to reproduce sexually, such as Bdelloidea, and some plants and animals that routinely reproduce asexually (by apomixis and parthenogenesis) without entirely losing sex. The evolution of sex contains two related, yet distinct, themes: its *origin* and its *maintenance*.

The maintenance of sexual reproduction in a highly competitive world had long been one of the major mysteries of biology given that asexual reproduction can

Figure 98: *Ladybirds mating*

Figure 99: *Pollen production is an essential step in sexual reproduction of seed plants.*

reproduce by budding, fission, or spore formation not involving union of gametes, which reproduce at a much faster rate compared to sexual reproduction. 50% of offspring from sexual reproduction are males, unable to produce offspring themselves.

Since hypotheses for the origins of sex are difficult to test experimentally (outside of evolutionary computation), most current work has focused on the maintenance of sexual reproduction. Sexual reproduction must offer significant fitness advantages to a species because despite the two-fold cost of sex, it dominates among multicellular forms of life, implying that the fitness of offspring produced outweighs the costs. Sexual reproduction derives from recombination, where parent genotypes are reorganized and shared with the offspring. This stands in contrast to single-parent asexual replication, where the offspring is identical to the parents. Recombination supplies two fault-tolerance mechanisms at the molecular level: *recombinational DNA repair* (promoted during meiosis because homologous chromosomes pair at that time) and *complementation* (also known as heterosis, hybrid vigor or masking of mutations).

Historical perspective

The issue features in the writings of Aristotle, and modern philosophical-scientific thinking on the problem dates from at least Erasmus Darwin (1731-1802) in the 18th century. August Weismann picked up the thread in 1889, arguing that sex served to generate genetic variation, as detailed in the majority of the explanations below. On the other hand, Charles Darwin (1809-1882) concluded that the effects of hybrid vigor (complementation) "is amply sufficient to account for the ... genesis of the two sexes".Wikipedia:Citation needed This is consistent with the repair and complementation hypothesis, described below. Biologists - including W. D. Hamilton, Alexey Kondrashov, George C. Williams, Harris Bernstein, Carol Bernstein, Michael M. Cox, Frederic A. Hopf and Richard E. Michod - have suggested several explanations for how a vast array of different living species maintain sexual reproduction.

Disadvantages of sex and sexual reproduction

This section will briefly focus on the ostensible disadvantages of sexual reproduction as compared to relative advantages in asexual reproduction. Given that sexual reproduction abounds in multicellular organisms, this section is followed by a lengthy overview of theories aiming to elucidate the advantages of sex and sexual reproduction.

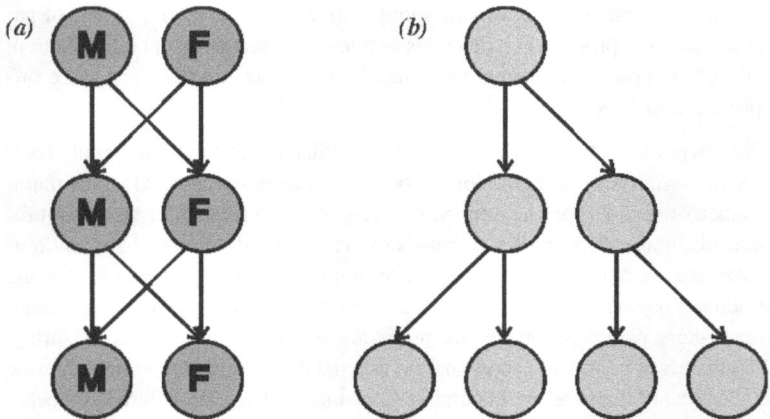

Figure 100: *This diagram illustrates the twofold cost of sex. If each individual were to contribute to the same number of offspring (two), (a) the sexual population remains the same size each generation, where the (b) asexual population doubles in size each generation.*

Population expansion cost of sex

An asexual population can grow much more rapidly with each generation. Assume the entire population of some theoretical species has 100 total organisms consisting of two sexes (i.e. males and females) with 50:50 male-to-female representation, and only the females of this species can bear offspring. If all capable members of this population procreated once, a total of 50 offspring would be produced (the *F1* generation). Contrast this outcome with an asexual species, where each member of the 100-organism population is capable of bearing young. If all capable members of this asexual population procreated once, a total of 100 offspring would be produced.

This idea is sometimes referred to as the *two-fold cost* of sexual reproduction. It was first described mathematically by John Maynard Smith.[195] In his manuscript Smith further speculated on the impact of an asexual mutant arising in a sexual population, which suppresses meiosis and allows eggs to develop by mitotic division into offspring genetically identical to the mother. The mutant-asexual lineage would double its representation in the population each generation, all else being equal.

Technically the problem above is not that of sexual reproduction but a problem of having a subset of organisms incapable of bearing offspring. Indeed some multicellular organisms (isogamous) engage in sexual reproduction but all members of the species are capable of bearing offspring.[196] The two-fold reproductive disadvantage assumes that males contribute only genes to their offspring and sexual females waste half their reproductive potential on sons. Thus, in this formulation, the principal costs of sex is that males and females must successfully copulate (which almost always involves expending energy to come together through time and space).

Selfish cytoplasmic genes

Sexual reproduction implies that chromosomes and alleles segregate and recombine in every generation, but not all genes transmitted together to the offspring. The chances of spreading mutants that cause unfair transmission at the expense of their non-mutant colleagues, these mutations are referred to as selfish because they promote their own spread at the cost of alternative alleles or host organism, these include; nuclear meiotic drivers and selfish cytoplasmic genes. Meiotic driver is defined as genes that distort meiosis to produce gametes containing themselves more than half the time and selfish cytoplasmic gene is a gene located in an organelle, plasmid, or intracellular parasite that modifies reproduction to cause its own increase at the expense of the cell or organism that carries it

Genetic heritability cost of sex

A sexually reproducing organism only passes on ∼50% of its own genetic material to each L2 offspring. This is a consequence of the fact that gametes from sexually reproducing species are haploid. Again however, this is not applicable to all sexual organisms. There are numerous species which are sexual but do not have a genetic-loss problem because they do not produce males or females. Yeast, for example, are isogamous sexual organisms which have two mating types which fuse and recombine their haploid genomes. Both sexes reproduce during the haploid and diploid stages of their life cycle and have a 100% chance of passing their genes into their offspring.

Some species avoid the cost of 50% of sexual reproduction, although they have "sex" (in the sense of genetic recombination). In these species (e.g., bacteria, ciliates, dinoflagellates and diatoms), "sex" and reproduction occurs separately.[197]

Advantages of sex and sexual reproduction

The concept of sex includes two fundamental phenomena: the sexual process (fusion of genetic information of two individuals) and sexual differentiation (separation of this information into two parts). Depending on the presence or absence of these phenomena, the existing ways of reproduction can be divided into asexual, hermaphrodite and dioecious forms. The sexual process and sexual differentiation are different phenomena, and, in essence, are diametrically opposed. The first creates (increases) diversity of genotypes, and the second decreases it by half.

Reproductive advantages of the asexual forms are in quantity of the progeny and the advantages of the hermaphrodite forms – in maximum diversity. Transition from the hermaphrodite to dioecious state leads to a loss of at least half of the diversity. So, the main question is to explain the advantages given by sexual differentiation, i.e. the benefits of two separate sexes compared to hermaphrodites rather than to explain benefits of sexual forms (hermaphrodite + dioecious) over asexual ones. It has already been understood that since sexual reproduction is not associated with any clear reproductive advantages, as compared with asexual, there should be some important advantages in evolution.[198]

Advantages due to genetic variation

For the advantage due to genetic variation, there are three possible reasons this might happen. First, sexual reproduction can combine the effects of two beneficial mutations in the same individual (i.e. sex aids in the spread of advantageous traits). Also, the necessary mutations do not have to have occurred one after another in a single line of descendants.Wikipedia:Identifying reliable sources Second, sex acts to bring together currently deleterious mutations to create severely unfit individuals that are then eliminated from the population (i.e. sex aids in the removal of deleterious genes). However, in organisms containing only one set of chromosomes, deleterious mutations would be eliminated immediately, and therefore removal of harmful mutations is an unlikely benefit for sexual reproduction. Lastly, sex creates new gene combinations that may be more fit than previously existing ones, or may simply lead to reduced competition among relatives.

For the advantage due to DNA repair, there is an immediate large benefit of removing DNA damage by recombinational DNA repair during meiosis, since this removal allows greater survival of progeny with undamaged DNA. The advantage of complementation to each sexual partner is avoidance of the bad effects of their deleterious recessive genes in progeny by the masking effect of normal dominant genes contributed by the other partner.

The classes of hypotheses based on the creation of variation are further broken down below. Any number of these hypotheses may be true in any given species (they are not mutually exclusive), and different hypotheses may apply in different species. However, a research framework based on creation of variation has yet to be found that allows one to determine whether the reason

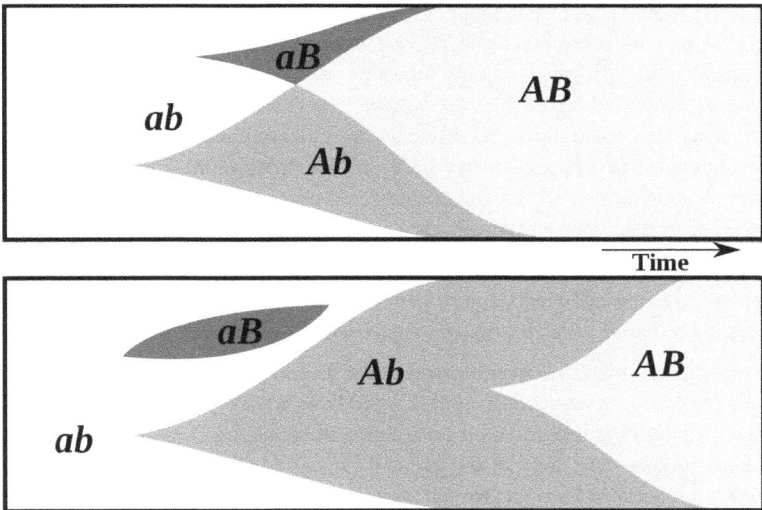

Figure 101: *This diagram illustrates how sex might create novel genotypes more rapidly. Two advantageous alleles A and B occur at random. The two alleles are recombined rapidly in a sexual population (top), but in an asexual population (bottom) the two alleles must independently arise because of clonal interference.*

for sex is universal for all sexual species, and, if not, which mechanisms are acting in each species.

On the other hand, the maintenance of sex based on DNA repair and complementation applies widely to all sexual species.

Protection from major genetic mutation

In contrast to the view that sex promotes genetic variation, Heng, and Gorelick and Heng reviewed evidence that sex actually acts as a constraint on genetic variation. They consider that sex acts as a coarse filter, weeding out major genetic changes, such as chromosomal rearrangements, but permitting minor variation, such as changes at the nucleotide or gene level (that are often neutral) to pass through the sexual sieve.

Novel genotypes

Sex could be a method by which novel genotypes are created. Because sex combines genes from two individuals, sexually reproducing populations can more easily combine advantageous genes than can asexual populations. If, in a sexual population, two different advantageous alleles arise at different loci on a chromosome in different members of the population, a chromosome containing the two advantageous alleles can be produced within a few generations by recombination. However, should the same two alleles arise in different members of an asexual population, the only way that one chromosome can develop the other allele is to independently gain the same mutation, which would take much longer. Several studies have addressed counterarguments, and the question of whether this model is sufficiently robust to explain the predominance of sexual versus asexual reproduction.[73–86]

Ronald Fisher also suggested that sex might facilitate the spread of advantageous genes by allowing them to better escape their genetic surroundings, if they should arise on a chromosome with deleterious genes.

Supporters of these theories respond to the balance argument that the individuals produced by sexual and asexual reproduction may differ in other respects too – which may influence the persistence of sexuality. For example, in the heterogamous water fleas of the genus *Cladocera*, sexual offspring form eggs which are better able to survive the winter versus those the fleas produce asexually.

Increased resistance to parasites

One of the most widely discussed theories to explain the persistence of sex is that it is maintained to assist sexual individuals in resisting parasites, also known as the Red Queen Hypothesis.[199:113–117][200]

When an environment changes, previously neutral or deleterious alleles can become favourable. If the environment changed sufficiently rapidly (i.e. between generations), these changes in the environment can make sex advantageous for the individual. Such rapid changes in environment are caused by the co-evolution between hosts and parasites.

Imagine, for example that there is one gene in parasites with two alleles p and P conferring two types of parasitic ability, and one gene in hosts with two alleles h and H, conferring two types of parasite resistance, such that parasites with allele p can attach themselves to hosts with the allele h, and P to H. Such a situation will lead to cyclic changes in allele frequency - as p increases in frequency, h will be disfavoured.

In reality, there will be several genes involved in the relationship between hosts and parasites. In an asexual population of hosts, offspring will only have the different parasitic resistance if a mutation arises. In a sexual population of hosts, however, offspring will have a new combination of parasitic resistance alleles.

In other words, like Lewis Carroll's Red Queen, sexual hosts are continually "running" (adapting) to "stay in one place" (resist parasites).

Evidence for this explanation for the evolution of sex is provided by comparison of the rate of molecular evolution of genes for kinases and immunoglobulins in the immune system with genes coding other proteins. The genes coding for immune system proteins evolve considerably faster.

Further evidence for the Red Queen hypothesis was provided by observing long-term dynamics and parasite coevolution in a "mixed" (sexual and asexual) population of snails (*Potamopyrgus antipodarum*). The number of sexuals, the number asexuals, and the rates of parasite infection for both were monitored. It was found that clones that were plentiful at the beginning of the study became more susceptible to parasites over time. As parasite infections increased, the once plentiful clones dwindled dramatically in number. Some clonal types disappeared entirely. Meanwhile, sexual snail populations remained much more stable over time.

However, Hanley et al. studied mite infestations of a parthenogenetic gecko species and its two related sexual ancestral species. Contrary to expectation based on the Red Queen hypothesis, they found that the prevalence, abundance and mean intensity of mites in sexual geckos was significantly higher than in asexuals sharing the same habitat.

In 2011, researchers used the microscopic roundworm *Caenorhabditis elegans* as a host and the pathogenic bacteria *Serratia marcescens* to generate a host-parasite coevolutionary system in a controlled environment, allowing them to conduct more than 70 evolution experiments testing the Red Queen Hypothesis. They genetically manipulated the mating system of *C. elegans*, causing populations to mate either sexually, by self-fertilization, or a mixture of both within the same population. Then they exposed those populations to the *S. marcescens* parasite. It was found that the self-fertilizing populations of *C. elegans* were rapidly driven extinct by the coevolving parasites while sex allowed populations to keep pace with their parasites, a result consistent with the Red Queen Hypothesis. In natural populations of *C. elegans*, self-fertilization is the predominant mode of reproduction, but infrequent out-crossing events occur at a rate of about 1%.

Critics of the Red Queen hypothesis question whether the constantly changing environment of hosts and parasites is sufficiently common to explain the evolution of sex. In particular, Otto and Nuismer presented results showing that species interactions (e.g. host vs parasite interactions) typically select against sex. They concluded that, although the Red Queen hypothesis favors sex under certain circumstances, it alone does not account for the ubiquity of sex. Otto and Gerstein further stated that "it seems doubtful to us that strong selection per gene is sufficiently commonplace for the Red Queen hypothesis to explain the ubiquity of sex." Parker reviewed numerous genetic studies on plant disease resistance and failed to uncover a single example consistent with the assumptions of the Red Queen hypothesis.

DNA repair and complementation

As discussed in the earlier part of this article, sexual reproduction is conventionally explained as an adaptation for producing genetic variation through allelic recombination. As acknowledged above, however, serious problems with this explanation have led many biologists to conclude that the benefit of sex is a major unsolved problem in evolutionary biology.

An alternative "informational" approach to this problem has led to the view that the two fundamental aspects of sex, genetic recombination and outcrossing, are adaptive responses to the two major sources of "noise" in transmitting genetic information. Genetic noise can occur as either physical damage to the genome (e.g. chemically altered bases of DNA or breaks in the chromosome) or replication errors (mutations) This alternative view is referred to as the repair and complementation hypothesis, to distinguish it from the traditional variation hypothesis.

The repair and complementation hypothesis assumes that genetic recombination is fundamentally a DNA repair process, and that when it occurs during meiosis it is an adaptation for repairing the genomic DNA which is passed on to progeny. Recombinational repair is the only repair process known which can accurately remove double-strand damages in DNA, and such damages are both common in nature and ordinarily lethal if not repaired. For instance, double-strand breaks in DNA occur about 50 times per cell cycle in human cells [see DNA damage (naturally occurring)]. Recombinational repair is prevalent from the simplest viruses to the most complex multicellular eukaryotes. It is effective against many different types of genomic damage, and in particular is highly efficient at overcoming double-strand damages. Studies of the mechanism of meiotic recombination indicate that meiosis is an adaptation for repairing DNA.[201] These considerations form the basis for the first part of the repair and complementation hypothesis.

In some lines of descent from the earliest organisms, the diploid stage of the sexual cycle, which was at first transient, became the predominant stage, because it allowed complementation — the masking of deleterious recessive mutations (i.e. hybrid vigor or heterosis). Outcrossing, the second fundamental aspect of sex, is maintained by the advantage of masking mutations and the disadvantage of inbreeding (mating with a close relative) which allows expression of recessive mutations (commonly observed as inbreeding depression). This is in accord with Charles Darwin,[202] who concluded that the adaptive advantage of sex is hybrid vigor; or as he put it, "the offspring of two individuals, especially if their progenitors have been subjected to very different conditions, have a great advantage in height, weight, constitutional vigor and fertility over the self fertilised offspring from either one of the same parents."

However, outcrossing may be abandoned in favor of parthenogenesis or selfing (which retain the advantage of meiotic recombinational repair) under conditions in which the costs of mating are very high. For instance, costs of mating are high when individuals are rare in a geographic area, such as when there has been a forest fire and the individuals entering the burned area are the initial ones to arrive. At such times mates are hard to find, and this favors parthenogenic species.

In the view of the repair and complementation hypothesis, the removal of DNA damage by recombinational repair produces a new, less deleterious form of informational noise, allelic recombination, as a by-product. This lesser informational noise generates genetic variation, viewed by some as the major effect of sex, as discussed in the earlier parts of this article.

Deleterious mutation clearance

Mutations can have many different effects upon an organism. It is generally believed that the majority of non-neutral mutations are deleterious, which means that they will cause a decrease in the organism's overall fitness.[203] If a mutation has a deleterious effect, it will then usually be removed from the population by the process of natural selection. Sexual reproduction is believed to be more efficient than asexual reproduction in removing those mutations from the genome.

There are two main hypotheses which explain how sex may act to remove deleterious genes from the genome.

Evading harmful mutation build-up

While DNA is able to recombine to modify alleles, DNA is also susceptible to mutations within the sequence that can affect an organism in a negative manner. Asexual organisms do not have the ability to recombine their genetic information to form new and differing alleles. Once a mutation occurs in the DNA or other genetic carrying sequence, there is no way for the mutation to be removed from the population until another mutation occurs that ultimately deletes the primary mutation. This is rare among organisms. Hermann Joseph Muller introduced the idea that mutations build up in asexual reproducing organisms. Muller described this occurrence by comparing the mutations that accumulate as a ratchet. Each mutation that arises in asexually reproducing organisms turns the ratchet once. The ratchet is unable to be rotated backwards, only forwards. The next mutation that occurs turns the ratchet once more. Additional mutations in a population continually turn the ratchet and the mutations, mostly deleterious, continually accumulate without recombination. These mutations are passed onto the next generation because the offspring are exact genetic clones of their parent. The genetic load of organisms and their populations will increase due to the addition of multiple deleterious mutations and decrease the overall reproductive success and fitness.

For sexually reproducing populations, studies have shown that single-celled bottlenecks are beneficial for resisting mutation build-up. Passaging a population through a single-celled bottleneck involves the fertilization event occurring with haploid sets of DNA, forming one fertilized cell. For example, humans undergo a single-celled bottleneck in that the haploid sperm fertilizes the haploid egg, forming the diploid zygote, which is unicellular. This passage through a single cell is beneficial in that it lowers the chance of mutations from being passed on through multiple individuals. Further studies using *Dictyostelium discoideum* suggest that this unicellular initial stage is important for resisting mutations due to the importance of high relatedness. Highly related individuals are more closely related, and more clonal, whereas less related individuals are less so, increasing the likelihood that an individual in a population of low relatedness may have a detrimental mutation. Highly related populations also tend to thrive better than lowly related because the cost of sacrificing an individual is greatly offset by the benefit gained by its relatives and in turn, its genes, according to kin selection. The studies with *D. discoideum* showed that conditions of high relatedness resisted mutant individuals more effectively than those of low relatedness, suggesting the importance of high relatedness to resist mutations from proliferating.

Removal of deleterious genes

This hypothesis was proposed by Alexey Kondrashov, and is sometimes known as the *deterministic mutation hypothesis*. It assumes that the majority of deleterious mutations are only slightly deleterious, and affect the individual such that the introduction of each additional mutation has an increasingly large effect on the fitness of the organism. This relationship between number of mutations and fitness is known as *synergistic epistasis*.

By way of analogy, think of a car with several minor faults. Each is not sufficient alone to prevent the car from running, but in combination, the faults combine to prevent the car from functioning.

Similarly, an organism may be able to cope with a few defects, but the presence of many mutations could overwhelm its backup mechanisms.

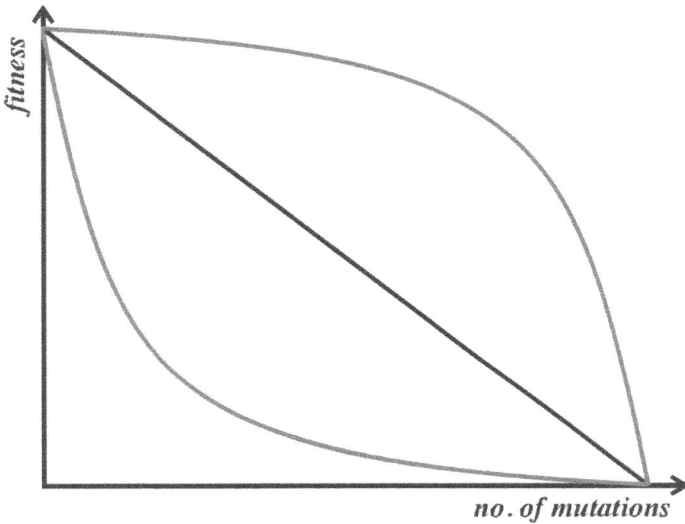

Figure 102: *Diagram illustrating different relationships between numbers of mutations and fitness. Kondrashov's model requires synergistic epistasis, which is represented by the red line[204,205] - each subsequent mutation has a disproportionately large effect on the organism's fitness.*

Kondrashov argues that the slightly deleterious nature of mutations means that the population will tend to be composed of individuals with a small number of mutations. Sex will act to recombine these genotypes, creating some individuals with fewer deleterious mutations, and some with more. Because there is a major selective disadvantage to individuals with more mutations, these individuals die out. In essence, sex compartmentalises the deleterious mutations.

There has been much criticism of Kondrashov's theory, since it relies on two key restrictive conditions. The first requires that the rate of deleterious mutation should exceed one per genome per generation in order to provide a substantial advantage for sex. While there is some empirical evidence for it (for example in Drosophila and E. coli), there is also strong evidence against it. Thus, for instance, for the sexual species *Saccharomyces cerevisiae* (yeast) and *Neurospora crassa* (fungus), the mutation rate per genome per replication are 0.0027 and 0.0030 respectively. For the nematode worm *Caenorhabditis elegans*, the mutation rate per effective genome per sexual generation is 0.036. Secondly, there should be strong interactions among loci (synergistic epistasis), a mutation-fitness relation for which there is only limited evidence. Conversely, there is also the same amount of evidence that mutations show no epistasis (purely additive model) or antagonistic interactions (each additional mutation has a disproportionally *small* effect).

Other explanations

Geodakyan's evolutionary theory of sex

Geodakyan suggested that sexual dimorphism provides a partitioning of a species' phenotypes into at least two functional partitions: a female partition that secures beneficial features of the species and a male partition that emerged in species with more variable and unpredictable environments. The male partition is suggested to be an "experimental" part of the species that allows the species to expand their ecological niche, and to have alternative configurations. This theory underlines the higher variability and higher mortality in males, in comparison to females. This functional partitioning also explains the higher susceptibility to disease in males, in comparison to females and therefore includes the idea of "protection against parasites" as another functionality of male sex. Geodakyan's evolutionary theory of sex was developed in Russia in 1960-80 and was not known to the West till the era of the Internet. Trofimova, who analysed psychological sex differences, hypothesised that the male sex might also provide a "redundancy pruning" function.

Speed of evolution

Ilan Eshel suggested that sex prevents rapid evolution. He suggests that recombination breaks up favourable gene combinations more often than it creates them, and sex is maintained because it ensures selection is longer-term than in asexual populations - so the population is less affected by short-term changes.[85-86] This explanation is not widely accepted, as its assumptions are very restrictive.

It has recently been shown in experiments with *Chlamydomonas* algae that sex can remove the speed limitWikipedia:Please clarify on evolution.

An information theoretic analysis using a simplified but useful model shows that in asexual reproduction, the information gain per generation of a species is limited to 1 bit per generation, while in sexual reproduction, the information gain is bounded by $\sqrt{G}$, where G is the size of the genome in bits.

Libertine bubble theory

The evolution of sex can alternatively be described as a kind of gene exchange that is independent from reproduction. According to the Thierry Lodé's "libertine bubble theory", sex originated from an archaic gene transfer process among prebiotic bubbles. The contact among the pre-biotic bubbles could, through simple food or parasitic reactions, promote the transfer of genetic material from one bubble to another. That interactions between two organisms be in balance appear to be a sufficient condition to make these interactions evolutionarily efficient, i.e. to select bubbles that tolerate these interactions ('libertine' bubbles) through a blind evolutionary process of self-reinforcing gene correlations and compatibility.

The "libertine bubble theory" proposes that meiotic sex evolved in proto-eukaryotes to solve a problem that bacteria did not have, namely a large amount of DNA material, occurring in an archaic step of proto-cell formation and genetic exchanges. So that, rather than providing selective advantages through reproduction, sex could be thought of as a series of separate events which combines step-by-step some very weak benefits of recombination, meiosis, gametogenesis and syngamy. Therefore, current sexual species could be descendants of primitive organisms that practiced more stable exchanges in the long term, while asexual species have emerged, much more recently in evolutionary history, from the conflict of interest resulting from anisogamy.Wikipedia:Please clarify

Origin of sexual reproduction

Life timeline

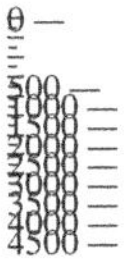

θ –
⁝
500
1000
1500
2000
2500
3000
3500
4000
4500

Axis scale: million years

🖐

Also see: *Human timeline* and *Nature timeline*

Many protists reproduce sexually, as do the multicellular plants, animals, and fungi. In the eukaryotic fossil record, sexual reproduction first appeared by 1.2 billion years ago in the Proterozoic Eon.[206] All sexually reproducing eukaryotic organisms likely derive from a single-celled common ancestor. It is probable that the evolution of sex was an integral part of the evolution of the first eukaryotic cell.[207] There are a few species which have secondarily lost this feature, such as Bdelloidea and some parthenocarpic plants.

Diploidy

Organisms need to replicate their genetic material in an efficient and reliable manner. The necessity to repair genetic damage is one of the leading theories explaining the origin of sexual reproduction. Diploid individuals can repair a damaged section of their DNA via homologous recombination, since there are two copies of the gene in the cell and if one copy is damaged, the other copy is unlikely to be damaged at the same site.

A mutation in a haploid individual, on the other hand, is more likely to become fixed, since any DNA repair mechanism would have no source to recover the original undamaged sequence from. The most primitive form of sex may have been one organism with damaged DNA replicating an undamaged strand from a similar organism in order to repair itself.

Meiosis

If, as evidence indicates, sexual reproduction arose very early in eukaryotic evolution, the essential features of meiosis may have already been present in the prokaryotic ancestors of eukaryotes. In extant organisms, proteins with central functions in meiosis are similar to key proteins in natural transformation in bacteria and DNA transfer in archaea.[208,209] For example, recA recombinase, that catalyses the key functions of DNA homology search and strand exchange in the bacterial sexual process of transformation, has orthologs in eukaryotes that perform similar functions in meiotic recombination (see Wikipedia articles RecA, RAD51 and DMC1).

Natural transformation in bacteria, DNA transfer in archaea, and meiosis in eukaryotic microorganisms are induced by stressful circumstances such as over-crowding, resource depletion, and DNA damaging conditions. This suggests that these sexual processes are adaptations for dealing with stress, particularly stress that causes DNA damage. In bacteria, these stresses induce an altered physiologic state, termed competence, that allows active take-up of DNA from a donor bacterium and the integration of this DNA into the recipient genome (see Natural competence) allowing recombinational repair of the recipients' damaged DNA.

If environmental stresses leading to DNA damage were a persistent challenge to the survival of early microorganisms, then selection would likely have been continuous through the prokaryote to eukaryote transition, and adaptive adjustments would have followed a course in which bacterial transformation or archaeal DNA transfer naturally gave rise to sexual reproduction in eukaryotes.

Virus-like RNA-based origin

Sex might also have been present even earlier, in the hypothesized RNA world that preceded DNA cellular life forms. One proposed origin of sex in the RNA world was based on the type of sexual interaction that is known to occur in extant single-stranded segmented RNA viruses, such as influenza virus, and in extant double-stranded segmented RNA viruses such as reovirus.

Exposure to conditions that cause RNA damage could have led to blockage of replication and death of these early RNA life forms. Sex would have allowed re-assortment of segments between two individuals with damaged RNA, permitting undamaged combinations of RNA segments to come together, thus allowing survival. Such a regeneration phenomenon, known as multiplicity reactivation, occurs in influenza virus and reovirus.

Parasitic DNA elements

Another theory is that sexual reproduction originated from selfish parasitic genetic elements that exchange genetic material (that is: copies of their own genome) for their transmission and propagation. In some organisms, sexual reproduction has been shown to enhance the spread of parasitic genetic elements (e.g.: yeast, filamentous fungi).

Bacterial conjugation is a form of genetic exchange that some sources describe as "sex", but technically is not a form of reproduction, even though it is a form of horizontal gene transfer. However, it does support the "selfish gene" part theory, since the gene itself is propagated through the F-plasmid.

A similar origin of sexual reproduction is proposed to have evolved in ancient haloarchaea as a combination of two independent processes: jumping genes and plasmid swapping.

Partial predation

A third theory is that sex evolved as a form of cannibalism: One primitive organism ate another one, but instead of completely digesting it, some of the "eaten" organism's DNA was incorporated into the DNA of the "eater".

Vaccination-like process

Sex may also be derived from another prokaryotic process. A comprehensive theory called "origin of sex as vaccination" proposes that eukaryan sex-as-syngamy (fusion sex) arose from prokaryan unilateral sex-as-infection, when infected hosts began swapping nuclearised genomes containing coevolved, vertically transmitted symbionts that provided protection against horizontal superinfection by other, more virulent symbionts.

Consequently, sex-as-meiosis (fission sex) would evolve as a host strategy for uncoupling from (and thereby render impotent) the acquired symbiotic/parasitic genes.

Mechanistic origin of sexual reproduction

While theories positing fitness benefits that led to the origin of sex are often problematicWikipedia:Citation needed, several theories addressing the emergence of the mechanisms of sexual reproduction have been proposed.

Viral eukaryogenesis

The viral eukaryogenesis (VE) theory proposes that eukaryotic cells arose from a combination of a lysogenic virus, an archaean, and a bacterium. This model suggests that the nucleus originated when the lysogenic virus incorporated genetic material from the archaean and the bacterium and took over the role of information storage for the amalgam. The archaeal host transferred much of its functional genome to the virus during the evolution of cytoplasm, but retained the function of gene translation and general metabolism. The bacterium transferred most of its functional genome to the virus as it transitioned into a mitochondrion.

For these transformations to lead to the eukaryotic cell cycle, the VE hypothesis specifies a pox-like virus as the lysogenic virus. A pox-like virus is a likely ancestor because of its fundamental similarities with eukaryotic nuclei. These include a double stranded DNA genome, a linear chromosome with short telomeric repeats, a complex membrane bound capsid, the ability to produce capped mRNA, and the ability to export the capped mRNA across the viral membrane into the cytoplasm. The presence of a lysogenic pox-like virus ancestor explains the development of meiotic division, an essential component of sexual reproduction.

Meiotic division in the VE hypothesis arose because of the evolutionary pressures placed on the lysogenic virus as a result of its inability to enter into the lytic cycle. This selective pressure resulted in the development of processes allowing the viruses to spread horizontally throughout the population. The outcome of this selection was cell-to-cell fusion. (This is distinct from the conjugation methods used by bacterial plasmids under evolutionary pressure, with important consequences.) The possibility of this kind of fusion is supported by the presence of fusion proteins in the envelopes of the pox viruses that allow them to fuse with host membranes. These proteins could have been transferred to the cell membrane during viral reproduction, enabling cell-to-cell fusion between the virus host and an uninfected cell. The theory proposes meiosis originated from the fusion between two cells infected with related but different viruses which recognised each other as uninfected. After the fusion of the two cells, incompatibilities between the two viruses result in a meiotic-like cell division.

The two viruses established in the cell would initiate replication in response to signals from the host cell. A mitosis-like cell cycle would proceed until the viral membranes dissolved, at which point linear chromosomes would be bound together with centromeres. The homologous nature of the two viral centromeres would incite the grouping of both sets into tetrads. It is speculated that this grouping may be the origin of crossing over, characteristic of the first division in modern meiosis. The partitioning apparatus of the mitotic-like cell cycle the cells used to replicate independently would then pull each set of chromosomes to one side of the cell, still bound by centromeres. These centromeres would prevent their replication in subsequent division, resulting in four daughter cells with one copy of one of the two original pox-like viruses. The process resulting from combination of two similar pox viruses within the same host closely mimics meiosis.

Neomuran revolution

An alternative theory, proposed by Thomas Cavalier-Smith, was labeled the Neomuran revolution. The designation "Neomuran revolution" refers to the appearances of the common ancestors of eukaryotes and archaea. Cavalier-Smith proposes that the first neomurans emerged 850 million years ago. Other molecular biologists assume that this group appeared much earlier, but Cavalier-Smith dismisses these claims because they are based on the "theoretically and empirically" unsound model of molecular clocks. Cavalier-Smith's theory of the Neomuran revolution has implications for the evolutionary history of the cellular machinery for recombination and sex. It suggests that this machinery evolved in two distinct bouts separated by a long period of stasis; first the appearance of recombination machinery in a bacterial ancestor which was maintained for 3 Gy, Wikipedia:Please clarify until the neomuran revolution when the mechanics were adapted to the presence of nucleosomes. The archaeal products of the revolution maintained recombination machinery that was essentially bacterial, whereas the eukaryotic products broke with this bacterial continuity. They introduced cell fusion and ploidy cycles into cell life histories. Cavalier-Smith argues that both bouts of mechanical evolution were motivated by similar selective forces: the need for accurate DNA replication without loss of viability.

Questions

Some questions biologists have attempted to answer include:

- Why sexual reproduction exists, if in many organisms it has a 50% cost (fitness disadvantage) in relation to asexual reproduction?[210]
- Did mating types (types of gametes, according to their compatibility) arise as a result of anisogamy (gamete dimorphism), or did mating types evolve before anisogamy?[211,212]
- Why do most sexual organisms use a binary mating system? Why do some organisms have gamete dimorphism?

Further reading

- Bell, Graham (1982). *The masterpiece of nature: the evolution and genetics of sexuality*. Berkeley: University of California Press. ISBN 0-520-04583-1.
- Bernstein, Carol; Harris Bernstein (1991). *Aging, sex, and DNA repair*. Boston: Academic Press. ISBN 0-12-092860-4.
- Hurst, L.D.; J.R. Peck (1996). "Recent advances in the understanding of the evolution and maintenance of sex". *Trends in Ecology and Evolution*. **11** (2): 46–52. doi: 10.1016/0169-5347(96)81041-X[213]. PMID 21237760[214].
- Levin, Bruce R.; Richard E. Michod (1988). *The Evolution of sex: an examination of current ideas*. Sunderland, Mass: Sinauer Associates. ISBN 0-87893-459-6.
- Maynard Smith, John (1978). *The evolution of sex*. Cambridge, UK: Cambridge University Press. ISBN 0-521-21887-X.
- Michod, Richard E. (1995). *Eros and evolution: a natural philosophy of sex*. Reading, Mass: Addison-Wesley Pub. Co. ISBN 0-201-40754-X.

- "Scientists put sex origin mystery to bed, Wild strawberry research provides evidence on when gender emerges"[215]. MSNBC. Retrieved 25 November 2008.
- Ridley, Mark (1993). *Evolution*. Oxford: Blackwell Scientific. ISBN 0-632-03481-5.
- Ridley, Mark (2000). *Mendel's demon: gene justice and the complexity of life*. London: Weidenfeld & Nicolson. ISBN 0-297-64634-6.
- Ridley, Matt (1995). *The Red Queen: sex and the evolution of human nature*. New York: Penguin Books. ISBN 0-14-024548-0.
- Szathmáry, Eörs; John Maynard Smith (1995). *The Major Transitions in Evolution*. Oxford: W.H. Freeman Spektrum. ISBN 0-7167-4525-9.
- Taylor, Timothy (1996). *The prehistory of sex: four million years of human sexual culture*. New York: Bantam Books. ISBN 0-553-09694-X.
- Williams, George (1975). *Sex and evolution*. Princeton, N.J: Princeton University Press. ISBN 0-691-08147-6.

External links

- Why Sex is Good[216]
- An essay summarising the different theories[217], dating from around 2001
- http://www.evolocus.com/Textbooks/Geodakian2012.pdf

Sexual reproduction

Sexual reproduction is a form of reproduction where two gametes fuse together. Each gamete contains half the number of chromosomes of normal cells. They are created by a specialized type of cell division, which only occurs in eukaryotic cells, known as meiosis. The two gametes fuse during fertilization to produce DNA replication and the creation of a single-celled zygote which includes genetic material from both gametes. In a process called genetic recombination, genetic material (DNA) joins up so that homologous chromosome sequences are aligned with each other, and this is followed by exchange of genetic information. Two rounds of cell division then produce four daughter cells with half the number of chromosomes from each original parent cell, and the same number of chromosomes as both parents. For instance, in human reproduction each human cell contains 46 chromosomes in 23 pairs. Meiosis in the parents' gonads produce gamete cells which only contain 23 chromosomes each. When the gametes are combined via sexual intercourse to form a fertilized egg, the resulting child will have 23 chromosomes from each parent genetically recombined into 23 chromosome pairs or 46 total.

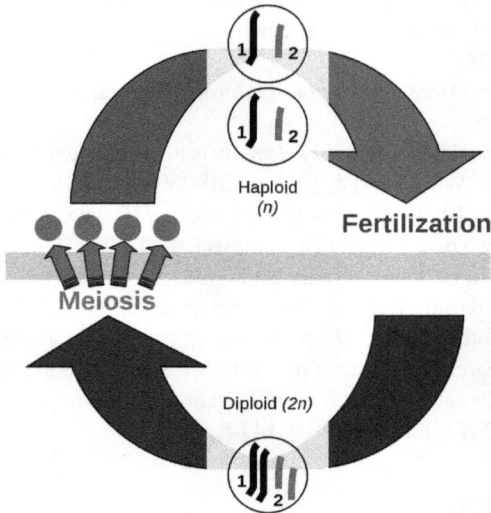

Figure 103: *In the first stage of sexual reproduction, "meiosis", the number of chromosomes is reduced from a diploid number (2n) to a haploid number (n). During "fertilization", haploid gametes come together to form a diploid zygote and the original number of chromosomes is restored.*

Cell division mitosis then initiates the development of a new individual organism in multicellular organisms, including animals and plants, for the vast majority of whom this is the primary method of reproduction.

The evolution of sexual reproduction is a major puzzle because asexual reproduction should be able to outcompete it as every young organism created can bear its own young. This implies that an asexual population has an intrinsic capacity to grow more rapidly with each generation.[218] This 50% cost is a fitness disadvantage of sexual reproduction.[219] The two-fold cost of sex includes this cost and the fact that any organism can only pass on 50% of its own genes to its offspring. One definite advantage of sexual reproduction is that it prevents the accumulation of genetic mutations.

Sexual selection is a mode of natural selection in which some individuals outreproduce others of a population because they are better at securing mates for sexual reproduction. It has been described as "a powerful evolutionary force that does not exist in asexual populations."

Prokaryotes, whose initial cell has additional or transformed genetic material, reproduce through asexual reproduction but may, in lateral gene transfer, display processes such as bacterial conjugation, transformation and transduction, which are similar to sexual reproduction although they do not lead to reproduction.

Evolution

The first fossilized evidence of sexual reproduction in eukaryotes is from the Stenian period, about 1 to 1.2 billion years ago.

Biologists studying evolution propose several explanations for why sexual reproduction developed and why it is maintained. These reasons include reducing the likelihood of the accumulation of deleterious mutations, increasing rate of adaptation to changing environments, dealing with competition, and masking deleterious mutations. All of these ideas about why sexual reproduction has been maintained are generally supported, but ultimately the size of the population determines if sexual reproduction is entirely beneficial. Larger populations appear to respond more quickly to benefits obtained through sexual reproduction than do smaller population sizes.

Maintenance of sexual reproduction has been explained by theories that work at several levels of selection, though some of these models remain controversial.Wikipedia:Citation needed However, newer models presented in recent years suggest a basic advantage for sexual reproduction in slowly reproducing complex organisms.

Sexual reproduction allows these species to exhibit characteristics that depend on the specific environment that they inhabit, and the particular survival strategies that they employ.

Sexual selection

In order to sexually reproduce, both males and females need to find a mate. Generally in animals mate choice is made by females while males compete to be chosen. This can lead organisms to extreme efforts in order to reproduce, such as combat and display, or produce extreme features caused by a positive feedback known as a Fisherian runaway. Thus sexual reproduction, as a form of natural selection, has an effect on evolution. Sexual dimorphism is where the basic phenotypic traits vary between males and females of the same species. Dimorphism is found in both sex organs and in secondary sex characteristics, body size, physical strength and morphology, biological ornamentation, behavior and other bodily traits. However, sexual selection is only implied over an extended period of time leading to sexual dimorphism.[220]

Figure 104: *Australian emperor laying egg, guarded by the male*

Sex ratio

Apart from some eusocial wasps, organisms which reproduce sexually have a 1:1 sex ratio of male and female births. The English statistician and biologist Ronald Fisher outlined why this is so in what has come to be known as Fisher's principle. This essentially says the following:

1. Suppose male births are less common than female.
2. A newborn male then has better mating prospects than a newborn female, and therefore can expect to have more offspring.
3. Therefore parents genetically disposed to produce males tend to have more than average numbers of grandchildren born to them.
4. Therefore the genes for male-producing tendencies spread, and male births become more common.
5. As the 1:1 sex ratio is approached, the advantage associated with producing males dies away.
6. The same reasoning holds if females are substituted for males throughout. Therefore 1:1 is the equilibrium ratio.

Animals

Insects

Insect species make up more than two-thirds of all extant animal species. Most insect species reproduce sexually, though some species are facultatively parthenogenetic. Many insects species have sexual dimorphism, while in others the sexes look nearly identical. Typically they have two sexes with males producing spermatozoa and females ova. The ova develop into eggs that have a covering called the chorion, which forms before internal fertilization. Insects have very diverse mating and reproductive strategies most often resulting in the male depositing spermatophore within the female, which she stores until she is ready for egg fertilization. After fertilization, and the formation of a zygote, and varying degrees of development, in many species the eggs are deposited outside the female; while in others, they develop further within the female and are born live.

Mammals

There are three extant kinds of mammals: monotremes, placentals and marsupials, all with internal fertilization. In placental mammals, offspring are born as juveniles: complete animals with the sex organs present although not reproductively functional. After several months or years, depending on the species, the sex organs develop further to maturity and the animal becomes sexually mature. Most female mammals are only fertile during certain periods during their estrous cycle, at which point they are ready to mate. Individual male and female mammals meet and carry out copulation.Wikipedia:Citation needed For most mammals, males and females exchange sexual partners throughout their adult lives.[221]

Fish

The vast majority of fish species lay eggs that are then fertilized by the male,[222] some species lay their eggs on a substrate like a rock or on plants, while others scatter their eggs and the eggs are fertilized as they drift or sink in the water column.

Some fish species use internal fertilization and then disperse the developing eggs or give birth to live offspring. Fish that have live-bearing offspring include the guppy and mollies or *Poecilia*. Fishes that give birth to live young can be ovoviviparous, where the eggs are fertilized within the female and the eggs simply hatch within the female body, or in seahorses, the male carries the developing young within a pouch, and gives birth to live young. Fishes can also be viviparous, where the female supplies nourishment to the internally growing offspring. Some fish are hermaphrodites, where a single fish is both male and female and can produce eggs and sperm. In hermaphroditic fish, some are male and female at the same time while in other fish they are serially hermaphroditic; starting as one sex and changing to the other. In at least one hermaphroditic species, self-fertilization occurs when the eggs and sperm are released together. Internal self-fertilization may occur in some other

Figure 105: *Common house geckos (Hemidactylus frenatus) mating*

species. One fish species does not reproduce by sexual reproduction but uses sex to produce offspring; *Poecilia formosa* is a unisex species that uses a form of parthenogenesis called gynogenesis, where unfertilized eggs develop into embryos that produce female offspring. *Poecilia formosa* mate with males of other fish species that use internal fertilization, the sperm does not fertilize the eggs but stimulates the growth of the eggs which develops into embryos.

Reptiles

Plants

Animals typically produce gametes directly by meiosis. Male gametes are called sperm, and female gametes are called eggs or ova. In animals, fertilization follows immediately after meiosis. Plants on the other hand have mitosis occurring in spores, which are produced by meiosis. The spores germinate into the gametophyte phase. The gametophytes of different groups of plants vary in size; angiosperms have as few as three cells in pollen, and mosses and other so called primitive plants may have several million cells. Plants have an alternation of generations where the sporophyte phase is succeeded by the gametophyte phase. The sporophyte phase produces spores within the sporangium by meiosis.

Figure 106: *Flowers are the sexual organs of flowering plants.*

Flowering plants

Flowering plants are the dominant plant form on land and they reproduce either sexually or asexually. Often their most distinguishing feature is their reproductive organs, commonly called flowers. The anther produces pollen grains which contain the male gametophytes (sperm). For pollination to occur, pollen grains must attach to the stigma of the female reproductive structure (carpel), where the female gametophytes (ovules) are located inside the ovary. After the pollen tube grows through the carpel's style, the sex cell nuclei from the pollen grain migrate into the ovule to fertilize the egg cell and endosperm nuclei within the female gametophyte in a process termed double fertilization. The resulting zygote develops into an embryo, while the triploid endosperm (one sperm cell plus two female cells) and female tissues of the ovule give rise to the surrounding tissues in the developing seed. The ovary, which produced the female gametophyte(s), then grows into a fruit, which surrounds the seed(s). Plants may either self-pollinate or cross-pollinate.

Nonflowering plants like ferns, moss and liverworts use other means of sexual reproduction.

In 2013, flowers dating from the Cretaceous (100 million years before present) were found encased in amber, the oldest evidence of sexual reproduction in a flowering plant. Microscopic images showed tubes growing out of pollen and penetrating the flower's stigma. The pollen was sticky, suggesting it was carried by insects.

Ferns

Ferns mostly produce large diploid sporophytes with rhizomes, roots and leaves; and on fertile leaves called sporangium, spores are produced. The spores are released and germinate to produce short, thin gametophytes that are typically heart shaped, small and green in color. The gametophytes or thallus, produce both motile sperm in the antheridia and egg cells in separate archegonia. After rains or when dew deposits a film of water, the motile sperm are splashed away from the antheridia, which are normally produced on the top side of the thallus, and swim in the film of water to the archegonia where they fertilize the egg. To promote out crossing or cross fertilization the sperm are released before the eggs are receptive of the sperm, making it more likely that the sperm will fertilize the eggs of different thallus. A zygote is formed after fertilization, which grows into a new sporophytic plant. The condition of having separate sporephyte and gametophyte plants is called alternation of generations. Other plants with similar reproductive means include the *Psilotum, Lycopodium, Selaginella* and *Equisetum*.

Bryophytes

The bryophytes, which include liverworts, hornworts and mosses, reproduce both sexually and vegetatively. They are small plants found growing in moist locations and like ferns, have motile sperm with flagella and need water to facilitate sexual reproduction. These plants start as a haploid spore that grows into the dominate form, which is a multicellular haploid body with leaf-like structures that photosynthesize. Haploid gametes are produced in antherida and archegonia by mitosis. The sperm released from the antherida respond to chemicals released by ripe archegonia and swim to them in a film of water and fertilize the egg cells thus producing a zygote. The zygote divides by mitotic division and grows into a sporophyte that is diploid. The multicellular diploid sporophyte produces structures called spore capsules, which are connected by seta to the archegonia. The spore capsules produce spores by meiosis, when ripe the capsules burst open and the spores are released. Bryophytes show considerable variation in their breeding structures and the above is a basic outline. Also in some species each plant is one sex while other species produce both sexes on the same plant.

Figure 107: *Puffballs emitting spores*

Fungi

Fungi are classified by the methods of sexual reproduction they employ. The outcome of sexual reproduction most often is the production of resting spores that are used to survive inclement times and to spread. There are typically three phases in the sexual reproduction of fungi: plasmogamy, karyogamy and meiosis. The cytoplasm of two parent cells fuse during plasmogamy and the nuclei fuse during karyogamy. New haploid gametes are formed during meiosis and develop into spores.

Bacteria and archaea

Three distinct processes in prokaryotes are regarded as similar to eukaryotic sex: bacterial transformation, which involves the incorporation of foreign DNA into the bacterial chromosome; bacterial conjugation, which is a transfer of plasmid DNA between bacteria, but the plasmids are rarely incorporated into the bacterial chromosome; and gene transfer and genetic exchange in archaea.

Bacterial transformation involves the recombination of genetic material and its function is mainly associated with DNA repair. Bacterial transformation is a complex process encoded by numerous bacterial genes, and is a bacterial adaptation for DNA transfer. This process occurs naturally in at least 40 bacterial species. For a bacterium to bind, take up, and recombine exogenous DNA into its chromosome, it must enter a special physiological state referred to as competence (see Natural competence). Sexual reproduction in early single-celled eukaryotes may have evolved from bacterial transformation,[223] or from a similar process in archaea (see below).

On the other hand, bacterial conjugation is a type of direct transfer of DNA between two bacteria through an external appendage called the conjugation pilus. Bacterial conjugation is controlled by plasmid genes that are adapted for spreading copies of the plasmid between bacteria. The infrequent integration of a plasmid into a host bacterial chromosome, and the subsequent transfer of a part of the host chromosome to another cell do not appear to be bacterial adaptations.

Exposure of hyperthermophilic archaeal Sulfolobus species to DNA damaging conditions induces cellular aggregation accompanied by high frequency genetic marker exchange. Ajon et al. hypothesized that this cellular aggregation enhances species-specific DNA repair by homologous recombination. DNA transfer in Sulfolobus may be an early form of sexual interaction similar to the more well-studied bacterial transformation systems that also involve species-specific DNA transfer leading to homologous recombinational repair of DNA damage.

Further reading

- Pang, K. "Certificate Biology: New Mastering Basic Concepts", Hong Kong, 2004
- Journal of Biology of Reproduction[224], accessed in August 2005.
- "Sperm Use Heat Sensors To Find The Egg; Weizmann Institute Research Contributes To Understanding Of Human Fertilization"[225], *Science Daily*, 3 February 2003
- Michod, RE; Levin, BE, eds. (1987). *The Evolution of sex: An examination of current ideas*. Sunderland, Massachusetts: Sinauer Associates. ISBN 978-0878934584.
- Michod, RE (1994). *Eros and Evolution: A Natural Philosophy of Sex*. Perseus Books. ISBN 978-0201407549.

External links

- Khan Academy, video lecture[226]

Multicellular organism

Multicellular organism
Temporal range: Mesoproterozoic–present
Had'nArchean   Proterozoic   Pha.

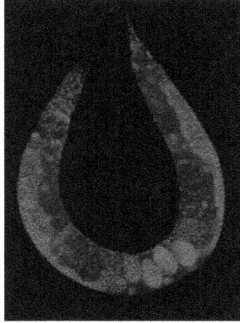

In this image, a wild-type *Caenorhabditis elegans* is stained to highlight the nuclei of its cells.

Scientific classification

Multicellular organisms are organisms that consist of more than one cell, in contrast to unicellular organisms.

All species of animals, land plants and most fungi are multicellular, as are many algae, whereas a few organisms are partially uni- and partially multicellular, like slime molds and social amoebae such as the genus *Dictyostelium*.

Multicellular organisms arise in various ways, for example by cell division or by aggregation of many single cells. Colonial organisms are the result of many identical individuals joining together to form a colony. However, it can often be hard to separate colonial protists from true multicellular organisms, because the two concepts are not distinct; colonial protists have been dubbed "pluricellular" rather than "multicellular".

Evolutionary history

Occurrence

Multicellularity has evolved independently at least 46 times in eukaryotes, and also in some prokaryotes, like cyanobacteria, myxobacteria, actinomycetes, *Magnetoglobus multicellularis* or *Methanosarcina*. However, complex multicellular organisms evolved only in six eukaryotic groups: animals, fungi, brown algae, red algae, green algae, and land plants.[227] It evolved repeatedly for Chloroplastida (green algae and land plants), once or twice for animals, once for brown algae, three times in the fungi (chytrids, ascomycetes and basidiomycetes) and perhaps several times for slime molds and red algae. The first evidence of multicellularity is from cyanobacteria-like organisms that lived 3–3.5 billion years ago. To reproduce, true multicellular organisms must solve the problem of regenerating a whole organism from germ cells (i.e. sperm and egg cells), an issue that is studied in evolutionary developmental biology. Animals have evolved a considerable diversity of cell types in a multicellular body (100–150 different cell types), compared with 10–20 in plants and fungi.[228]

Loss of multicellularity

Loss of multicellularity occurred in some groups.[229] Fungi are predominantly multicellular, though early diverging lineages are largely unicellular (e.g. Microsporidia) and there have been numerous reversions to unicellularity across fungi (e.g. *Saccharomycotina*, *Cryptococcus*, and other yeasts).[230] It may also have occurred in some red algae (e.g. *Porphyridium*), but it is possible that they are primitively unicellular.[231] Loss of multicellularity is also considered probable in some green algae (e.g. *Chlorella vulgaris* and some Ulvophyceae).[232] In other groups, generally parasites, a reduction of multicellularity occurred, in number or types of cells (e.g. the myxozoans, multicellular organisms, earlier thought to be unicellular, are probably extremely reduced cnidarians).

Cancer

Multicellular organisms, especially long-living animals, face the challenge of cancer, which occurs when cells fail to regulate their growth within the normal program of development. Changes in tissue morphology can be observed during this process. Cancer in animals (metazoans) has often been described as a loss of multicellularity. There is a discussion about the possibility of existence of cancer in other multicellular organisms[233] or even in protozoa.[234] For example, plant galls have been characterized as tumors but some authors argue that plants do not develop cancer.

Figure 108: *Tetrabaena socialis consists of four cells.*

Separation of somatic and germ cells

In some multicellular groups, which are called Weismannists, a separation between a sterile somatic cell line and a germ cell line evolved. However, Weismannist development is relatively rare (e.g. vertebrates, arthropods, *Volvox*), as great part of species have the capacity for somatic embryogenesis (e.g. land plants, most algae, many invertebrates).[235,236]

Hypotheses for origin

One hypothesis for the origin of multicellularity is that a group of function-specific cells aggregated into a slug-like mass called a grex, which moved as a multicellular unit. This is essentially what slime molds do. Another hypothesis is that a primitive cell underwent nucleus division, thereby becoming a coenocyte. A membrane would then form around each nucleus (and the cellular space and organelles occupied in the space), thereby resulting in a group of connected cells in one organism (this mechanism is observable in Drosophila). A third hypothesis is that as a unicellular organism divided, the daughter cells failed to separate, resulting in a conglomeration of identical cells in one organism, which could later develop specialized tissues. This is what plant and animal embryos do as well as colonial choanoflagellates.[237]

Because the first multicellular organisms were simple, soft organisms lacking bone, shell or other hard body parts, they are not well preserved in the fossil record.[238] One exception may be the demosponge, which may have left a chemical signature in ancient rocks. The earliest fossils of multicellular organisms include the contested Grypania spiralis and the fossils of the black shales of the Palaeoproterozoic Francevillian Group Fossil B Formation in Gabon (Gabonionta). The Doushantuo Formation has yielded 600 million year old microfossils with evidence of multicellular traits.

Until recently, phylogenetic reconstruction has been through anatomical (particularly embryological) similarities. This is inexact, as living multicellular organisms such as animals and plants are more than 500 million years removed from their single-cell ancestors. Such a passage of time allows both divergent and convergent evolution time to mimic similarities and accumulate differences between groups of modern and extinct ancestral species. Modern phylogenetics uses sophisticated techniques such as alloenzymes, satellite DNA and other molecular markers to describe traits that are shared between distantly related lineages.Wikipedia:Citation needed

The evolution of multicellularity could have occurred in a number of different ways, some of which are described below:

The symbiotic theory

This theory suggests that the first multicellular organisms occurred from symbiosis (cooperation) of different species of single-cell organisms, each with different roles. Over time these organisms would become so dependent on each other they would not be able to survive independently, eventually leading to the incorporation of their genomes into one multicellular organism. Each respective organism would become a separate lineage of differentiated cells within the newly created species.

This kind of severely co-dependent symbiosis can be seen frequently, such as in the relationship between clown fish and Riterri sea anemones. In these cases, it is extremely doubtful whether either species would survive very long if the other became extinct. However, the problem with this theory is that it is still not known how each organism's DNA could be incorporated into one single genome to constitute them as a single species. Although such symbiosis is theorized to have occurred (e.g. mitochondria and chloroplasts in animal and plant cells—endosymbiosis), it has happened only extremely rarely and, even then, the genomes of the endosymbionts have retained an element of distinction, separately replicating their DNA during mitosis of the host species. For instance, the two or three symbiotic organisms forming the composite lichen, although dependent on each other for survival, have to separately reproduce and then re-form to create one individual organism once more.

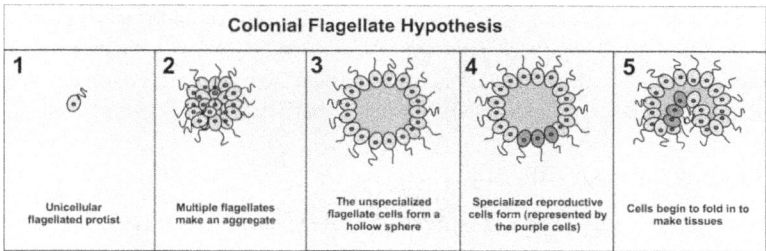

Colonial Flagellate Hypothesis				
1	**2**	**3**	**4**	**5**
Unicellular flagellated protist	Multiple flagellates make an aggregate	The unspecialized flagellate cells form a hollow sphere	Specialized reproductive cells form (represented by the purple cells)	Cells begin to fold in to make tissues

The cellularization (syncytial) theory

This theory states that a single unicellular organism, with multiple nuclei, could have developed internal membrane partitions around each of its nuclei. Many protists such as the ciliates or slime molds can have several nuclei, lending support to this hypothesis. However, the simple presence of multiple nuclei is not enough to support the theory. Multiple nuclei of ciliates are dissimilar and have clear differentiated functions. The macronucleus serves the organism's needs, whereas the micronucleus is used for sexual reproduction with exchange of genetic material. Slime molds syncitia form from individual amoeboid cells, like syncitial tissues of some multicellular organisms, not the other way round. To be deemed valid, this theory needs a demonstrable example and mechanism of generation of a multicellular organism from a pre-existing syncytium.

The colonial theory

The Colonial Theory of Haeckel, 1874, proposes that the symbiosis of many organisms of the same species (unlike the symbiotic theory, which suggests the symbiosis of different species) led to a multicellular organism. At least some, it is presumed land-evolved, multicellularity occurs by cells separating and then rejoining (e.g. cellular slime molds) whereas for the majority of multicellular types (those that evolved within aquatic environments), multicellularity occurs as a consequence of cells failing to separate following division. The mechanism of this latter colony formation can be as simple as incomplete cytokinesis, though multicellularity is also typically considered to involve cellular differentiation.

The advantage of the Colonial Theory hypothesis is that it has been seen to occur independently in 16 different protoctistan phyla. For instance, during food shortages the amoeba Dictyostelium groups together in a colony that moves as one to a new location. Some of these amoeba then slightly differentiate from each other. Other examples of colonial organisation in protista are Volvocaceae, such as Eudorina and Volvox, the latter of which consists of up to 500–50,000 cells (depending on the species), only a fraction of which reproduce.[239] For example, in one species 25–35 cells reproduce, 8 asexually and around 15–25 sexually. However, it can often be hard to separate colonial protists from true multicellular organisms, as the two concepts are not distinct; colonial protists have been dubbed "pluricellular" rather than "multicellular".

The Synzoospore theory

Some authors suggest that the origin of multicellularity, at least in Metazoa, occurred due to a transition from temporal to spatial cell differentiation, rather than through a gradual evolution of cell differentiation, as affirmed in Haeckel's Gastraea theory.[240]

GK-PID

About 800 million years ago, a minor genetic change in a single molecule called guanylate kinase protein-interaction domain (GK-PID) may have allowed organisms to go from a single cell organism to one of many cells.

The role of viruses

Genes borrowed from viruses have recently been identified as playing a crucial role in the differentiation of multicellular tissues and organs and even in sexual reproduction, in the fusion of egg cell and sperm. Such fused cells are also involved in metazoan membranes such as those that prevent chemicals crossing the placenta and the brain body separation. Two viral components have been identified. The first is syncytin, which came from a virus. The second identified in 2007 is called EFF1, which helps form the skin of *Caenorhabditis elegans*, part of a whole family of FF proteins. Felix Rey, of the Pasteur Institute in Paris has constructed the 3D structure of the EFF1 protein[241] and shown it does the work of linking one cell to another, in viral infections. The fact that all known cell fusion molecules are viral in origin suggests that they have been vitally important to the inter-cellular communication systems that enabled multicellularity. Without the ability of cellular fusion, colonies could have formed, but anything even as complex as a sponge would not have been possible.[242]

Advantages

Multicellularity allows an organism to exceed the size limits normally imposed by diffusion: single cells with increased size have a decreased surface-to-volume ratio and have difficulty absorbing sufficient nutrients and transporting them throughout the cell. Multicellular organisms thus have the competitive advantages of an increase in size without its limitations. They can have longer lifespans as they can continue living when individual cells die. Multicellularity also permits increasing complexity by allowing differentiation of cell types within one organism.

External links

- Tree of Life Eukaryotes[243]

Emergence of animals

Ediacaran biota

<indicator name="featured-star"> ⭐ </indicator>

Part of a series on
The Cambrian explosion
• v • t • e[244]

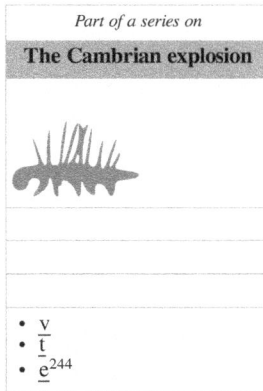

The **Ediacaran** (/ˌiːdiˈækərən/; formerly **Vendian**) **biota** consisted of enigmatic tubular and frond-shaped, mostly sessile organisms that lived during the Ediacaran Period (ca. 635–542 Mya). Trace fossils of these organisms have been found worldwide, and represent the earliest known complex multicellular organisms.[245] The Ediacaran biota may have radiated in a proposed event called the Avalon explosion, 575[246] million years ago,[247] after the Earth had thawed from the Cryogenian period's extensive glaciation. The biota largely disappeared with the rapid increase in biodiversity known as the Cambrian explosion. Most of the currently existing body plans of animals first appeared in the fossil record of the Cambrian rather than the Ediacaran. For macroorganisms, the Cambrian biota appears to have completely replaced the organisms that dominated the Ediacaran fossil record, although relationships are still a matter of debate.

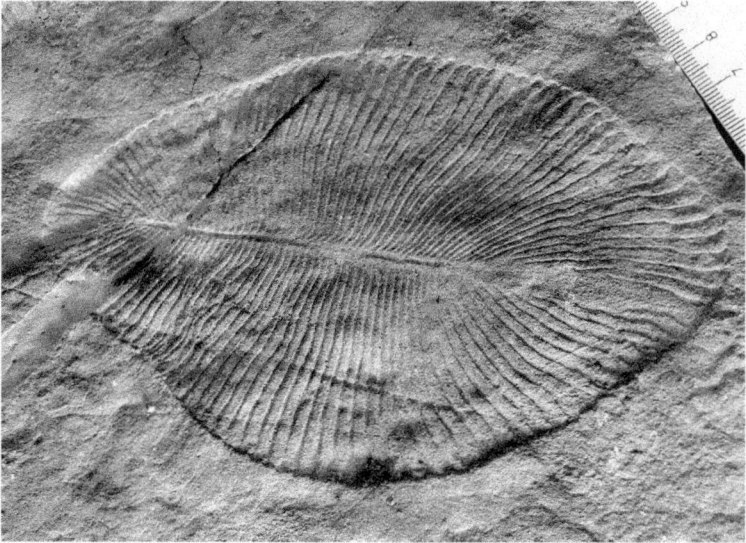

Figure 109: *Dickinsonia costata, an iconic Ediacaran organism, displays the characteristic quilted appearance of Ediacaran enigmata*

The organisms of the Ediacaran Period first appeared around 600[248] million years ago and flourished until the cusp of the Cambrian 542[249] million years ago, when the characteristic communities of fossils vanished. The earliest reasonably diverse Ediacaran community was discovered in 1995 in Sonora, Mexico, and is approximately 600 million years in age, pre-dating the Gaskiers glaciation of about 580 million years ago. While rare fossils that may represent survivors have been found as late as the Middle Cambrian (510 to 500 million years ago), the earlier fossil communities disappear from the record at the end of the Ediacaran leaving only curious fragments of once-thriving ecosystems. Multiple hypotheses exist to explain the disappearance of this biota, including preservation bias, a changing environment, the advent of predators and competition from other life-forms.

Determining where Ediacaran organisms fit in the tree of life has proven challenging; it is not even established that they were animals, with suggestions that they were lichens (fungus-alga symbionts), algae, protists known as foraminifera, fungi or microbial colonies, or hypothetical intermediates between plants and animals. The morphology and habit of some taxa (e.g. *Funisia dorothea*) suggest relationships to Porifera or Cnidaria. *Kimberella* may show a similarity to molluscs, and other organisms have been thought to possess bilateral symmetry, although this is controversial. Most macroscopic fossils are morphologically distinct from later life-forms: they resemble discs, tubes, mud-filled bags or quilted mattresses. Due to the difficulty of deducing evolutionary relationships among these organisms, some palaeontologists have suggested that these represent completely extinct lineages that do not resemble any living organism. One palaeontologist proposed a separate kingdom level category **Vendozoa** (now renamed **Vendobionta**) in the Linnaean hierarchy for the Ediacaran biota. If these enigmatic organisms left no descendants, their strange forms might be seen as a "failed experiment" in multicellular life, with later multicellular life evolving independently from unrelated single-celled organisms.

The Ediacara biota in context

Neoproterozoic

(last era of the Precambrian)

Palaeozoic

(first era of the Phanerozoic)

Axis scale: millions of years ago.

References: Waggoner 1998, Hofmann 1990

The concept of "Ediacaran Biota" is somewhat artificial as it can not be defined geographically, stratigraphically, taphonomically, or biologically.

History

Life timeline

θ —
ⁱⁱ
500
1000
1500
2000
2500
3000
3500
4000
4500

Figure 110: *Palaeontologist Guy Narbonne examining Ediacaran fossils in Newfoundland*

Axis scale: million years

Also see: *Human timeline* and *Nature timeline*

The first Ediacaran fossils discovered were the disc-shaped *Aspidella terranovica* in 1868. Their discoverer, Scottish geologist Alexander Murray, found them useful aids for correlating the age of rocks around Newfoundland. However, since they lay below the "Primordial Strata" of the Cambrian that was then thought to contain the very first signs of animal life, a proposal four years after their discovery by Elkanah Billings that these simple forms represented fauna was dismissed by his peers. Instead, they were interpreted as gas escape structures or inorganic concretions. No similar structures elsewhere in the world were then known and the one-sided debate soon fell into obscurity. In 1933, Georg Gürich discovered specimens in Namibia but the firm belief that complex life originated in the Cambrian led to them being assigned to the Cambrian Period and no link to *Aspidella* was made. In 1946, Reg Sprigg noticed "jellyfishes" in the Ediacara Hills of Australia's Flinders Ranges but these rocks were believed to be Early Cambrian so, while the discovery sparked some interest, little serious attention was garnered.

It was not until the British discovery of the iconic *Charnia* in 1957 that the pre-Cambrian was seriously considered as containing life. This frond-shaped fossil was found in England's Charnwood Forest, and due to the detailed geological mapping of the British Geological Survey there was no doubt these fossils sat in Precambrian rocks. Palaeontologist Martin Glaessner finally, in

1959, made the connection between this and the earlier finds and with a combination of improved dating of existing specimens and an injection of vigour into the search many more instances were recognised.

All specimens discovered until 1967 were in coarse-grained sandstone that prevented preservation of fine details, making interpretation difficult. S.B. Misra's discovery of fossiliferous ash-beds at the Mistaken Point assemblage in Newfoundland changed all this as the delicate detail preserved by the fine ash allowed the description of features that were previously undiscernible.

Poor communication, combined with the difficulty in correlating globally distinct formations, led to a plethora of different names for the biota. In 1960 the French name "Ediacarien" – after the Ediacara Hills – was added to the competing terms "Sinian" and "Vendian" for terminal-Precambrian rocks, and these names were also applied to the life-forms. "Ediacaran" and "Ediacarian" were subsequently applied to the epoch or period of geological time and its corresponding rocks. In March 2004, the International Union of Geological Sciences ended the inconsistency by formally naming the terminal period of the Neoproterozoic after the Australian locality.[250]

The term "Ediacaran biota" and similar ("Ediacara"/"Ediacaran"/"Ediacarian"/"Vendian", "fauna"/"biota") has, at various times, been used in a geographic, stratigraphic, taphonomic, or biological sense, with the latter the most common in modern literature.[251]

Preservation

Microbial mats

Microbial mats are areas of sediment stabilised by the presence of colonies of microbes that secrete sticky fluids or otherwise bind the sediment particles. They appear to migrate upwards when covered by a thin layer of sediment but this is an illusion caused by the colony's growth; individuals do not, themselves, move. If too thick a layer of sediment is deposited before they can grow or reproduce through it, parts of the colony will die leaving behind fossils with a characteristically wrinkled ("elephant skin") and tubercular texture.

Some Ediacaran strata with the texture characteristics of microbial mats contain fossils, and Ediacaran fossils are almost always found in beds that contain these microbial mats. Although microbial mats were once widespread, the evolution of grazing organisms in the Cambrian vastly reduced their numbers. These communities are now limited to inhospitable refugia, such as the stromatolites found in Hamelin Pool Marine Nature Reserve in Shark Bay, Western Australia where the salt levels can be twice those of the surrounding sea.

Figure 111: *Modern cyanobacterial-algal mat, salty lake on the White Sea seaside*

Figure 112: *The fossil Charniodiscus is barely distinguishable from the "elephant skin" texture on this cast.*

Fossilization

The preservation of these fossils is one of their great fascinations to science. As soft-bodied organisms, they would normally not fossilize and, unlike later soft-bodied fossil biota such as the Burgess Shale or Solnhofen Limestone, the Ediacaran biota is not found in a restricted environment subject to unusual local conditions: they were a global phenomenon. The processes that were operating must have been systemic and worldwide. There was something very different about the Ediacaran Period that permitted these delicate creatures to be left behind and it is thought the fossils were preserved by virtue of rapid covering by ash or sand, trapping them against the mud or microbial mats on which they lived. Their preservation was possibly enhanced by the high concentration of silica in the oceans before silica-secreting organisms such as sponges and diatoms became prevalent. Ash beds provide more detail and can readily be dated to the nearest million years or better using radiometric dating. However, it is more common to find Ediacaran fossils under sandy beds deposited by storms or high-energy bottom-scraping ocean currents known as turbidites. Soft-bodied organisms today rarely fossilize during such events, but the presence of widespread microbial mats probably aided preservation by stabilising their impressions in the sediment below.

Scale of preservation

The rate of cementation of the overlying substrate relative to the rate of decomposition of the organism determines whether the top or bottom surface of an organism is preserved. Most disc-shaped fossils decomposed before the overlying sediment was cemented, whereupon ash or sand slumped in to fill the void, leaving a cast of the organism's underside.

Conversely, quilted fossils tended to decompose *after* the cementation of the overlying sediment; hence their upper surfaces are preserved. Their more resistant nature is reflected in the fact that, in rare occasions, quilted fossils are found *within* storm beds as the high-energy sedimentation did not destroy them as it would have the less-resistant discs. Further, in some cases, the bacterial precipitation of minerals formed a "death mask", ultimately leaving a positive, cast-like impression of the organism.

Morphology

Forms of Ediacaran fossil	
The earliest discovered potential embryo, preserved within an acantho-morphic acritarch. The term 'acritarch' describes a range of unclassi-fied cell-like fossils.	
Tateana inflata (= *'Cyclomedusa' radiata*) is the attachment disk of an unknown organism.	
A cast of the quilted *Charnia*, the first accepted complex Precambrian organism. *Charnia* was once interpreted as a relative of the sea pens.	
Spriggina was originally interpreted as annelid or arthropod. How-ever, lack of known limbs, and glide reflected isomers instead of true segments, rejects any such classification despite some superficial re-semblance.	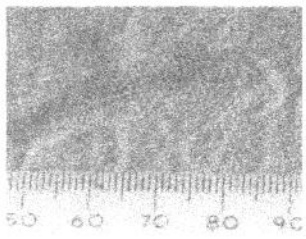
Late Ediacaran *Archaeonassa*-type trace fossils are commonly pre-served on the top surfaces of sandstone strata.	
Epibaion waggoneris, chain of trace platforms and the imprint of the body of *Yorgia waggoneri* (right), which created these traces on microbial mat.	

The Ediacaran biota exhibited a vast range of morphological characteristics. Size ranged from millimetres to metres; complexity from "blob-like" to intri-cate; rigidity from sturdy and resistant to jelly-soft. Almost all forms of sym-metry were present. These organisms differed from earlier fossils by display-ing an organised, differentiated multicellular construction and centimetre-plus sizes.

These disparate morphologies can be broadly grouped into form taxa:

"Embryos"

Recent discoveries of Precambrian multicellular life have been dominated by reports of embryos, particularly from the Doushantuo Formation in China. Some finds generated intense media excitement[252] though some have claimed they are instead inorganic structures formed by the precipitation of minerals on the inside of a hole. Other "embryos" have been interpreted as the remains of the giant sulfur-reducing bacteria akin to *Thiomargarita*,[253] a view that, while it had enjoyed a notable gain of supporters[254] as of 2007, has since suffered following further research comparing the potential Doushantuo embryos' morphologies with those of *Thiomargarita* specimens, both living and in various stages of decay.

Microfossils dating from 632.5^{255} million years ago – just 3 million years after the end of the Cryogenian glaciations – may represent embryonic 'resting stages' in the life cycle of the earliest known animals. An alternative proposal is that these structures represent adult stages of the multicellular organisms of this period.

Discs

Circular fossils, such as *Ediacaria*, *Cyclomedusa* and *Rugoconites* led to the initial identification of Ediacaran fossils as cnidaria, which include jellyfish and corals. Further examination has provided alternative interpretations of all disc-shaped fossils: not one is now confidently recognised as a jellyfish. Alternate explanations include holdfasts and protists; the patterns displayed where two meet have led to many 'individuals' being identified as microbial colonies, and yet others may represent scratch marks formed as stalked organisms spun around their holdfasts. Useful diagnostic characters are often lacking because only the underside of the organism is preserved by fossilisation.

Bags

Fossils such as *Pteridinium* preserved within sediment layers resemble "mud-filled bags". The scientific community is a long way from reaching a consensus on their interpretation.[256]

Toroids

The fossil *Vendoglossa tuberculata* from the Nama Group, Namibia, has been interpreted as a dorso-ventrally compressed stem-group metazoan, with a large gut cavity and a transversely ridged ectoderm. The organism is in the shape of a flattened torus, with the long axis of its toroidal body running through the approximate center of the presumed gut cavity.

Quilted organisms

The organisms considered in Seilacher's revised definition of the Vendobionta share a "quilted" appearance and resembled an inflatable mattress. Sometimes these quilts would be torn or ruptured prior to preservation: such damaged specimens provide valuable clues in the reconstruction process. For example, the three (or more) petaloid fronds of *Swartpuntia germsi* could only be recognised in a posthumously damaged specimen – usually multiple fronds were hidden as burial squashed the organisms flat.

These organisms appear to form two groups: the fractal rangeomorphs and the simpler erniettomorphs. Including such fossils as the iconic *Charnia* and *Swartpuntia*, the group is both the most iconic of the Ediacaran biota and the most difficult to place within the existing tree of life. Lacking any mouth, gut, reproductive organs, or indeed any evidence of internal anatomy, their lifestyle was somewhat peculiar by modern standards; the most widely accepted hypothesis holds that they sucked nutrients out of the surrounding seawater by osmotrophy or osmosis. However, others argue against this.

Non-Ediacarans

Some Ediacaran organisms have more complex details preserved, which has allowed them to be interpreted as possible early forms of living phyla excluding them from some definitions of the Ediacaran biota.

The earliest such fossil is the reputed bilaterian *Vernanimalcula* claimed by some, however, to represent the infilling of an egg-sac or acritarch. Later examples are almost universally accepted as bilaterians and include the mollusc-like *Kimberella*, *Spriggina* (pictured) and the shield-shaped *Parvancorina* whose affinities are currently debated.[257]

A suite of fossils known as the Small shelly fossils are represented in the Ediacaran, most famously by *Cloudina* a shelly tube-like fossil that often shows evidence of predatory boring, suggesting that, while predation may not have been common in the Ediacaran Period, it was at least present.

Representatives of modern taxa existed in the Ediacaran, some of which are recognisable today. Sponges, red and green algæ, protists and bacteria are all easily recognisable with some pre-dating the Ediacaran by nearly three billion years. Possible arthropods have also been described.

Trace fossils

With the exception of some very simple vertical burrows the only Ediacaran burrows are horizontal, lying on or just below the surface of the seafloor. Such burrows have been taken to imply the presence of motile organisms with heads, which would probably have had a bilateral symmetry. This could place them in the bilateral clade of animals but they could also have been made by simpler organisms feeding as they slowly rolled along the sea floor. Putative "burrows" dating as far back as $1,100^{258}$ million years may have been made by animals that fed on the undersides of microbial mats, which would have shielded them from a chemically unpleasant ocean; however their uneven width and tapering ends make a biological origin so difficult to defend that even the original proponent no longer believes they are authentic.

The burrows observed imply simple behaviour, and the complex efficient feeding traces common from the start of the Cambrian are absent. Some Ediacaran fossils, especially discs, have been interpreted tentatively as trace fossils but this hypothesis has not gained widespread acceptance. As well as burrows, some trace fossils have been found directly associated with an Ediacaran fossil. *Yorgia* and *Dickinsonia* are often found at the end of long pathways of trace fossils matching their shape; these fossils are thought to be associated with ciliary feeding but the precise method of formation of these disconnected and overlapping fossils largely remains a mystery.[259] The potential mollusc *Kimberella* is associated with scratch marks, perhaps formed by a radula.[260]

Classification and interpretation

Classification of the Ediacarans is difficult, and hence a variety of theories exist as to their placement on the tree of life.

Martin Glaessner proposed in *The Dawn of Animal Life* (1984) that the Ediacaran biota were recognizable crown group members of modern phyla, but were unfamiliar because they had yet to evolve the characteristic features we use in modern classification.

In 1998 Mark McMenamin claimed Ediacarans did not possess an embryonic stage, and thus could not be animals. He believed that they independently evolved a nervous system and brains, meaning that "the path toward intelligent life was embarked upon more than once on this planet".

Figure 113: *A sea pen, a modern cnidarian bearing a passing resemblance to Charnia*

Cnidarians

Since the most primitive eumetazoans—multi-cellular animals with tissues—are cnidarians, the first attempt to categorise these fossils designated them as jellyfish and sea pens. However, more recent discoveries have established that many of the circular forms formerly considered "cnidarian medusa" are actually holdfasts – sand-filled vesicles occurring at the base of the stem of upright frond-like Ediacarans. A notable example is the form known as *Charniodiscus*, a circular impression later found to be attached to the long 'stem' of a frond-like organism that now bears the name.[261]

The link between certain frond-like Ediacarans and sea pens has been thrown into doubt by multiple lines of evidence; chiefly the derived nature of the most frond-like pennatulacean octocorals, their absence from the fossil record before the Tertiary, and the apparent cohesion between segments in Ediacaran frond-like organisms.[262] Some researchers have suggested that an analysis of "growth poles" discredits the pennatulacean nature of Ediacaran fronds.

Figure 114: *A single-celled xenophyophore in the Galapagos Rift*

Protozoans

Adolf Seilacher has suggested the Ediacaran sees animals usurping giant protists as the dominant life form. The modern xenophyophores are giant single-celled protozoans found throughout the world's oceans, largely on the abyssal plain. A recent genetic study suggested that the xenophyophores are a specialised group of Foraminifera. There are approximately 42 recognised species in 13 genera and 2 orders; one of which, *Syringammina fragilissima*, is among the largest known protozoans at up to 20 centimetres in diameter.

New phylum

Seilacher has suggested that the Ediacaran organisms represented a unique and extinct grouping of related forms descended from a common ancestor (clade) and created the kingdom Vendozoa, named after the now-obsolete Vendian era. He later excluded fossils identified as metazoans and relaunched the phylum "Vendobionta".

He described the Vendobionta as quilted cnidarians lacking stinging cells. This absence precludes the current cnidarian method of feeding, so Seilacher suggested that the organisms may have survived by symbiosis with photosynthetic or chemoautotrophic organisms. Mark McMenamin saw such feeding strategies as characteristic for the entire biota, and referred to the marine biota of this period as a "Garden of Ediacara".

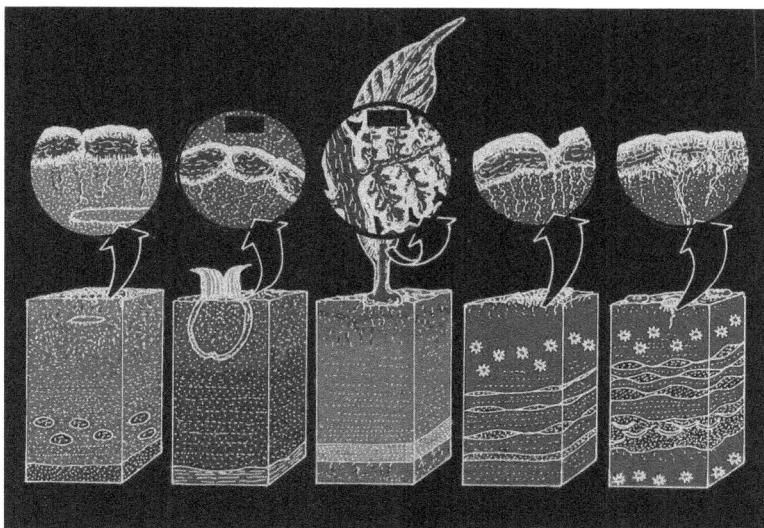

Figure 115: *Thin sections and substrates of a variety of Ediacaran fossils*

Lichen hypothesis

Greg Retallack's hypothesis that Ediacaran organisms were lichens has been controversial He argues that the fossils are not as squashed as known fossil jellyfish, and their relief is closer to compressed woody branches whose compaction can be estimated as compressed cylinders. He points out the chitinous walls of lichen colonies would provide a similar resistance to compaction, and claims the large size of the organisms (up to 1.5 metres long, far larger than any of the preserved burrows) also hints against classification with animals. Thin sections of Ediacaran fossils show lichen-like compartments and hypha-like wisps of ferruginized clay . Finally, Ediacaran fossils from classic localities of the Flinders Ranges have been found in growth position within red calcareous and gypsiferous paleosols, interpreted as soils of well-drained temperate desert soils. Such habitats limit interpretive options for fractal Ediacaran fossils such as *Dickinsonia* to lichenised or unlichenised fungi, but other Ediacaran fossils could have been slime moulds or microbial colonies.

Other interpretations

Several classifications have been used to accommodate the Ediacaran biota at some point, from algae, to protozoans, to fungi to bacterial or microbial colonies, to hypothetical intermediates between plants and animals.

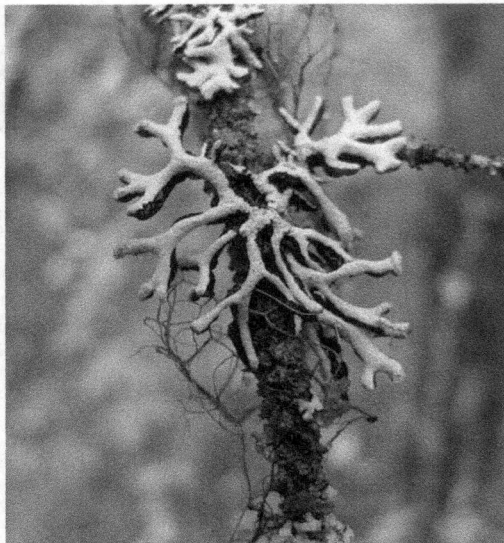

Figure 116: *A modern lichen, Hypogymnia. The lichen hypothesis addresses unusual features in the fossil record from this epoch.*

A new extant genus discovered in 2014, *Dendrogramma*, which appears to be a basal metazoan but of unknown taxonomic placement, has been noted to have similarities with the Ediacaran fauna. It has since been found to be a siphonophore, possibly even sections of a more complex species,[263] though this in turn has raised suspicions for a similar status for at least some ediacaran organisms.

Origin

It took almost 4 billion years from the formation of the Earth for the Ediacaran fossils to first appear, 655 million years ago. While putative fossils are reported from 3,460[264] million years ago, the first uncontroversial evidence for life is found 2,700[265] million years ago, and cells with nuclei certainly existed by 1,200[266] million years ago: The reason why it took so long for forms with an Ediacaran grade of organisation to appear is uncertain.

It could be that no special explanation is required: the slow process of evolution simply required 4 billion years to accumulate the necessary adaptations. Indeed, there does seem to be a slow increase in the maximum level of complexity seen over this time, with more and more complex forms of life evolving as time progresses, with traces of earlier semi-complex life such as *Nimbia*,

Figure 117: *Global ice sheets may have delayed or prevented the establishment of multicellular life.*

found in the 610^{267} million year old Twitya formation, (and possibly older rocks dating to 770^{268} million years ago) possibly displaying the most complex morphology of the time.

The alternative train of thought is that it was simply not advantageous to be large until the appearance of the Ediacarans: the environment favoured the small over the large. Examples of such scenarios today include plankton, whose small size allows them to reproduce rapidly to take advantage of ephemerally abundant nutrients in algal blooms. But for large size *never* to be favourable, the environment would have to be very different indeed.

A primary size-limiting factor is the amount of atmospheric oxygen. Without a complex circulatory system, low concentrations of oxygen cannot reach the centre of an organism quickly enough to supply its metabolic demand.

On the early Earth, reactive elements, such as iron and uranium, existed in a reduced form that would react with any free oxygen produced by photosynthesising organisms. Oxygen would not be able to build up in the atmosphere until all the iron had rusted (producing banded iron formations), and all the other reactive elements had been oxidised. Donald Canfield detected records of the first significant quantities of atmospheric oxygen just before the first Ediacaran fossils appeared – and the presence of atmospheric oxygen was soon heralded as a possible trigger for the Ediacaran radiation. Oxygen seems to have accumulated in two pulses; the rise of small, sessile (stationary) organisms seems to correlate with an early oxygenation event, with larger and mobile organisms appearing around the second pulse of oxygenation. However, the assumptions underlying the reconstruction of atmospheric composition have attracted some criticism, with widespread anoxia having little effect on life where it occurs in the Early Cambrian and the Cretaceous.

Periods of intense cold have also been suggested as a barrier to the evolution of multicellular life. The earliest known embryos, from China's Doushantuo Formation, appear just a million years after the Earth emerged from a global glaciation, suggesting that ice cover and cold oceans may have prevented the emergence of multicellular life. Potentially, complex life may have evolved before these glaciations, and been wiped out. However, the diversity of life in modern Antarctica has sparked disagreement over whether cold temperatures increase or decrease the rate of evolution.

In early 2008 a team analysed the range of basic body structures ("disparity") of Ediacaran organisms from three different fossil beds: Avalon in Canada, 575^{246} to 565^{269} million years ago; White Sea in Russia, 560^{270} to 550^{271} million years ago; and Nama in Namibia, 550^{271} to 542^{249} million years ago, immediately before the start of the Cambrian. They found that, while the White Sea assemblage had the most species, there was no significant difference in disparity between the three groups, and concluded that before the beginning of the Avalon timespan these organisms must have gone through their own evolutionary "explosion", which may have been similar to the famous Cambrian explosion .

Preservation bias

The paucity of Ediacaran fossils after the Cambrian could simply be due to conditions that no longer favoured the fossilisation of Ediacaran organisms, which may have continued to thrive unpreserved. However, if they were common, more than the occasional specimen might be expected in exceptionally preserved fossil assemblages (Konservat-Lagerstätten) such as the Burgess Shale and Chengjiang. There are at present no widely accepted reports of Ediacara-type organisms in the Cambrian period, though there are a few disputed reports, as well as unpublished observations of 'vendobiont' fossils from 535 Ma Orsten-type deposits in China.

Predation and grazing

It is suggested that by the Early Cambrian, organisms higher in the food chain caused the microbial mats to largely disappear. If these grazers first appeared as

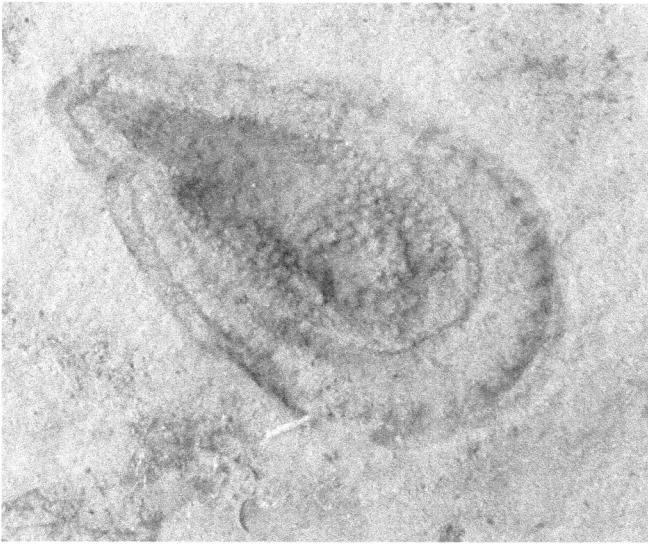

Figure 118: *Kimberella may have had a predatory or grazing lifestyle.*

the Ediacaran biota started to decline, then it may suggest that they destabilised the microbial substrate, leading to displacement or detachment of the biota; or that the destruction of the mat destabilised the ecosystem, causing extinctions.

Alternatively, skeletonised animals could have fed directly on the relatively undefended Ediacaran biota. However, if the interpretation of the Ediacaran age *Kimberella* as a grazer is correct then this suggests that the biota had already had limited exposure to "predation".

There is however little evidence for any trace fossils in the Ediacaran Period, which may speak against the active grazing theory. Further, the onset of the Cambrian Period is defined by the appearance of a worldwide trace fossil assemblage, quite distinct from the activity-barren Ediacaran Period.

Competition

It is possible that increased competition due to the evolution of key innovations among other groups, perhaps as a response to predation, drove the Ediacaran biota from their niches. However, this argument has not successfully explained similar phenomena. For instance, the bivalve molluscs' "competitive exclusion" of brachiopods was eventually deemed to be a coincidental result of two unrelated trends.

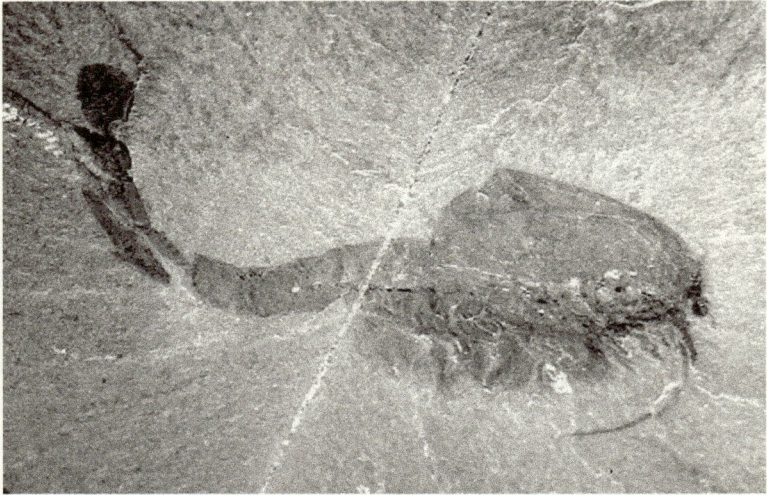

Figure 119: *Cambrian animals such as Waptia may
have competed with, or fed upon, Ediacaran life-forms.*

Change in environmental conditions

While it is difficult to infer the effect of changing planetary conditions on organisms, communities and ecosystems, great changes were happening at the end of the Precambrian and the start of the Early Cambrian. The breakup of the supercontinents, rising sea levels (creating shallow, "life-friendly" seas), a nutrient crisis, fluctuations in atmospheric composition, including oxygen and carbon dioxide levels, and changes in ocean chemistry (promoting biomineralisation) could all have played a part.

Assemblages

Ediacaran-type fossils are recognised globally in 25 localities and a variety of depositional conditions, and are commonly grouped into three main types, known as assemblages and named after typical localities. Each assemblage tends to occupy its own region of morphospace, and after an initial burst of diversification changes little for the rest of its existence.

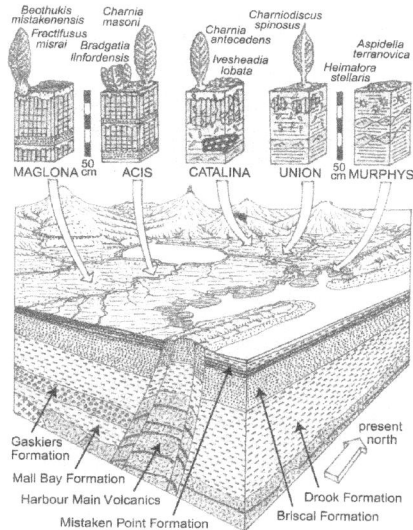

Figure 120: *Reconstruction of fossil soils and their biota in the Mistaken Point Formation of Newfoundland*

Avalon-type assemblage

The Avalon-type assemblage is defined at Mistaken Point in Canada, the oldest locality with a large quantity of Ediacaran fossils. The assemblage is easily dated because it contains many fine ash-beds, which are a good source of zircons used in the uranium-lead method of radiometric dating. These fine-grained ash beds also preserve exquisite detail. Constituents of this biota appear to survive through until the extinction of all Ediacarans at the base of the Cambrian.

One interpretation of the biota is as deep-sea-dwelling rangeomorphs such as *Charnia*, all of which share a fractal growth pattern. They were probably preserved *in situ* (without post-mortem transportation), although this point is not universally accepted. The assemblage, while less diverse than the Ediacara- or Nama-types, resembles Carboniferous suspension-feeding communities, which may suggest filter feeding – by most interpretations, the assemblage is found in water too deep for photosynthesis. The low diversity may reflect the depth of water – which would restrict speciation opportunities – or it may just be too young for a rich biota to have evolved. Opinion is currently divided between these conflicting hypotheses.

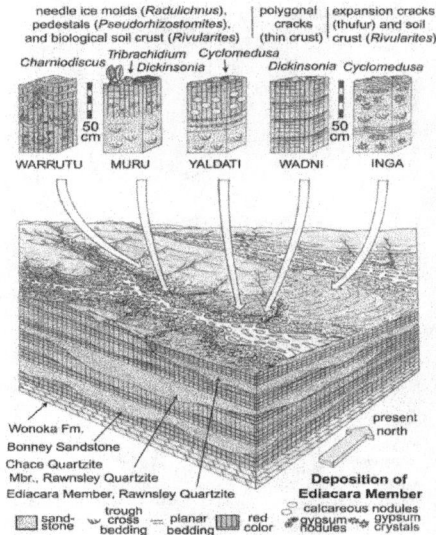

Figure 121: *Reconstruction of Ediacaran biota and their soils in the Ediacara Member of the Rawnsley Quartzite in the Flinders Ranges, South Australia*

An alternative explanation for the distinct composition of the Avalon-type assemblage is that it was a terrestrial assemblage of volcaniclastic coastal soils near a continental volcanic arc . This view is based on geochemical studies of the substrates of Mistaken Point fossils and associated matrix supported tuffs and volcanic bombs that could only form on land . Some of these fossils such as Fractofusus and Charniodiscus were found in red well drained paleosols of coastal plains, but others such as Aspidella were found in pyritic intertidal paleosols.

Ediacara-type assemblage

The Ediacara-type assemblage is named after Australia's Ediacara Hills, and consists of fossils preserved in facies of coastal lagoons and rivers. They are typically found in red gypsiferous and calcareous paleosols formed on loess and flood deposits in an arid cool temperate paleoclimate. Most fossils are preserved as imprints in microbial earths, but a few are preserved *within* sandy units.[272]

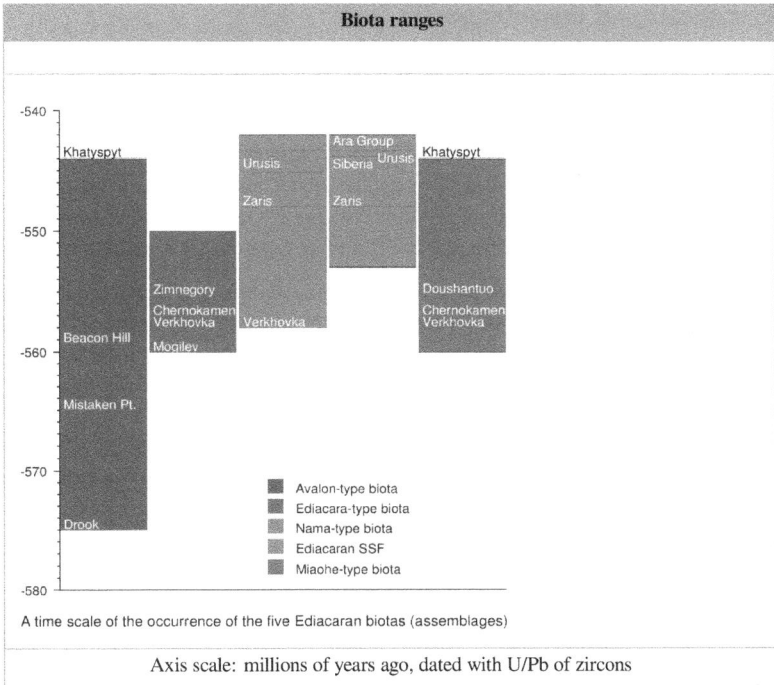

A time scale of the occurrence of the five Ediacaran biotas (assemblages)

Axis scale: millions of years ago, dated with U/Pb of zircons

Nama-type assemblage

The Nama assemblage is best represented in Namibia. Three-dimensional preservation is most common, with organisms preserved in sandy beds containing internal bedding. Dima Grazhdankin believes that these fossils represent burrowing organisms, while Guy Narbonne maintains they were surface dwellers. These beds are sandwiched between units comprising interbedded sandstones, siltstones and shales – with microbial mats, where present, usually containing the fossils. The environment is interpreted as sand bars formed at the mouth of a delta's distributaries. Matress-like vendobionts (Ernietta, Pteridinium, Rangea) in these sandstones form a very differnt assemblage from vermiform fossils (Cloudina, Namacalathus) of Ediacaran "wormworld" in marine dolostones of Namibia .

Significance of assemblages

In the White Sea region of Russia, all three assemblage types have been found in close proximity. This, and the faunas' considerable temporal overlap, makes it unlikely that they represent evolutionary stages or temporally distinct communities. Since they are globally distributed – described on all continents except Antarctica – geographical boundaries do not appear to be a factor; the same fossils are found at all palaeolatitudes (the latitude where the fossil was created, accounting for continental drift) and in separate sedimentary basins.

It is most likely that the three assemblages mark organisms adapted to survival in different environments, and that any apparent patterns in diversity or age are in fact an artefact of the few samples that have been discovered – the timeline (right) demonstrates the paucity of Ediacaran fossil-bearing assemblages. An analysis of one of the White Sea fossil beds, where the layers cycle from continental seabed to inter-tidal to estuarine and back again a few times, found that a specific set of Ediacaran organisms was associated with each environment.

As the Ediacaran biota represent an early stage in multicellular life's history, it is unsurprising that not all possible modes of life are occupied. It has been estimated that of 92 potentially possible modes of life – combinations of feeding style, tiering and motility — no more than a dozen are occupied by the end of the Ediacaran. Just four are represented in the Avalon assemblage. The lack of large-scale predation and vertical burrowing are perhaps the most significant factors limiting the ecological diversity; the emergence of these during the Early Cambrian allowed the number of lifestyles occupied to rise to 30.

Further reading

- Mark McMenamin (1998). *The Garden of Ediacara: Discovering the First Complex Life*. New York: Columbia University Press. pp. 368pp. ISBN 0-231-10558-4. OCLC 3758852[273]. A popular science account of these fossils, with a particular focus on the Namibian fossils.
- Derek Briggs; Peter Crowther, eds. (2001). *Palæobiology II: A synthesis*. Malden, MA: Blackwell Science. pp. Chapter 1. ISBN 0-632-05147-7. OCLC 43945263[274]. Excellent further reading for the keen – includes many interesting chapters with macroevolutionary theme.

External links

<indicator name="spoken-icon"> �))) </indicator>

- Ediacara Biota[275] on *In Our Time* at the BBC
- "The oldest complex animal fossils"[276] – Queen's University, Canada

- "Ediacaran fossils of Canada"[277] – Queen's University, Canada
- "The Ediacaran Assemblage"[278] – Thorough, though slightly out-of-date, description
- "Database of Ediacaran Biota"[279] Advent of Complex Life
- Earth's oldest animal ecosystem held in fossils at Nilpena Station in SA outback[280] *ABC News*, 5 August 2013. Accessed 6 August 2013.
- Meet the fossils[281] ABC *Landline* TV program on Ediacaran fossils at Nilpena (audio + transcript). First broadcast 3 August 2013. Accessed 11 August 2013.

Cambrian explosion

<indicator name="pp-default"> 🔧 </indicator>

Part of a series on
The Cambrian explosion

- v
- t
- e[282]

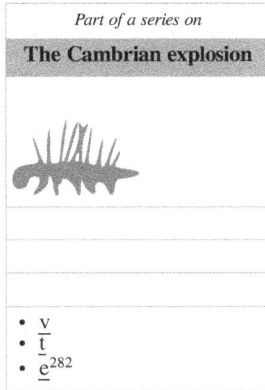

The **Cambrian explosion** or **Cambrian radiation** was an event approximately 541[283] million years ago in the Cambrian period when most major animal phyla appeared in the fossil record. It lasted for about 20–25 million years. It resulted in the divergence of most modern metazoan phyla. The event was accompanied by major diversification of other organisms.[284]

Before the Cambrian explosion,[285] most organisms were simple, composed of individual cells occasionally organized into colonies. Over the following 70 to 80 million years, the rate of diversification accelerated, and the variety of life began to resemble that of today. Almost all present animal phyla appeared during this period.

The Cambrian explosion has generated extensive scientific debate.

Key Cambrian explosion events

490 —
500 —
510 —
520 —
530 —
540 —
550 —
560 —
570 —
580 —
590 —

History and significance

Life timeline

θ —
500 —
1000 —
1500 —
2000 —
2500 —
3000 —
3500 —
4000 —
4500 —

Axis scale: million years

Also see: *Human timeline* and *Nature timeline*

The seemingly rapid appearance of fossils in the "Primordial Strata" was noted by William Buckland in the 1840s, and in his 1859 book *On the Origin of Species*, Charles Darwin discussed the then inexplicable lack of earlier fossils as one of the main difficulties for his theory of descent with slow modification through natural selection. The long-running puzzlement about the appearance of the Cambrian fauna, seemingly abruptly, without precursor, centers on three key points: whether there really was a mass diversification of complex organisms over a relatively short period of time during the early Cambrian; what might have caused such rapid change; and what it would imply about the origin of animal life. Interpretation is difficult due to a limited supply of evidence, based mainly on an incomplete fossil record and chemical signatures remaining in Cambrian rocks.

The first discovered Cambrian fossils were trilobites, described by Edward Lhuyd, the curator of Oxford Museum, in 1698. Although their evolutionary importance was not known, on the basis of their old age, William Buckland (1784–1856) realised that a dramatic step-change in the fossil record had occurred around the base of what we now call the Cambrian. Nineteenth-century geologists such as Adam Sedgwick and Roderick Murchison used the fossils for dating rock strata, specifically for establishing the Cambrian and Silurian periods. By 1859, leading geologists including Roderick Murchison, were convinced that what was then called the lowest Silurian stratum showed the origin of life on Earth, though others, including Charles Lyell, differed. In *On the Origin of Species*, Charles Darwin considered this sudden appearance of a solitary group of trilobites, with no apparent antecedents, and absence of other fossils, to be "undoubtedly of the gravest nature" among the difficulties in his theory of natural selection. He reasoned that earlier seas had swarmed with living creatures, but that their fossils had not been found due to the imperfections of the fossil record. In the sixth edition of his book, he stressed his problem further as:

> *To the question why we do not find rich fossiliferous deposits belonging to these assumed earliest periods prior to the Cambrian system, I can give no satisfactory answer.*

American paleontologist Charles Walcott, who studied the Burgess Shale fauna, proposed that an interval of time, the "Lipalian", was not represented in the fossil record or did not preserve fossils, and that the ancestors of the Cambrian animals evolved during this time.

Earlier fossil evidence has since been found. The earliest claim is that the history of life on earth goes back 3,850[286] million years: Rocks of that age at Warrawoona, Australia, were claimed to contain fossil stromatolites, stubby pillars formed by colonies of microorganisms. Fossils (*Grypania*) of more complex eukaryotic cells, from which all animals, plants, and fungi are built, have been found in rocks from 1,400[287] million years ago, in China and Montana. Rocks dating from 580 to 543[288] million years ago contain fossils of the Ediacara biota, organisms so large that they are likely multicelled, but very

Figure 122: *Opabinia made the largest single contribution to modern interest in the Cambrian explosion.*

unlike any modern organism. In 1948, Preston Cloud argued that a period of "eruptive" evolution occurred in the Early Cambrian, but as recently as the 1970s, no sign was seen of how the 'relatively' modern-looking organisms of the Middle and Late Cambrian arose.

The intense modern interest in this "Cambrian explosion" was sparked by the work of Harry B. Whittington and colleagues, who, in the 1970s, reanalysed many fossils from the Burgess Shale and concluded that several were as complex as, but different from, any living animals.[289] The most common organism, *Marrella*, was clearly an arthropod, but not a member of any known arthropod class. Organisms such as the five-eyed *Opabinia* and spiny slug-like *Wiwaxia* were so different from anything else known that Whittington's team assumed they must represent different phyla, seemingly unrelated to anything known today. Stephen Jay Gould's popular 1989 account of this work, *Wonderful Life*, brought the matter into the public eye and raised questions about what the explosion represented. While differing significantly in details, both Whittington and Gould proposed that all modern animal phyla had appeared almost simultaneously in a rather short span of geological period. This view led to the modernization of Darwin's tree of life and the theory of punctuated equilibrium, which Eldredge and Gould developed in the early 1970s and which views evolution as long intervals of near-stasis "punctuated" by short periods of rapid change.

Other analyses, some more recent and some dating back to the 1970s, argue that complex animals similar to modern types evolved well before the start of the Cambrian.

Dating the Cambrian

Radiometric dates for much of the Cambrian, obtained by analysis of radioactive elements contained within rocks, have only recently become available, and for only a few regions.

Relative dating (*A* was before *B*) is often assumed sufficient for studying processes of evolution, but this, too, has been difficult, because of the problems involved in matching up rocks of the same age across different continents.[290]

Therefore, dates or descriptions of sequences of events should be regarded with some caution until better data become available.

Body fossils

Fossils of organisms' bodies are usually the most informative type of evidence. Fossilization is a rare event, and most fossils are destroyed by erosion or metamorphism before they can be observed. Hence, the fossil record is very incomplete, increasingly so as earlier times are considered. Despite this, they are often adequate to illustrate the broader patterns of life's history.[291] Also, biases exist in the fossil record: different environments are more favourable to the preservation of different types of organism or parts of organisms. Further, only the parts of organisms that were already mineralised are usually preserved, such as the shells of molluscs. Since most animal species are soft-bodied, they decay before they can become fossilised. As a result, although 30-plus phyla of living animals are known, two-thirds have never been found as fossils.

The Cambrian fossil record includes an unusually high number of lagerstätten, which preserve soft tissues. These allow paleontologists to examine the internal anatomy of animals, which in other sediments are only represented by shells, spines, claws, etc. – if they are preserved at all. The most significant Cambrian lagerstätten are the early Cambrian Maotianshan shale beds of Chengjiang (Yunnan, China) and Sirius Passet (Greenland); the middle Cambrian Burgess Shale (British Columbia, Canada); and the late Cambrian Orsten (Sweden) fossil beds.

While lagerstätten preserve far more than the conventional fossil record, they are far from complete. Because lagerstätten are restricted to a narrow range of environments (where soft-bodied organisms can be preserved very quickly,

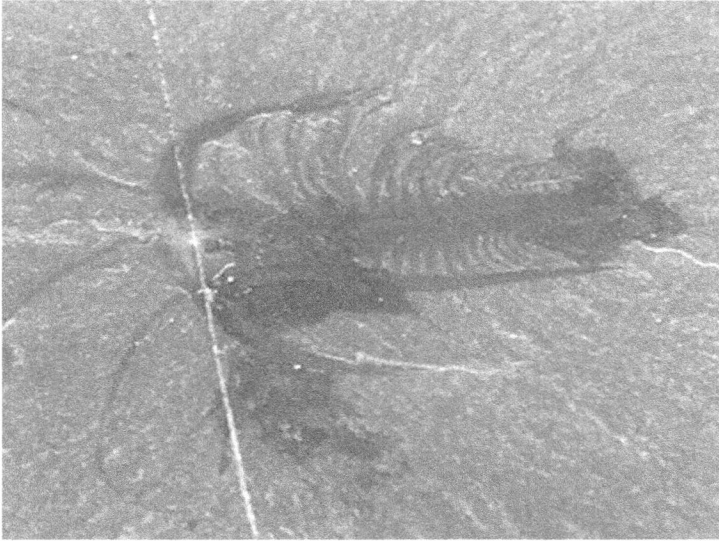

Figure 123: *This Marrella specimen illustrates how clear and detailed the fossils from the Burgess Shale Lagerstätte are.*

e.g. by mudslides), most animals are probably not represented; further, the exceptional conditions that create lagerstätten probably do not represent normal living conditions. In addition, the known Cambrian lagerstätten are rare and difficult to date, while Precambrian lagerstätten have yet to be studied in detail.

The sparseness of the fossil record means that organisms usually exist long before they are found in the fossil record – this is known as the Signor–Lipps effect.

Trace fossils

Trace fossils consist mainly of tracks and burrows, but also include coprolites (fossil feces) and marks left by feeding. Trace fossils are particularly significant because they represent a data source that is not limited to animals with easily fossilized hard parts, and reflects organisms' behaviour. Also, many traces date from significantly earlier than the body fossils of animals that are thought to have been capable of making them.[292] While exact assignment of trace fossils to their makers is generally impossible, traces may, for example, provide the earliest physical evidence of the appearance of moderately complex animals (comparable to earthworms).

Figure 124: *Rusophycus and other trace fossils from the Gog Group, Middle Cambrian, Lake Louise, Alberta, Canada*

Geochemical observations

Several chemical markers indicate a drastic change in the environment around the start of the Cambrian. The markers are consistent with a mass extinction, or with a massive warming resulting from the release of methane ice. Such changes may reflect a cause of the Cambrian explosion, although they may also have resulted from an increased level of biological activity – a possible result of the explosion. Despite these uncertainties, the geochemical evidence helps by making scientists focus on theories that are consistent with at least one of the likely environmental changes.

Phylogenetic techniques

Cladistics is a technique for working out the "family tree" of a set of organisms. It works by the logic that, if groups B and C have more similarities to each other than either has to group A, then B and C are more closely related to each other than either is to A. Characteristics that are compared may be anatomical, such as the presence of a notochord, or molecular, by comparing sequences of DNA or protein. The result of a successful analysis is a hierarchy of clades – groups whose members are believed to share a common ancestor. The cladistic technique is sometimes problematic, as some features, such as wings or camera eyes, evolved more than once, convergently – this must be taken into account in analyses.

From the relationships, it may be possible to constrain the date that lineages first appeared. For instance, if fossils of B or C date to X million years ago and the calculated "family tree" says A was an ancestor of B and C, then A must have evolved more than X million years ago.

It is also possible to estimate how long ago two living clades diverged – i.e. about how long ago their last common ancestor must have lived – by assuming that DNA mutations accumulate at a constant rate. These "molecular clocks", however, are fallible, and provide only a very approximate timing: they are not sufficiently precise and reliable for estimating when the groups that feature in the Cambrian explosion first evolved, and estimates produced by different techniques vary by a factor of two. However, the clocks can give an indication of branching rate, and when combined with the constraints of the fossil record, recent clocks suggest a sustained period of diversification through the Ediacaran and Cambrian.

Explanation of key scientific terms

Phylum

A phylum is the highest level in the Linnaean system for classifying organisms. Phyla can be thought of as groupings of animals based on general body plan.[293] Despite the seemingly different external appearances of organisms, they are classified into phyla based on their internal and developmental organizations.[294] For example, despite their obvious differences, spiders and barnacles both belong to the phylum Arthropoda, but earthworms and tapeworms, although similar in shape, belong to different phyla. As chemical and genetic testing becomes more accurate, previously hypothesised phyla are often entirely reworked.

A phylum is not a fundamental division of nature, such as the difference between electrons and protons. It is simply a very high-level grouping in a classification system created to describe all currently living organisms. This system is imperfect, even for modern animals: different books quote different numbers of phyla, mainly because they disagree about the classification of a huge number of worm-like species. As it is based on living organisms, it accommodates extinct organisms poorly, if at all.[295]

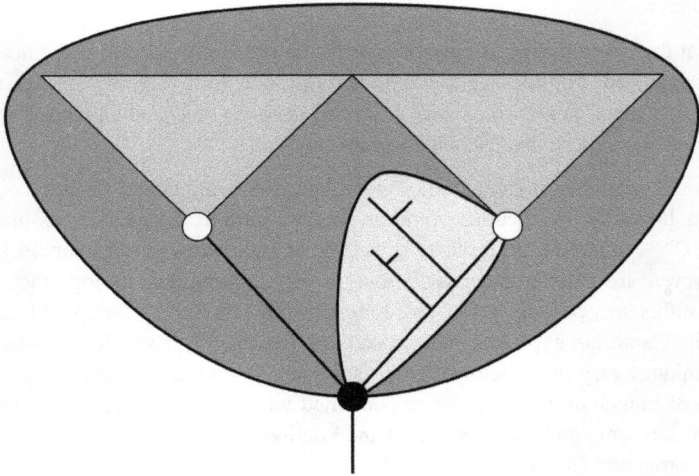

Figure 125:
Stem groups

- ——= *Lines of descent*
- = *Basal node*
- = *Crown node*
- = *Total group*
- = *Crown group*
- = *Stem group*

Stem group

The concept of stem groups was introduced to cover evolutionary "aunts" and "cousins" of living groups, and have been hypothesized based on this scientific theory. A crown group is a group of closely related living animals plus their last common ancestor plus all its descendants. A stem group is a set of offshoots from the lineage at a point earlier than the last common ancestor of the crown group; it is a relative concept, for example tardigrades are living animals that form a crown group in their own right, but Budd (1996) regarded them as also being a stem group relative to the arthropods.

File:Coelomate.svg

A coelomate animal is basically a set of concentric tubes, with a gap between the gut and the outer tubes.

Triploblastic

The term *Triploblastic* means consisting of three layers, which are formed in the embryo, quite early in the animal's development from a single-celled egg to a larva or juvenile form. The innermost layer forms the digestive tract (gut); the outermost forms skin; and the middle one forms muscles and all the internal organs except the digestive system. Most types of living animal are triploblastic – the best-known exceptions are Porifera (sponges) and Cnidaria (jellyfish, sea anemones, etc.).

Bilaterian

The bilaterians are animals that have right and left sides at some point in their life histories. This implies that they have top and bottom surfaces and, importantly, distinct front and back ends. All known bilaterian animals are triploblastic, and all known triploblastic animals are bilaterian. Living echinoderms (sea stars, sea urchins, sea cucumbers, etc.) 'look' radially symmetrical (like wheels) rather than bilaterian, but their larvae exhibit bilateral symmetry and some of the earliest echinoderms may have been bilaterally symmetrical. Porifera and Cnidaria are radially symmetrical, not bilaterian, and not triploblastic.

Coelomate

The term *Coelomate* means having a body cavity (coelom) containing the internal organs. Most of the phyla featured in the debate about the Cambrian explosion are coelomates: arthropods, annelid worms, molluscs, echinoderms, and chordates – the noncoelomate priapulids are an important exception. All known coelomate animals are triploblastic bilaterians, but some triploblastic bilaterian animals do not have a coelom – for example flatworms, whose organs are surrounded by unspecialized tissues.

Precambrian life

Understanding of the Cambrian explosion relies upon knowing what was there beforehand – did the event herald the sudden appearance of a wide range of animals and behaviours, or did such things exist beforehand?

Phylogenetic analysis has been used to support the view that during the Cambrian explosion, metazoans (multi-celled animals) evolved monophyletically from a single common ancestor: flagellated colonial protists similar to modern choanoflagellates.

Figure 126: *Stromatolites (Pika Formation, Middle Cambrian) near Helen Lake, Banff National Park, Canada*

Figure 127: *Modern stromatolites in Hamelin Pool Marine Nature Reserve, Western Australia*

Evidence of animals around 1 billion years ago

For further information, see Acritarch and Stromatolite

Changes in the abundance and diversity of some types of fossil have been interpreted as evidence for "attacks" by animals or other organisms. Stromatolites, stubby pillars built by colonies of microorganisms, are a major constituent of the fossil record from about 2,700[296] million years ago, but their abundance and diversity declined steeply after about $1,250^{297}$ million years ago. This decline has been attributed to disruption by grazing and burrowing animals.

Precambrian marine diversity was dominated by small fossils known as acritarchs. This term describes almost any small organic walled fossil – from the egg cases of small metazoans to resting cysts of many different kinds of green algae. After appearing around $2,000^{298}$ million years ago, acritarchs underwent a boom around $1,000^{299}$ million years ago, increasing in abundance, diversity, size, complexity of shape, and especially size and number of spines. Their increasingly spiny forms in the last 1 billion years may indicate an increased need for defence against predation. Other groups of small organisms from the Neoproterozoic era also show signs of antipredator defenses. A consideration of taxon longevity appears to support an increase in predation pressure around this time. In general, the fossil record shows a very slow appearance of these lifeforms in the Precambrian, with many cyanobacterial species making up much of the underlying sediment.

Fossils of the Doushantuo formation

The layers of the Doushantuo formation from around 580^{300} million year old harbour microscopic fossils that may represent early bilaterians. Some have been described as animal embryos and eggs, although some may represent the remains of giant bacteria. Another fossil, *Vernanimalcula*, has been interpreted as a coelomate bilaterian, but may simply be an infilled bubble.

These fossils form the earliest hard-and-fast evidence of animals, as opposed to other predators.

Burrows

The traces of organisms moving on and directly underneath the microbial mats that covered the Ediacaran sea floor are preserved from the Ediacaran period, about 565^{301} million years ago.^{302} have since been recognised as nonbiogenic.</ref> They were probably made by organisms resembling earthworms in shape, size, and how they moved. The burrow-makers have never been found preserved, but, because they would need a head and a tail, the burrowers probably had bilateral symmetry – which would in all probability make them bilaterian animals. They fed above the sediment surface, but were forced to burrow to avoid predators.

Around the start of the Cambrian (about 542^{303} million years ago), many new types of traces first appear, including well-known vertical burrows such as *Diplocraterion* and *Skolithos*, and traces normally attributed to arthropods,

Figure 128: *An Ediacaran trace fossil, made when an organism burrowed below a microbial mat.*

such as *Cruziana* and *Rusophycus*. The vertical burrows indicate that worm-like animals acquired new behaviours, and possibly new physical capabilities. Some Cambrian trace fossils indicate that their makers possessed hard exoskeletons, although they were not necessarily mineralised.

Burrows provide firm evidence of complex organisms; they are also much more readily preserved than body fossils, to the extent that the absence of trace fossils has been used to imply the genuine absence of large, motile, bottom-dwelling organisms.Wikipedia:Citation needed They provide a further line of evidence to show that the Cambrian explosion represents a real diversification, and is not a preservational artefact.

This new habit changed the seafloor's geochemistry, and led to decreased oxygen in the ocean and increased CO_2-levels in the seas and the atmosphere, resulting in global warming for tens of millions years, and could be responsible for mass extinctions.[304] But as burrowing became established, it allowed an explosion of its own, for as burrowers disturbed the sea floor, they aerated it, mixing oxygen into the toxic muds. This made the bottom sediments more hospitable, and allowed a wider range of organisms to inhabit them – creating new niches and the scope for higher diversity.

Figure 129: *Dickinsonia costata, an Ediacaran organism of unknown affinity, with a quilted appearance*

Ediacaran organisms

At the start of the Ediacaran period, much of the acritarch fauna, which had remained relatively unchanged for hundreds of millions of years, became extinct, to be replaced with a range of new, larger species, which would prove far more ephemeral. This radiation, the first in the fossil record, is followed soon after by an array of unfamiliar, large, fossils dubbed the Ediacara biota, which flourished for 40 million years until the start of the Cambrian. Most of this "Ediacara biota" were at least a few centimeters long, significantly larger than any earlier fossils. The organisms form three distinct assemblages, increasing in size and complexity as time progressed.

Many of these organisms were quite unlike anything that appeared before or since, resembling discs, mud-filled bags, or quilted mattresses – one palæontologist proposed that the strangest organisms should be classified as a separate kingdom, Vendozoa.

At least some may have been early forms of the phyla at the heart of the "Cambrian explosion" debate, having been interpreted as early molluscs (*Kimberella*), echinoderms (*Arkarua*); and arthropods (*Spriggina, Parvancorina*). Still, debate exists about the classification of these specimens, mainly because the diagnostic features that allow taxonomists to classify more recent

Figure 130: *Fossil of Kimberella, a triploblastic bilaterian, and possibly a mollusc*

organisms, such as similarities to living organisms, are generally absent in the ediacarans. However, there seems little doubt that *Kimberella* was at least a triploblastic bilaterian animal. These organisms are central to the debate about how abrupt the Cambrian explosion was. If some were early members of the animal phyla seen today, the "explosion" looks a lot less sudden than if all these organisms represent an unrelated "experiment", and were replaced by the animal kingdom fairly soon thereafter (40M years is "soon" by evolutionary and geological standards).

Beck Spring Dolomite

Paul Knauth, a geologist at Arizona State University, maintains that photosynthesizing organisms such as algae may have grown over a 750- to 800-million-year-old formation in Death Valley known as the Beck Spring Dolomite. In the early 1990s, samples from this 1,000-foot thick layer of dolomite revealed that the region housed flourishing mats of photosynthesizing, unicellular life forms which antedated the Cambrian explosion.

Microfossils have been unearthed from holes riddling the otherwise barren surface of the dolomite. These geochemical and microfossil findings support the idea that during the Precambrian period, complex life evolved both in the oceans and on land. Knauth contends that animals may well have had their origins in freshwater lakes and streams, and not in the oceans.

Some 30 years later, a number of studies have documented an abundance of geochemical and microfossil evidence showing that life covered the continents as far back as 2.2 billion years ago. Many paleobiologists now accept the idea that simple life forms existed on land during the Precambrian, but are opposed to the more radical idea that multicellular life thrived on land more than 600 million years ago.[305]

Ediacaran–Early Cambrian skeletonisation

The first Ediacaran and lowest Cambrian (Nemakit-Daldynian) skeletal fossils represent tubes and problematic sponge spicules. The oldest sponge spicules are monaxon siliceous, aged around 580[300] million years ago, known from the Doushantou Formation in China and from deposits of the same age in Mongolia, although the interpretation of these fossils as spicules has been challenged. In the late Ediacaran-lowest Cambrian, numerous tube dwellings of enigmatic organisms appeared. It was organic-walled tubes (e.g. *Saarina*) and chitinous tubes of the sabelliditids (e.g. *Sokoloviina, Sabellidites, Paleolina*) that prospered up to the beginning of the Tommotian. The mineralized tubes of *Cloudina, Namacalathus, Sinotubulites*, and a dozen more of the other organisms from carbonate rocks formed near the end of the Ediacaran period from 549 to 542[306] million years ago, as well as the triradially symmetrical mineralized tubes of anabaritids (e.g. *Anabarites, Cambrotubulus*) from uppermost Ediacaran and lower Cambrian. Ediacaran mineralized tubes are often found in carbonates of the stromatolite reefs and thrombolites, i.e. they could live in an environment adverse to the majority of animals.

Although they are as hard to classify as most other Ediacaran organisms, they are important in two other ways. First, they are the earliest known calcifying organisms (organisms that built shells from calcium carbonate). Secondly, these tubes are a device to rise over a substrate and competitors for effective feeding and, to a lesser degree, they serve as armor for protection against predators and adverse conditions of environment. Some *Cloudina* fossils show small holes in shells. The holes possibly are evidence of boring by predators sufficiently advanced to penetrate shells. A possible "evolutionary arms race" between predators and prey is one of the hypotheses that attempt to explain the Cambrian explosion.

In the lowest Cambrian, the stromatolites were decimated. This allowed animals to begin colonization of warm-water pools with carbonate sedimentation. At first, it was anabaritids and *Protohertzina* (the fossilized grasping spines of chaetognaths) fossils. Such mineral skeletons as shells, sclerites, thorns, and plates appeared in uppermost Nemakit-Daldynian; they were the earliest species of halkierids, gastropods, hyoliths and other rare organisms. The beginning of the Tommotian has historically been understood to mark an explosive increase of the number and variety of fossils of molluscs, hyoliths, and sponges, along with a rich complex of skeletal elements of unknown animals, the first archaeocyathids, brachiopods, tommotiids, and others. This sudden increase is partially an artefact of missing strata at the Tommotian type section, and most of this fauna in fact began to diversify in a series of pulses through the Nemakit-Daldynian and into the Tommotian.

Some animals may already have had sclerites, thorns, and plates in the Ediacaran (e.g. *Kimberella* had hard sclerites, probably of carbonate), but thin carbonate skeletons cannot be fossilized in siliciclastic deposits. Older ($\sim$750 Ma) fossils indicate that mineralization long preceded the Cambrian, probably defending small photosynthetic algae from single-celled eukaryotic predators.

Cambrian life

Trace fossils

Trace fossils (burrows, etc.) are a reliable indicator of what life was around, and indicate a diversification of life around the start of the Cambrian, with the freshwater realm colonized by animals almost as quickly as the oceans.

Small shelly fauna

Fossils known as "small shelly fauna" have been found in many parts on the world, and date from just before the Cambrian to about 10 million years after the start of the Cambrian (the Nemakit-Daldynian and Tommotian ages; see timeline). These are a very mixed collection of fossils: spines, sclerites (armor plates), tubes, archeocyathids (sponge-like animals), and small shells very like those of brachiopods and snail-like molluscs – but all tiny, mostly 1 to 2 mm long.

While small, these fossils are far more common than complete fossils of the organisms that produced them; crucially, they cover the window from the start of the Cambrian to the first lagerstätten: a period of time otherwise lacking in fossils. Hence, they supplement the conventional fossil record and allow the fossil ranges of many groups to be extended.

Figure 131: *A fossilized trilobite, an ancient type of arthropod: This specimen, from the Burgess Shale, preserves "soft parts" – the antennae and legs.*

Early Cambrian trilobites and echinoderms

The earliest trilobite fossils are about 530 million years old, but the class was already quite diverse and worldwide, suggesting they had been around for quite some time. The fossil record of trilobites began with the appearance of trilobites with mineral exoskeletons – not from the time of their origin.

The earliest generally accepted echinoderm fossils appeared a little bit later, in the Late Atdabanian; unlike modern echinoderms, these early Cambrian echinoderms were not all radially symmetrical.

These provide firm data points for the "end" of the explosion, or at least indications that the crown groups of modern phyla were represented.

Burgess Shale type faunas

The Burgess Shale and similar lagerstätten preserve the soft parts of organisms, which provide a wealth of data to aid in the classification of enigmatic fossils. It often preserved complete specimens of organisms only otherwise known from dispersed parts, such as loose scales or isolated mouthparts. Further, the majority of organisms and taxa in these horizons are entirely soft-bodied, hence absent from the rest of the fossil record. Since a large part of the ecosystem is preserved, the ecology of the community can also be tentatively reconstructed.Wikipedia:Verifiability However, the assemblages may represent a "museum": a deep-water ecosystem that is evolutionarily "behind" the rapidly diversifying fauna of shallower waters.

Because the lagerstätten provide a mode and quality of preservation that is virtually absent outside of the Cambrian, many organisms appear completely different from anything known from the conventional fossil record. This led early workers in the field to attempt to shoehorn the organisms into extant phyla; the shortcomings of this approach led later workers to erect a multitude of new phyla to accommodate all the oddballs. It has since been realised that most oddballs diverged from lineages before they established the phyla known todayWikipedia:Please clarify – slightly different designs, which were fated to perish rather than flourish into phyla, as their cousin lineages did.

The preservational mode is rare in the preceding Ediacaran period, but those assemblages known show no trace of animal life – perhaps implying a genuine absence of macroscopic metazoans.

Early Cambrian crustaceans

Crustaceans, one of the four great modern groups of arthropods, are very rare throughout the Cambrian. Convincing crustaceans were once thought to be common in Burgess Shale-type biotas, but none of these individuals can be shown to fall into the crown group of "true crustaceans". The Cambrian record of crown-group crustaceans comes from microfossils. The Swedish Orsten horizons contain later Cambrian crustaceans, but only organisms smaller than 2 mm are preserved. This restricts the data set to juveniles and miniaturised adults.

A more informative data source is the organic microfossils of the Mount Cap formation, Mackenzie Mountains, Canada. This late Early Cambrian assemblage (510 to 515[307] million years ago) consists of microscopic fragments of arthropods' cuticle, which is left behind when the rock is dissolved with hydrofluoric acid. The diversity of this assemblage is similar to that of modern crustacean faunas. Analysis of fragments of feeding machinery found in the formation shows that it was adapted to feed in a very precise and refined fashion. This contrasts with most other early Cambrian arthropods, which fed messily by shovelling anything they could get their feeding appendages on into their mouths. This sophisticated and specialised feeding machinery belonged to a large (about 30 cm) organism, and would have provided great potential for diversification; specialised feeding apparatus allows a number of different approaches to feeding and development, and creates a number of different approaches to avoid being eaten.

Early Ordovician radiation

After an extinction at the Cambrian–Ordovician boundary, another radiation occurred, which established the taxa that would dominate the Palaeozoic.

During this radiation, the total number of orders doubled, and families tripled, increasing marine diversity to levels typical of the Palaeozoic, and disparity to levels approximately equivalent to today's.

Stages

The event lasted for about the next 20–25 million years. Different authors break the explosion down into stages in different ways.

Ed Landing recognizes three stages: Stage 1, spanning the Ediacaran-Cambrian boundary, corresponds to a diversification of biomineralizing animals and of deep and complex burrows; Stage 2, corresponding to the radiation of molluscs and stem-group Brachiopods (hyoliths and tommotiids), which apparently arose in intertidal waters; and Stage 3, seeing the Atdabanian diversification of trilobites in deeper waters, but little change in the intertidal realm.

Graham Budd synthesises various schemes to produce a compatible view of the SSF record of the Cambrian explosion, divided slightly differently into four intervals: a "Tube world", lasting from 550 to 536[308] million years ago, spanning the Ediacaran-Cambrian boundary, dominated by Cloudina, Namacalathus ans pseudoconodont-type element; a "Sclerite world", seeing the rise of halkieriids, tommotiids, and hyoliths, lasting to the end of the Fortunian (c. 525 Ma); a brachiopod world, perhaps corresponding to the as yet unratified Cambrian Stage 2; and Trilobite World, kicking off in Stage 3.

Complementary to the shelly fossil record, trace fossils can be divided into five subdivisions: "Flat world" (late Ediacaran), with traces restricted to the sediment surface; Protreozoic III (after Jensen), with increasing complexity; *pedum* world, initiated at the base of the Cambrian with the base of the *T.pedum* zone (see discussion at Cambrian#Dating the Cambrian); *Rusophycus* world, spanning 536 to 521[309] million years ago and thus corresponding exactly to the periods of Sclerite World and Brachiopod World under the SSF paradigm; and *Cruziana* world, with an obvious correspondence to Trilobite World.

Validity

There is strong evidence for species of Cnidaria and Porifera existing in the Ediacaran and possible members of Porifera even before that during the Cryogenian. Bryozoans don't appear in the fossil record until after the Cambrian, in the Lower Ordovician.

The fossil record as Darwin knew it seemed to suggest that the major metazoan groups appeared in a few million years of the early to mid-Cambrian, and even in the 1980s, this still appeared to be the case.

However, evidence of Precambrian Metazoa is gradually accumulating. If the Ediacaran *Kimberella* was a mollusc-like protostome (one of the two main groups of coelomates), the protostome and deuterostome lineages must have split significantly before 550[310] million years ago (deuterostomes are the other main group of coelomates). Even if it is not a protostome, it is widely accepted as a bilaterian. Since fossils of rather modern-looking cnidarians (jellyfish-like organisms) have been found in the Doushantuo lagerstätte, the cnidarian and bilaterian lineages must have diverged well over 580[300] million years ago.

Trace fossils and predatory borings in *Cloudina* shells provide further evidence of Ediacaran animals. Some fossils from the Doushantuo formation have been interpreted as embryos and one (*Vernanimalcula*) as a bilaterian coelomate, although these interpretations are not universally accepted. Earlier still, predatory pressure has acted on stromatolites and acritarchs since around 1,250[297] million years ago.

Some say that the evolutionary change was accelerated by an order of magnitude,[311] but the presence of Precambrian animals somewhat dampens the "bang" of the explosion; not only was the appearance of animals gradual, but their evolutionary radiation ("diversification") may also not have been as rapid as once thought. Indeed, statistical analysis shows that the Cambrian explosion was no faster than any of the other radiations in animals' history.[312]</ref> However, it does seem that some innovations linked to the explosion – such as resistant armour – only evolved once in the animal lineage; this makes a lengthy Precambrian animal lineage harder to defend. Further, the conventional view that all the phyla arose in the Cambrian is flawed; while the phyla may have diversified in this time period, representatives of the crown groups of many phyla do not appear until much later in the Phanerozoic. Further, the mineralised phyla that form the basis of the fossil record may not be representative of other phyla, since most mineralised phyla originated in a benthic setting. The fossil record is consistent with a Cambrian explosion that was limited to the benthos, with pelagic phyla evolving much later.

Ecological complexity among marine animals increased in the Cambrian, as well later in the Ordovician. However, recent research has overthrown the once-popular idea that disparity was exceptionally high throughout the Cambrian, before subsequently decreasing. In fact, disparity remains relatively low throughout the Cambrian, with modern levels of disparity only attained after the early Ordovician radiation.

The diversity of many Cambrian assemblages is similar to today's, and at a high (class/phylum) level, diversity is thought by some to have risen relatively smoothly through the Cambrian, stabilizing somewhat in the Ordovician. This interpretation, however, glosses over the astonishing and fundamental pattern of basal polytomy and phylogenetic telescoping at or near the Cambrian boundary, as seen in most major animal lineages. Thus Harry Blackmore Whittington's questions regarding the abrupt nature of the Cambrian explosion remain, and have yet to be satisfactorily answered.

Possible causes

Despite the evidence that moderately complex animals (triploblastic bilaterians) existed before and possibly long before the start of the Cambrian, it seems that the pace of evolution was exceptionally fast in the early Cambrian. Possible explanations for this fall into three broad categories: environmental, developmental, and ecological changes. Any explanation must explain both the timing and magnitude of the explosion.

Changes in the environment

Increase in oxygen levels

Earth's earliest atmosphere contained no free oxygen (O_2); the oxygen that animals breathe today, both in the air and dissolved in water, is the product of billions of years of photosynthesis. Cyanobacteria were the first organisms to evolve the ability to photosynthesize, introducing a steady supply of oxygen into the environment. Initially, oxygen levels did not increase substantially in the atmosphere. The oxygen quickly reacted with iron and other minerals in the surrounding rock and ocean water. Once a saturation point was reached for the reactions in rock and water, oxygen was able to exist as a gas in its diatomic form. Oxygen levels in the atmosphere increased substantially afterward. As a general trend, the concentration of oxygen in the atmosphere has risen gradually over about the last 2.5 billion years.

Oxygen levels seem to have a positive correlation with diversity in eukaryotes well before the Cambrian period. The last common ancestor of all extant eukaryotes is thought to have lived around 1.8 billion years ago. Around 800 million years ago, there was a notable increase in the complexity and number of eukaryotes species in the fossil record. Before the spike in diversity, eukaryotes are thought to have lived in highly sulfuric environments. Sulfide interferes with mitochondrial function in aerobic organisms, limiting the amount of oxygen that could be used to drive metabolism. Oceanic sulfide levels decreased around 800 million years ago, which supports the importance of oxygen in eukaryotic diversity.

The shortage of oxygen might well have prevented the rise of large, complex animals. The amount of oxygen an animal can absorb is largely determined by the area of its oxygen-absorbing surfaces (lungs and gills in the most complex animals; the skin in less complex ones); but, the amount needed is determined by its volume, which grows faster than the oxygen-absorbing area if an animal's size increases equally in all directions. An increase in the concentration of oxygen in air or water would increase the size to which an organism could grow without its tissues becoming starved of oxygen. However, members of the Ediacara biota reached metres in length tens of millions of years before the Cambrian explosion.[313] Other metabolic functions may have been inhibited by lack of oxygen, for example the construction of tissue such as collagen, required for the construction of complex structures, or to form molecules for the construction of a hard exoskeleton. However, animals are not affected when similar oceanographic conditions occur in the Phanerozoic; there is no convincing correlation between oxygen levels and evolution, so oxygen may have been no more a prerequisite to complex life than liquid water or primary productivity.

Ozone formation

The amount of ozone (O_3) required to shield Earth from biologically lethal UV radiation, wavelengths from 200 to 300 nanometers (nm), is believed to have been in existence around the Cambrian explosion. The presence of the ozone layer may have enabled the development of complex life and life on land, as opposed to life being restricted in the water.

Snowball Earth

In the late Neoproterozoic (extending into the early Ediacaran period), the Earth suffered massive glaciations in which most of its surface was covered by ice. This may have caused a mass extinction, creating a genetic bottleneck; the resulting diversification may have given rise to the Ediacara biota, which appears soon after the last "Snowball Earth" episode. However, the snowball episodes occurred a long time before the start of the Cambrian, and it is hard to see how so much diversity could have been caused by even a series of bottlenecks; the cold periods may even have *delayed* the evolution of large size organisms.

Increase in the calcium concentration of the Cambrian seawater

Newer research suggests that volcanically active midocean ridges caused a massive and sudden surge of the calcium concentration in the oceans, making it possible for marine organisms to build skeletons and hard body parts.[314] Alternatively a high influx of ions could have been provided by the widespread erosion that produced Powell's Great Unconformity.

An increase of calcium may also have been caused by erosion of the Transgondwanan Supermountain that existed at the time the explosion. The roots of the mountain are preserved in present-day East Africa as an orogen.

Developmental explanations

A range of theories are based on the concept that minor modifications to animals' development as they grow from embryo to adult may have been able to cause very large changes in the final adult form. The Hox genes, for example, control which organs individual regions of an embryo will develop into. For instance, if a certain *Hox* gene is expressed, a region will develop into a limb; if a different Hox gene is expressed in that region (a minor change), it could develop into an eye instead (a phenotypically major change).

Such a system allows a large range of disparity to appear from a limited set of genes, but such theories linking this with the explosion struggle to explain why the origin of such a development system should by itself lead to increased diversity or disparity. Evidence of Precambrian metazoans combines with molecular data to show that much of the genetic architecture that could feasibly have played a role in the explosion was already well established by the Cambrian.

This apparent paradox is addressed in a theory that focuses on the physics of development. It is proposed that the emergence of simple multicellular forms provided a changed context and spatial scale in which novel physical processes and effects were mobilized by the products of genes that had previously evolved to serve unicellular functions. Morphological complexity (layers, segments, lumens, appendages) arose, in this view, by self-organization.

Horizontal gene transfer has also been identified as a possible factor in the rapid acquisition of the biochemical capability of biomineralization among organisms during this period, based on evidence that the gene for a critical protein in the process was originally transferred from a bacterium into sponges.

Ecological explanations

These focus on the interactions between different types of organism. Some of these hypotheses deal with changes in the food chain; some suggest arms races between predators and prey, and others focus on the more general mechanisms of coevolution. Such theories are well suited to explaining why there was a rapid increase in both disparity and diversity, but they must explain why the "explosion" happened when it did.

End-Ediacaran mass extinction

Evidence for such an extinction includes the disappearance from the fossil record of the Ediacara biota and shelly fossils such as *Cloudina*, and the accompanying perturbation in the $\delta^{13}C$ record.

Mass extinctions are often followed by adaptive radiations as existing clades expand to occupy the ecospace emptied by the extinction. However, once the dust had settled, overall disparity and diversity returned to the pre-extinction level in each of the Phanerozoic extinctions.

Evolution of eyes

Andrew Parker has proposed that predator-prey relationships changed dramatically after eyesight evolved. Prior to that time, hunting and evading were both close-range affairs – smell, vibration, and touch were the only senses used. When predators could see their prey from a distance, new defensive strategies were needed. Armor, spines, and similar defenses may also have evolved in response to vision. He further observed that, where animals lose vision in unlighted environments such as caves, diversity of animal forms tends to decrease. Nevertheless, many scientists doubt that vision could have caused the explosion. Eyes may well have evolved long before the start of the Cambrian. It is also difficult to understand why the evolution of eyesight would have caused an explosion, since other senses, such as smell and pressure detection, can detect things at a greater distance in the sea than sight can; but the appearance of these other senses apparently did not cause an evolutionary explosion.

Arms races between predators and prey

The ability to avoid or recover from predation often makes the difference between life and death, and is therefore one of the strongest components of natural selection. The pressure to adapt is stronger on the prey than on the predator: if the predator fails to win a contest, it loses a meal; if the prey is the loser, it loses its life.

But, there is evidence that predation was rife long before the start of the Cambrian, for example in the increasingly spiny forms of acritarchs, the holes drilled in *Cloudina* shells, and traces of burrowing to avoid predators. Hence, it is unlikely that the *appearance* of predation was the trigger for the Cambrian "explosion", although it may well have exhibited a strong influence on the body forms that the "explosion" produced. However, the intensity of predation does appear to have increased dramatically during the Cambrian as new predatory "tactics" (such as shell-crushing) emerged. This rise of predation during the Cambrian was confirmed by the temporal pattern of the median predator ratio at the scale of genus, in fossil communities covering the Cambrian and Ordovician periods, but this pattern is not correlated to diversification rate. This lack of correlation between predator ratio and diversification over the Cambrian and Ordovician suggests that predators did not trigger the large evolutionary radiation of animals during this interval. Thus the role of predators as triggerers of diversification may have been limited to the very beginning of the "Cambrian explosion".

Increase in size and diversity of planktonic animals

Geochemical evidence strongly indicates that the total mass of plankton has been similar to modern levels since early in the Proterozoic. Before the start of the Cambrian, their corpses and droppings were too small to fall quickly towards the seabed, since their drag was about the same as their weight. This meant they were destroyed by scavengers or by chemical processes before they reached the sea floor.

Mesozooplankton are plankton of a larger size. Early Cambrian specimens filtered microscopic plankton from the seawater. These larger organisms would have produced droppings and corpses that were large enough to fall fairly quickly. This provided a new supply of energy and nutrients to the mid-levels and bottoms of the seas, which opened up a huge range of new possible ways of life. If any of these remains sank uneaten to the sea floor they could be buried; this would have taken some carbon out of circulation, resulting in an increase in the concentration of breathable oxygen in the seas (carbon readily combines with oxygen).

The initial herbivorous mesozooplankton were probably larvae of benthic (seafloor) animals. A larval stage was probably an evolutionary innovation driven by the increasing level of predation at the seafloor during the Ediacaran period.

Metazoans have an amazing ability to increase diversity through coevolution. This means that an organism's traits can lead to traits evolving in other organisms; a number of responses are possible, and a different species can potentially emerge from each one. As a simple example, the evolution of predation may have caused one organism to develop a defence, while another developed motion to flee. This would cause the predator lineage to split into two species: one that was good at chasing prey, and another that was good at breaking through defences. Actual coevolution is somewhat more subtle, but, in this fashion, great diversity can arise: three quarters of living species are animals, and most of the rest have formed by coevolution with animals.

Ecosystem engineering

Evolving organisms inevitably change the environment they evolve in. The Devonian colonization of land had planet-wide consequences for sediment cycling and ocean nutrients, and was likely linked to the Devonian mass extinction. A similar process may have occurred on smaller scales in the oceans, with, for example, the sponges filtering particles from the water and depositing them in the mud in a more digestible form; or burrowing organisms making previously unavailable resources available for other organisms.

Complexity threshold

The explosion may not have been a significant evolutionary event. It may represent a threshold being crossed: for example a threshold in genetic complexity that allowed a vast range of morphological forms to be employed. This genetic threshold may have a correlation to the amount of oxygen available to organisms. Using oxygen for metabolism produces much more energy than anaerobic processes. Organisms that use more oxygen have the opportunity to produce more complex proteins, providing a template for further evolution. These proteins translate into larger, more complex structures that allow organisms better to adapt to their environments. With the help of oxygen, genes that code for these proteins could contribute to the expression of complex traits more efficiently. Access to a wider range of structures and functions would allow organisms to evolve in different directions, increasing the number of niches that could be inhabited. Furthermore, organisms had the opportunity to become more specialized in their own niches.

Uniqueness of the explosion

The "Cambrian explosion" can be viewed as two waves of metazoan expansion into empty niches: first, a coevolutionary rise in diversity as animals explored niches on the Ediacaran sea floor, followed by a second expansion in the early Cambrian as they became established in the water column. The rate of diversification seen in the Cambrian phase of the explosion is unparalleled among marine animals: it affected all metazoan clades of which Cambrian fossils have been found. Later radiations, such as those of fish in the Silurian and Devonian periods, involved fewer taxa, mainly with very similar body plans. Although the recovery from the Permian-Triassic extinction started with about as few animal species as the Cambrian explosion, the recovery produced far fewer significantly new types of animals.

Whatever triggered the early Cambrian diversification opened up an exceptionally wide range of previously unavailable ecological niches. When these were all occupied, limited space existed for such wide-ranging diversifications to occur again, because strong competition existed in all niches and incumbents usually had the advantage. If a wide range of empty niches had continued, clades would be able to continue diversifying and become disparate enough for us to recognise them as different phyla; when niches are filled, lineages will continue to resemble one another long after they diverge, as limited opportunity exists for them to change their life-styles and forms.

There were two similar explosions in the evolution of land plants: after a cryptic history beginning about 450[315] million years ago, land plants underwent a uniquely rapid adaptive radiation during the Devonian period, about 400[316] million years ago. Furthermore, Angiosperms (flowering plants) originated and rapidly diversified during the Cretaceous period.

Notes

Further reading

- Budd, G. E.; Jensen, J. (2000). "A critical reappraisal of the fossil record of the bilaterian phyla". *Biological Reviews*. **75** (2): 253–295. doi: 10.1111/j.1469-185X.1999.tb00046.x[317]. PMID 10881389[318].
- Collins, Allen G. "Metazoa: Fossil record"[319]. Retrieved Dec. 14, 2005.
- Conway Morris, S. (1997). *The Crucible of Creation: the Burgess Shale and the rise of animals*. Oxford University Press. ISBN 0-19-286202-2.
- Conway Morris, S. (June 2006). "Darwin's dilemma: the realities of the Cambrian 'explosion'"[320]. *Philosophical Transactions of the Royal Society B: Biological Sciences*. **361** (1470): 1069–1083. doi: 10.1098/rstb.2006.1846[321]. ISSN 0962-8436[322]. PMC 1578734[320] ⊚ . PMID 16754615[323]. An enjoyable account.
- Gould, S.J. (1989). *Wonderful Life: The Burgess Shale and the Nature of History*. W.W. Norton & Company.
- Kennedy, M.; M. Droser; L. Mayer.; D. Pevear & D. Mrofka (2006). "Clay and Atmospheric Oxygen". *Science*. **311** (5766): 1341. doi: 10.1126/science.311.5766.1341c[324].

- Knoll, A.H.; Carroll, S.B. (1999-06-25). "Early Animal Evolution: Emerging Views from Comparative Biology and Geology". *Science*. **284** (5423): 2129–37. doi: 10.1126/science.284.5423.2129[325]. PMID 10381872[326].

- Markov, Alexander V.; Korotayev, Andrey V. (2007). "Phanerozoic marine biodiversity follows a hyperbolic trend"[327]. *Palaeoworld*. **16** (4): 311–318. doi: 10.1016/j.palwor.2007.01.002[328].

- Montenari, M.; Leppig, U. (2003). "The Acritarcha: their classification morphology, ultrastructure and palaeoecological/palaeogeographical distribution". *Paläontologische Zeitschrift*. **77**: 173–194. doi: 10.1007/bf03004567[329].

- Wang, D. Y.-C.; S. Kumar; S. B. Hedges (January 1999). "Divergence time estimates for the early history of animal phyla and the origin of plants, animals and fungi"[330]. *Proceedings of the Royal Society B*. **266** (1415): 163–71. doi: 10.1098/rspb.1999.0617[331]. ISSN 0962-8452[332]. PMC 1689654[330] ⊚. PMID 10097391[333].

- Xiao, S.; Y. Zhang & A. Knoll (January 1998). "Three-dimensional preservation of algae and animal embryos in a Neoproterozoic phosphorite". *Nature*. **391** (1): 553–58. Bibcode: 1998Natur.391..553X[334]. doi: 10.1038/35318[335]. ISSN 0090-9556[336].

Timeline References:

- Martin, M.W; Grazhdankin, D.V; Bowring, S.A; Evans, D.A.D; Fedonkin, M.A; Kirschvink, J.L (2000). "Age of Neoproterozoic Bilaterian Body and Trace Fossils, White Sea, Russia: Implications for Metazoan Evolution". *Science*. **288** (5467): 841–845. Bibcode: 2000Sci...288..841M[337]. doi: 10.1126/science.288.5467.841[338]. PMID 10797002[339].

External links

- The Cambrian "explosion" of metazoans and molecular biology: would Darwin be satisfied?[340]
- On embryos and ancestors[341] by Stephen Jay Gould
- Conway Morris, S. (April 2000). "The Cambrian "explosion": Slow-fuse or megatonnage?"[342]. *Proceedings of the National Academy of Sciences*. **97**: 4426–4429. Bibcode: 2000PNAS...97.4426C[343]. doi: 10.1073/pnas.97.9.4426[344]. PMC 34314[345] ⊚. PMID 10781036[346].
- The Cambrian Explosion[347] – *In Our Time*[348], BBC Radio 4 broadcast, 17 February 2005
- "Burgess Shale"[349]. Virtual Museum of Canada. 2011., exhaustive details about the Burgess Shale, its fossils, and its significance for the Cambrian explosion

- Utah's Cambrian life[350] – new (2008) website with good images of a range of Burgess-shale-type and other Cambrian fossils
- Smithsonian National Museum[351]

Burgess Shale type fauna

Part of a series on
The Burgess Shale

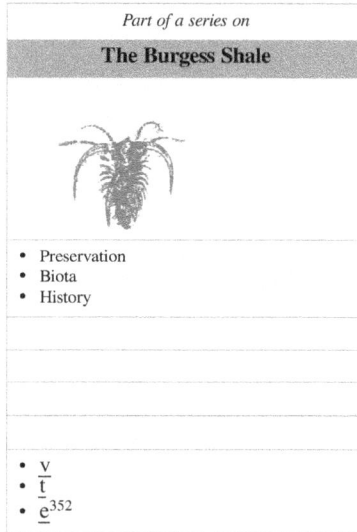

- Preservation
- Biota
- History

- $\underline{v}$
- $\underline{t}$
- $\underline{e}$[352]

A number of assemblages bear **fossil assemblages similar in character to that of the Burgess Shale**. While many are also preserved in a similar fashion to the Burgess Shale, the term "**Burgess Shale type fauna**" covers assemblages based on taxonomic criteria only.

Extent

The fauna of the middle Cambrian has a cosmopolitan range. All assemblages preserving soft-part anatomy have a very similar fauna, even though they span almost every continent. The wide distribution has been attributed to the advent of pelagic larvae.

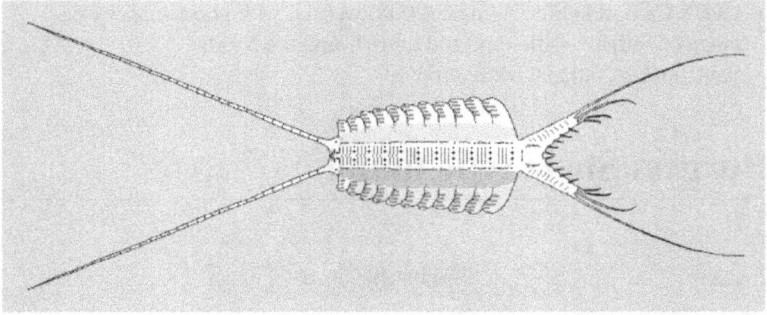

Figure 132: *Reconstruction of Kerygmachela from Sirius Passet, viewed from the top, with the head to the right. The shaded areas on the lobes are thought to have functioned as gills.*

Composition

The fauna is composed of a range of soft bodied organisms; creatures with hard, mineralised skeletons are rare, although trilobites are quite commonly found. The major soft-bodied groups are sponges, palaeoscolecid worms, lobopods, arthropods and anomalocaridids. Assemblages are typically diverse, with the most famous localities each containing in the region of 150 described species. The fauna of the Burgess Shale lived in the photic zone, as bottom-dwelling photosynthesisers are present in the assemblage.

Example faunas

Sirius Passet fauna

Sirius Passet is a lagerstätte in Greenland which was formed about 527 million years ago. Its most common fossils are arthropods, but there is only a handful of trilobite species. There are also very few species with hard parts: trilobites, hyoliths, sponges, brachiopods, and no echinoderms or molluscs.

Halkieria has features associated with more than one living phylum, and is discussed below.

The strangest-looking animals from Sirius Passet are *Pambdelurion* and *Kerygmachela*. They are generally regarded as anomalocarids because they have long, soft, segmented bodies with a pair of broad fin-like flaps on most segments and a pair of segmented appendages at the rear. The outer parts of the top surfaces of the flaps have grooved areas which are thought to have acted as gills. Under each flap there is a short, fleshy leg. This arrangement suggests the animals are related to biramous arthropods.

Chengjiang fauna

There are several Cambrian fossil sites in the Chengjiang county of China's Yunnan province. The most significant is the Maotianshan shale, a lagerstätte which preserves soft tissues very well. The Chengjiang fauna date to between 525 million and 520 million years ago, about the middle of the early Cambrian epoch, a few million years after Sirius Passet and at least 10 million years earlier than the Burgess Shale.

The Chengjiang sediments provide what are currently the oldest known chordates, the phylum to which all vertebrates belong. The 8 chordate species include *Myllokunmingia*, possibly a very primitive agnathid and *Haikouichthys*, which may be related to lampreys. *Yunnanozoon* may be the oldest known hemichordate.

Anomalocaris was a mainly soft-bodied swimming predator which was gigantic for its time (up to 70 cm = 2¼ feet long; some later species were 3 times as long); the soft, segmented body had a pair of broad fin-like flaps along each side, except that the last 3 segments had a pair of fans arranged in a V shape. Unlike *Kerygmachela* and *Pambdelurion* (see above), *Anomalocaris* apparently had no legs, and the grooved patches which are thought to have acted as gills were at the bases of the flaps, or even overlapping on to its back. The two eyes were on relatively long horizontal stalks; the mouth lay under the head and was a round-cornered square of plates which could not close completely; and in front of the mouth were two jointed appendages which were shaped like a shrimp's body, curved backwards and with short spines on the inside of the curve. *Amplectobelua*, also found at Chengjiang, was similar, smaller than *Anomalocaris* but considerably larger than most other Chengjiang animals. Both are thought to have been powerful predators.

Hallucigenia looks like a long-legged caterpillar with spines on its back, and almost certainly crawled on the seabed.

Nearly half of the Chengjiang fossil species are arthropods, few of which had the hard, mineral-reinforced exoskeletons found in most later marine arthropods; only about 3% of the organisms known from Chengjiang have hard shells, and most of those are trilobites (although *Misszhouia* is a *soft-bodied* trilobite). Many other phyla are found there: Porifera (sponges) and Priapulida (burrowing "worms" which were ambush predators), Brachiopoda (these had bivalve-like shells, but fed by means of a lophophore, a fan-like filter which occupied about of half of the internal space), Chaetognatha (arrow worms), Cnidaria, Ctenophora (comb jellies), Echinodermata, Hyolitha (Lophophorata with small conical shells), Nematomorpha, Phoronida (horseshoe worms), and Protista.

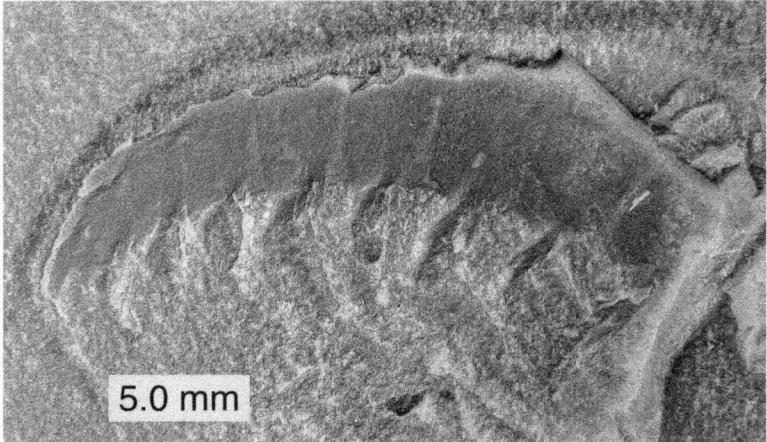

Figure 133: *Anomalocaridid "arm" from the Walcott Quarry,
Burgess Shale, Middle Cambrian, British Columbia, Canada.*

Burgess Shale

The Burgess Shale was the first of the Cambrian lagerstätten to be discovered
(by Walcott in 1909), and the re-analysis of the Burgess Shale by Whittington
and others in the 1970s was the basis of Gould's book *Wonderful Life*, which
was largely responsible for non-scientists' awareness of the Cambrian explosion. The fossils date from the mid Cambrian, about 515 million years ago
and 10 million years later than the Chengjiang fauna.

The shelled fossils in the Burgess Shale are similar in proportions to other shelly
fossil deposits; however, they are a minor component of the biota, accounding
for only 14% of the Burgess Shale fossils. When organisms that were not
preserved are entered into the equation, the shelly fossils probably represent
about 2% of the animals that were alive at the time.

Arthropods are the most abundant and diverse group of organisms in the
Burgess Shale, followed closely by sponges. Many Burgess Shale fossils are
unusual and difficult to classify, for example:

- *Marrella* is the most common fossil, but Whittington's re-analysis showed
  that it belonged to none of the known marine arthropod groups (trilobites,
  crustaceans, chelicerates).
- *Yohoia* was a tiny animal (7 mm to 23 mm long) with: a head shield; a
  slim, segmented body covered on top by armor plates; a paddle-like tail;
  3 pairs of legs under the head shield; a *single* flap-like appendage fringed
  with setae under each body segment, probably used for swimming and/or

Figure 134: *Reconstruction of Opabinia, one of the strangest animals from the Burgess Shale*

respiration; a pair of relatively large appendages at the front of the head shield, each with a pronounced "elbow" and ending in four long spines which may have functioned as "fingers". *Yohoia* is assumed to have been a mainly benthic (bottom-dwelling) creature that swam just above the ocean floor and used its appendages to scavenge or capture prey. It may be a member of the arachnomorphs, a group of arthropods that includes the chelicerates and trilobites.

- *Naraoia* was a soft-bodied animal which is classified as a trilobite because its appendages (legs, mouth-parts) are very similar.
- *Waptia*, *Canadaspis* and *Plenocaris* had bivalve-like carapaces. It is uncertain whether these animals are related or acquired bivalve-like carapaces by convergent evolution.
- *Molaria* was a chelicerate-like arthropod with a long, narrow tail.
- *Pikaia* resembled the modern lancelet, and was the earliest known chordate until the discovery of the fish-like *Myllokunmingia* and *Haikouichthys* among the Chengjiang fauna.

But the "weird wonders", creatures that resembled nothing known in the 1970s, attracted the most publicity, for example:

- Whittington's first presentation about *Opabinia* made the audience laugh.[353] The reconstruction showed a soft-bodied animal with: a slim, segmented body; a pair of flap-like appendages on each segment with gills above the flaps, except that the last 3 segments had no gills and the flaps formed a tail; *five* stalked eyes; a *backward*-facing mouth under the head; a long, flexible, hose-like proboscis which extended from under the front of the head and ended in a "claw" fringed with spines. Subsequent research has concluded that *Opabinia* is a lobopod, closely related to the arthropods and possibly even closer to ancestors of the arthropods.

- *Anomalocaris* and *Hallucigenia* were first found in the Burgess Shale, but older specimens have been found in the Chengjiang fauna. They are now regarded as lobopods, and *Anomalocaris* is very similar to *Opabinia* in most respects (except the eyes and feeding mechanisms) – see above.
- *Odontogriphus* is currently regarded as either a mollusc or a lophotro-chozoan, i.e. fairly closely related to the ancestors of molluscs (see above).

Other fauna

Other fauna include the Middle Cambrian Wheeler Shale Formation of Utah.

Ichnofauna

Trace fossils are associated with many Burgess Shale-type deposits. They are often associated with the innards of soft-bodied organisms,[354] and are particularly prevalent under the carapaces of bivalved arthropods. Burrowing organisms seem to have used the high-sulfur decay fluids as a nutrient source when farming bacteria in the microenvironment under the carapaces, indicated by their repeated uses of individual burrows.

Further sources

- Conway Morris, S. (1981). Taylor, M. E, ed. "The Burgess Shale fauna as a mid-Cambrian community"[355] (djvu). *Short papers for the Second International Symposium on the Cambrian System, 1981*. Open-File Report. U.S. Geological Survey,. 81-743: 47–49. USGS Library Call Number (200) R29o no.81-743.

Deuterostomes and the first vertebrates

Chordate

<table>
<tr><td colspan="2" align="center">Chordates
Temporal range:
Terreneuvian – Holocene, 542–0 Ma</td></tr>
</table>

Chordates
Temporal range: Terreneuvian – Holocene, 542–0 Ma PreЄ OSD C P T J K PgN

The Glass catfish (*Kryptopterus vitreolus*) is one of the few chordates with a visible backbone. The spinal cord is housed within its backbone.

Scientific classification 🖉	
Kingdom:	Animalia
Subkingdom:	Eumetazoa
Clade:	Bilateria
Clade:	Nephrozoa
Superphylum:	Deuterostomia
Phylum:	**Chordata** Haeckel, 1874[356]
Subgroups	

- Cephalochordata
  - †Pikaiidae
  - Leptocardii
- **Olfactores**
  - Tunicata
  - Craniata
  - †*Palaeospondylus*
  - †*Zhongxiniscus*
- †Vetulicolia

And see text

A **chordate** (/kɔːrdeɪt/) is an animal belonging to the phylum **Chordata**; chordates possess a notochord, a hollow dorsal nerve cord, pharyngeal slits, an endostyle, and a post-anal tail, for at least some period of their life cycle. Chordates are deuterostomes, as during the embryo development stage the anus forms before the mouth. They are also bilaterally symmetric coelomates with metameric segmentation and a circulatory system. In the case of vertebrate chordates, the notochord is usually replaced by a vertebral column during development.

Taxonomically, the phylum includes the following subphyla: the Vertebrata, which includes fish, amphibians, reptiles, birds, and mammals; the Tunicata, which includes salps and sea squirts; and the Cephalochordata, which include the lancelets. There are also additional extinct taxa such as the Vetulicolia. The Vertebrata are sometimes considered as a subgroup of the clade Craniata, consisting of chordates with a skull; the Craniata and Tunicata compose the clade Olfactores.

Of the more than 65,000 living species of chordates, about half are bony fish of the superclass Osteichthyes. The world's largest and fastest animals, the blue whale and peregrine falcon respectively, are chordates, as are humans. Fossil chordates are known from at least as early as the Cambrian explosion.

Hemichordata, which includes the acorn worms, has been presented as a fourth chordate subphylum, but it now is usually treated as a separate phylum. The Hemichordata, along with the Echinodermata (which includes starfish, sea urchins, sea cucumbers, and crinoids), form the Ambulacraria, the sister taxon of the Chordates. The Chordata and Ambulacraria form the superphylum Deuterostomia, composed of the deuterostomes.

Overview of affinities

Attempts to work out the evolutionary relationships of the chordates have produced several hypotheses. The current consensus is that chordates are monophyletic, meaning that the Chordata include all and only the descendants of a single common ancestor, which is itself a chordate, and that craniates' nearest relatives are tunicates.

All of the earliest chordate fossils have been found in the Early Cambrian Chengjiang fauna, and include two species that are regarded as fish, which implies that they are vertebrates. Because the fossil record of early chordates is poor, only molecular phylogenetics offers a reasonable prospect of dating their emergence. However, the use of molecular phylogenetics for dating evolutionary transitions is controversial.

It has also proved difficult to produce a detailed classification within the living chordates. Attempts to produce evolutionary "family trees" shows that many of the traditional classes are paraphyletic.

While this has been well known since the 19th century, an insistence on only monophyletic taxa has resulted in vertebrate classification being in a state of flux.

Origin of name

Although the name Chordata is attributed to William Bateson (1885), it was already in prevalent use by 1880. Ernst Haeckel described a taxon comprising tunicates, cephalochordates, and vertebrates in 1866. Though he used the German vernacular form, it is allowed under the ICZN code because of its subsequent latinization.

File:BranchiostomaLanceolatum PioM.svg

Anatomy of the cephalochordate *Amphioxus*. Bolded items are components of all chordates at some point in their lifetimes, and distinguish them from other phyla.

Definition

Chordates form a phylum of animals that are defined by having at some stage in their lives all of the following:

- A notochord, a fairly stiff rod of cartilage that extends along the inside of the body. Among the vertebrate sub-group of chordates the notochord develops into the spine, and in wholly aquatic species this helps the animal to swim by flexing its tail.
- A dorsal neural tube. In fish and other vertebrates, this develops into the spinal cord, the main communications trunk of the nervous system.
- Pharyngeal slits. The pharynx is the part of the throat immediately behind the mouth. In fish, the slits are modified to form gills, but in some other chordates they are part of a filter-feeding system that extracts particles of food from the water in which the animals live.
- Post-anal tail. A muscular tail that extends backwards behind the anus.

Figure 135: *Craniate: Hagfish*

- An endostyle. This is a groove in the ventral wall of the pharynx. In filter-feeding species it produces mucus to gather food particles, which helps in transporting food to the esophagus. It also stores iodine, and may be a precursor of the vertebrate thyroid gland.

There are soft constraints that separate chordates from certain other biological lineages, but have not yet been made part of the formal definition:

- All chordates are deuterostomes. This means that, during the embryo development stage, the anus forms before the mouth.
- All chordates are based on a bilateral body plan.[357]
- All chordates are coelomates, and have a fluid filled body cavity called a coelom with a complete lining called peritoneum derived from mesoderm (see Brusca and Brusca).[358]

There is still much ongoing differential (DNA sequence based) comparison research that is trying to separate out the simplest forms of chordates. As some lineages of the 90% of species that lack a backbone or notochord might have lost these structures over time, this complicates the classification of chordates. Some chordate lineages may only be found by DNA analysis, when there is no physical trace of any chordate-like structures.

Subdivisions

Craniata (Vertebrata)

Craniates, one of the three subdivisions of chordates, all have distinct skulls. They include the hagfish, which have no vertebrae. Michael J. Benton commented that "craniates are characterized by their heads, just as chordates, or possibly all deuterostomes, are by their tails".

Most craniates are vertebrates, in which the notochord is replaced by the vertebral column. These consist of a series of bony or cartilaginous cylindrical vertebrae, generally with neural arches that protect the spinal cord, and with projections that link the vertebrae. However hagfish have incomplete braincases and no vertebrae, and are therefore not regarded as vertebrates, but as members of the craniates, the group from which vertebrates are thought to have evolved. However the cladistic exclusion of hagfish from the vertebrates is controversial, as they may be degenerate vertebrates who have lost their vertebral columns.

The position of lampreys is ambiguous. They have complete braincases and rudimentary vertebrae, and therefore may be regarded as vertebrates and true fish. However, molecular phylogenetics, which uses biochemical features to classify organisms, has produced both results that group them with vertebrates and others that group them with hagfish. If lampreys are more closely related to the hagfish than the other vertebrates, this would suggest that they form a clade, which has been named the Cyclostomata.

Tunicata

Comparison of two invertebrate chordates

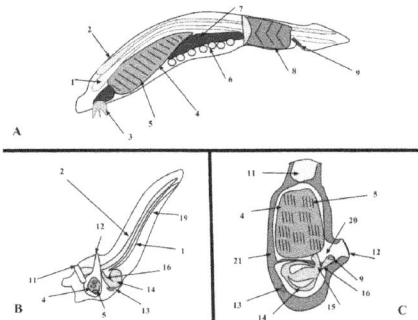

A. Lancelet, B. Larval tunicate, C. Adult tunicate

1. Notochord, 2. Nerve chord, 3. Buccal cirri, 4. Pharynx, 5. Gill slit, 6. Gonad, 7. Gut, 8. V-shaped muscles, 9. Anus, 10. Inhalant syphon, 11. Exhalant syphon, 12. Heart, 13. Stomach, 14. Esophagus, 15. Intestines, 16. Tail, 17. Atrium, 18. Tunic

Most tunicates appear as adults in two major forms, known as "sea squirts" and salps, both of which are soft-bodied filter-feeders that lack the standard features of chordates. Sea squirts are sessile and consist mainly of water pumps and filter-feeding apparatus; salps float in mid-water, feeding on plankton, and have a two-generation cycle in which one generation is solitary and the next

Figure 136: *Tunicates: sea squirts*

forms chain-like colonies. However, all tunicate larvae have the standard chordate features, including long, tadpole-like tails; they also have rudimentary brains, light sensors and tilt sensors. The third main group of tunicates, Appendicularia (also known as Larvacea), retain tadpole-like shapes and active swimming all their lives, and were for a long time regarded as larvae of sea squirts or salps. The etymology of the term Urochorda(ta) (Balfour 1881) is from the ancient Greek οὐρά (oura, "tail") + Latin chorda ("cord"), because the notochord is only found in the tail.[359] The term **Tunicata** (Lamarck 1816) is recognised as having precedence and is now more commonly used.

Cephalochordata: Lancelets

Cephalochordates are small, "vaguely fish-shaped" animals that lack brains, clearly defined heads and specialized sense organs. These burrowing filter-feeders compose the earliest-branching chordate sub-phylum.

Origins

The majority of animals more complex than jellyfish and other Cnidarians are split into two groups, the protostomes and deuterostomes, the latter of which contains chordates. It seems very likely the 555[360] million-year-old

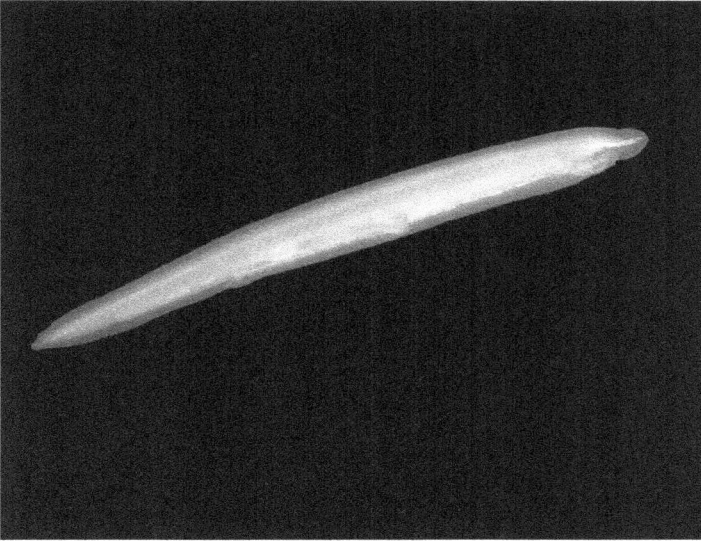

Figure 137: *Cephalochordate: Lancelet*

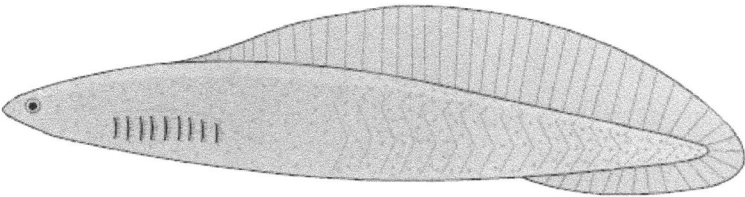

Figure 138: *Haikouichthys, from about 518[365] million years ago in China, may be the earliest known fish.*

Kimberella was a member of the protostomes. If so, this means the protostome and deuterostome lineages must have split some time before *Kimberella* appeared—at least 558[361] million years ago, and hence well before the start of the Cambrian 541[362] million years ago. The Ediacaran fossil *Ernietta*, from about 549 to 543[363] million years ago, may represent a deuterostome animal.[364]

Fossils of one major deuterostome group, the echinoderms (whose modern members include starfish, sea urchins and crinoids), are quite common from the start of the Cambrian, 542[366] million years ago. The Mid Cambrian fossil *Rhabdotubus johanssoni* has been interpreted as a pterobranch hemichordate. Opinions differ about whether the Chengjiang fauna fossil *Yunnanozoon*, from the earlier Cambrian, was a hemichordate or chordate. Another fossil, *Haikouella lanceolata*, also from the Chengjiang fauna, is interpreted as a chordate and possibly a craniate, as it shows signs of a heart, arteries, gill filaments, a tail, a neural chord with a brain at the front end, and possibly eyes—although it also had short tentacles round its mouth. *Haikouichthys* and *Myllokunmingia*, also from the Chengjiang fauna, are regarded as fish. *Pikaia*, discovered much earlier (1911) but from the Mid Cambrian Burgess Shale (505 Ma), is also regarded as a primitive chordate. On the other hand, fossils of early chordates are very rare, since invertebrate chordates have no bones or teeth, and only one has been reported for the rest of the Cambrian.

<templatestyles src="Template:Clade/styles.css" />

Deuteros-   <templatestyles src="Template:Clade/styles.css" />
tomes

      <templatestyles src="Template:Clade/styles.css" />
      Ambu-   <templatestyles src="Template:Clade/styles.css" />
      lacraria

             Hemichordates

             Echinoderms

Chordates   <templatestyles src="Template:Clade/styles.css" />
             Cephalochordates

      Olfactores   <templatestyles src="Template:Clade/styles.css" />
             Tunicates

             Craniates

A consensus family tree of the chordates

The evolutionary relationships between the chordate groups and between chordates as a whole and their closest deuterostome relatives have been debated since 1890. Studies based on anatomical, embryological, and paleontological data have produced different "family trees". Some closely linked chordates and hemichordates, but that idea is now rejected. Combining such analyses with data from a small set of ribosome RNA genes eliminated some older ideas, but opened up the possibility that tunicates (urochordates) are "basal deuterostomes", surviving members of the group from which echinoderms, hemichordates and chordates evolved. Some researchers believe that, within the chordates, craniates are most closely related to cephalochordates, but there are also reasons for regarding tunicates (urochordates) as craniates' closest relatives.

Since early chordates have left a poor fossil record, attempts have been made to calculate the key dates in their evolution by molecular phylogenetics techniques—by analyzing biochemical differences, mainly in RNA. One such study

Figure 139: *A skeleton of the blue whale, the world's largest animal, outside the Long Marine Laboratory at the University of California, Santa Cruz*

suggested that deuterostomes arose before 900[367] million years ago and the earliest chordates around 896[368] million years ago. However, molecular estimates of dates often disagree with each other and with the fossil record, and their assumption that the molecular clock runs at a known constant rate has been challenged.

Classification

Taxonomy

Traditionally, Cephalochordata and Craniata were grouped into the proposed clade "Euchordata", which would have been the sister group to Tunicata/Urochordata. More recently, Cephalochordata has been thought of as a sister group to the "Olfactores", which includes the craniates and tunicates. The matter is not yet settled.

The following schema is from the third edition of *Vertebrate Palaeontology*.[369] The invertebrate chordate classes are from Fishes of the World. While it is structured so as to reflect evolutionary relationships (similar to a cladogram), it also retains the traditional ranks used in Linnaean taxonomy.

- **Phylum Chordata**
  - †Vetulicolia
  - Subphylum **Cephalochordata** (Acraniata) – (lancelets; 30 species)
    - Class **Leptocardii** (lancelets)
  - Clade **Olfactores**
    - Subphylum **Tunicata** (Urochordata) – (tunicates; 3,000 species)

Figure 140: *A peregrine falcon, the world's fastest animal*

- Class **Ascidiacea** (sea squirts)
- Class **Thaliacea** (salps)
- Class **Appendicularia** (larvaceans)
- Class **Sorberacea**
- Subphylum **Vertebrata** (Craniata) (vertebrates – animals with back-bones; 57,674 species)
  - Infraphylum incertae sedis **Cyclostomata**
    - Superclass **'Agnatha'** paraphyletic (jawless vertebrates; 100+ species)
      - Class Myxini (hagfish; 65 species)
      - Class Petromyzontida or Hyperoartia (lampreys)
      - Class †Conodonta
      - Class †Myllokunmingiida
      - Class †Pteraspidomorphi
      - Class †Thelodonti
      - Class †Anaspida
      - Class †Cephalaspidomorphi
  - Infraphylum **Gnathostomata** (jawed vertebrates)
    - Class †**Placodermi** (Paleozoic armoured forms; paraphyletic in relation to all other gnathostomes)
    - Class **Chondrichthyes** (cartilaginous fish; 900+ species)

- Class †**Acanthodii** (Paleozoic "spiny sharks"; paraphyletic in relation to Chondrichthyes)
- Superclass **Osteichthyes** (bony fish; 30,000+ species)
  - Class **Actinopterygii** (ray-finned fish; about 30,000 species)
  - Class **Sarcopterygii** (lobe-finned fish: 8 species)
- Superclass **Tetrapoda** (four-limbed vertebrates; 28,000+ species) (The classification below follows Benton 2004, and uses a synthesis of rank-based Linnaean taxonomy and also reflects evolutionary relationships. Benton included the Superclass Tetrapoda in the Subclass Sarcopterygii in order to reflect the direct descent of tetrapods from lobe-finned fish, despite the former being assigned a higher taxonomic rank.)
  - Class **Amphibia** (amphibians; 7,000+ species)
  - Class **Sauropsida** (reptiles (including birds); 19,000+ species - 10,000+ species of birds and 9,500+ species of reptiles)[370,371]
  - Class **Synapsida** (mammals; 5,700+ species)

Phylogeny

Chordates

Phylogenetic tree of the Chordate phylum. Lines show probable evolutionary relationships, including extinct taxa, which are denoted with a dagger, †. Some are invertebrates. The positions (relationships) of the Lancelet, Tunicate, and Craniata clades are as reported <templatestyles src="Template:Clade/styles.css" />

Chordata `<templatestyles src="Template:Clade/styles.css" />`

Cephalochordata `<templatestyles src="Template:Clade/styles.css" />`
Amphioxus

Olfactores `<templatestyles src="Template:Clade/styles.css" />`
Haikouellat

Tunicata `<templatestyles src="Template:Clade/styles.css" />`
Appendicularia (formerly Larvacea)

Thaliacea

Ascidiacea

Vertebrata `<templatestyles src="Template:Clade/styles.css" />`
Agnatha `<templatestyles src="Template:Clade/styles.css" />`
Myxini (hagfish)

Hyperoartia (Petromyzontida) (Lampreys)

Cyclostomata

`<templatestyles src="Template:Clade/styles.css" />`
Myllokunmingia fengjiaoat

Zhongjianichthys rostratust

Conodontat

Cephalaspidomorphit

Pteraspidomorphit

osteostracant

Gnathostomata `<templatestyles src="Template:Clade/styles.css" />`
Antiarchit

`<templatestyles src="Template:Clade/styles.css" />`

Closest nonchordate relatives

Hemichordates

Hemichordates ("half chordates") have some features similar to those of chordates: branchial openings that open into the pharynx and look rather like gill slits; stomochords, similar in composition to notochords, but running in a circle round the "collar", which is ahead of the mouth; and a dorsal nerve cord—but also a smaller ventral nerve cord.

There are two living groups of hemichordates. The solitary enteropneusts, commonly known as "acorn worms", have long proboscises and worm-like bodies with up to 200 branchial slits, are up to 2.5 metres (8.2 ft) long, and burrow though seafloor sediments. Pterobranchs are colonial animals, often less than 1 millimetre (0.039 in) long individually, whose dwellings are interconnected. Each filter feeds by means of a pair of branched tentacles, and has a short, shield-shaped proboscis. The extinct graptolites, colonial animals whose fossils look like tiny hacksaw blades, lived in tubes similar to those of pterobranchs.

Echinoderms

Echinoderms differ from chordates and their other relatives in three conspicuous ways: they possess bilateral symmetry only as larvae - in adulthood they have radial symmetry, meaning that their body pattern is shaped like a wheel; they have tube feet; and their bodies are supported by skeletons made of calcite, a material not used by chordates. Their hard, calcified shells keep their bodies well protected from the environment, and these skeletons enclose their bodies, but are also covered by thin skins. The feet are powered by another unique feature of echinoderms, a water vascular system of canals that also functions as a "lung" and surrounded by muscles that act as pumps. Crinoids look rather like flowers, and use their feather-like arms to filter food particles out of the water; most live anchored to rocks, but a few can move very slowly. Other echinoderms are mobile and take a variety of body shapes, for example starfish, sea urchins and sea cucumbers.

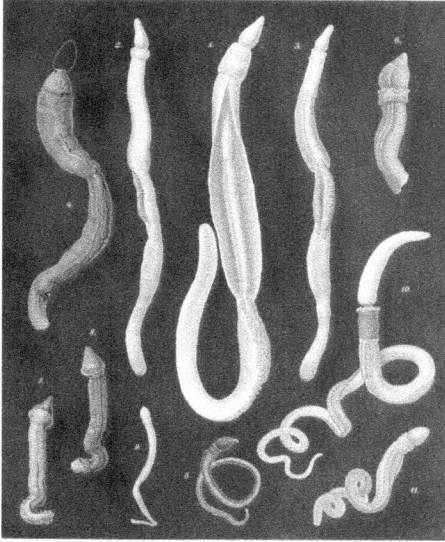

Figure 141: *Acorn worms or Enteropneusts are example of hemichordates.*

Figure 142: *A red knob sea star, Protoreaster linckii is an example of Asterozoan Echinoderm.*

External links

> Wikispecieshas information related to *Chordata*

> The Wikibook *Dichotomous Key*has a page on the topic of: *Chordata*

- *Chordate*[372] at the Encyclopedia of Life ✐
- Chordate on GlobalTwitcher.com[373]
- Chordate node at Tree Of Life[374]
- Chordate node at NCBI Taxonomy[375]

Evolution of fish

The **evolution of fish** began about 530 million years ago during the Cambrian explosion. It was during this time that the early chordates developed the skull and the vertebral column, leading to the first craniates and vertebrates. The first fish lineages belong to the Agnatha, or jawless fish. Early examples include *Haikouichthys*. During the late Cambrian, eel-like jawless fish called the conodonts, and small mostly armoured fish known as ostracoderms, first appeared. Most jawless fish are now extinct; but the extant lampreys may approximate ancient pre-jawed fish. Lampreys belong to the Cyclostomata, which includes the extant hagfish, and this group may have split early on from other agnathans.

The first jawed vertebrates probably developed during the late Ordovician period. They are first represented in the fossil record from the Silurian by two groups of fish: the armoured fish known as placoderms, which evolved from the ostracoderms; and the Acanthodii (or spiny sharks). The jawed fish that are still extant in modern days also appeared in late Silurian: the Chondrichthyes (or cartilaginous fish) and the Osteichthyes (or bony fish). The bony fish evolved into two separate groups: the Actinopterygii (or ray-finned fish) and Sarcopterygii (which includes the lobe-finned fish).

During the Devonian period a great increase in fish variety occurred, especially that of the ostracoderms and placoderms, as well as lobe-finned fish and early sharks. This has led to the Devonian being known as the *age of fishes*. It was from the lobe-finned fish that the tetrapods evolved, the four-limbed vertebrates, represented today by amphibians, mammals, reptiles and birds. Transitional tetrapods first appeared during the early Devonian, and by the

Figure 143: *The Devonian period 419–359 Ma (Age of Fishes) saw the development of early sharks, armoured placoderms and various lobe-finned fishes including the tetrapod transitional species*

late Devonian the first tetrapods appeared. The diversity of jawed vertebrates may indicate the evolutionary advantage of a jawed mouth; but it is unclear if the advantage of a hinged jaw is greater biting force, improved respiration, or a combination of factors. Fish do not represent a monophyletic group, but a paraphyletic one, as they exclude the tetrapods.[376]

Fish, like many other organisms, have been greatly affected by extinction events throughout natural history. The Ordovician–Silurian extinction events led to the loss of many species. The late Devonian extinction led to the extinction of the ostracoderms and placoderms by the end of the Devonian, as well as other fish. The spiny sharks became extinct at the Permian–Triassic extinction event; the conodonts became extinct at the Triassic–Jurassic extinction event. The Cretaceous–Paleogene extinction event, and the present day Holocene extinction, have also affected fish variety and fish stocks.

Overview

Vertebrate classes

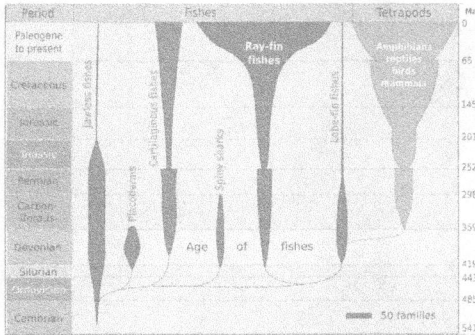

Spindle diagram for the evolution of fish and other vertebrate classes. The diagram is based on Michael Benton, 2005.[377] Conventional classification has living vertebrates as a subphylum grouped into eight classes based on traditional interpretations of gross anatomical and physiological traits. In turn, these classes are grouped into the vertebrates that have four limbs (the tetrapods) and those that do not: fishes. The extant vertebrate classes are:[378]

Fish:

- jawless fishes (Agnatha)
- cartilaginous fishes (Chondrichthyes)
- ray-finned fishes (Actinopterygii)
- lobe-finned fishes (Sarcopterygii)

Tetrapods:

- amphibians (Amphibia)
- reptiles (Reptilia)
- birds (Aves)
- mammals (Mammalia)

In addition to these are two classes of extinct jawed fishes, the armoured placoderms and the spiny sharks.

Fish may have evolved from an animal similar to a coral-like sea squirt (a tunicate), whose larvae resemble early fish in important ways. The first ancestors of fish may have kept the larval form into adulthood (as some sea squirts do today), although this path cannot be proven.

Vertebrates, among them the first fishes, originated about 530 million years ago during the Cambrian explosion, which saw the rise in organism diversity.[379]

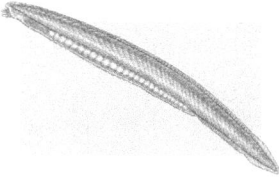

The lancelet, a small, translucent, fish-like animal, is the closest living invertebrate relative of the olfactoreans (vertebrates and tunicates).[380]

The early vertebrate *Haikouichthys*, from about 518[381] million years ago in China, may be the "ancestor to all vertebrates" and is one of the earliest known fish.

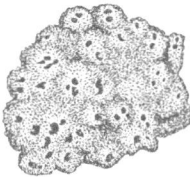

Somewhat dated view of continuous evolutionary gradation (click to animate)

The first ancestors of fish, or animals that were probably closely related to fish, were *Pikaia*, *Haikouichthys* and *Myllokunmingia*.[379] These three genera all appeared around 530 Ma. *Pikaia* had a primitive notochord, a structure that could have developed into a vertebral column later. Unlike the other fauna that dominated the Cambrian, these groups had the basic vertebrate body plan: a notochord, rudimentary vertebrae, and a well-defined head and tail. All of these early vertebrates lacked jaws in the common sense and relied on filter feeding close to the seabed.[382]

These were followed by indisputable fossil vertebrates in the form of heavily armoured fishes discovered in rocks from the Ordovician Period 500–430 Ma.

The first jawed vertebrates appeared in the late Ordovician and became common in the Devonian, often known as the "Age of Fishes".[383] The two groups of bony fishes, the actinopterygii and sarcopterygii, evolved and became common.[384] The Devonian also saw the demise of virtually all jawless fishes, save for lampreys and hagfish, as well as the Placodermi, a group of armoured fish that dominated much of the late Silurian. The Devonian also saw the rise of the first labyrinthodonts, which was a transitional between fishes and amphibians.

Figure 144: *A modern jawless fish, the lamprey, attached to a modern jawed fish*

The colonisation of new niches resulted in diversification of body plans and sometimes an increase in size. The Devonian Period (395 to 345 Ma) brought in such giants as the placoderm *Dunkleosteus*, which could grow up to seven meters long, and early air-breathing fish that could remain on land for extended periods. Among this latter group were ancestral amphibians.

The reptiles appeared from labyrinthodonts in the subsequent Carboniferous period. The anapsid and synapsid reptiles were common during the late Paleozoic, while the diapsids became dominant during the Mesozoic. In the sea, the bony fishes became dominant.

The later radiations, such as those of fish in the Silurian and Devonian periods, involved fewer taxa, mainly with very similar body plans. The first animals to venture onto dry land were arthropods. Some fish had lungs and strong, bony fins and could crawl onto the land also.

Jawless fish

Jawless fishes belong to the superclass Agnatha in the phylum Chordata, subphylum Vertebrata. Agnatha comes from the Greek, and means "no jaws".[385] It excludes all vertebrates with jaws, known as gnathostomes. Although a minor element of modern marine fauna, jawless fish were prominent among the early fish in the early Paleozoic. Two types of Early Cambrian animal apparently having fins, vertebrate musculature, and gills are known from the early Cambrian Maotianshan shales of China: *Haikouichthys* and *Myllokunmingia*. They have been tentatively assigned to Agnatha by Janvier. A third possible agnathid from the same region is *Haikouella*. A possible agnathid that has not been formally described was reported by Simonetti from the Middle Cambrian Burgess Shale of British Columbia.Wikipedia:Citation needed

Figure 145:
Lamprey mouth

Many Ordovician, Silurian, and Devonian agnathians were armoured with heavy bony-spiky plates. The first armoured agnathans—the Ostracoderms, precursors to the bony fish and hence to the tetrapods (including humans)—are known from the middle Ordovician, and by the Late Silurian the agnathans had reached the high point of their evolution. Most of the ostracoderms, such as thelodonts, osteostracans, and galeaspids, were more closely related to the gnathostomes than to the surviving agnathans, known as cyclostomes. Cyclostomes apparently split from other agnathans before the evolution of dentine and bone, which are present in many fossil agnathans, including conodonts. Agnathans declined in the Devonian and never recovered.

The agnathans as a whole are paraphyletic, because most extinct agnathans belong to the stem group of gnathostomes. Recent molecular data, both from rRNA and from mtDNA strongly supports the theory that living agnathans, known as cyclostomes, are monophyletic.[386] In phylogenetic taxonomy, the relationships between animals are not typically divided into ranks, but illustrated as a nested "family tree" known as a cladogram. Phylogenetic groups are given definitions based on their relationship to one another, rather than purely on physical traits such as the presence of a backbone. This nesting pattern is often combined with traditional taxonomy, in a practice known as evolutionary taxonomy.

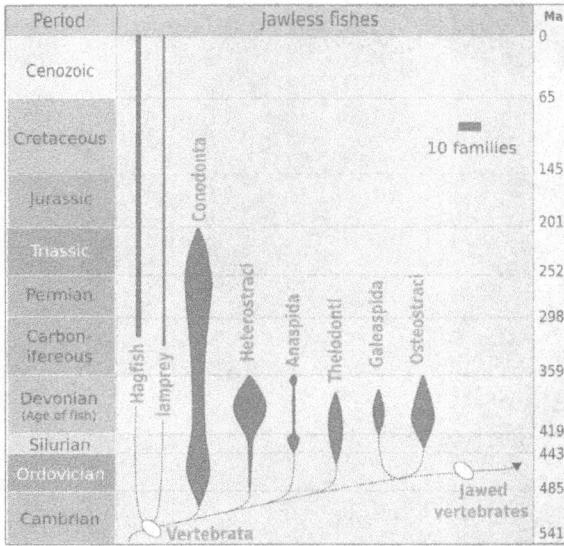

Figure 146: *Evolution of jawless fishes. The diagram is based on Michael Benton, 2005.*[387]

The cladogram below for jawless fish is based on studies compiled by Philippe Janvier and others for the *Tree of Life Web Project*.[388] († = group is extinct)

<templatestyles src="Template:Clade/styles.css" />

Jaw-   <templatestyles src="Template:Clade/styles.css" />
less fish

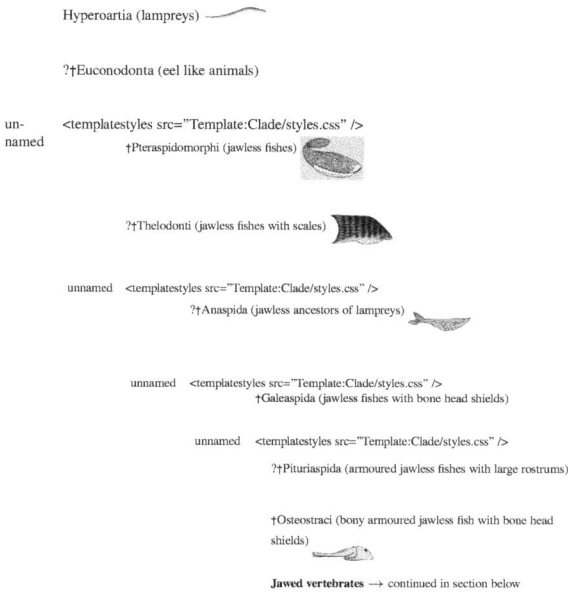

Hyperoartia (lampreys) ⎯⎯⎯⎯

?†Euconodonta (eel like animals)

un-   <templatestyles src="Template:Clade/styles.css" />
named   †Pteraspidomorphi (jawless fishes)

?†Thelodonti (jawless fishes with scales)

unnamed   <templatestyles src="Template:Clade/styles.css" />
                ?†Anaspida (jawless ancestors of lampreys)

unnamed   <templatestyles src="Template:Clade/styles.css" />
                †Galeaspida (jawless fishes with bone head shields)

unnamed   <templatestyles src="Template:Clade/styles.css" />
                ?†Pituriaspida (armoured jawless fishes with large rostrums)

                †Osteostraci (bony armoured jawless fish with bone head
                shields)

Jawed vertebrates ⟶ continued in section below

†Conodonts

Conodonts resembled primitive jawless eels. They appeared 495 Ma and were wiped out 200 Ma. Initially they were known only from tooth-like microfossils called *conodont elements*. These "teeth" have been variously interpreted as filter-feeding apparatuses or as a "grasping and crushing array". Conodonts ranged in length from a centimeter to the 40 cm *Promissum*. Their large eyes had a lateral position, which makes a predatory role unlikely. The preserved musculature hints that some conodonts (*Promissum* at least) were efficient cruisers but incapable of bursts of speed. In 2012 researchers classified the conodonts in the phylum Chordata on the basis of their fins with fin rays, chevron-shaped muscles and notochord. Some researchers see them as vertebrates similar in appearance to modern hagfish and lampreys, though phylogenetic analysis suggests that they are more derived than either of these groups.

†Ostracoderms

Ostracoderms *(shell-skinned)* are armoured jawless fishes of the Paleozoic. The term does not often appear in classifications today because it is paraphyletic or polyphyletic, and has no phylogenetic meaning.[389] However, the term is still used informally to group together the armoured jawless fishes.

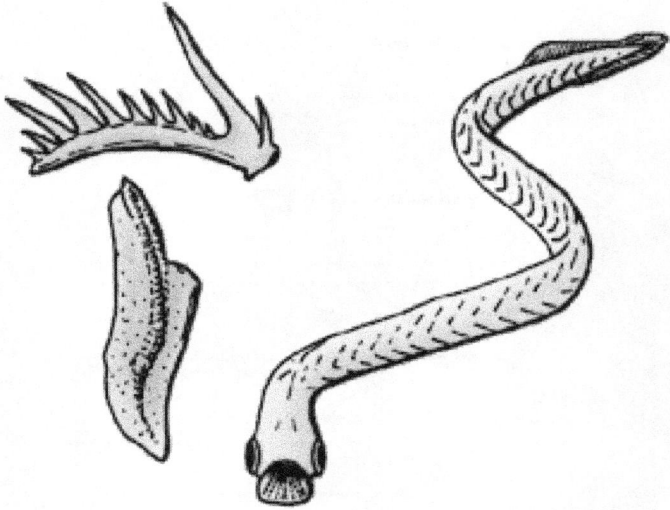

Figure 147:
†Conodonts (extinct) resembled primitive jawless eels

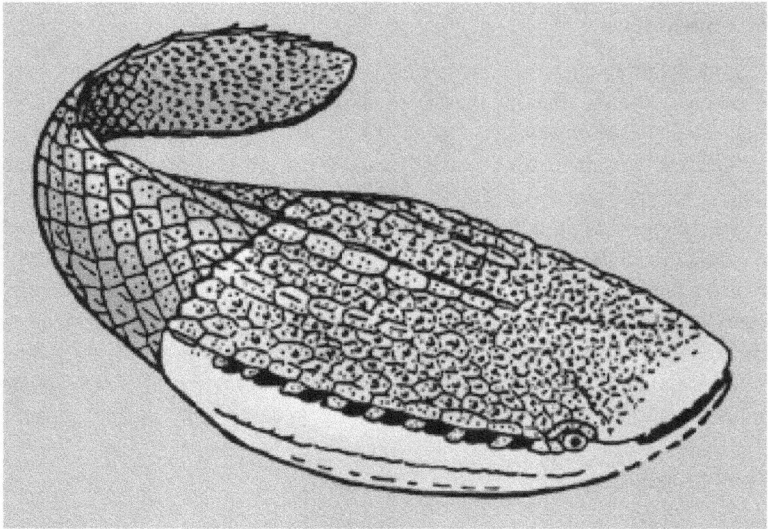

Figure 148:
†Ostracoderms (extinct) were armoured jawless fishes

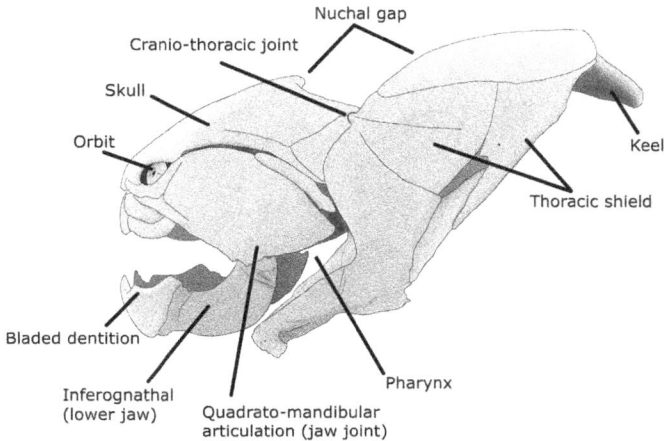

The ostracoderm armour consisted of 3–5 mm polygonal plates that shielded the head and gills, and then overlapped further down the body like scales. The eyes were particularly shielded. Earlier chordates used their gills for both respiration and feeding, whereas ostracoderms used their gills for respiration only. They had up to eight separate pharyngeal gill pouches along the side of the head, which were permanently open with no protective operculum. Unlike invertebrates that use ciliated motion to move food, ostracoderms used their muscular pharynx to create a suction that pulled small and slow moving prey into their mouths.

The first fossil fishes that were discovered were ostracoderms. The Swiss anatomist Louis Agassiz received some fossils of bony armored fish from Scotland in the 1830s. He had a hard time classifying them as they did not resemble any living creature. He compared them at first with extant armored fish such as catfish and sturgeons but later realizing that they had no movable jaws, classified them in 1844 into a new group "ostracoderms".

Ostracoderms existed in two major groups, the more primitive heterostracans and the cephalaspids. Later, about 420 million years ago, the jawed fish evolved from one of the ostracoderms. After the appearance of jawed fish, most ostracoderm species underwent a decline, and the last ostracoderms became extinct at the end of the Devonian period.[390]

Jawed fish

The vertebrate jaw probably originally evolved in the Silurian period and appeared in the Placoderm fish, which further diversified in the Devonian. The two most anterior pharyngeal arches are thought to have become the jaw itself and the hyoid arch, respectively. The hyoid system suspends the jaw from the braincase of the skull, permitting great mobility of the jaws. Already long assumed to be a paraphyletic assemblage leading to more derived gnathostomes, the discovery of *Entelognathus* suggests that placoderms are directly ancestral to modern bony fish.

As in most vertebrates, fish jaws are bony or cartilaginous and oppose vertically, comprising an *upper jaw* and a *lower jaw*. The jaw is derived from the most anterior two pharyngeal arches supporting the gills, and usually bears numerous teeth. The skull of the last common ancestor of today's jawed vertebrates is assumed to have resembled sharks.[391]

It is thought that the original selective advantage garnered by the jaw was not related to feeding, but to increased respiration efficiency. The jaws were used in the buccal pump (observable in modern fish and amphibians) that pumps water across the gills of fish or air into the lungs in the case of amphibians. Over evolutionary time the more familiar use of jaws (to humans), in feeding, was selected for and became a very important function in vertebrates. Many teleost fish have substantially modified their jaws for suction feeding and jaw protrusion, resulting in highly complex jaws with dozens of bones involved.

Jawed vertebrates and jawed fish evolved from earlier jawless fish, and the cladogram below for jawed vertebrates is a continuation of the cladogram in the section above. († = group is extinct)

<templatestyles src="Template:Clade/styles.css" />

Jawed vertebrates

<templatestyles src="Template:Clade/styles.css" />

unnamed †Placodermi (armoured fishes)

<templatestyles src="Template:Clade/styles.css" />

Bony fishesAcanthodians and Chondrichthyes (cartilaginous fishes)

<templatestyles src="Template:Clade/styles.css" />

Actinopterygii (ray-finned fishes) <dominant class of fish today

Lobe-finned fishes
<templatestyles src="Template:Clade/styles.css" />

?†Onychodontiformes (lobe-finned)

Actinistia (coelacanths)

un-named
<templatestyles src="Template:Clade/styles.css" />

†Porolepiformes (lobe-finned)

Dipnoi (lungfishes)

un-named
<templatestyles src="Template:Clade/styles.css" />

†Rhizodontimorpha (predatory lobe-finned)

<templatestyles src="Template:Clade/styles.css" />

†Tristichopteridae (tetrapodomorphs)

Tetrapods (four-legged animals)

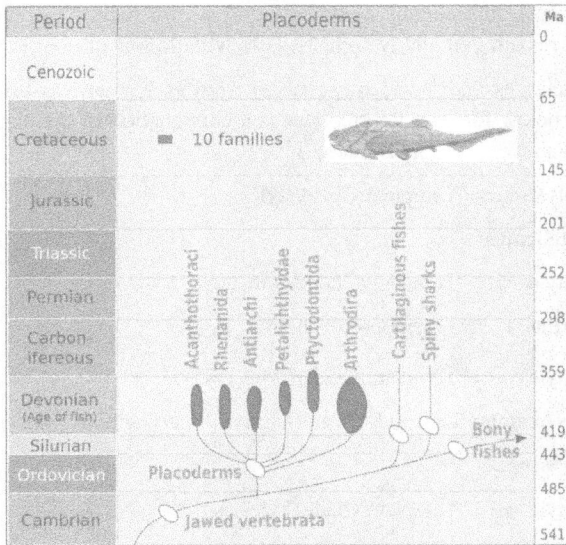

Figure 149: *Evolution of the (now extinct) placo-derms. The diagram is based on Michael Benton, 2005.*

Figure 150: †*Placoderms (extinct) were armoured jawed fishes (compare with the ostracoderms above)*

†Placoderms

Placoderms, class Placodermi *(plate skinned)*, are extinct armoured prehistoric fish, which appeared about 430 Ma in the Early to Middle Silurian. They were mostly wiped out during the Late Devonian Extinction event, 378 Ma, though some survived and made a slight recovery in diversity during the Famennian epoch before dying out entirely at the close of the Devonian, 360 mya; they are ultimately ancestral to modern vertebrates. Their head and thorax were covered with massive and often ornamented armoured plates. The rest of the body was scaled or naked, depending on the species. The armour shield was articulated, with the head armour hinged to the thoratic armour. This allowed placoderms

Figure 151: †*Spiny sharks (extinct) were the earliest known jawed fishes. They resembled sharks and were indeed ancestral to them.*

to lift their heads, unlike ostracoderms. Placoderms were the first jawed fish; their jaws likely evolved from the first of their gill arches. The chart on the right shows the rise and demise of the separate placoderm lineages: Acanthothoraci, Rhenanida, Antiarchi, Petalichthyidae, Ptyctodontida and Arthrodira.

†Spiny sharks

Spiny sharks, class Acanthodii, are extinct fishes that share features with both bony and cartilaginous fishes, though ultimately more closely related to and ancestral to the latter. Despite being called "spiny sharks", acanthodians pre-date sharks, though they gave rise to them. They evolved in the sea at the beginning of the Silurian Period, some 50 million years before the first sharks appeared. Eventually competition from bony fishes proved too much, and the spiny sharks died out in Permian times about 250 Ma. In form they resembled sharks, but their epidermis was covered with tiny rhomboid platelets like the scales of holosteans (gars, bowfins).

Cartilaginous fishes

Cartilaginous fishes, class Chondrichthyes, consisting of sharks, rays and chimaeras, appeared by about 395 million years ago, in the middle Devonian, evolving from acanthodians. The class contains the sub classes Holocephali (chimaera) and Elasmobranchii (sharks and rays). The radiation of elasmobranches in the chart on the right is divided into the taxa: Cladoselache, Eugeneodontiformes, Symmoriida, Xenacanthiformes, Ctenacanthiformes, Hybodontiformes, Galeomorphi, Squaliformes and Batoidea.

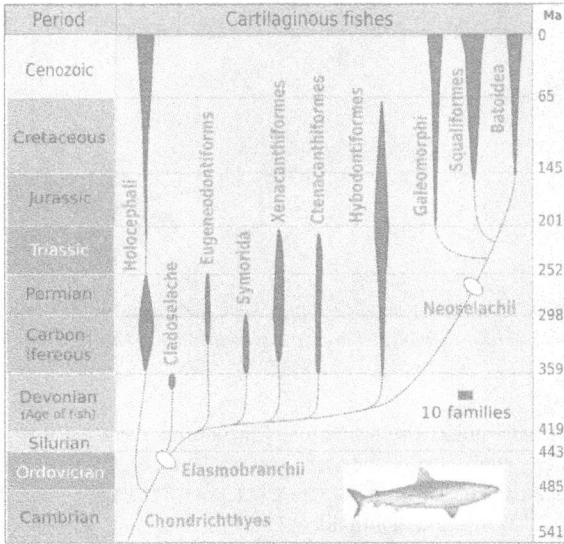

Figure 152: *Radiation of cartilaginous fishes, derived from work by Michael Benton, 2005.*[392]

Bony fishes

🎞 Chordate Evolution and Bony Fish[393] *YouTube*

Bony fishes, class Osteichthyes, are characterised by bony skeleton rather than cartilage. They appeared in the late Silurian, about 419 million years ago. The recent discovery of *Entelognathus* strongly suggests that bony fishes (and possibly cartilaginous fishes, via acanthodians) evolved from early placoderms. A subclass of the Osteichthyes, the ray-finned fishes (Actinopterygii), have become the dominant group of fishes in the post-Paleozoic and modern world, with some 30,000 living species.

The bony (and cartilaginous) fish groups that emerged after the Devonian, were characterised by steady improvements in foraging and locomotion.[394]

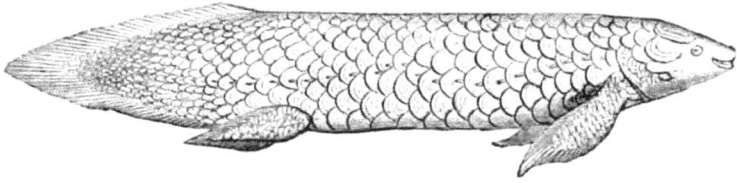

Figure 153:
The Queensland lungfish is a lobe-finned fish that is a living fossil. Lung-
fish evolved the first proto-lungs and proto-limbs. They developed the abil-
ity to live outside a water environment in the middle Devonian (397-385
Ma), and have remained virtually the same for over 100 million years.

Phylogenomic analysis has shown that "the closest living fish
to the tetrapod ancestor is the lungfish, not the coelacanth".

Lobe-finned fishes

Lobe-finned fishes, fish belonging to the class Sarcopterygii, are mostly extinct bony fishes, basally characterised by robust and stubby lobe fins containing a robust internal skeleton, cosmoid scales and internal nostrils. Their fins are fleshy, lobed, paired fins, joined to the body by a single bone.[395] The fins of lobe-finned fish differ from those of all other fish in that each is borne on a fleshy, lobelike, scaly stalk extending from the body. The pectoral and pelvic fins are articulated in ways resembling the tetrapod limbs they were the pre-cursors to. The fins evolved into the legs of the first tetrapod land vertebrates, amphibians. They also possess two dorsal fins with separate bases, as opposed to the single dorsal fin of ray-finned fish. The braincase of lobe-finned fishes primitively has a hinge line, but this is lost in tetrapods and lungfish. Many early lobe-finned fishes have a symmetrical tail. All lobe-finned fishes possess teeth covered with true enamel.

Lobe-finned fishes, such as coelacanths and lungfish, were the most diverse group of bony fishes in the Devonian. Taxonomists who subscribe to the cladistic approach include the grouping Tetrapoda within the Sarcopterygii, and the tetrapods in turn include all species of four-limbed vertebrates.[396] The fin-limbs of lobe-finned fishes such as the coelacanths show a strong simi-larity to the expected ancestral form of tetrapod limbs. The lobe-finned fish apparently followed two different lines of development and are accordingly separated into two subclasses, the Rhipidistia (including the lungfish, and the Tetrapodomorpha, which include the Tetrapoda) and the Actinistia (coela-canths). The first lobe-finned fishes, found in the uppermost Silurian (ca 418 Ma), closely resembled spiny sharks, which became extinct at the end of the

Figure 154: *The coelacanth is another lobe-finned fish that is a living fossil. It is thought to have evolved into roughly its current form about 408 million years ago, during the early Devonian, and has not essentially evolved further from its ancient form.*[397]

Paleozoic. In the early–middle Devonian (416 - 385 Ma), while the predatory placoderms dominated the seas, some lobe-finned fishes came into freshwater habitats.

In the Early Devonian (416-397 Ma), the lobe-finned fishes split into two main lineages — the coelacanths and the rhipidistians. The former never left the oceans and their heyday was the Late Devonian and Carboniferous, from 385 to 299 Ma, as they were more common during those periods than in any other period in the Phanerozoic; coelacanths still live today in the oceans (genus *Latimeria*). The Rhipidistians, whose ancestors probably lived in estuaries, migrated into freshwater habitats. They in turn split into two major groups: the lungfish and the tetrapodomorphs. The lungfish's greatest diversity was in the Triassic period; today there are fewer than a dozen genera left. The lungfish evolved the first proto-lungs and proto-limbs; developing the ability to live outside a water environment in the middle Devonian (397-385 Ma). The first tetrapodomorphs, which included the gigantic rhizodonts, had the same general anatomy as the lungfish, who were their closest kin, but they appear not to have left their water habitat until the late Devonian epoch (385 - 359 Ma), with the appearance of tetrapods (four-legged vertebrates). Tetrapods are the only tetrapodomorphs that survived after the Devonian. Lobe-finned fishes continued until towards the end of Paleozoic era, suffering heavy losses during the Permian-Triassic extinction event (251 Ma).

Evolution of ray-finned fishes

Ray-finned fishes

Ray-finned fishes, class Actinopterygii, differ from lobe-finned fishes in that their fins consist of webs of skin supported by spines ("rays") made of bone or horn. There are other differences in respiratory and circulatory structures. Ray-finned fishes normally have skeletons made from true bone, though this is not true of sturgeons and paddlefishes.[398]

Ray-finned fishes are the dominant vertebrate group, containing half of all known vertebrate species. They inhabit abyssal depths in the sea, coastal inlets and freshwater rivers and lakes, and are a major source of food for humans.

Timeline

Pre Devonian: Origin of fish

Cam-brian	Cambrian (541–485 Ma): The beginning of the Cambrian was marked by the Cambrian explosion, the sudden appearance of nearly all of the invertebrate animal phyla (molluscs, jellyfish, worms and arthropods, such as crustaceans) in great abundance. The first vertebrates appeared in the form of primitive fish, which were subsequently greatly diversified in the Silurian and Devonian.

	Pikaia	🎞 *Pikaia gracilens* animation[399] *The Burgess Shale* *Pikaia*, along with *Myllokunmingia* and *Haikouichthys ercaicunensis* immediately below, are all candidates in the fossil record for the titles of "first vertebrate" and "first fish". *Pikaia* is a genus that appeared about 530 Ma during the Cambrian explosion of multicellular life. *Pikaia gracilens (pictured)* is a transitional fossil between invertebrates and vertebrates,[400] and may be the earliest known chordate. In this sense it may have been the original ancestor of fishes. It was a primitive creature with no evidence of eyes, without a well defined head, and less than 2 inches (5 centimetres) long. *Pikaia* was a sideways-flattened, leaf-shaped animal that swam by throwing its body into a series of S-shaped, zig-zag curves, similar to movement of snakes. Fish inherited the same swimming movement, but they generally have stiffer backbones. It had a pair of large head tentacles and a series of short appendages, which may be linked to gill slits, on either side of its head. *Pikaia* shows the essential prerequisites for vertebrates. The flattened body is divided into pairs of segmented muscle blocks, seen as faint vertical lines. The muscles lie on either side of a flexible structure resembling a rod that runs from the tip of the head to the tip of the tail.[401]
	Haik- ouichthys	*Haikouichthys (fish from Haikou)* is another genus that also appears in the fossil record about 530 Ma, and also marks the transition from invertebrate to vertebrates.[402] Haikouichthys are craniates (animals with backbones and distinct heads). Unlike *Pikaia*, they had eyes. They also had a defined skull and other characteristics that have led paleontologists to label it a true craniate, and even to be popularly characterized as one of the earliest fishes. Cladistic analysis indicates that the animal is probably a basal chordate or a basal craniate; but it does not possess sufficient features to be included uncontroversially even in either stem group.[402]
	Myllokun- mingia	*Myllokunmingia* is a genus that appeared about 530 Ma. It is a chordate, and it has been argued that it is a vertebrate, It is 28 mm long and 6 mm high, and is among the oldest possible craniates.
	Conodont	Conodonts *(cone-teeth)* resembled primitive eels. They appeared 495 Ma and were wiped out 200 Ma. Initially they were known only from tooth-like microfossils called *conodont elements*. These "teeth" have been variously interpreted as filter-feeding apparatuses or as a "grasping and crushing array". Conodonts ranged in length from a centimeter to the 40 cm *Promissum.* Their large eyes had a lateral position of which makes a predatory role unlikely. The preserved musculature hints that some conodonts (*Promissum* at least) were efficient cruisers but incapable of bursts of speed. In 2012 researchers classify the conodonts in the phylum Chordata on the basis of their fins with fin rays, chevron-shaped muscles and notochord. Some researchers see them as vertebrates similar in appearance to modern hagfish and lampreys, though phylogenetic analysis suggests that they are more derived than either of these groups.
	Ostraco- derms	Ostracoderms *(shell-skinned)* are any of several groups of extinct, primitive, jawless fishes that were covered in an armour of bony plates. They appeared in the Cambrian, about 510 million years ago, and became extinct towards the end of the Devonian, about 377 million years ago. Initially Ostracoderms had poorly formed fins, and paired fins, or limbs, first evolved within this group. They were covered with a bony armour or scales and were often less than 30 cm (0.98 ft) long.

Ordov-ician		Ordovician (485–443 Ma): Fish, the world's first true vertebrates, continued to evolve, and those with jaws (Gnathostomata) may have first appeared late in this period. Life had yet to diversify on land.
	Aran-daspis	*Arandaspis* are jawless fish that lived in the early Ordovician period, about 480–470 Ma. It was about 15 cm (6 in) long, with a streamlined body covered in rows of knobbly armoured scutes. The front of the body and the head were protected by hard plates with openings for the eyes, nostrils and gills. Although it was jawless, *Arandaspis* might have had some moveable plates in its mouth, serving as lips, sucking in food particles. The low position of its mouth suggests it foraged the ocean floor. It lacked fins and its only method of propulsion was its horizontally flattened tail. As a result, it probably swam in a fashion similar to a modern tadpole.
	Astraspis	*Astraspis (star shield)* is an extinct genus of primitive jawless fish related to other Ordovician fishes, such as *Sacabambaspis* and *Arandaspis*. Fossils show clear evidence of a sensory structure (lateral line system). The arrangement of these organs in regular lines allows the fish to detect the direction and distance from which a disturbance in the water is coming. *Arandaspis* are thought to have had a mobile tail covered with small protective plates and a head region covered with larger plates. A specimen described by Sansom *et al.* had relatively large, lateral eyes and a series of eight gill openings on each side.
	Pteraspido-morphi	Pteraspidomorphi is an extinct class of early jawless fish. The fossils show extensive shielding of the head. Many had hypocercal tails to generate lift to increase ease of movement through the water for their armoured bodies, which were covered in dermal bone. They also had sucking mouth parts and some species may have lived in fresh water.
	Thelodonts	Thelodonts *(nipple teeth)* are a class of small, extinct jawless fishes with distinctive scales instead of large plates of armour. There is debate over whether these represent a monophyletic grouping, or disparate stem groups to the major lines of jawless and jawed fish. Thelodonts are united by their characteristic "thelodont scales". This defining character is not necessarily a result of shared ancestry, as it may have been evolved independently by different groups. Thus the thelodonts are generally thought to represent a polyphyletic group.[403] If they are monophyletic, there is no firm evidence on what their ancestral state was.[404] These scales were easily dispersed after death; their small size and resilience makes them the most common vertebrate fossil of their time.[405] The fish lived in both freshwater and marine environments, first appearing during the Ordovician, and perishing during the Frasnian–Famennian extinction event of the Late Devonian. They were predominantly deposit-feeding bottom dwellers, although some species may have been pelagic.
		The Ordovician ended with the Ordovician–Silurian extinction event (450–440 Ma). Two events occurred that killed off 27% of all families, 57% of all genera and 60% to 70% of all species. Together they are ranked by many scientists as the second largest of the five major extinctions in Earth's history in terms of percentage of genera that became extinct.
Sil-urian		Silurian (443–419 Ma): Many evolutionary milestones occurred during this period, including the appearance of armoured jawless fish, jawed fish, spiny sharks and ray-finned fish.
		While it is traditional to refer to the Devonian as the age of fishes, recent findings have shown the Silurian was also a period of considerable diversification. Jawed fish developed movable jaws, adapted from the supports of the front two or three gill arches

	Anaspida	Anaspida *(without shield)* is an extinct class of primitive jawless vertebrates that lived during the Silurian and Devonian periods.[406] They are classically regarded as the ancestors of lampreys.[407] Anaspids were small, primarily marine agnathans that lacked heavy bony shield and paired fins, but have highly exaggerated hypocercal tails. They first appeared in the early Silurian, and flourished until the Late Devonian extinction,[408] where most species, save for lampreys, became extinct. Unusually for an agnathan, anaspids did not possess a bony shield or armour. The head is instead covered in an array of smaller, weakly mineralised scales.[409]
	Osteostraci	Osteostraci ("bony shields") was a class of bony-armored jawless fish that lived from the Middle Silurian to Late Devonian. Anatomically speaking, the osteostracans, especially the Devonian species, were among the most advanced of all known agnathans. This is due to the development of paired fins, and their complicated cranial anatomy. The osteostracans were more similar to lampreys than to jawed vertebrates in possessing two pairs of semicircular canals in the inner ear, as opposed to the three pairs found in the inner ears of jawed vertebrates. Most osteostracans had a massive cephalothorac shield, but all Middle and Late Devonian species appear to have had a reduced, thinner, and often micromeric dermal skeleton. They were probably relatively good swimmers, possessing dorsal fins, paired pectoral fins, and a strong tail.
	Spiny sharks	Spiny sharks, more formally called "Acanthodians" *(having spines)*, constitute the class Acanthodii. They first appeared by the late Silurian ~420 Ma, and were among the first fishes to evolve jaws. They share features with both cartilaginous fish and bony fish, but they are not true sharks, though leading to them. They became extinct before the end of the Permian ~250 Ma. However, scales and teeth attributed to this group, as well as more derived jawed fish, such as cartilaginous and bony fish, date from the Ordovician ~460 Ma. Acanthodians were generally small shark-like fishes varying from toothless filter-feeders to toothed predators. They were once classified as an order of the class Placodermi, but recent authorities tend to place the acanthodians as a paraphyletic assemblage leading to modern cartilaginous fish. They are distinguished in two respects: they were the earliest known jawed vertebrates, and they had stout spines supporting all their fins, fixed in place and non-movable (like a shark's dorsal fin), an important defensive adaptation. Their fossils are extremely rare.
	Placo-derms	Placoderms, *(plate-like skin)*, are a group of armoured jawed fishes, of the class Placodermi. The oldest fossils appeared during the late Silurian, and became extinct at the end of the Devonian. Recent studies suggest that the placoderms are possibly a paraphyletic group of basal jawed fishes, and the closest relatives of all living jawed vertebrates. Some placoderms were small, flattened bottom-dwellers, such as antiarchs. However many, particularly the arthrodires, were active midwater predators. *Dunkleosteus*, which appeared later in the Devonian below, was the largest and most famous of these. The upper jaw was firmly fused to the skull, but there was a hinge joint between the skull and the bony plating of the trunk region. This allowed the upper part of the head to be thrown back and, in arthrodires, allowed them to take larger bites.
	Guiyu oneiros	*Guiyu oneiros*, the earliest known bony fish. It has the combination of both ray-finned and lobe-finned features, although analysis of the totality of its features place it closer to lobe-finned fish.

		An-dreolepis	The extinct genus *Andreolepis* includes the earliest known ray finned fish *Andreolepis hedei*, which appeared in the late Silurian, around 420 Ma.

Devonian: Age of fishes

Events of the Devonian Period

Key events of the Devonian Period.

Axis scale: millions of years ago.

The Devonian Period is broken into the Early, Middle and Late Devonian. By the start of the Early Devonian 419 mya, jawed fishes had divided into four

distinct clades: the placoderms and spiny sharks, both of which are now extinct, and the cartilaginous and bony fishes, both of which are still extant. The
modern bony fishes, class Osteichthyes, appeared in the late Silurian or early
Devonian, about 416 million years ago. Both the cartilaginous and bony fishes
may have arisen from either the placoderms or the spiny sharks. A subclass
of bony fishes, the ray-finned fishes (Actinopterygii), have become the dominant group in the post-Paleozoic and modern world, with some 30,000 living
species.

Sea levels in the Devonian were generally high. Marine faunas were dominated
by bryozoa, diverse and abundant brachiopods, the enigmatic hederelloids, mi-
croconchids and corals. Lily-like crinoids were abundant, and trilobites were
still fairly common. Among vertebrates, jawless armoured fish (ostracoderms)
declined in diversity, while the jawed fish (gnathostomes) simultaneously increased in both the sea and fresh water. Armoured placoderms were numerous during the lower stages of the Devonian Period but became extinct in the
Late Devonian, perhaps because of competition for food against the other fish
species. Early cartilaginous (Chondrichthyes) and bony fishes (Osteichthyes)
also become diverse and played a large role within the Devonian seas. The
first abundant genus of shark, *Cladoselache*, appeared in the oceans during the
Devonian Period. The great diversity of fish around at the time have led to the
Devonian being given the name "The Age of Fish" in popular culture.

The first ray-finned and lobe-finned bony fish appeared in the Devonian, while
the placoderms began dominating almost every known aquatic environment.
However, another subclass of Osteichthyes, the Sarcopterygii, including lobe-
finned fishes including coelacanths and lungfish) and tetrapods, was the most
diverse group of bony fishes in the Devonian. Sarcopterygians are basally
characterized by internal nostrils, lobe fins containing a robust internal skeleton, and cosmoid scales.

During the Middle Devonian 393–383 Ma, the armoured jawless ostracoderm
fishes were declining in diversity; the jawed fish were thriving and increasing
in diversity in both the oceans and freshwater. The shallow, warm, oxygen-
depleted waters of Devonian inland lakes, surrounded by primitive plants, provided the environment necessary for certain early fish to develop essential characteristics such as well developed lungs and the ability to crawl out of the water
and onto the land for short periods of time. Cartilaginous fishes, class Chondrichthyes, consisting of sharks, rays and chimaeras, appeared by about 395
million years ago, in the middle Devonian

During the Late Devonian the first forests were taking shape on land. The
first tetrapods appear in the fossil record over a period, the beginning and end
of which are marked with extinction events. This lasted until the end of the
Devonian 359 mya. The ancestors of all tetrapods began adapting to walking

on land, their strong pectoral and pelvic fins gradually evolved into legs (see *Tiktaalik*). In the oceans, primitive sharks became more numerous than in the Silurian and the late Ordovician. The first ammonite mollusks appeared. Trilobites, the mollusk-like brachiopods and the great coral reefs, were still common.

The Late Devonian extinction occurred at the beginning of the last phase of the Devonian period, the Famennian faunal stage, (the Frasnian-Famennian boundary), about 372.2 Ma. Many fossil agnathan fishes, save for the psammosteid heterostracans, make their last appearance shortly before this event. The Late Devonian extinction crisis primarily affected the marine community, and selectively affected shallow warm-water organisms rather than cool-water organisms. The most important group affected by this extinction event were the reef-builders of the great Devonian reef-systems.

A second extinction pulse, the Hangenberg event closed the Devonian period and had a dramatic impact on vertebrate faunas. Placoderms mostly became extinct during this event, as did most members of other groups including lobe-finned fishes, acanthodians and early tetrapods in both marine and terrestrial habitats, leaving only a handful of survivors. This event has been related to glaciation in the temperate and polar zones as well as euxinia and anoxia in the seas.

Devonian (419–359 mya): The start of Devonian saw the first appearance of lobe-finned fish, precursors to the tetrapods (animals with four limbs). Major groups of fish evolved during this period, often referred to as the **age of fishes**. See Category:Devonian fish.			
D e v o n i a n	Early Devonian		Early Devonian (419–393 Ma):
		Psarolepis	*Psarolepis (speckled scale)* is a genus of extinct lobe-finned fish that lived around 397 to 418 Ma. Fossils of *Psarolepis* have been found mainly in South China and described by paleontologist Xiaobo Yu in 1998. It is not known for certain which group *Psarolepis* belongs, but paleontologists agree that it probably is a basal genus and seems to be close to the common ancestor of lobe-finned and ray-finned fishes.[410]
		Holoptychius	*Holoptychius* is an extinct genus from the order of porolepiform lobe-finned fish, extant from 416 to 359 Ma. It was a stream-lined predator about 50 centimetres (20 in) long (though it could grow up to 2.5 m), which fed on other bony fish. Its rounded scales and body form indicate that it could have swum quickly through the water to catch prey.[411] Similar to other rhipidistians, it had fang-like teeth on its palate in addition to smaller teeth on the jaws. Its asymmetrical tail sported a caudal fin on its lower end. To compensate for the downward push caused by this fin placement, *Holoptychius*'s pectoral fins were placed high on the body.

		Ptyctodon-tida	The ptyctodontids (*beak-teeth*) are an extinct monotypic order of unarmored placoderms, containing only one family. They were extant from the start to the end of the Devonian. With their big heads, big eyes, and long bodies, the ptyctodontids bore a strong resemblance to modern day chimaeras (Holocephali). Their armor was reduced to a pattern of small plates around the head and neck. Like the extinct and related acanthothoracids, and the living and unrelated holocephalians, most of the ptyctodontids are thought to have lived near the sea bottom and preyed on shellfish.
		Petalichthyida	The Petalichthyida was an order of small, flattened placoderms that existed from the beginning of the Devonian to the Late Devonian. They were typified by splayed fins and numerous tubercles that decorated all of the plates and scales of their armour. They reached a peak in diversity during the Early Devonian and were found throughout the world. Because they had compressed body forms, it is supposed they were bottom-dwellers that chased after or ambushed smaller fish. Their diet is not clear, as none of the fossil specimens found have preserved mouth parts.
		Laccog-nathus	*Laccognathus* (*pitted jaw*) was a genus of amphibious lobe-finned fish that existed 398–360 Ma. They were characterized by the three large pits (fossae) on the external surface of the lower jaw, which may have had sensory functions. *Laccognathus* grew to 1–2 metres (3–7 ft) in length. They had very short dorsoventrally flattened heads, less than one-fifth the length of the body. The skeleton was structured so large areas of skin were stretched over solid plates of bone. This bone was composed of particularly dense fibers – so dense that exchange of oxygen through the skin was unlikely. Rather, the dense ossifications served to retain water inside the body as *Laccognathus* traveled on land between bodies of water.
Mid-dle De-vo-nian		Middle Devonian (393–383 Ma): Cartilaginous fishes, consisting of sharks, rays and chimaeras, appeared about 395 Ma.	
		Dipterus	*Dipterus* (*two wings*) is an extinct genus of lungfish from 376–361 Ma. It was about 35 centimetres (14 in) long, mostly ate invertebrates, and had lungs, not an air bladder. Like its ancestor *Dipnorhynchus* it had tooth-like plates on its palate instead of real teeth. However, unlike its modern relatives, in which the dorsal, caudal, and anal fin are fused into one, its fins were still separated. Otherwise *Dipterus* closely resembled modern lungfish.[412]
		Cheirolepis	*Cheirolepis* (*hand fin*) was a genus of ray-finned fishes. It was among the most basal of the Devonian ray-finned fishes and is considered the first to possess the "standard" dermal cranial bones seen in later ray-finned fish. It was a predatory freshwater fish about 55 centimetres (22 in) long, and based on the size of its eyes it hunted by sight.

	Cladoselache	*Cladoselache* was the first abundant genus of primitive shark, appearing about 370 Ma.[413] It grew to 6 feet (1.8 m) long, with anatomical features similar to modern mackerel sharks. It had a streamlined body almost entirely devoid of scales, with five to seven gill slits and a short, rounded snout that had a terminal mouth opening at the front of the skull.[413] It had a very weak jaw joint compared with modern-day sharks, but it compensated for that with very strong jaw-closing muscles. Its teeth were multi-cusped and smooth-edged, making them suitable for grasping, but not tearing or chewing. *Cladoselache* therefore probably seized prey by the tail and swallowed it whole.[413] It had powerful keels that extended onto the side of the tail stalk and a semi-lunate tail fin, with the superior lobe about the same size as the inferior. This combination helped with its speed and agility, which was useful when trying to outswim its probable predator, the heavily armoured 10 metres (33 ft) long placoderm fish *Dunkleosteus*.[413]
	Coccosteus	*Coccosteus (seed bone)* is an extinct genus of arthrodire placoderm. The majority of fossils have been found in freshwater sediments, though they may have been able to enter saltwater. They grew up to 40 centimetres (16 in) long. Like all other arthrodires, *Coccosteus* had a joint between the armour of the body and skull. In addition, it also had an internal joint between its neck vertebrae and the back of the skull, allowing it to open its mouth even wider. Along with the longer jaws, this allowed *Coccosteus* to feed on fairly large prey. As with all other arthrodires, *Coccosteus* had bony dental plates embedded in its jaws, forming a beak. The beak was kept sharp by having the edges of the dental plates grind away at each other.[414]
	Bothriolepis	▣ *Bothriolepis*[415] – *Animal Planet* *Bothriolepis (pitted scale)* was the most successful genus of antiarch placoderms, if not the most successful genus of any placoderm, with over 100 species spread across Middle to Late Devonian strata across every continent.
	Pituriaspida	Pituriaspida *(hallucinogenic shield)* is a class containing two bizarre species of armoured jawless fishes with tremendous nose-like rostrums. They lived in estuaries around 390 Ma. The paleontologist Gavin Young, named the class after the hallucinogenic drug pituri, since he thought he might be hallucinating upon viewing the bizarre forms.[416] The better studied species looked like a throwing-dart-like, with an elongate headshield and spear-like rostrum. The other species looked like a guitar pick with a tail, with a smaller and shorter rostrum and a more triangular headshield.

Late Devonian extinction: 375–360 Ma. A prolonged series of extinctions eliminated about 19% of all families, 50% of all genera and 70% of all species. This extinction event lasted perhaps as long as 20 Ma, and there is evidence for a series of extinction pulses within this period.

Late Devonian	Late Devonian (383–359 Ma):

	Dunkleosteus	🎞 *Dunkleosteus*[417] – *Animal Planet* *Dunkleosteus* is a genus of arthrodire placoderms that existed from 380 to 360 Ma. It grew up to 10 metres (33 ft) long[418] and weighed up to 3.6 tonnes. It was a hypercarnivorous apex predator. Apart from its contemporary *Titanichthys* (below), no other placoderm rivalled it in size. Instead of teeth, *Dunkleosteus* had two pairs of sharp bony plates, which formed a beak-like structure. Apart from megalodon, it had the most powerful bite of any fish, generating bite forces in the same league as *Tyrannosaurus rex* and the modern crocodile.
	Titanichthys	*Titanichthys* is a genus of giant, aberrant marine placoderm that lived in shallow seas. Many of the species approached *Dunkleosteus* in size and build. Unlike its relative, however, the various species of *Titanichys* had small, ineffective-looking mouth-plates that lacked a sharp cutting edge. It is assumed that *Titanichthys* was a filter feeder that used its capacious mouth to swallow or inhale schools of small, anchovy-like fish, or possibly krill-like zooplankton, and that the mouth-plates retained the prey while allowing the water to escape as it closed its mouth.
	Materpiscis	🎞 The mother fish[419] – *Nature* *Materpiscis (mother fish)* is a genus of ptyctodontid placoderm from about 380 Ma. Known from only one specimen, it is unique in having an unborn embryo present inside, and with remarkable preservation of a mineralised placental feeding structure (umbilical cord). This makes *Materpiscis* the first known vertebrate to show viviparity, or giving birth to live young. The specimen was named *Materpiscis attenboroughi* in honour of David Attenborough.
	Hyneria	Hyneria is a genus of predatory lobe-finned fish, about 2.5 m (8.2 ft) long, that lived 360 million years ago.
	Rhizodonts	Rhizodonts were an order of lobe-finned fish that survived to the end of the Carboniferous, 377–310 Ma. They reached huge sizes. The largest known species, *Rhizodus hibberti* grew up to 7 metres in length, making it the largest freshwater fish known.

Fish to tetrapods

The first tetrapods are four-legged, air-breathing, terrestrial animals from which the land vertebrates descended, including humans. They evolved from lobe-finned fish of the clade Sarcopterygii, appearing in coastal water in the middle Devonian, and giving rise to the first amphibians.

The group of lobe-finned fishes that were the ancestors of the tetrapod are grouped together as the Rhipidistia, and the first tetrapods evolved from these fish over the relatively short timespan 385–360 Ma. The early tetrapod groups themselves are grouped as Labyrinthodontia. They retained aquatic, fry-like tadpoles, a system still seen in modern amphibians. From the 1950s to the early 1980s it was thought that tetrapods evolved from fish that had already acquired the ability to crawl on land, possibly so they could go from a pool that was drying out to one that was deeper. However, in 1987, nearly complete

Figure 155:
A cladogram of the evolution of tetrapods showing some of the best-known transitional fossils. It starts with Eusthenopteron at the bottom, indisputably still a fish, through Panderichthys, Tiktaalik, Acanthostega and Ichthyostega to Pederpes at the top, indisputably a tetrapod

fossils of *Acanthostega* from about 363[420] Ma showed that this Late Devonian transitional animal had legs and both lungs and gills, but could never have survived on land: its limbs and its wrist and ankle joints were too weak to bear its weight; its ribs were too short to prevent its lungs from being squeezed flat by its weight; its fish-like tail fin would have been damaged by dragging on the ground. The current hypothesis is that *Acanthostega*, which was about 1 metre (3.3 ft) long, was a wholly aquatic predator that hunted in shallow water. Its skeleton differed from that of most fish, in ways that enabled it to raise its head to breathe air while its body remained submerged, including: its jaws show modifications that would have enabled it to gulp air; the bones at the back of its skull are locked together, providing strong attachment points for muscles that raised its head; the head is not joined to the shoulder girdle and it has a distinct neck.

The Devonian proliferation of land plants may help to explain why air-breathing would have been an advantage: leaves falling into streams and rivers would have encouraged the growth of aquatic vegetation; this would have attracted grazing invertebrates and small fish that preyed on them; they would

Figure 156:
*Until the 1980s early transitional lobe-finned fishes, such as the Eu-
sthenopteron shown here, were depicted as emerging onto land. Paleontologists now widely agree this did not happen, and they were strictly aquatic.*

have been attractive prey but the environment was unsuitable for the big marine
predatory fish; air-breathing would have been necessary because these waters
would have been short of oxygen, since warm water holds less dissolved oxygen than cooler marine water and since the decomposition of vegetation would
have used some of the oxygen.

There are three major hypotheses as to how tetrapods evolved their stubby fins
(proto-limbs). The traditional explanation is the "shrinking waterhole hypothesis" or "desert hypothesis" posited by the American paleontologist Alfred
Romer. He believed limbs and lungs may have evolved from the necessity of
having to find new bodies of water as old waterholes dried up.

The second hypothesis is the "inter-tidal hypothesis" put forward in 2010 by
a team of Polish paleontologists led by Grzegorz Niedźwiedzki. They argued
that sarcopterygians may have first emerged unto land from intertidal zones
rather than inland bodies of water. Their hypothesis is based on the discovery
of the 395 million-year-old Zachełmie tracks in Zachełmie, Poland, the oldest
ever discovered fossil evidence of tetrapods.

The third hypothesis, the "woodland hypothesis", was proposed by the American paleontologist Gregory J. Retallack in 2011. He argues that limbs may

have developed in shallow bodies of water in woodlands as a means of navigating in environments filled with roots and vegetation. He based his conclusions on the evidence that transitional tetrapod fossils are consistently found in habitats that were formerly humid and wooded floodplains.

Research by Jennifer A. Clack and her colleagues showed that the very earliest tetrapods, animals similar to *Acanthostega*, were wholly aquatic and quite unsuited to life on land. This is in contrast to the earlier view that fish had first invaded the land — either in search of prey (like modern mudskippers) or to find water when the pond they lived in dried out — and later evolved legs, lungs, etc.

Two ideas about the homology of arms, hands and digits have existed in the past 130 years. First that digits are unique to tetrapods and second that antecedents were present in the fins of early sarcopterygian fish. Until recently it was believed that "genetic and fossil data support the hypothesis that digits are evolutionary novelties".[p. 640] However new research that created a three-dimensional reconstruction of Panderichthys, a coastal fish from the Devonian period 385 million years ago, shows that these animals already had many of the homologous bones present in the forelimbs of limbed vertebrates. For example, they had radial bones similar to rudimentary fingers but positioned in the arm-like base of their fins. Thus there was in the evolution of tetrapods a shift such that the outermost part of the fins were lost and eventually replaced by early digits. This change is consistent with additional evidence from the study of actinopterygians, sharks and lungfish that the digits of tetrapods arose from pre-existing distal radials present in more primitive fish. Controversy still exists since Tiktaalik, a vertebrate often considered the missing link between fishes and land-living animals, had stubby leg-like limbs that lacked the finger-like radial bones found in the Panderichthys. The researchers of the paper commented that it "is difficult to say whether this character distribution implies that Tiktaalik is autapomorphic, that Panderichthys and tetrapods are convergent, or that Panderichthys is closer to tetrapods than Tiktaalik. At any rate, it demonstrates that the fish–tetrapod transition was accompanied by significant character incongruence in functionally important structures".[p. 638]

From the end of the Devonian to the Mid Carboniferous a 30 million year gap occurs in the fossil record. This gap, called Romer's gap, is marked by the absence of ancestral tetrapod fossils and fossils of other vertebrates that look well-adapted for life on land.

Transition from lobe-finned fishes to tetrapods		
~385 Ma	Eu-sthenopteron	🎞 *Eusthenopteron*[421] – *Animal Planet* Genus of extinct lobe-finned fishes that has attained an iconic status from its close relationships to tetrapods. Early depictions of this animal show it emerging onto land, however paleontologists now widely agree that it was a strictly aquatic animal. The genus *Eusthenopteron* is known from several species that lived during the Late Devonian period, about 385 Ma. It was the object of intense study from the 1940s to the 1990s by the paleoichthyologist Erik Jarvik.
	Gogonasus	*Gogonasus (snout from Gogo)* was a lobe-finned fish known from 3-dimensionally preserved 380 million-year-old fossils found in the Gogo Formation. It was a small fish reaching 30–40 cm (0.98–1.31 ft) in length.[422] Its skeleton shows several tetrapod-like features. They included the structure of its middle ear, and its fins show the precursors of the forearm bones, the radius and ulna. Researchers believe it used its forearm-like fins to dart out of the reef to catch prey. *Gogonasus* was first described in 1985 by John A. Long. For almost 100 years *Eusthenopteron* has been the role model for demonstrating stages in the evolution of lobe-finned fishes to tetrapods. *Gogonasus* now replaces *Eusthenopteron* in being a better preserved representative without any ambiguity in interpreting its anatomy.
~385 Ma	Pan-derichthys	Adapted to muddy shallows, and capable of some kind of shallow water or terrestrial body flexion locomotion. Had the ability to prop itself up.[423] They had large tetrapod-like heads, and are thought to be the most crownward stem fish-tetrapod with paired fins.
~375 Ma	Tiktaalik	A fish with limb-like fins that could take it onto land. It is an example from several lines of ancient sarcopterygian fish developing adaptations to the oxygen-poor shallow-water habitats of its time, which led to the evolution of tetrapods.[424] Paleontologists suggest that it is representative of the transition between non-tetrapod vertebrates (fish) such as *Panderichthys*, known from fossils 380 million years old, and early tetrapods such as *Acanthostega* and *Ichthyostega*, known from fossils about 365 million years old. Its mixture of primitive fish and derived tetrapod characteristics led one of its discoverers, Neil Shubin, to characterize *Tiktaalik* as a "fishapod".[425]
365 Ma	Acan-thostega	A fish-like early labyrinthodont that occupied swamps and changed views about the early evolution of tetrapods. It had eight digits on each hand (the number of digits on the feet is unclear) linked by webbing, it lacked wrists, and was generally poorly adapted to come onto land.[426] Subsequent discoveries revealed earlier transitional forms between *Acanthostega* and completely fish-like animals.

 374–359 Ma	*Ichthyostega*	🖼 *Ichthyostega*[427] – *Animal Planet*
		🖼 Pierce et al. Vertebral Architecture 1^{428} 2^{429} 3^{430} 4^{431} 5^{432} – *YouTube*
		Until finds of other early tetrapods and closely related fishes in the late 20th century, *Ichthyostega* stood alone as the transitional fossil between fish and tetrapods, combining a fishlike tail and gills with an amphibian skull and limbs. It possessed lungs and limbs with seven digits that helped it navigate through shallow water in swamps.
 359–345 Ma	*Pederpes*	*Pederpes* is the earliest known fully terrestrial tetrapod. It is included here to complete the transition of lobe-finned fishes to tetrapods, even though *Pederpes* is no longer a fish.

By the late Devonian, land plants had stabilized freshwater habitats, allowing the first wetland ecosystems to develop, with increasingly complex food webs that afforded new opportunities. Freshwater habitats were not the only places to find water filled with organic matter and choked with plants with dense vegetation near the water's edge. Swampy habitats like shallow wetlands, coastal lagoons and large brackish river deltas also existed at this time, and there is much to suggest that this is the kind of environment in which the tetrapods evolved. Early fossil tetrapods have been found in marine sediments, and because fossils of primitive tetrapods in general are found scattered all around the world, they must have spread by following the coastal lines — they could not have lived in freshwater only.

- Fossil Illuminates Evolution of Limbs from Fins[433] *Scientific American*, 2 2 April 2004.

Post Devonian

- The Mesozoic Era began about 250 million years ago in the wake of the Permian-Triassic event, the largest mass extinction in Earth's history, and ended about 66 million years ago with the Cretaceous–Paleogene extinction event, another mass extinction that killed off non-avian dinosaurs, as well as other plant and animal species. It is often referred to as the *Age of Reptiles* because reptiles were the dominant vertebrates of the time. The Mesozoic witnessed the gradual rifting of the supercontinent Pangaea into separate landmasses. The climate alternated between warming and cooling periods; overall the Earth was hotter than it is today.

Carbon-iferous	Carboniferous (359–299 Ma): Sharks underwent a major evolutionary radiation during the Carboniferous. It is believed that this evolutionary radiation occurred because the decline of the placoderms at the end of the Devonian period caused many environmental niches to become unoccupied and allowed new organisms to evolve and fill these niches.

		Coastal seas during the Carboniferous c. 300 Ma	The first 15 million years of the Carboniferous has very few terrestrial fossils. This gap in the fossil record, is called Romer's gap after the American palaentologist Alfred Romer. While it has long been debated whether the gap is a result of fossilisation or relates to an actual event, recent work indicates the gap period saw a drop in atmospheric oxygen levels, indicating some sort of ecological collapse. The gap saw the demise of the Devonian fish-like ichthyostegalian labyrinthodonts, and the rise of the more advanced temnospondyl and reptiliomorphan amphibians that so typify the Carboniferous terrestrial vertebrate fauna. The Carboniferous seas were inhabited by many fish, mainly Elasmobranchs (sharks and their relatives). These included some, like *Psammodus*, with crushing pavement-like teeth adapted for grinding the shells of brachiopods, crustaceans, and other marine organisms. Other sharks had piercing teeth, such as the Symmoriida; some, the petalodonts, had peculiar cycloid cutting teeth. Most of the sharks were marine, but the Xenacanthida invaded fresh waters of the coal swamps. Among the bony fish, the Palaeonisciformes found in coastal waters also appear to have migrated to rivers. Sarcopterygian fish were also prominent, and one group, the Rhizodonts, reached very large size. Most species of Carboniferous marine fish have been described largely from teeth, fin spines and dermal ossicles, with smaller freshwater fish preserved whole. Freshwater fish were abundant, and include the genera *Ctenodus*, *Uronemus*, *Acanthodes*, *Cheirodus*, and *Gyracanthus*.
		Stethacanthidae	
			As a result of the evolutionary radiation, carboniferous sharks assumed a wide variety of bizarre shapes—including sharks of the family Stethacanthidae, which possessed a flat brush-like dorsal fin with a patch of denticles on its top. *Stethacanthus'* unusual fin may have been used in mating rituals. Apart from the fins, Stethacanthidae resembled *Falcatus* (below).

		Falcatus	*Falcatus* is a genus of small cladodont-toothed sharks that lived 335–318 Ma. They were about 25–30 cm (10–12 in) long.[434] They are characterised by the prominent fin spines that curved anteriorly over their heads.
		Orodus	*Orodus* is another shark of the Carboniferous, a genus from the family Orodontidae that lived into the early Permian from 303 to 295 Ma. It grew to 2 m (6.6 ft) in length.
Permian	Permian (298–252 Ma):		
		Acanthodes	*Acanthodes* are an extinct genus of spiny shark. It had gills but no teeth, and was presumably a filter feeder. *Acanthodes* had only two skull bones and were covered in cubical scales. Each paired pectoral and pelvic fins had one spine, as did the single anal and dorsal fins, giving it a total of six spines, less than half that of many other spiny sharks. Acanthodians share qualities of both bony fish (osteichthyes) and cartilaginous fish (chondrichthyes), and it has been suggested that they may have been stem chondrichthyans and stem gnathostomes.[435,436]
	The Permian ended with the most extensive extinction event recorded in paleontology: the Permian-Triassic extinction event. 90% to 95% of marine species became extinct, as well as 70% of all land organisms. It is also the only known mass extinction of insects.[437,438] Recovery from the Permian-Triassic extinction event was protracted; land ecosystems took 30M years to recover, and marine ecosystems took even longer.		
Triassic	Triassic (252–201 Ma): The fish fauna of the Triassic was remarkably uniform, reflecting the fact that very few families survived the Permian extinction. A considerable radiation of ray-finned fishes occurred during the Triassic, laying the foundation for many modern fishes.[439] *See Category:Triassic fish.*		
		Perleidus	*Perleidus* was a ray-finned fish from the Early Triassic. About 15 centimetres (5.9 in) in length, it was a freshwater predatory fish with jaws that hung vertically under the braincase, allowing them to open wide. *Perleidus* had highly flexible dorsal and anal fins, with a reduced number of fin rays, which would have made the fish more agile in the water.
		Pachycormiformes	Pachycormiformes are an extinct order of ray-finned fish that existed from the Middle Triassic to the K-Pg extinction (below). They were characterized by serrated pectoral fins, reduced pelvic fins and a bony rostrum. Their relations with other fish are unclear.

		*Pholidophorus*	*Pholidophorus* was an extinct genus of teleost, around 40 centimetres (16 in) long, from about 240–140 Ma. Although not closely related to the modern herring, it was somewhat like them. It had a single dorsal fin, a symmetrical tail, and an anal fin placed towards the rear of the body. It had large eyes and was probably a fast swimming predator, hunting planktonic crustaceans and smaller fish.[440] A very early teleost, *Pholidophoris* had many primitive characteristics such as ganoid scales and a spine that was partially composed of cartilage, rather than bone.[440]

The Triassic ended with the Triassic–Jurassic extinction event. About 23% of all families, 48% of all genera (20% of marine families and 55% of marine genera) and 70% to 75% of all species became extinct. Non-dinosaurian archosaurs continued to dominate aquatic environments, while non-archosaurian diapsids continued to dominate marine environments.

Jurassic	Jurassic (201–145 Ma): During the Jurassic period, the primary vertebrates living in the seas were fish and marine reptiles. The latter include ichthyosaurs who were at the peak of their diversity, plesiosaurs, pliosaurs, and marine crocodiles of the families Teleosauridae and Metriorhynchidae.[441] Numerous turtles could be found in lakes and rivers. *See Category:Jurassic fish.*

		*Leedsichthys*	Along with its close pachycormid relatives *Bonnerichthys* and *Rhinconichthys*, *Leedsichthys* is part of a lineage of large-sized filter-feeders that swam the Mesozoic seas for over 100 million years, from the middle Jurassic until the end of the Cretaceous period. Pachycormids might represent an early branch of Teleostei, the group most modern bony fishes belong to; in that case *Leedsichthys* is the largest known teleost fish.[442] In 2003, a fossil specimen 22 meters (72 feet) long was unearthed.[443]
		Ichthyodectidae	

Figure 157: *This fossil Ichthyodectidae from the Lower Jurassic is one of the best conserved fossil fishes worldwide*

		The family Ichthyodecti-dae (literally "fish-biters") was a family of marine actinopterygian fish. They first appeared 156 Ma during the Late Jurassic and disappeared during the K-Pg extinction event 66 Ma. They were most diverse throughout the Cretaceous period. Sometimes classified in the primitive bony fish order Pachycormiformes, they are today generally regarded as members of the "bulldog fish" order Ichthyodectiformes in the far more advanced Osteoglossomorpha. Most ichthyodectids ranged between 1 and 5 meters (3.3 and 16.4 ft) in length. All known taxa were predators, feeding on smaller fish; in several cases, larger Ichthyodectidae preyed on smaller members of the family. Some species had remarkably large teeth, though others, such as *Gillicus arcuatus*, had small ones and sucked in their prey. The largest *Xiphactinus* was 20 feet long, and appeared in the Late Cretaceous (below).	
Cretaceous		Cretaceous (145–66 Ma): *See Category:Cretaceous fish.*	
		Sturgeon	True sturgeons appear in the fossil record during the Upper Cretaceous. Since that time, sturgeons have undergone remarkably little morphological change, indicating their evolution has been exceptionally slow and earning them informal status as living fossils.[444] This is explained in part by the long generation interval, tolerance for wide ranges of temperature and salinity, lack of predators due to size, and the abundance of prey items in the benthic environment.

	Cretoxyrhina	*Cretoxyrhina mantelli* was a large shark that lived about 100 to 82 million years ago, during the mid Cretaceous period. It is commonly known as the Ginsu Shark. This shark was first identified by a famous Swiss Naturalist, Louis Agassiz in 1843, as *Cretoxyrhina mantelli*. However, the most complete specimen of this shark was discovered in 1890, by the fossil hunter Charles H. Sternberg, who published his findings in 1907. The specimen consisted of a nearly complete associated vertebral column and over 250 associated teeth. This kind of exceptional preservation of fossil sharks is rare because a shark's skeleton is made of cartilage, which is not prone to fossilization. Charles dubbed the specimen *Oxyrhina mantelli*. This specimen represented a 20-foot-long (6.1 m) shark.
	Enchodus	*Enchodus* is an extinct genus of bony fish. It flourished during the Upper Cretaceous and was small to medium in size. One of the genus' most notable attributes are the large "fangs" at the front of the upper and lower jaws and on the palatine bones, leading to its misleading nickname among fossil hunters and paleoichthyologists, "the saber-toothed herring". These fangs, along with a long sleek body and large eyes, suggest *Enchodus* was a predatory species.
	Xiphactinus	📑 *Xiphactinus*[445] – *YouTube* *Xiphactinus* is an extinct genus of large predatory marine bony fish of the Late Cretaceous. They grew more than 4.5 metres (15 feet) long.[446]
	Ptychodus	*Ptychodus* is a genus of extinct hybodontiform shark that lived from the late Cretaceous to the Paleogene.[447,448] *Ptychodus mortoni (pictured)* was about 32 feet (9.8 metres) long and was unearthed in Kansas, United States.[449]

The end of the Cretaceous was marked by the Cretaceous–Paleogene extinction event (K–Pg extinction). There are substantial fossil records of jawed fishes across the K–T boundary, which provides good evidence of extinction patterns of these classes of marine vertebrates. Within cartilaginous fish, approximately 80% of the sharks, rays, and skates families survived the extinction event, and more than 90% of teleost fish (bony fish) families survived. There is evidence of a mass kill of bony fishes at a fossil site immediately above the K–T boundary layer on Seymour Island near Antarctica, apparently precipitated by the K–Pg extinction event. However, the marine and freshwater environments of fishes mitigated environmental effects of the extinction event, and evidence shows that there was a major increase in size and abundance of teleosts immediately after the extinction, apparently due to the elimination of their ammonite competitors (there was no similar change in shark populations across the boundary).

Ceno-zoic Era	Cenozoic Era (66 Ma to present): The current era has seen great diversification of bony fishes. Over half of all living vertebrate species (about 32,000 species) are fishes (non-tetrapod craniates), a diverse set of lineages that inhabit all the world's aquatic ecosystems, from snow minnows (Cypriniformes) in Himalayan lakes at elevations over 4,600 metres (15,100 feet) to flatfishes (order Pleuronectiformes) in the Challenger Deep, the deepest ocean trench at about 11,000 metres (36,000 feet). Fishes of myriad varieties are the main predators in most of the world's water bodies, both freshwater and marine.

	Amphistium	Amphistium is a 50-million-year-old fossil fish that has been identified as an early relative of the flatfish, and as a transitional fossil. In a typical modern flatfish, the head is asymmetric with both eyes on one side of the head. In *Amphistium*, the transition from the typical symmetric head of a vertebrate is incomplete, with one eye placed near the top of the head.
	Megalodon	Megalodon Giant Shark[450] – *National Geographic* (full documentary)
		Megalodon battle[451] *History Channel*
		The Nightmarish Megalodon[452] *Discovery*
		Megalodon is an extinct species of shark that lived about 28 to 1.5 Ma. It looked much like a stocky version of the great white shark, but was much larger with fossil lengths reaching 20.3 metres (67 ft). Found in all oceans it was one of the largest and most powerful predators in vertebrate history, and probably had a profound impact on marine life.

Prehistoric fish

Fossil Fishes[453] *American Museum of Natural History*

Prehistoric fish are early fish that are known only from fossil records. They are the earliest known vertebrates, and include the first and extinct fish that lived through the Cambrian to the Tertiary. The study of prehistoric fish is called *paleoichthyology*. A few living forms, such as the coelacanth are also referred to as prehistoric fish, or even living fossils, due to their current rarity and similarity to extinct forms. Fish that have become recently extinct are not usually referred to as prehistoric fish.

Living fossils

Bony fishes

- Arowana and Arapaima
- Bowfin
- Coelacanth
- Gar
- Queensland lungfish
- Sturgeons and paddlefish
- Bichir
- Polypterus retropinnis

Figure 158: *The jawless hagfish is a living fossil that essentially has not changed for 300 million years.*[454]

Sharks

- Blind shark
- Bullhead shark
- Elephant shark
- Frilled shark
- Goblin shark
- Gulper shark

Jawless fishes

- Hagfish
- Northern brook lamprey

Eels

- Protoanguilla palau

The coelacanth was thought to have gone extinct 66[455] million years ago, until a living specimen belonging to the order was discovered in 1938

Fossil sites

Some fossil sites that have produced notable fish fossils

- Abbey Wood SSSI
- Bracklesham Beds
- Bear Gulch Limestone
- Burgess Shale
- Canowindra

Figure 159: *Miguasha National Park: outcrop of Devonian beds rich in fossil fish*

- Crato Formation
- Dura Den
- Feltville Formation
- Fossil Butte National Monument
- Fur Formation
- Gogo Formation
- Green's Creek
- Green River Formation
- Kakwa Provincial Park
- Land Grove Quarry
- Maotianshan Shales
- Matanuska Formation
- McAbee Fossil Beds
- Miguasha National Park
- MoClay
- Monte Bolca
- Mount Ritchie
- Orcadian Basin
- Portishead Pier to Black Nore SSSI
- Santana Formation
- Southerham Grey Pit

- Thanet Formation
- Towaco Formation
- Weydale
- Zhoukoudian

Fossil collections

Part of a series on
Paleontology
Paleontology Portal
Category
• v
• t
• e[456]

Some notable fossil fish collections.

- Fossil fish collection[457] *Natural History Museum*, Britain.
- Collection and expertise[458] *Museum für Naturkunde*, Germany.
- Fossil fishes[459] *The Field Museum*, United States.

Paleoichthyologists

Paleoichthyology is the scientific study of the prehistoric life of fish. Listed below are some researchers who have made notable contributions to paleoichthyology.

- Louis Agassiz
- Mary Anning

- Michael Benton
- Derek Briggs
- Hans C. Bjerring
- John Samuel Budgett
- Frederick Chapman
- Jenny Clack
- Ted Daeschler
- Bashford Dean
- Robert Dick
- Philip Grey Egerton
- Edwin Sherbon Hills
- Jeffrey A. Hutchings
- Thomas Henry Huxley
- Johan Aschehoug Kiær
- Philippe Janvier
- Erik Jarvik
- George V. Lauder
- John A. Long
- Hugh Miller
- Charles Moore
- Paul E. Olsen
- Heinz Christian Pander
- Elizabeth Philpot
- Jean Piveteau
- Colin Patterson
- Alfred Romer
- Ira Rubinoff
- Neil Shubin
- Franz Steindachner
- Erik Stensiö
- Ramsay Heatley Traquair
- Thomas Stanley Westoll
- Tiberius Cornelis Winkler
- Arthur Smith Woodward

References

Bibliography

<templatestyles src="Template:Refbegin/styles.css" />

- Ahlberg, Per Erik (2001). *Major events in early vertebrate evolution: palaeontology, phylogeny, genetics, and development*[460]. Washington, DC: Taylor & Francis. ISBN 0-415-23370-4.
- Benton, Michael J (2005): *Vertebrate Palaeontology*
- Benton, Michael (2009) *Vertebrate Palaeontology*[461] Edition 3, John Wiley & Sons. ISBN 9781405144490.
- Berg, Linda R.; Eldra Pearl Solomon; Diana W. Martin (2004). *Biology*. Cengage Learning. ISBN 978-0-534-49276-2.
- Haines, Tim; Chambers, Paul (2005). "The Complete Guide to Prehistoric Life". Firefly Books.
- Cloudsley-Thompson, J. L. (2005). *Ecology and behaviour of Mesozoic reptiles*. 9783540224211: Springer.
- Dawkins, Richard (2004). *The Ancestor's Tale: A Pilgrimage to the Dawn of Life*. Boston: Houghton Mifflin Company. ISBN 0-618-00583-8.
- Donoghue, P. C., P. L. Forey & R. J. Aldridge (2000). "Conodont affinity and chordate phylogeny". *Biological Reviews of the Cambridge Philosophical Society*. **75** (2): 191–251. doi: 10.1111/j.1469-185X.1999.tb00045.x[462]. PMID 10881388[463].
- Encyclopædia Britannica (1954). *Encyclopædia Britannica: A new survey of universal knowledge*. **17**.
- Forey, Peter L (1998) *History of the Coelacanth Fishes*. London: Chapman & Hall.
- Hall, Brian Keith; Hanken, James (1993). *The Skull*[464]. Chicago: University of Chicago Press. ISBN 0-226-31568-1.
- Helfman G, Collette BB, Facey DH and Bowen BW (2009) *The Diversity of Fishes: Biology, Evolution, and Ecology*[465] Wiley-Blackwell. ISBN 978-1-4051-2494-2
- Janvier, Philippe (2003). *Early Vertebrates*. Oxford University Press. ISBN 978-0-19-852646-9.
- Lecointre, G; Le Guyader, H (2007). "The Tree of Life: A Phylogenetic Classification". Harvard University Press Reference Library.
- Long, John A. *The Rise of Fishes: 500 Million Years of Evolution*. Baltimore: The Johns Hopkins University Press, 1996. ISBN 0-8018-5438-5
- Nelson, Joseph S. (2006). *Fishes of the World*. John Wiley & Sons, Inc. ISBN 0-471-25031-7.

- Palmer, D., ed. (1999). *The Marshall Illustrated Encyclopedia of Dinosaurs and Prehistoric Animals*. London: Marshall Editions. p. 26. ISBN 1-84028-152-9.
- Palmer, D. (2000). "The Atlas of the Prehistoric World". Philadelphia: Marshall Publishing Ltd.
- Patterson, Colin (1987). *Molecules and morphology in evolution: conflict or compromise?*[466]. Cambridge, UK: Cambridge University Press. ISBN 0-521-32271-5.
- Sarjeant, William Antony S.; L. B. Halstead (1995). *Vertebrate fossils and the evolution of scientific concepts: writings in tribute to Beverly Halstead*[467]. ISBN 978-2-88124-996-9.
- Romer, AS (1970). "The Vertebrate Body" (4 ed.). London: W.B. Saunders.
- Turner, S. (1999). "Early Silurian to Early Devonian thelodont assemblages and their possible ecological significance". In A. J. Boucot; J. Lawson. *Palaeocommunities, International Geological Correlation Programme 53, Project Ecostratigraphy, Final Report*. Cambridge University Press.

Further reading

External video

📹 Feeding Mechanism of Conodonts[468] – *YouTube*
📹 Chordate evolution[469] – *YouTube*

- Benton MJ (1998) "The quality of the fossil record of the vertebrates"[470] Pages 269–303 in Donovan, SK and Paul CRC (eds), *The adequacy of the fossil record*. Wiley. ISBN 9780471969884.
- Cloutier R (2010). "The fossil record of fish ontogenies: Insights into developmental patterns and processes"[471] (PDF). *Seminars in Cell & Developmental Biology*. **21** (4): 400–413. doi: 10.1016/j.semcdb.2009.11.004[472].Wikipedia:Link rot
- Janvier, Philippe (1998) *Early Vertebrates*, Oxford, New York: Oxford University Press. ISBN 0-19-854047-7
- Long, John A. (1996) *The Rise of Fishes: 500 Million Years of Evolution*[473] Johns Hopkins University Press. ISBN 0-8018-5438-5
- McKenzie DJ, Farrell AP and Brauner CJ (2011) *Fish Physiology: Primitive Fishes*[474] Academic Press. ISBN 9780080549521.
- Maisey JG (1996) fossil fishes[475] Holt. ISBN 9780805043662.

- Near, T.J.; Dornburg, A.; Eytan, R.I.; Keck, B.P.; Smith, W.L.; Kuhn, K.L.; Moore, J.A.; Price, S.A.; Burbrink, F.T.; Friedman, M. (2013). "Phylogeny and tempo of diversification in the superradiation of spiny-rayed fishes"[476]. *Proceedings of the National Academy of Sciences*. **110** (31): 12738–12743. doi: 10.1073/pnas.1304661110[477]. PMC 3732986[476] ∂ . PMID 23858462[478].
- Shubin, Neil (2009) *Your inner fish: A journey into the 3.5-billion-year history of the human body*[479] Vintage Books. ISBN 9780307277459.
- Introduction to the Vertebrates[480] *Museum of Palaeontology*, University of California.

External links

- Fossil Fish[481]
- Origins of Fish[482]
- Overview of evolution[483] – Carl Sagan
- The Origin of Vertebrates[484] Marc W. Kirschner, *iBioSeminars*.
- 150 Million Years of Fish Evolution in One Handy Figure[485] *ScientificAmerican*, 29 August 2013.

Colonization of land

Evolutionary history of plants

The evolution of plants has resulted in a wide range of complexity, from the earliest algal mats, through multicellular marine and freshwater green algae, terrestrial bryophytes, lycopods and ferns, to the complex gymnosperms and angiosperms of today. While many of the earliest groups to appear continue to thrive, as exemplified by red and green algae in marine environments, more recently derived groups have displaced previously ecologically dominant ones, e.g. the ascendance of flowering plants over gymnosperms in terrestrial environments.[498]

There is evidence that cyanobacteria and multicellular photosynthetic eukaryotes lived in freshwater communities on land as early as 1 billion years ago, and that communities of complex, mutilcellular photosynthesizing organisms existed on land in the late Precambrian, around 850[490] million years ago.

Evidence of the emergence of embryophyte land plants first occurs in the mid-Ordovician ($\sim$ 470[491] million years ago), and by the middle of the Devonian ($\sim$ 390[492] million years ago), many of the features recognised in land plants today were present, including roots and leaves. Late Devonian ($\sim$ 370[493] million years ago) free-sporing plants such as *Archaeopteris* had secondary vascular tissue that produced wood and had formed forests of tall trees. Also by late Devonian, *Elkinsia*, an early seed fern, had evolved seeds. Evolutionary innovation continued into the Carboniferous and still continues today. Most plant groups were relatively unscathed by the Permo-Triassic extinction event, although the structures of communities changed. This may have set the scene for the appearance of the flowering plants in the Triassic ($\sim$ 200[494] million years ago), and their later diversification in the Cretaceous and Paleogene. The latest major group of plants to evolve were the grasses, which became important in the mid-Paleogene, from around 40[495] million years ago. The grasses,

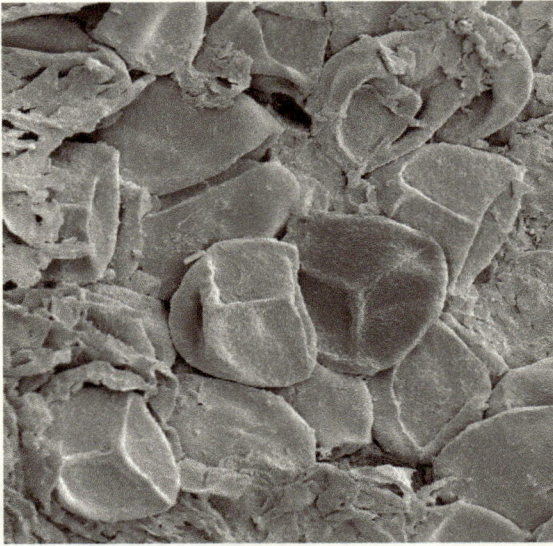

Figure 160: *A Late Silurian sporangium, artificially colored.*
Green: *A spore tetrad.* ***Blue:*** *A spore bearing a trilete mark –*
the Y-shaped scar. The spores are about 30–35 μm across

as well as many other groups, evolved new mechanisms of metabolism to survive the low CO_2 and warm, dry conditions of the tropics over the last 10^{496} million years.

Colonization of land

Life timeline

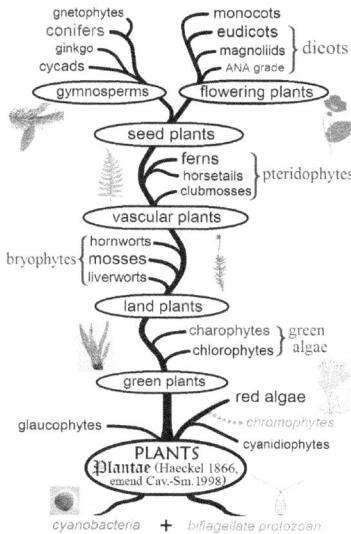

Figure 161: *Phylogenetic plant tree, showing the major clades and traditional groups. Monophyletic groups are in black and paraphyletics in blue. Diagram according to symbiogenetic origin of plant cells,*[486] *and phylogeny of algae, bryophytes,*[487] *vascular plants,*[488] *and flowering plants.*[489]

Axis scale: million years

Also see: *Human timeline* and *Nature timeline*

Land plants evolved from a group of green algae, perhaps as early as 850 mya, but algae-like plants might have evolved as early as 1 billion years ago. The closest living relatives of land plants are the charophytes, specifically Charales; assuming that the Charales' habit has changed little since the divergence of lineages, this means that the land plants evolved from a branched, filamentous alga dwelling in shallow fresh water, perhaps at the edge of seasonally desiccating pools. However, some recent evidence suggests that land plants might have originated from unicellular terrestrial charophytes similar to extant Klebsormidiophyceae. The alga would have had a haplontic life cycle: it would only very briefly have had paired chromosomes (the diploid condition) when the egg and sperm first fused to form a zygote; this would have immediately divided by meiosis to produce cells with half the number of unpaired chromosomes (the haploid condition). Co-operative interactions with fungi may have helped early plants adapt to the stresses of the terrestrial realm.

Figure 162: *The Devonian marks the beginning of extensive land colonization by plants, which – through their effects on erosion and sedimentation – brought about significant climatic change.*

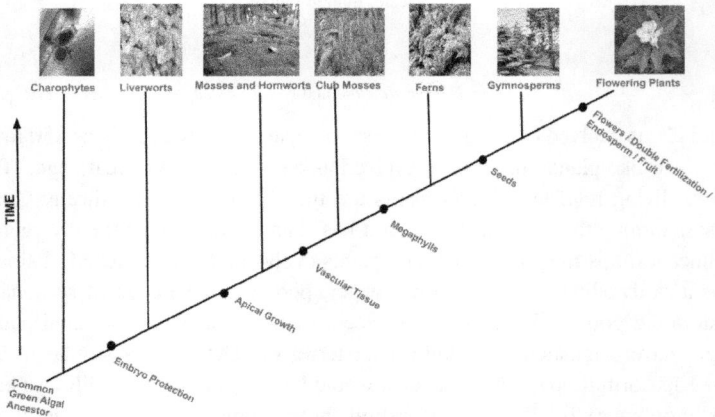

Figure 163: *Cladogram of plant evolution*

Plants were not the first photosynthesisers on land. Weathering rates suggest that organisms capable of photosynthesis were already living on the land 1,200[497] million years ago, and microbial fossils have been found in freshwater lake deposits from 1,000[498] million years ago, but the carbon isotope record suggests that they were too scarce to impact the atmospheric composition until around 850[490] million years ago. These organisms, although phylogenetically diverse, were probably small and simple, forming little more than an algal scum.

However, evidence of the earliest land plants occurs much later at about 470Ma, in lower middle Ordovician rocks from Saudi Arabia and Gondwana in the form of spores with decay-resistant walls. These spores, known as cryptospores, were produced either singly (monads), in pairs (dyads) or groups of four (tetrads), and their microstructure resembles that of modern liverwort spores, suggesting they share an equivalent grade of organisation. Their walls contain sporopollenin – further evidence of an embryophytic affinity. It could be that atmospheric 'poisoning' prevented eukaryotes from colonising the land prior to this, or it could simply have taken a great time for the necessary complexity to evolve.

Trilete spores similar to those of vascular plants appear soon afterwards, in Upper Ordovician rocks. Depending exactly when the tetrad splits, each of the four spores may bear a "trilete mark", a Y-shape, reflecting the points at which each cell squashed up against its neighbours. However, this requires that the spore walls be sturdy and resistant at an early stage. This resistance is closely associated with having a desiccation-resistant outer wall—a trait only of use when spores must survive out of water. Indeed, even those embryophytes that have returned to the water lack a resistant wall, thus don't bear trilete marks. A close examination of algal spores shows that none have trilete spores, either because their walls are not resistant enough, or in those rare cases where it is, the spores disperse before they are squashed enough to develop the mark, or don't fit into a tetrahedral tetrad.

The earliest megafossils of land plants were thalloid organisms, which dwelt in fluvial wetlands and are found to have covered most of an early Silurian flood plain. They could only survive when the land was waterlogged. There were also microbial mats.

Once plants had reached the land, there were two approaches to dealing with desiccation. Modern bryophytes either avoid it or give in to it, restricting their ranges to moist settings, or drying out and putting their metabolism "on hold" until more water arrives, as in the liverwort genus *Targionia*. Tracheophytes resist desiccation, by controlling the rate of water loss. They all bear a waterproof outer cuticle layer wherever they are exposed to air (as do some bryophytes), to reduce water loss, but since a total covering would cut them off from CO_2 in the atmosphere early tracheophytes used variable openings, the stomata, to regulate the rate of gas exchange. Tracheophytes also developed vascular tissue to aid in the movement of water within the organisms (see below), and moved away from a gametophyte dominated life cycle (see below). Vascular tissue ultimately also facilitated upright growth without the support of water and paved the way for the evolution of larger plants on land.

A snowball earth, from around 850-630 mya, is believed to have been caused by early photosynthetic organisms, which reduced the concentration of carbon dioxide and increased the amount of oxygen in the atmosphere. The establishment of a land-based flora increased the rate of accumulation of oxygen in the atmosphere, as the land plants produced oxygen as a waste product. When this concentration rose above 13%,Wikipedia:Manual of Style/Dates and numbers#Chronological items wildfires became possible, evident from charcoal in the fossil record. Apart from a controversial gap in the Late Devonian, charcoal is present ever since.

Charcoalification is an important taphonomic mode. Wildfire or burial in hot volcanic ash drives off the volatile compounds, leaving only a residue of pure carbon. This is not a viable food source for fungi, herbivores or detritovores, so is prone to preservation. It is also robust, so can withstand pressure and display exquisite, sometimes sub-cellular, detail.

Evolution of life cycles

All multicellular plants have a life cycle comprising two generations or phases. The gametophyte phase has a single set of chromosomes (denoted $1n$), and produces gametes (sperm and eggs). The sporophyte phase has paired chromosomes (denoted $2n$), and produces spores. The gametophyte and sporophyte phases may be homomorphic, appearing identical in some algae, such as *Ulva lactuca*, but are very different in all modern land plants, a condition known as heteromorphy.

The pattern in plant evolution has been a shift from homomorphy to heteromorphy. The algal ancestors of land plants were almost certainly haplobiontic, being haploid for all their life cycles, with a unicellular zygote providing the 2N stage. All land plants (i.e. embryophytes) are diplobiontic – that is, both the haploid and diploid stages are multicellular. Two trends are apparent: bryophytes (liverworts, mosses and hornworts) have developed the gametophyte as the dominant phase of the life cycle, with the sporophyte becoming almost entirely dependent on it; vascular plants have developed the sporophyte as the dominant phase, with the gametophytes being particularly reduced in the seed plants.

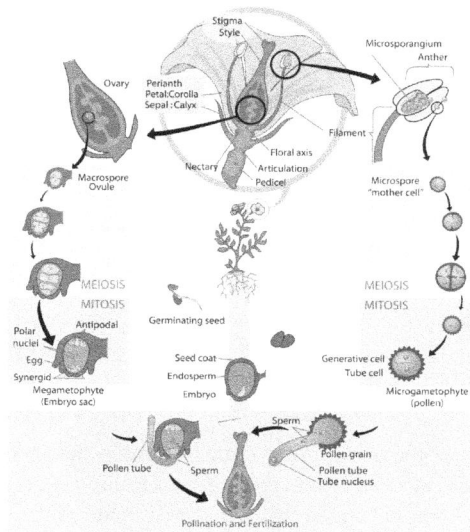

Figure 164: *Angiosperm life cycle*

It has been proposed that the basis for the emergence of the diploid phase of the life cycle as the dominant phase, is that diploidy allows masking of the expression of deleterious mutations through genetic complementation. Thus if one of the parental genomes in the diploid cells contains mutations leading to defects in one or more gene products, these deficiencies could be compensated for by the other parental genome (which nevertheless may have its own defects in other genes). As the diploid phase was becoming predominant, the masking effect likely allowed genome size, and hence information content, to increase without the constraint of having to improve accuracy of replication. The opportunity to increase information content at low cost is advantageous because it permits new adaptations to be encoded. This view has been challenged, with evidence showing that selection is no more effective in the haploid than in the diploid phases of the lifecycle of mosses and angiosperms.

There are two competing theories to explain the appearance of a diplobiontic lifecycle.

The **interpolation theory** (also known as the antithetic or intercalary theory) holds that the interpolation of a multicellular sporophyte phase between two successive gametophyte generations was an innovation caused by preceding meiosis in a freshly germinated zygote with one or more rounds of mitotic division, thereby producing some diploid multicellular tissue before finally meiosis produced spores. This theory implies that the first sporophytes bore a very different and simpler morphology to the gametophyte they depended on. This seems to fit well with what is known of the bryophytes, in which a vegetative thalloid gametophyte nurtures a simple sporophyte, which consists of little more than an unbranched sporangium on a stalk. Increasing complexity of the ancestrally simple sporophyte, including the eventual acquisition of photosynthetic cells, would free it from its dependence on a gametophyte, as seen in some hornworts (*Anthoceros*), and eventually result in the sporophyte developing organs and vascular tissue, and becoming the dominant phase, as in the tracheophytes (vascular plants). This theory may be supported by observations that smaller *Cooksonia* individuals must have been supported by a gametophyte generation. The observed appearance of larger axial sizes, with room for photosynthetic tissue and thus self-sustainability, provides a possible route for the development of a self-sufficient sporophyte phase.

The alternative hypothesis, called the **transformation theory** (or homologous theory), posits that the sporophyte might have appeared suddenly by delaying the occurrence of meiosis until a fully developed multicellular sporophyte had formed. Since the same genetic material would be employed by both the haploid and diploid phases they would look the same. This explains the behaviour of some algae, such as *Ulva lactuca*, which produce alternating phases of identical sporophytes and gametophytes. Subsequent adaption to the desiccating land environment, which makes sexual reproduction difficult, might have resulted in the simplification of the sexually active gametophyte, and elaboration of the sporophyte phase to better disperse the waterproof spores. The tissue of sporophytes and gametophytes of vascular plants such as *Rhynia* preserved in the Rhynie chert is of similar complexity, which is taken to support this hypothesis. By contrast, with the exception of *Psilotum* modern vascular plants have heteromorphic sporophytes and gametophytes in which the gametophytes rarely have any vascular tissue.

Evolution of plant anatomy

Arbuscular mycorrhizal symbiosis

There is no evidence that early land plants of the Silurian and early Devonian had roots, although fossil evidence of rhizoids occurs for several species, such as *Horneophyton*. The earliest land plants did not have vascular systems for transport of water and nutrients either. *Aglaophyton*, a rootless vascular plant known from Devonian fossils in the Rhynie chert was the first land plant discovered to have had a mycorrhizal relationship with fungi which formed arbuscules, literally "tree-like fungal roots", in a well-defined cylinder of cells (ring in cross section) in the cortex of its stems. The fungi fed on the plant's sugars, in exchange for nutrients generated or extracted from the soil (especially phosphate), to which the plant would otherwise have had no access. Like other rootless land plants of the Silurian and early Devonian *Aglaophyton* may have relied on arbuscular mycorrhizal fungi for acquisition of water and nutrients from the soil.

The fungi were of the phylum Glomeromycota, a group that probably first appeared 1 billion years ago and still forms arbuscular mycorrhizal associations today with all major land plant groups from bryophytes to pteridophytes, gymnosperms and angiosperms and with more than 80% of vascular plants.

Evidence from DNA sequence analysis indicates that the arbuscular mycorrhizal mutualism arose in the common ancestor of these land plant groups during their transition to land and it may even have been the critical step that enabled them to colonise the land. Appearing as they did before these plants had evolved roots, mycorrhizal fungi would have assisted plants in the acquisition of water and mineral nutrients such as phosphorus, in exchange for organic compounds which they could not synthesize themselves. Such fungi increase the productivity even of simple plants such as liverworts.

Xylem

To photosynthesise, plants must absorb CO_2 from the atmosphere. However, this comes at a price, since making the tissues available for CO_2 to enter allows water to evaporate. Water is lost much faster than CO_2 is absorbed, so plants need to replace it. Early land plants transported water apoplastically, within the porous walls of their cells. Later, they evolved the ability to control water loss (and CO_2 acquisition) through the use of a waterproof outer covering or cuticle perforated by stomata, variable apertures that could open and close to regulate evapotranspiration. Specialised water transport vascular tissues subsequently evolved, first in the form of hydroids, then tracheids and secondary xylem, followed by vessels in flowering plants. As water transport mechanisms and waterproof cuticles evolved, plants could survive without being continually covered by a film of water. This transition from poikilohydry to homoiohydry opened up new potential for colonisation.

The high CO_2 concentrations of the Silurian and early Devonian, when plants were first colonising land, meant that the need for water was relatively low. As CO_2 was withdrawn from the atmosphere by plants, more water was lost in its capture, and more elegant water acquisition and transport mechanisms evolved. Plants then needed a robust internal structure that contained long narrow channels for transporting water from the soil to all the different parts of the above-soil plant, especially to the parts where photosynthesis occurred. By the end of the Carboniferous, when CO_2 concentrations had been reduced

Figure 165: *A banded tube from the Late Silurian/Early Devo-nian. The bands are difficult to see on this specimen, as an opaque carbonaceous coating conceals much of the tube. Bands are just visible in places on the left half of the image. Scale bar: 20 μm*

to something approaching today's, around 17 times more water was lost per unit of CO_2 uptake. However, even in these "easy" early days, water was at a premium, and had to be transported to parts of the plant from the wet soil to avoid desiccation. Even today, water transport takes advantage of the cohesion-tension property of water. Water can be wicked along a fabric with small spaces, and in narrow columns of water, such as those within the plant cell walls or in tracheids, when molecules evaporate from one end, they pull the molecules behind them along the channels. Therefore, transpiration alone provides the driving force for water transport in plants. However, without ded-icated transport vessels, the cohesion-tension mechanism can cause negative pressures sufficient to collapse the water conducting cells, limiting the trans-port water to no more than a few cm, and therefore limiting the size of the earliest plants.

To be free from the constraints of small size and constant moisture that the parenchymatic transport system inflicted, plants needed a more efficient water transport system. During the early Silurian, they developed specialized xylem cells, with walls that were strengthened by bands of lignification (or similar chemical compounds) This process was followed by cell death, allowing the cell contents to be emptied and water to be passed through them. These wider, dead, empty cells, the xylem tracheids were much more conductive than the inter-cell pathway, and more resistant to collapse under the tension caused by water stress, giving the potential for transport over longer distances.

The early Devonian pretracheophytes *Aglaophyton* and *Horneophyton* have unreinforced water transport tubes with wall structures very similar to the hy-droids of modern moss sporophytes, but they grew alongside several species of tracheophytes, such as *Rhynia gwynne-vaughanii* that had well-reinforced xylem tracheids. The earliest macrofossils known to have xylem tracheids are small, mid-Silurian plants of the genus *Cooksonia*. Plants continued to innovate ways of reducing the resistance to flow within their cells, thereby increasing the efficiency of their water transport. Thickened bands on the walls of tubes are apparent from the early Silurian onwards are adaptations to increase the resistance to collapse under tension. and, when they form single celled conduits, are referred to as tracheids. These, the "next generation" of transport cell design, have a more rigid structure than hydroids, preventing their collapse at higher levels of water tension. Tracheids may have a single evolutionary origin, possibly within the hornworts, uniting all tracheophytes (but they may have evolved more than once).

Water transport requires regulation, and dynamic control is provided by stomata.[521] By adjusting the rate of gas exchange, they can restrict the amount of water lost through transpiration. This is an important role where water supply is not constant, and indeed stomata appear to have evolved before tracheids, since they are present in the sporophytes of mosses and the non-vascular hornworts.

An endodermis may have evolved in the earliest plant roots during the Devonian, but the first fossil evidence for such a structure is Carboniferous. The endodermis in the roots surrounds the water transport tissue and regulates ion exchange (and prevents unwanted pathogens etc. from entering the water transport system). The endodermis can also provide an upwards pressure, forcing water out of the roots when transpiration is not enough of a driver.

Once plants had evolved this level of controlled water transport, they were truly homoiohydric, able to extract water from their environment through root-like organs rather than relying on a film of surface moisture, enabling them to grow to much greater size. As a result of their independence from their surroundings, they lost their ability to survive desiccation – a costly trait to retain.

During the Devonian, maximum xylem diameter increased with time, with the minimum diameter remaining pretty constant. By the Middle Devonian, the tracheid diameter of some plant lineages such as the Zosterophyllophytes had plateaued. Wider tracheids allow water to be transported faster, but the overall transport rate depends also on the overall cross-sectional area of the xylem bundle itself.

While wider tracheids with robust walls make it possible to achieve higher water transport flow rates, this increases the problem of cavitation that occurs when the cohesive tension of the water column is broken, resulting in the formation of a bubble. Pits in tracheid walls have very small diameters, preventing air bubbles from passing through to adjacent tracheids., but at the cost of restricted flow rates. By the Carboniferous, Gymnosperms had developed bordered pits,[499] valve-like structures that seal the pits when one side of a tracheid is depressurized.

Growing to height also employed another trait of tracheids – the support offered by their lignified walls. Defunct tracheids were retained to form a strong, woody stem, produced in most instances by a secondary xylem. However, in early plants, tracheids were too mechanically vulnerable, and retained a central position, with a layer of tough sclerenchyma on the outer rim of the stems. Even when tracheids do take a structural role, they are supported by sclerenchymatic tissue.

Tracheids end with walls, which impose a great deal of resistance on flow; vessel members have perforated end walls, and are arranged in series to operate as if they were one continuous vessel. The function of end walls, which were the default state in the Devonian, was probably to avoid embolisms. An embolism is where an air bubble is created in a tracheid. This may happen as a result of freezing, or by gases dissolving out of solution. Once an embolism is formed, it usually cannot be removed (but see later); the affected cell cannot pull water up, and is rendered useless.

End walls excluded, the tracheids of prevascular plants were able to operate under the same hydraulic conductivity as those of the first vascular plant, *Cooksonia*.

The size of tracheids is limited as they comprise a single cell; this limits their length, which in turn limits their maximum useful diameter to 80 μm. Conductivity grows with the fourth power of diameter, so increased diameter has huge rewards; **vessel elements**, consisting of a number of cells, joined at their ends, overcame this limit and allowed larger tubes to form, reaching diameters of up to 500 μm, and lengths of up to 10 m.

Vessels first evolved during the dry, low CO_2 periods of the Late Permian, in the horsetails, ferns and Selaginellales independently, and later appeared in the mid Cretaceous in angiosperms and gnetophytes. Vessels allow the same cross-sectional area of wood to transport around a hundred times more water than tracheids! This allowed plants to fill more of their stems with structural fibres, and also opened a new niche to vines, which could transport water without being as thick as the tree they grew on. Despite these advantages, tracheid-based wood is a lot lighter, thus cheaper to make, as vessels need to be much more reinforced to avoid cavitation.

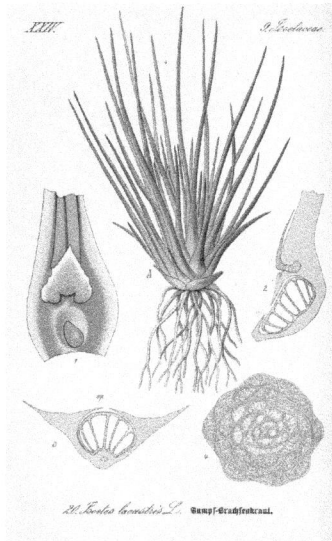

Figure 166: *The lycopod Isoetes bears micro-phylls (leaves with a single vascular trace).*

Evolution of plant morphology

Leaves

Leaves are the primary photosynthetic organs of a modern plant. The origin of leaves was almost certainly triggered by falling concentrations of atmospheric CO_2 during the Devonian period, increasing the efficiency with which carbon dioxide could be captured for photosynthesis.

Leaves certainly evolved more than once. Based on their structure, they are classified into two types: microphylls, which lack complex venation and may have originated as spiny outgrowths known as enations, and megaphylls, which are large and have complex venation that may have arisen from the modification of groups of branches. It has been proposed that these structures arose independently. Megaphylls, according to Walter Zimmerman's telome theory, have evolved from plants that showed a three-dimensional branching architecture, through three transformations—**overtopping**, which led to the lateral position typical of leaves, **planation**, which involved formation of a planar architecture, **webbing** or **fusion**, which united the planar branches, thus leading to the formation of a proper leaf lamina. All three steps happened multiple times in the evolution of today's leaves.

Figure 167: *The branching pattern of megaphyll veins may indicate their origin as webbed, dichotomising branches.*

Figure 168: *Leaf lamina. The megaphyllous leaf architecture arose multiple times in different plant lineages*

It is widely believed that the telome theory is well supported by fossil evidence. However, Wolfgang Hagemann questioned it for morphological and ecological reasons and proposed an alternative theory.[500] Whereas according to the telome theory the most primitive land plants have a three-dimensional branching system of radially symmetrical axes (telomes), according to Hagemann's alternative the opposite is proposed: the most primitive land plants that gave rise to vascular plants were flat, thalloid, leaf-like, without axes, somewhat like a liverwort or fern prothallus. Axes such as stems and roots evolved later as new organs. Rolf Sattler proposed an overarching process-oriented view that leaves some limited room for both the telome theory and Hagemann's alternative and in addition takes into consideration the whole continuum between dorsiventral (flat) and radial (cylindrical) structures that can be found in fossil and living land plants.[501] This view is supported by research in molecular genetics. Thus, James (2009) concluded that "it is now widely accepted that... radiality [characteristic of axes such as stems] and dorsiventrality [characteristic of leaves] are but extremes of a continuous spectrum. In fact, it is simply the timing of the KNOX gene expression!"

From the point of view of the telome theory, it has been proposed that before the evolution of leaves, plants had the photosynthetic apparatus on the stems. Today's megaphyll leaves probably became commonplace some 360mya, about 40my after the simple leafless plants had colonized the land in the Early Devonian. This spread has been linked to the fall in the atmospheric carbon dioxide concentrations in the Late Paleozoic era associated with a rise in density of stomata on leaf surface. This would have resulted in greater transpiration rates and gas exchange, but especially at high CO_2 concentrations, large leaves with fewer stomata would have heated to lethal temperatures in full sunlight. Increasing the stomatal density allowed for a better-cooled leaf, thus making its spread feasible, but increased CO2 uptake at the expense of decreased water use efficiency.[502]

The rhyniophytes of the Rhynie chert consisted of nothing more than slender, unornamented axes. The early to middle Devonian trimerophytes may be considered leafy. This group of vascular plants are recognisable by their masses of terminal sporangia, which adorn the ends of axes which may bifurcate or trifurcate. Some organisms, such as *Psilophyton*, bore enations. These are small, spiny outgrowths of the stem, lacking their own vascular supply.

Around the same time, the zosterophyllophytes were becoming important. This group is recognisable by their kidney-shaped sporangia, which grew on short lateral branches close to the main axes. They sometimes branched in a distinctive H-shape. The majority of this group bore pronounced spines on their axes. However, none of these had a vascular trace, and the first evidence of vascularised enations occurs in the Rhynie genus *Asteroxylon*. The spines of *Asteroxylon* had a primitive vascular supply – at the very least, leaf traces could be seen departing from the central protostele towards each individual "leaf". A fossil clubmoss known as *Baragwanathia* had already appeared in the fossil record about 20 million years earlier, in the Late Silurian. In this organism, these leaf traces continue into the leaf to form their mid-vein. One theory, the "enation theory", holds that the leaves developed by outgrowths of the protostele connecting with existing enations, but it is also possible that microphylls evolved by a branching axis forming "webbing".

Asteroxylon and *Baragwanathia* are widely regarded as primitive lycopods, a group still extant today, represented by the quillworts *Isoetes*, the spikemosses and the club mosses. Lycopods bear distinctive microphylls, defined as leaves with a single vascular trace. Microphylls could grow to some size, those of Lepidodendrales reaching over a meter in length, but almost all just bear the one vascular bundle. An exception is the rare branching in some *Selaginella* species.

The more familiar leaves, megaphylls, are thought to have originated four times independently, in the ferns, horsetails, progymnosperms and seed plants. They appear to have originated by modifying dichotomising branches, which first overlapped (or "overtopped") one another, became flattened or planated and eventually developed "webbing" and evolved gradually into more leaf-like structures. Megaphylls, by Zimmerman's telome theory, are composed of a group of webbed branches and hence the "leaf gap" left where the leaf's vascular bundle leaves that of the main branch resembles two axes splitting. In each of the four groups to evolve megaphylls, their leaves first evolved during the Late Devonian to Early Carboniferous, diversifying rapidly until the designs settled down in the mid Carboniferous.

The cessation of further diversification can be attributed to developmental constraints, but why did it take so long for leaves to evolve in the first place? Plants had been on the land for at least 50 million years before megaphylls became significant. However, small, rare mesophylls are known from the early Devonian genus *Eophyllophyton* – so development could not have been a barrier to their appearance. The best explanation so far incorporates observations that atmospheric CO_2 was declining rapidly during this time – falling by around 90% during the Devonian. This required an increase in stomatal density by 100 times to maintain rates of photosynthesis. When stomata open to allow water to evaporate from leaves it has a cooling effect, resulting from the loss of latent heat of evaporation. It appears that the low stomatal density in the early Devonian meant that evaporation and evaporative cooling were limited, and that leaves would have overheated if they grew to any size. The stomatal density could not increase, as the primitive steles and limited root systems would not be able to supply water quickly enough to match the rate of transpiration. Clearly, leaves are not always beneficial, as illustrated by the frequent occurrence of secondary loss of leaves, famously exemplified by cacti and the "whisk fern" *Psilotum*.

Secondary evolution can also disguise the true evolutionary origin of some leaves. Some genera of ferns display complex leaves which are attached to the pseudostele by an outgrowth of the vascular bundle, leaving no leaf gap. Further, horsetail (*Equisetum*) leaves bear only a single vein, and appear to be microphyllous; however, both the fossil record and molecular evidence indicate that their forebears bore leaves with complex venation, and the current state is a result of secondary simplification.

Deciduous trees deal with another disadvantage to having leaves. The popular belief that plants shed their leaves when the days get too short is misguided; evergreens prospered in the Arctic circle during the most recent greenhouse earth. The generally accepted reason for shedding leaves during winter is to cope with the weather – the force of wind and weight of snow are much more comfortably weathered without leaves to increase surface area. Seasonal leaf loss has evolved independently several times and is exhibited in the ginkgoales, some pinophyta and certain angiosperms. Leaf loss may also have arisen as a response to pressure from insects; it may have been less costly to lose leaves entirely during the winter or dry season than to continue investing resources in their repair.

Factors influencing leaf architectures

Various physical and physiological factors such as light intensity, humidity, temperature, wind speeds etc. have influenced evolution of leaf shape and size. High trees rarely have large leaves, because they are damaged by high winds. Similarly, trees that grow in temperate or taiga regions have pointed leaves,Wikipedia:Citation needed presumably to prevent nucleation of ice onto the leaf surface and reduce water loss due to transpiration. Herbivory, by mammals and insects, has been a driving force in leaf evolution. An example is that plants of the New Zealand genus *Aciphylla* have spines on their laminas, which probably functioned to discourage the extinct Moas from feeding on them. Other members of *Aciphylla*, which did not co-exist with the moas, do not have these spines.

At the genetic level, developmental studies have shown that repression of KNOX genes is required for initiation of the leaf primordium. This is brought about by *ARP* genes, which encode transcription factors. Repression of KNOX genes in leaf primordia seems to be quite conserved, while expression of KNOX genes in leaves produces complex leaves. The *ARP* function appears to have arisen early in vascular plant evolution, because members of

Figure 169: *The diversity of leaves*

the primitive group Lycophytes also have a functionally similar gene. Other players that have a conserved role in defining leaf primordia are the phytohormones auxin, gibberelin and cytokinin.

The arrangement of leaves or phyllotaxy on the plant body can maximally harvest light and might be expected to be genetically robust. However, in maize, a mutation in only one gene called *ABPHYL (ABnormal PHYLlotaxy)* is enough to change the phyllotaxy of the leaves, implying that mutational adjustment of a single locus on the genome is enough to generate diversity.

Once the leaf primordial cells are established from the SAM cells, the new axes for leaf growth are defined, among them being the abaxial-adaxial (lower-upper surface) axes. The genes involved in defining this, and the other axes seem to be more or less conserved among higher plants. Proteins of the *HD-ZIPIII* family have been implicated in defining the adaxial identity. These proteins deviate some cells in the leaf primordium from the default abaxial state, and make them adaxial. In early plants with leaves, the leaves probably just had one type of surface — the abaxial one, the underside of today's leaves. The definition of the adaxial identity occurred some 200 million years after the abaxial identity was established.

How the wide variety of observed plant leaf morphology is generated is a subject of intense research. Some common themes have emerged. One of the most significant is the involvement of KNOX genes in generating compound leaves, as in the tomato *(see above)*. But, this is not universal. For example, the pea uses a different mechanism for doing the same thing. Mutations in genes affecting leaf curvature can also change leaf form, by changing the leaf from flat, to a crinkly shape, like the shape of cabbage leaves. There also exist different morphogen gradients in a developing leaf which define the leaf's axis and may also affect the leaf form. Another class of regulators of leaf development are the microRNAs.

Roots

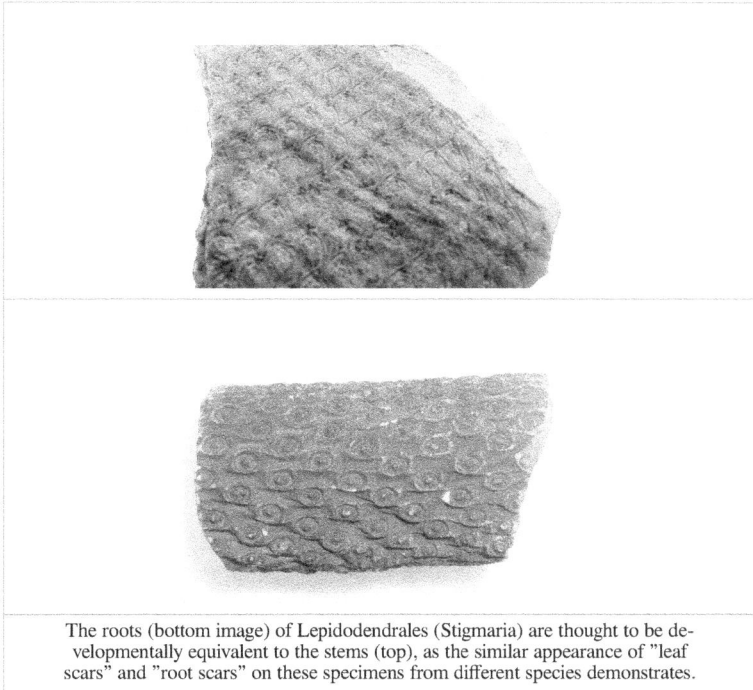

The roots (bottom image) of Lepidodendrales (Stigmaria) are thought to be developmentally equivalent to the stems (top), as the similar appearance of "leaf scars" and "root scars" on these specimens from different species demonstrates.

Roots are important to plants for two main reasons: Firstly, they provide anchorage to the substrate; more importantly, they provide a source of water and nutrients from the soil. Roots allowed plants to grow taller and faster.

The evolution of roots had consequences on a global scale. By disturbing the soil and promoting its acidification (by taking up nutrients such as nitrate and phosphate), they enabled it to weather more deeply, injecting carbon compounds deeper into soils with huge implications for climate. These effects may have been so profound they led to a mass extinction.

While there are traces of root-like impressions in fossil soils in the Late Silurian, body fossils show the earliest plants to be devoid of roots. Many had prostrate branches that sprawled along the ground, with upright axes or thalli dotted here and there, and some even had non-photosynthetic subterranean branches which lacked stomata. The distinction between root and specialised branch is developmental.Wikipedia:Please clarify differing in their branching pattern and in possession of a root cap. So while Siluro-Devonian plants such as *Rhynia* and *Horneophyton* possessed the physiological equivalent of roots, roots – defined as organs differentiated from stems – did not arrive until later. Unfortunately, roots are rarely preserved in the fossil record, and our understanding of their evolutionary origin is sparse.

Rhizoids – small structures performing the same role as roots, usually a cell in diameter – probably evolved very early, perhaps even before plants colonised the land; they are recognised in the Characeae, an algal sister group to land plants. That said, rhizoids probably evolved more than once; the rhizines of lichens, for example, perform a similar role. Even some animals (*Lamellibrachia*) have root-like structures. Rhizoids are clearly visible in the Rhynie chert fossils, and were present in most of the earliest vascular plants, and on this basis seem to have presaged true plant roots.

More advanced structures are common in the Rhynie chert, and many other fossils of comparable early Devonian age bear structures that look like, and acted like, roots. The rhyniophytes bore fine rhizoids, and the trimerophytes and herbaceous lycopods of the chert bore root-like structure penetrating a few centimetres into the soil. However, none of these fossils display all the features borne by modern roots. Roots and root-like structures became increasingly common and deeper penetrating during the Devonian, with lycopod trees forming roots around 20 cm long during the Eifelian and Givetian. These were joined by progymnosperms, which rooted up to about a metre deep, during the ensuing Frasnian stage. True gymnosperms and zygopterid ferns also formed shallow rooting systems during the Famennian.

The rhizophores of the lycopods provide a slightly different approach to rooting. They were equivalent to stems, with organs equivalent to leaves performing the role of rootlets. A similar construction is observed in the extant lycopod *Isoetes*, and this appears to be evidence that roots evolved independently at least twice, in the lycophytes and other plants, a proposition supported by studies showing that roots are initiated and their growth promoted by different mechanisms in lycophytes and euphyllophytes.

A vascular system is indispensable to rooted plants, as non-photosynthesising roots need a supply of sugars, and a vascular system is required to transport water and nutrients from the roots to the rest of the plant. Rooted

Figure 170: *The trunk of early tree fern Psaronius, showing internal structure. The top of the plant would have been to the left of the image*

plantsWikipedia:Avoid weasel words are little more advanced than their Silurian forebears, without a dedicated root system; however, the flat-lying axes can be clearly seen to have growths similar to the rhizoids of bryophytes today.

By the Middle to Late Devonian, most groups of plants had independently developed a rooting system of some nature. As roots became larger, they could support larger trees, and the soil was weathered to a greater depth. This deeper weathering had effects not only on the aforementioned drawdown of CO_2, but also opened up new habitats for colonisation by fungi and animals.

Roots today have developed to the physical limits. They penetrate many-Wikipedia:Manual of Style/Dates and numbers metres of soil to tap the water table. The narrowest roots are a mere 40 µm in diameter, and could not physically transport water if they were any narrower. The earliest fossil roots recovered, by contrast, narrowed from 3 mm to under 700 µm in diameter; of course, taphonomy is the ultimate control of what thickness can be seen.

Tree form

The early Devonian landscape was devoid of vegetation taller than waist height. Greater height provided a competitive advantage in the harvesting of

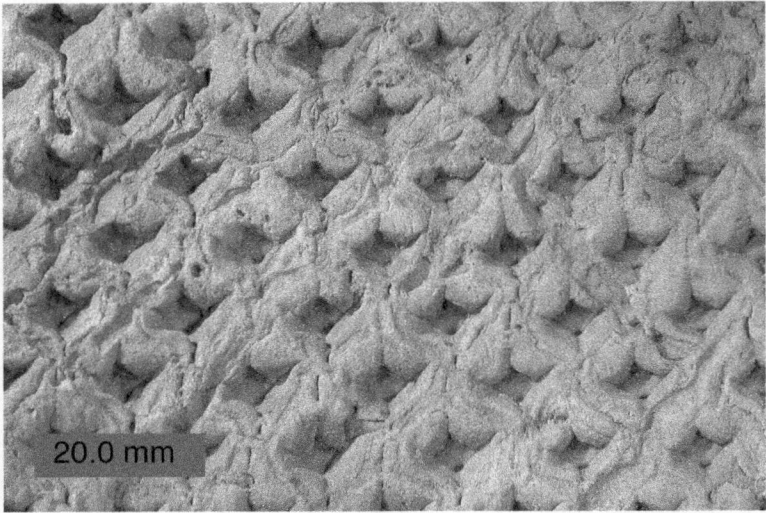

Figure 171: *External mold of Lepidodendron trunk showing leaf scars from the Upper Carboniferous of Ohio*

sunlight for photosynthesis , overshadowing of competitors and in spore distribution, as spores (and, later, seeds) could be blown for greater distances
if they started higher. An effective vascular system was required in order to
achieve greater heights. To attain arborescence, plants had to develop woody
tissue that provided both support and water transport, and thus needed to
evolve the capacity for secondary growth. The stele of plants undergoing secondary growth is surrounded by a vascular cambium, a ring of meristematic
cells which produces more xylem on the inside and phloem on the outside.
Since xylem cells comprise dead, lignified tissue, subsequent rings of xylem
are added to those already present, forming wood.

The first plants to develop secondary growth and a woody habit, were apparently the ferns, and as early as the Middle Devonian one species, *Wattieza*,
had already reached heights of 8 m and a tree-like habit.

Other clades did not take long to develop a tree-like stature. The Late Devonian *Archaeopteris*, a precursor to gymnosperms which evolved from the
trimerophytes, reached 30 m in height. The progymnosperms were the first
plants to develop true wood, grown from a bifacial cambium. The first appearance of one of them, *Rellimia*, was in the Middle Devonian. True wood is
only thought to have evolved once, giving rise to the concept of a "lignophyte"
clade.Wikipedia:Citation needed

Archaeopteris forests were soon supplemented by arborescent lycopods, in the form of Lepidodendrales, which exceeded 50m in height and 2m across at the base. These arborescent lycopods rose to dominate Late Devonian and Carboniferous forests that gave rise to coal deposits. Lepidodendrales differ from modern trees in exhibiting determinate growth: after building up a reserve of nutrients at a lower height, the plants would "bolt" as a single trunk to a genetically determined height, branch at that level, spread their spores and die. They consisted of "cheap" wood to allow their rapid growth, with at least half of their stems comprising a pith-filled cavity. Their wood was also generated by a unifacial vascular cambium – it did not produce new phloem, meaning that the trunks could not grow wider over time.Wikipedia:Verifiability

The horsetail *Calamites* appeared in the Carboniferous. Unlike the modern horsetail *Equisetum*, *Calamites* had a unifacial vascular cambium, allowing them to develop wood and grow to heights in excess of 10 m and to branch repeatedly.

While the form of early trees was similar to that of today's, the Spermatophytes or seed plants, the group that contain all modern trees, had yet to evolve. The dominant tree groups today are all seed plants, the gymnosperms, which include the coniferous trees, and the angiosperms, which contain all fruiting and flowering trees. No free-sporing trees like *Archaeopteris* exist in the extant flora. It was long thought that the angiosperms arose from within the gymnosperms, but recent molecular evidence suggests that their living representatives form two distinct groups. The molecular data has yet to be fully reconciled with morphological data, but it is becoming accepted that the morphological support for paraphyly is not especially strong. This would lead to the conclusion that both groups arose from within the pteridosperms, probably as early as the Permian.

The angiosperms and their ancestors played a very small role until they diversified during the Cretaceous. They started out as small, damp-loving organisms in the understorey, and have been diversifying ever since the mid-Wikipedia:Verifiability-Cretaceous, to become the dominant member of non-boreal forests today.

Seeds

Early land plants reproduced in the fashion of ferns: spores germinated into small gametophytes, which produced eggs and/or sperm. These sperm would swim across moist soils to find the female organs (archegonia) on the same or another gametophyte, where they would fuse with an egg to produce an embryo, which would germinate into a sporophyte.

Figure 172: *The fossil seed Trigonocarpus*

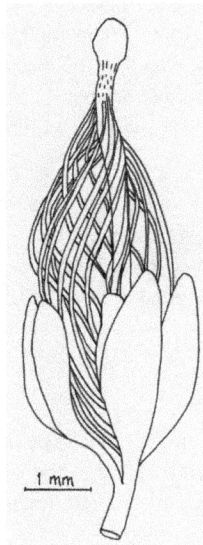

Figure 173: *The transitional fossil Runcaria*

Heterosporic plants, as their name suggests, bear spores of two sizes – microspores and megaspores. These would germinate to form microgametophytes and megagametophytes, respectively. This system paved the way for ovules and seeds: taken to the extreme, the megasporangia could bear only a single megaspore tetrad, and to complete the transition to true ovules, three of the megaspores in the original tetrad could be aborted, leaving one megaspore per megasporangium.

The transition to ovules continued with this megaspore being "boxed in" to its sporangium while it germinates. Then, the megagametophyte is contained within a waterproof integument, which forms the bulk of the seed. The microgametophyte – a pollen grain which has germinated from a microspore – is employed for dispersal, only releasing its desiccation-prone sperm when it reaches a receptive megagametophyte.

Lycopods and sphenopsids got a fair way down the path to the seed habit without ever crossing the threshold. Fossil lycopod megaspores reaching 1 cm in diameter, and surrounded by vegetative tissue, are known (Lepidocarpon, Achlamydocarpon);– these even germinate into a megagametophyte *in situ*. However, they fall short of being ovules, since the nucellus, an inner spore-covering layer, does not completely enclose the spore. A very small slit (micropyle) remains, meaning that the megasporangium is still exposed to the atmosphere. This has two consequences – firstly, it means it is not fully resistant to desiccation, and secondly, sperm do not have to "burrow" to access the archegonia of the megaspore.

A Middle Devonian precursor to seed plants from Belgium has been identified predating the earliest seed plants by about 20 million years. *Runcaria*, small and radially symmetrical, is an integumented megasporangium surrounded by a cupule. The megasporangium bears an unopened distal extension protruding above the multilobed integument. It is suspected that the extension was involved in anemophilous pollination. *Runcaria* sheds new light on the sequence of character acquisition leading to the seed. *Runcaria* has all of the qualities of seed plants except for a solid seed coat and a system to guide the pollen to the ovule.

The first spermatophytes (literally: "seed plants") – that is, the first plants to bear true seeds – are called **pteridosperms**: literally, "seed ferns", so called because their foliage consisted of fern-like fronds, although they were not closely related to ferns. The oldest fossil evidence of seed plants is of Late Devonian age, and they appear to have evolved out of an earlier group known as the progymnosperms. These early seed plants ranged from trees to small, rambling shrubs; like most early progymnosperms, they were woody plants with fern-like foliage. They all bore ovules, but no cones, fruit or similar. While it is difficult to track the early evolution of seeds, the lineage of the seed ferns may be traced from the simple trimerophytes through homosporous Aneurophytes.

This seed model is shared by basically all gymnosperms (literally: "naked seeds"), most of which encase their seeds in a woody cone or fleshy aril (the yew, for example), but none of which fully enclose their seeds. The angiosperms ("vessel seeds") are the only group to fully enclose the seed, in a carpel.

Fully enclosed seeds opened up a new pathway for plants to follow: that of seed dormancy. The embryo, completely isolated from the external atmosphere and hence protected from desiccation, could survive some years of drought before germinating. Gymnosperm seeds from the Late Carboniferous have been found to contain embryos, suggesting a lengthy gap between fertilisation and germination. This period is associated with the entry into a greenhouse earth period, with an associated increase in aridity. This suggests that dormancy arose as a response to drier climatic conditions, where it became advantageous to wait for a moist period before germinating. This evolutionary breakthrough appears to have opened a floodgate: previously inhospitable areas, such as dry mountain slopes, could now be tolerated, and were soon covered by trees.

Seeds offered further advantages to their bearers: they increased the success rate of fertilised gametophytes, and because a nutrient store could be "packaged" in with the embryo, the seeds could germinate rapidly in inhospitable environments, reaching a size where it could fend for itself more quickly. For example, without an endosperm, seedlings growing in arid environments would not have the reserves to grow roots deep enough to reach the water table before they expired from dehydration. Likewise, seeds germinating in a gloomy understory require an additional reserve of energy to quickly grow high enough to capture sufficient light for self-sustenance. A combination of these advantages gave seed plants the ecological edge over the previously dominant genus *Archaeopteris*, thus increasing the biodiversity of early forests.

Despite these advantages, it is common for fertilized ovules to fail to mature as seeds. Also during seed dormancy (often associated with unpredictable and stressful conditions) DNA damage accumulates. Thus DNA damage appears to be a basic problem for survival of seed plants, just as DNA damage is a major problem for life in general.[503]

Flowers

Flowers are modified leaves possessed only by the angiosperms, which are relatively late to appear in the fossil record. The group originated and diversified during the Early Cretaceous and became ecologically significant thereafter.

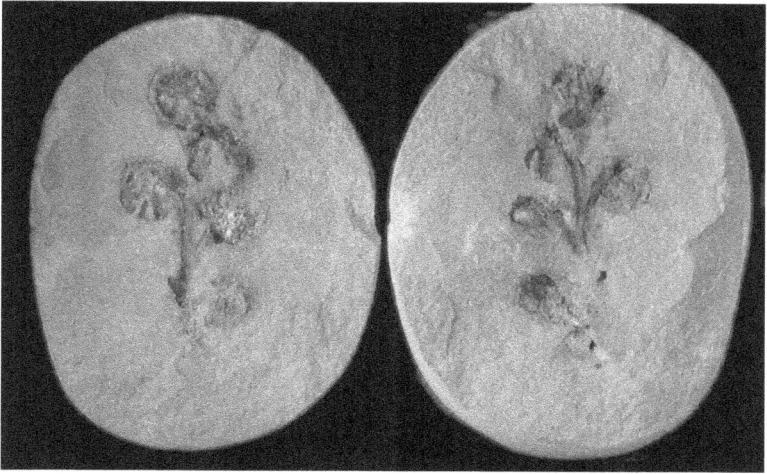

Figure 174: *The pollen bearing organs of the early "flower" Crossotheca*

Flower-like structures first appear in the fossil records some ~130 mya, in the Cretaceous.

Colorful and/or pungent structures surround the cones of plants such as cycads and Gnetales, making a strict definition of the term "flower" elusive.

The main function of a flower is reproduction, which, before the evolution of the flower and angiosperms, was the job of microsporophylls and megasporophylls. A flower can be considered a powerful evolutionary innovation, because its presence allowed the plant world to access new means and mechanisms for reproduction.

The flowering plants have long been assumed to have evolved from within the gymnosperms; according to the traditional morphological view, they are closely allied to the Gnetales. However, as noted above, recent molecular evidence is at odds with this hypothesis, and further suggests that Gnetales are more closely related to some gymnosperm groups than angiosperms, and that extant gymnosperms form a distinct clade to the angiosperms, the two clades diverging some 300[504] million years ago.

The relationship of stem groups to the angiosperms is important in determining the evolution of flowers. Stem groups provide an insight into the state of earlier "forks" on the path to the current state. Convergence increases the risk of misidentifying stem groups. Since the protection of the megagametophyte is evolutionarily desirable, probably many separate groups evolved protective encasements independently. In flowers, this protection takes the form of a

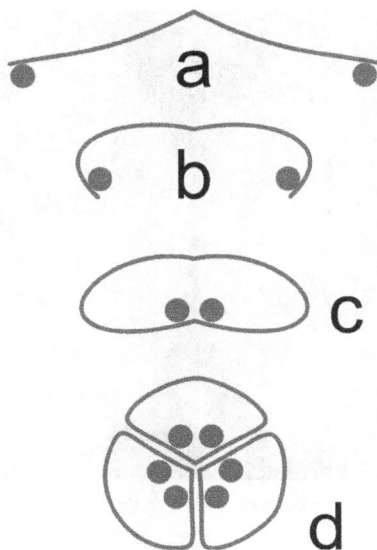

Figure 175: *The evolution of syncarps.*
a: sporangia borne at tips of leaf
b: Leaf curls up to protect sporangia
c: leaf curls to form enclosed roll
d: grouping of three rolls into a syncarp

carpel, evolved from a leaf and recruited into a protective role, shielding the ovules. These ovules are further protected by a double-walled integument.

Penetration of these protective layers needs something more than a free-floating microgametophyte. Angiosperms have pollen grains comprising just three cells. One cell is responsible for drilling down through the integuments, and creating a conduit for the two sperm cells to flow down. The megagametophyte has just seven cells; of these, one fuses with a sperm cell, forming the nucleus of the egg itself, and another joins with the other sperm, and dedicates itself to forming a nutrient-rich endosperm. The other cells take auxiliary roles.Wikipedia:Please clarify This process of "double fertilisation" is unique and common to all angiosperms.

In the fossil record, there are three intriguing groups which bore flower-like structures. The first is the Permian pteridosperm *Glossopteris*, which already bore recurved leaves resembling carpels. The Mesozoic *Caytonia* is more flower-like still, with enclosed ovules – but only a single integument. Further, details of their pollen and stamens set them apart from true flowering plants.

Figure 176: *The inflorescences of the Bennettitales are strikingly similar to flowers*

The Bennettitales bore remarkably flower-like organs, protected by whorls of bracts which may have played a similar role to the petals and sepals of true flowers; however, these flower-like structures evolved independently, as the Bennettitales are more closely related to cycads and ginkgos than to the angiosperms.

However, no true flowers are found in any groups save those extant today. Most morphological and molecular analyses place *Amborella*, the nymphaeales and Austrobaileyaceae in a basal clade called "ANA". This clade appear to have diverged in the early Cretaceous, around 130[505] million years ago – around the same time as the earliest fossil angiosperm,[506] and just after the first angiosperm-like pollen, 136 million years ago. The magnoliids diverged soon after, and a rapid radiation had produced eudicots and monocots by 125[507] million years ago. By the end of the Cretaceous 66[508] million years ago, over 50% of today's angiosperm orders had evolved, and the clade accounted for 70% of global species. It was around this time that flowering trees became dominant over conifers.[498]

The features of the basal "ANA" groups suggest that angiosperms originated in dark, damp, frequently disturbed areas. It appears that the angiosperms remained constrained to such habitats throughout the Cretaceous – occupying the niche of small herbs early in the successional series. This may have restricted their initial significance, but given them the flexibility that accounted for the rapidity of their later diversifications in other habitats.

<templatestyles src="Template:Clade/-styles.css" />
 	
Cycads
 	
<templatestyles src="Template:Clade/-styles.css" />
  Ginkgo
 	
    <templatestyles src="Template:Clade/styles.css" />
    Conifers
 	
      <templatestyles src="Template:Clade/styles.css" />
  Antho-   <templatestyles src="Template:Clade/-
  phytes   styles.css" />
        Bennettitales
 	
        Gnetales
 	
        Angiosperms
	<templatestyles src="Template:Clade/-styles.css" />
	Angiosperms
	
	  <templatestyles src="Template:Clade/styles.css
	  />
	Gym- <templatestyles src="Template:Clade/-
	nosperms styles.css" />
	        <templatestyles src="Template:Clade/-
	        styles.css" />
	          <templatestyles src="Template:Clade/-
	          styles.css" />
	          Cycads
	
	          Bennettitales
	
	        Ginkgo
	
	          <templatestyles src="Template:Clade/-
	          styles.css" />
	          Conifers
	
	          Gnetales

Traditional view	**+Phylogeny of anthophytes and gymnosperms, from**

Some propose that the Angiosperms arose from an unknown Seed Fern, Pteridophyte, and view Cycads as living Seed Ferns with both Seed-Bearing and sterile leaves (Cycas revoluta)

In August 2017, scientists presented a detailed description and 3D image of a reconstruction of possibly the first flower that lived about 140 million years ago.

Origins of the flower

The family Amborellaceae is regarded as being the sister clade to all other living flowering plants. The complete genome of *Amborella trichopoda* is still being sequenced as of March 2012[509]. By comparing its genome with those of all other living flowering plants, it will be possible to work out the most likely characteristics of the ancestor of *A. trichopoda* and all other flowering plants, i.e. the ancestral flowering plant.

It seems that on the level of the organ, the leaf may be the ancestor of the flower, or at least some floral organs. When some crucial genes involved in

Figure 177: *A male flower of Amborella trichopoda. Amborellaceae is considered the sister family of all other flowering plants.*

flower development are mutated, clusters of leaf-like structures arise in place of flowers. Thus, sometime in history, the developmental program leading to formation of a leaf must have been altered to generate a flower. There probably also exists an overall robust framework within which the floral diversity has been generated. An example of that is a gene called *LEAFY (LFY)*, which is involved in flower development in *Arabidopsis thaliana*. The homologs of this gene are found in angiosperms as diverse as tomato, snapdragon, pea, maize and even gymnosperms. Expression of *Arabidopsis thaliana* LFY in distant plants like poplar and citrus also results in flower-production in these plants. The *LFY* gene regulates the expression of some genes belonging to the MADS-box family. These genes, in turn, act as direct controllers of flower development.Wikipedia:Citation needed

Evolution of the MADS-box family

The members of the MADS-box family of transcription factors play a very important and evolutionarily conserved role in flower development. According to the ABC Model of flower development, three zones — A, B and C — are generated within the developing flower primordium, by the action of some transcription factors, that are members of the MADS-box family. Among these, the functions of the B and C domain genes have been evolutionarily more conserved than the A domain gene. Many of these genes have arisen through gene duplications of ancestral members of this family. Quite a few of them show redundant functions.

The evolution of the MADS-box family has been extensively studied. These genes are present even in pteridophytes, but the spread and diversity is many times higher in angiosperms. There appears to be quite a bit of pattern into how this family has evolved. Consider the evolution of the C-region gene *AG-AMOUS (AG)*. It is expressed in today's flowers in the stamens, and the carpel, which are reproductive organs. Its ancestor in gymnosperms also has the same expression pattern. Here, it is expressed in the strobili, an organ that produces pollen or ovules. Similarly, the B-genes' *(AP3 and PI)* ancestors are expressed only in the male organs in gymnosperms. Their descendants in the modern angiosperms also are expressed only in the stamens, the male reproductive organ. Thus, the same, then-existing components were used by the plants in a novel manner to generate the first flower. This is a recurring pattern in evolution.

Factors influencing floral diversity

🏛 IIII	Wikiversityhas bloom time data for *Linaria vulgaris*on the Bloom Clock

There is enormous variation in floral structure in plants, typically due to changes in the MADS-box genes and their expression pattern. For example, grasses possess unique floral structures. The carpels and stamens are surrounded by scale-like lodicules and two bracts, the lemma and the palea, but genetic evidence and morphology suggest that lodicules are homologous to eudicot petals.[510] The palea and lemma may be homologous to sepals in other groups, or may be unique grass structures.Wikipedia:Citation needed

Another example is that of *Linaria vulgaris*, which has two kinds of flower symmetries-radial and bilateral. These symmetries are due to epigenetic changes in just one gene called *CYCLOIDEA*.

Arabidopsis thaliana has a gene called *AGAMOUS* that plays an important role in defining how many petals and sepals and other organs are generated. Mutations in this gene give rise to the floral meristem obtaining an indeterminate fate, and proliferation of floral organs in double-flowered forms of roses, carnations and morning glory. These phenotypes have been selected by horticulturists for their increased number of petals. Several studies on diverse plants like petunia, tomato, Impatiens, maize etc. have suggested that the enormous diversity of flowers is a result of small changes in genes controlling their development.

Figure 178: *Large number of petals in roses is the result of human selection*

The Floral Genome Project confirmed that the ABC Model of flower development is not conserved across all angiosperms. Sometimes expression domains change, as in the case of many monocots, and also in some basal angiosperms like *Amborella*. Different models of flower development like the *Fading boundaries model*, or the *Overlapping-boundaries model* which propose non-rigid domains of expression, may explain these architectures. There is a possibility that from the basal to the modern angiosperms, the domains of floral architecture have become more and more fixed through evolution.

Flowering time

Another floral feature that has been a subject of natural selection is flowering time. Some plants flower early in their life cycle, others require a period of vernalization before flowering. This outcome is based on factors like temperature, light intensity, presence of pollinators and other environmental signals: genes like *CONSTANS (CO)*, *Flowering Locus C (FLC)* and *FRIGIDA* regulate integration of environmental signals into the pathway for flower development. Variations in these loci have been associated with flowering time variations between plants. For example, *Arabidopsis thaliana* ecotypes that grow in the cold, temperate regions require prolonged vernalization before they flower, while the tropical varieties, and the most common lab strains, don't. This variation is due to mutations in the *FLC* and *FRIGIDA* genes, rendering them non-functional.

Many of the genes involved in this process are conserved across all the plants studied. Sometimes though, despite genetic conservation, the mechanism of action turns out to be different. For example, rice is a short-day plant, while *Arabidopsis thaliana* is a long-day plant. Both plants have the proteins *CO* and *FLOWERING LOCUS T (FT)*, but, in *Arabidopsis thaliana*, *CO* enhances *FT* production, while in rice, the *CO* homolog represses *FT* production, resulting in completely opposite downstream effects.

Theories of flower evolution

The Anthophyte theory was based on the observation that a gymnospermic group Gnetales has a flower-like ovule. It has partially developed vessels as found in the angiosperms, and the megasporangium is covered by three envelopes, like the ovary structure of angiosperm flowers. However, many other lines of evidence show that Gnetales is not related to angiosperms.

The *Mostly Male theory* has a more genetic basis. Proponents of this theory point out that the gymnosperms have two very similar copies of the gene *LFY*, while angiosperms just have one. Molecular clock analysis has shown that the other *LFY* paralog was lost in angiosperms around the same time as flower fossils become abundant, suggesting that this event might have led to floral evolution. According to this theory, loss of one of the *LFY* paralog led to flowers that were more male, with the ovules being expressed ectopically. These ovules initially performed the function of attracting pollinators, but sometime later, may have been integrated into the core flower.

Mechanisms and players in evolution of plant morphology

While environmental factors are significantly responsible for evolutionary change, they act merely as agents for natural selection. Change is inherently brought about via phenomena at the genetic level: mutations, chromosomal rearrangements, and epigenetic changes. While the general types of mutations hold true across the living world, in plants, some other mechanisms have been implicated as highly significant.

Genome doubling is a relatively common occurrence in plant evolution and results in polyploidy, which is consequently a common feature in plants. It is estimated that at least half (and probably all) plants have seen genome doubling in their history. Genome doubling entails gene duplication, thus generating functional redundancy in most genes. The duplicated genes may attain new function, either by changes in expression pattern or changes in activity. Polyploidy and gene duplication are believed to be among the most powerful forces in evolution of plant form; though it is not known why genome doubling is such a frequent process in plants. One probable reason is the production of large amounts of secondary metabolites in plant cells. Some of them might

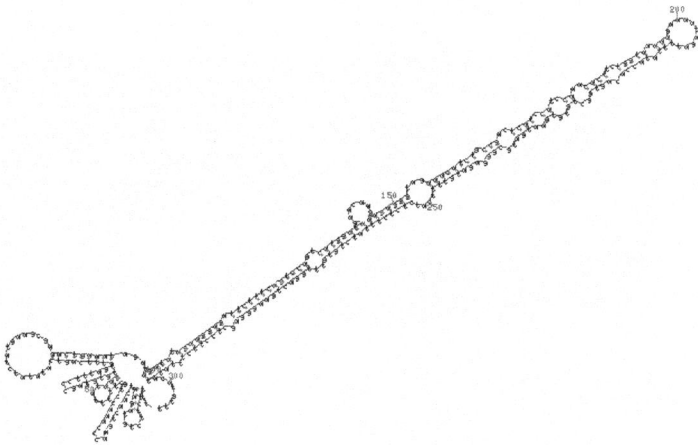

Figure 179: *The stem-loop secondary structure of a pre-microRNA from Brassica oleracea*

interfere in the normal process of chromosomal segregation, causing genome duplication.

In recent times, plants have been shown to possess significant microRNA families, which are conserved across many plant lineages. In comparison to animals, while the number of plant miRNA families are lesser than animals, the size of each family is much larger. The miRNA genes are also much more spread out in the genome than those in animals, where they are more clustered. It has been proposed that these miRNA families have expanded by duplications of chromosomal regions. Many miRNA genes involved in regulation of plant development have been found to be quite conserved between plants studied.

Domestication of plants like maize, rice, barley, wheat etc. has also been a significant driving force in their evolution. Research concerning the origin of maize has found that it is a domesticated derivative of a wild plant from Mexico called teosinte. Teosinte belongs to the genus *Zea*, just as maize, but bears very small inflorescence, 5–10 hard cobs and a highly branched and spread out stem.

Crosses between a particular teosinte variety and maize yields fertile offspring that are intermediate in phenotype between maize and teosinte. QTL analysis has also revealed some loci that, when mutated in maize, yield a teosinte-like stem or teosinte-like cobs. Molecular clock analysis of these genes estimates their origins to some 9,000 years ago, well in accordance with other records of maize domestication. It is believed that a small group of farmers must have

Figure 180: *Top: teosinte, bottom: maize, middle: maize-teosinte hybrid*

Figure 181: *Cauliflower – Brassica oleracea var. botrytis*

selected some maize-like natural mutant of teosinte some 9,000 years ago in Mexico, and subjected it to continuous selection to yield the familiar maize plant of today.

The edible cauliflower is a domesticated version of the wild plant *Brassica oleracea*, which does not possess the dense undifferentiated inflorescence, called the curd, that cauliflower possesses.

> Wikispecieshas information related to ***Brassicaceae***

Cauliflower possesses a single mutation in a gene called *CAL*, controlling meristem differentiation into inflorescence. This causes the cells at the floral meristem to gain an undifferentiated identity and, instead of growing into a flower, they grow into a dense mass of inflorescence meristem cells in arrested development. This mutation has been selected through domestication since at least the time of the Greek empire.

Evolution of photosynthetic pathways

The C_4 metabolic pathway is a valuable recent evolutionary innovation in plants, involving a complex set of adaptive changes to physiology and gene expression patterns.

Photosynthesis is a complex chemical pathway facilitated by a range of enzymes and co-enzymes. The enzyme RuBisCO is responsible for "fixing" CO_2 – that is, it attaches it to a carbon-based molecule to form a sugar, which can be used by the plant, releasing an oxygen molecule. However, the enzyme is notoriously inefficient, and, as ambient temperature rises, will increasingly fix oxygen instead of CO_2 in a process called photorespiration. This is energetically costly as the plant has to use energy to turn the products of photorespiration back into a form that can react with CO_2.

Concentrating carbon

C_4 plants evolved carbon concentrating mechanisms that work by increasing the concentration of CO_2 around RuBisCO, and excluding oxygen, thereby increasing the efficiency of photosynthesis by decreasing photorespiration. The process of concentrating CO_2 around RuBisCO requires more energy than allowing gases to diffuse, but under certain conditions – i.e. warm temperatures (>25 °C), low CO_2 concentrations, or high oxygen concentrations – pays off in terms of the decreased loss of sugars through photorespiration.

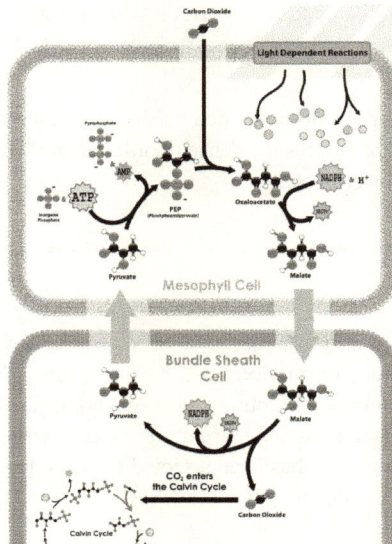

Figure 182: *The C_4 carbon concentrating mechanism*

One type of C_4 metabolism employs a so-called Kranz anatomy. This transports CO_2 through an outer mesophyll layer, via a range of organic molecules, to the central bundle sheath cells, where the CO_2 is released. In this way, CO_2 is concentrated near the site of RuBisCO operation. Because RuBisCO is operating in an environment with much more CO_2 than it otherwise would be, it performs more efficiently.

A second mechanism, CAM photosynthesis, temporally separates photosynthesis from the action of RuBisCO. RuBisCO only operates during the day, when stomata are sealed and CO_2 is provided by the breakdown of the chemical malate. More CO_2 is then harvested from the atmosphere when stomata open, during the cool, moist nights, reducing water loss.

Evolutionary record

These two pathways, with the same effect on RuBisCO, evolved a number of times independently – indeed, C_4 alone arose 62 times in 18 different plant families. A number of 'pre-adaptations' seem to have paved the way for C4, leading to its clustering in certain clades: it has most frequently been innovated in plants that already had features such as extensive vascular bundle sheath tissue. Many potential evolutionary pathways resulting in the C_4 phenotype are possible and have been characterised using Bayesian inference, confirming that non-photosynthetic adaptations often provide evolutionary stepping stones for the further evolution of C_4.

The C_4 construction is used by a subset of grasses, while CAM is employed by many succulents and cacti. The C_4 trait appears to have emerged during the Oligocene, around 25 to 32[511] million years ago; however, they did not become ecologically significant until the Miocene, 6 to 7[512] million years ago. Remarkably, some charcoalified fossils preserve tissue organised into the Kranz anatomy, with intact bundle sheath cells, allowing the presence C_4 metabolism to be identified. Isotopic markers are used to deduce their distribution and significance. C_3 plants preferentially use the lighter of two isotopes of carbon in the atmosphere, ^{12}C, which is more readily involved in the chemical pathways involved in its fixation. Because C_4 metabolism involves a further chemical step, this effect is accentuated. Plant material can be analysed to deduce the ratio of the heavier ^{13}C to ^{12}C. This ratio is denoted $\delta^{13}C$. C_3 plants are on average around 14‰ (parts per thousand) lighter than the atmospheric ratio, while C_4 plants are about 28‰ lighter. The $\delta^{13}C$ of CAM plants depends on the percentage of carbon fixed at night relative to what is fixed in the day, being closer to C_3 plants if they fix most carbon in the day and closer to C_4 plants if they fix all their carbon at night.

Original fossil material in sufficient quantity to analyse the grass itself is scarce, but horses provide a good proxy. They were globally widespread in the period of interest, and browsed almost exclusively on grasses. There's an old phrase in isotope paleontology, "you are what you eat (plus a little bit)" – this refers to the fact that organisms reflect the isotopic composition of whatever they eat, plus a small adjustment factor. There is a good record of horse teeth throughout the globe, and their $\delta^{13}C$ record shows a sharp negative inflection around 6 to 7[512] million years ago, during the Messinian that is interpreted as resulting from the rise of C_4 plants on a global scale.

When is C_4 an advantage?

While C_4 enhances the efficiency of RuBisCO, the concentration of carbon is highly energy intensive. This means that C_4 plants only have an advantage over C_3 organisms in certain conditions: namely, high temperatures and low rainfall. C_4 plants also need high levels of sunlight to thrive. Models suggest that, without wildfires removing shade-casting trees and shrubs, there would be no space for C_4 plants. But, wildfires have occurred for 400 million years – why did C_4 take so long to arise, and then appear independently so many times? The Carboniferous (~ 300[504] million years ago) had notoriously high oxygen levels – almost enough to allow spontaneous combustion[513] – and very low CO_2, but there is no C_4 isotopic signature to be found. And there doesn't seem to be a sudden trigger for the Miocene rise.

During the Miocene, the atmosphere and climate were relatively stable. If anything, CO_2 increased gradually from 14 to 9[514] million years ago before settling down to concentrations similar to the Holocene. This suggests that it did not have a key role in invoking C_4 evolution. Grasses themselves (the group which would give rise to the most occurrences of C_4) had probably been around for 60 million years or more, so had had plenty of time to evolve C_4, which, in any case, is present in a diverse range of groups and thus evolved independently. There is a strong signal of climate change in South Asia; increasing aridity – hence increasing fire frequency and intensity – may have led to an increase in the importance of grasslands. However, this is difficult to reconcile with the North American record. It is possible that the signal is entirely biological, forced by the fire- (and elephant?)- driven acceleration of grass evolution – which, both by increasing weathering and incorporating more carbon into sediments, reduced atmospheric CO_2 levels. Finally, there is evidence that the onset of C_4 from 9 to 7[515] million years ago is a biased signal, which only holds true for North America, from where most samples originate; emerging evidence suggests that grasslands evolved to a dominant state at least 15Ma earlier in South America.

Evolution of transcriptional regulation

Transcription factors and transcriptional regulatory networks play key roles in plant development and stress responses, as well as their evolution. During plant landing, many novel transcription factor families emerged and are preferentially wired into the networks of multicellular development, reproduction, and organ development, contributing to more complex morphogenesis of land plants.

Evolution of secondary metabolism

Secondary metabolites are essentially low molecular weight compounds, sometimes having complex structures, that are not essential for the normal processes of growth, development, or reproduction. They function in processes as diverse as immunity, anti-herbivory, pollinator attraction, communication between plants, maintaining symbiotic associations with soil flora, or enhancing the rate of fertilization, and hence are significant from the evo-devo perspective. Secondary metabolites are structurally and functionally diverse, and it is estimated that hundreds of thousands of enzymes might be involved in the process of producing them, with about 15–25% of the genome coding for these enzymes, and every species having its unique arsenal of secondary metabolites. Many of these metabolites, such as salicylic acid are of medical significance to humans.

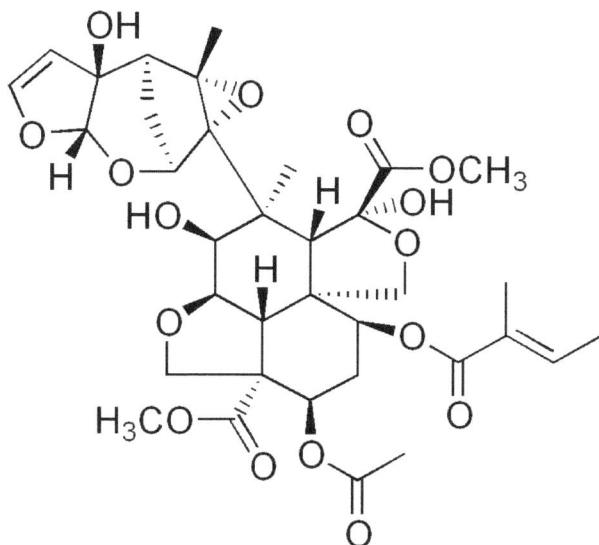

Figure 183: *Structure of Azadirachtin, a terpenoid produced by the Neem plant, which helps ward off microbes and insects. Many secondary metabolites have complex structures*

The purpose of producing so many secondary metabolites, with a significant proportion of the metabolome devoted to this activity is unclear. It is postulated that most of these chemicals help in generating immunity and, in consequence, the diversity of these metabolites is a result of a constant arms race between plants and their parasites. Some evidence supports this case. A central question involves the reproductive cost to maintaining such a large inventory of genes devoted to producing secondary metabolites. Various models have been suggested that probe into this aspect of the question, but a consensus on the extent of the cost has yet to be established; as it is still difficult to predict whether a plant with more secondary metabolites increases its survival or reproductive success compared to other plants in its vicinity.

Secondary metabolite production seems to have arisen quite early during evolution. In plants, they seem to have spread out using mechanisms including gene duplications or the evolution of novel genes. Furthermore, research has shown that diversity in some of these compounds may be positively selected for. Although the role of novel gene evolution in the evolution of secondary metabolism is clear, there are several examples where new metabolites have been formed by small changes in the reaction. For example, cyanogen glycosides have been proposed to have evolved multiple times in different plant lineages. There are several such instances of convergent evolution. For example, enzymes for synthesis of limonene – a terpene – are more similar between angiosperms and gymnosperms than to their own terpene synthesis enzymes. This suggests independent evolution of the limonene biosynthetic pathway in these two lineages.

Coevolution of plants and fungal parasites

An additional contributing factor in some plants leading to evolutionary change is the force due to coevolution with fungal parasites. In an environment with a fungal parasite, which is common in nature, the plants must make adaptation in an attempt to evade the harmful effects of the parasite.

Whenever a parasitic fungus is siphoning limited resources away from a plant, there is selective pressure for a phenotype that is better able to prevent parasitic attack from fungi. At the same time, fungi that are better equipped to evade the defenses of the plant will have greater fitness level. The combination of these two factors leads to an endless cycle of evolutionary change in the host-pathogen system.

Because each species in the relationship is influenced by a constantly changing symbiont, evolutionary change usually occurs at a faster pace than if the other species was not present. This is true of most instances of coevolution. This makes the ability of a population to quickly evolve vital to its survival. Also, if the pathogenic species is too successful and threatens the survival and reproductive success of the host plants, the pathogenic fungi risk losing their nutrient source for future generations. These factors create a dynamic that shapes the evolutionary changes in both species generation after generation.

Genes that code for defense mechanisms in plants must keep changing to keep up with the parasite that constantly works to evade the defenses. Genes that code for attachment mechanisms are the most dynamic and are directly related to the evading ability of the fungi. The greater the changes in these genes, the more change in the attachment mechanism. After selective forces on the resulting phenotypes, evolutionary change that promotes evasion of host defenses occurs.

Fungi not only evolve to avoid the defenses of the plants, but they also attempt to prevent the plant from enacting the mechanisms to improve its defenses. Anything the fungi can do to slow the evolution process of the host plants will improve the fitness of future generations because the plant will not be able to keep up with the evolutionary changes of the parasite. One of the main processes by which plants quickly evolve in response to the environment is sexual reproduction. Without sexual reproduction, advantageous traits could not be spread through the plant population as quickly allowing the fungi to gain a competitive advantage. For this reason, the sexual reproductive organs of plants are targets for attacks by fungi. Studies have shown that many different current types of obligate parasitic plant fungi have developed mechanisms to disable or otherwise affect the sexual reproduction of the plants. If successful, the sexual reproduction process slows for the plant, thus slowing down evolutionary change or in extreme cases, the fungi can render the plant sterile creating an advantage for the pathogens. It is unknown exactly how this adaptive trait developed in fungi, but it is clear that the relationship to the plant forced the development of the process.

Some researchers are also studying how a range of factors affect the rate of evolutionary change and the outcomes of change in different environments. For example, as with most evolution, increases in heritability in a population allow for a greater evolutionary response in the presence of selective pressure. For traits specific to the plant-fungi coevolution, researchers have studied how the virulence of the invading pathogen affects the coevolution. Studies involving *Mycosphaerella graminicola* have consistently showed that virulence of a pathogen does not have a significant impact on the evolutionary track of the host plant.

There can be other factors in that can affect the process of coevolution. For example, in small populations, selection is a relatively weaker force on the population due to genetic drift. Genetic drift increases the likelihood of having fixed alleles which decreases the genetic variance in the population. Therefore, if there is only a small population of plants in an area with the ability to reproduce together, genetic drift may counteract the effects of selection putting the plant in a disadvantageous position to fungi which can evolve at a normal rate. The variance in both the host and pathogen population is a major determinant of evolutionary success compared to the other species. The greater the genetic variance, the faster the species can evolve to counteract the other organism's avoidance or defensive mechanisms.

Due to the process of pollination for plants, the effective population size is normally larger than for fungi because pollinators can link isolated populations in a way that the fungus is not able. This means positive traits that evolve in non-adjacent but close areas can be passed to nearby areas. Fungi must individually evolve to evade host defenses in each area. This is obviously a clear competitive advantage for the host plants. Sexual reproduction with a broad, high variance population leads to fast evolutionary change and higher reproductive success of offspring.

Environment and climate patterns also play a role in evolutionary outcomes. Studies with oak trees and an obligate fungal parasite at different altitudes clearly show this distinction. For the same species, different altitudinal positions had drastically different rates of evolution and changes in the response to the pathogens due to the organism also in a selective environment due to their surroundings.

Coevolution is a process that is related to the red queen hypothesis. Both the host plant and parasitic fungi have to continue to survive to stay in their ecological niche. If one of the two species in the relationship evolves at a significantly faster rate than the other, the slower species will be at a competitive disadvantage and risk the loss of nutrients. Because the two species in the system are so closely linked, they respond to external environment factors together and each species affects the evolutionary outcome of the other. In other words, each species exerts selective pressure on the other. Population size is also a major factor in the outcome because differences in gene flow and genetic drift could cause evolutionary changes that do not match the direction of selection expected by forces due to the other organism. Coevolution is an important phenomenon necessary for understanding the vital relationship between plants and their fungal parasites.

Early land vertebrates

Tetrapod

Tetrapods	
Temporal range: Late Devonian–Present, 367.5–0 Ma PreЄЄ OSD C P T J K PgN	

Representatives of extant tetrapod groups, (clockwise from upper left): a frog (a lissamphibian), a hoatzin, a skink (two sauropsids), and a mouse (a synapsid)

Scientific classification 🖉	
Kingdom:	Animalia
Phylum:	Chordata
Infraphylum:	Gnathostomata
Clade:	Eugnathostomata
Clade:	Teleostomi
Superclass:	**Tetrapoda** Goodrich, 1930
Subgroups	

- Batrachomorpha
  - various extinct clades
  - Amphibia
- Reptiliomorpha
  - various extinct clades
  - Amniota (Crown group)
    - Synapsida
    - Sauropsida

The superclass **Tetrapoda** (from Greek: τετρα- "four" and πούς "foot") contains the four-limbed vertebrates known as **tetrapods** (/ˈtɛtrəpɒd/); it includes living and extinct amphibians, reptiles (including dinosaurs and thus birds) and mammals (including primates, and all hominid subgroups including humans), as well as earlier extinct groups. Tetrapods evolved from a group of animals known as the Tetrapodomorpha, who in turn evolved from ancient Sarcopterygii lobe-finned fishes around 390 million years ago in the middle Devonian period; their forms were transitional between lobe-finned fishes and the four-limbed tetrapods.

The first tetrapods appeared by the late Devonian, 367.5 million years ago; the specific aquatic ancestors of the tetrapods, and the process by which they colonized Earth's land after emerging from water remains unclear, and is an area of research and debate among palaeontologists.

The first tetrapods were primarily aquatic. Modern amphibians, which evolved from earlier groups, are generally semiaquatic; the first stage of their lives is as fish-like tadpoles, and later stages are partly terrestrial and partly aquatic. However, most tetrapod species today are amniotes, most of those are terrestrial tetrapods whose branch evolved from earlier tetrapods about 340 million years ago (crown amniotes evolved 318 million years ago) the key innovation in amniotes over amphibians is laying of eggs on land or having further evolved to retain the fertilized egg(s) within the mother.

Amniote tetrapods began to dominate and drove most amphibian tetrapods to extinction. One group of amniotes diverged into the reptiles, which includes lepidosaurs, dinosaurs (which includes birds), crocodilians, turtles, and extinct relatives; while another group of amniotes diverged into the mammals and their extinct relatives. Amniotes include the tetrapods that further evolved for flight—such as birds from among the dinosaurs, and bats from among the mammals.

The change from a body plan for breathing and navigating in water to a body plan enabling the animal to move on land is one of the most profound evolutionary changes known.[516] It is becoming increasingly understood thanks to additional transitional fossil finds and improved phylogenetic analysis.

Several groups of tetrapods, such as the caecilians, snakes, cetaceans, sirenians, and moas have lost some or all of their limbs through further speciation and evolution; some have only concealed vestigial bones as a remnant of the limbs of their distant ancestors.

Many tetrapods have returned to partially or fully aquatic lives throughout the history of the group. Modern examples of tetrapods that evolved back to aquatic life include mammalian species such as cetaceans (like whales and dolphins), sirenians (like the sea cow), and pinnipeds (like seals); and reptilian species such as certain avian dinosaurs (aquatic birds like penguins) and various venomous elapid snakes (sea snakes like coral snakes, and other species of Hydrophiinae). The first tetrapods to return aquatic life may have done so as early as the Carboniferous period; others returned as recently as the Cenozoic, as in cetaceans and pinnipeds and several modern amphibians.

Definitions

Tetrapods can be defined in cladistics as the nearest common ancestor of all living amphibians (the lissamphibians) and all living amniotes (reptiles, birds, and mammals), along with all of the descendants of that ancestor. This is a node-based definition (the node being the nearest common ancestor). The group so defined is the crown group, or crown tetrapods. The term tetrapodomorph is used for the stem-based definition: any animal that is more closely related to living amphibians, reptiles, birds, and mammals than to living dipnoi (lungfishes). The group so defined is known as the tetrapod total group.

Stegocephalia is a larger group equivalent to some broader uses of the word *tetrapod*, used by scientists who prefer to reserve *tetrapod* for the crown group (based on the nearest common ancestor of living forms). Such scientists use the term "stem-tetrapod" to refer to those tetrapod-like vertebrates that are not members of the crown group, including the tetrapodomorph fishes.

The two subclades of crown tetrapods are Batrachomorpha and Reptiliomorpha. Batrachomorphs are all animals sharing a more recent common ancestry with living amphibians than with living amniotes (reptiles, birds, and mammals). Reptiliomorphs are all animals sharing a more recent common ancestry with living amniotes than with living amphibians.

Biodiversity

Tetrapoda includes four classes: amphibians, reptiles, mammals, and birds. Overall, the biodiversity of lissamphibians, as well as of tetrapods generally, has grown exponentially over time; the more than 30,000 species living today are descended from a single amphibian group in the Early to Middle Devonian. However, that diversification process was interrupted at least a few times by major biological crises, such as the Permian–Triassic extinction event, which at least affected amniotes. The overall composition of biodiversity was driven primarily by amphibians in the Palaeozoic, dominated by reptiles in the Mesozoic and expanded by the explosive growth of birds and mammals in the Cenozoic. As biodiversity has grown, so has the number of niches that tetrapods have occupied. The first tetrapods were aquatic and fed primarily on fish. Today, the Earth supports a great diversity of tetrapods that live in many habitats and subsist on a variety of diets. The following table shows summary estimates for each tetrapod class from the *IUCN Red List of Threatened Species*, 2014.3, for the number of extant species that have been described in the literature, as well as the number of threatened species.[517]

IUCN global summary estimates for extant tetrapod species as of 2014						
Tetrapod group	Image	Class	Estimated number of described species	Threatened species in Red List	Species evaluated as percent of described species	Best estimate of percent of threatened species
Anamniotes lay eggs in water		Amphibians	7,302	1,957	88%	41%
Amniotes adapted to lay eggs on land		Sauropsids (traditional reptilia+birds)	20,463	2,300	75%	13%
		Mammals	5,513	1,199	100%	26%
		Overall	33,278	5,456	80%	?

Evolution

Ancestry

Tetrapods evolved from early bony fishes (Osteichthyes), specifically from the tetrapodomorph branch of lobe-finned fishes (Sarcopterygii), living in the early to middle Devonian period.

The first tetrapods probably evolved in the Emsian stage of the Early Devonian from Tetrapodomorph fish living in shallow water environments. The very earliest tetrapods would have been animals similar to *Acanthostega*, with legs and lungs as well as gills, but still primarily aquatic and unsuited to life on land.

The earliest tetrapods inhabited saltwater, brackish-water, and freshwater environments, as well as environments of highly variable salinity. These traits

Figure 184: *Devonian fishes, including an early shark Cladoselache, Eusthenopteron and other lobe-finned fishes,
and the placoderm Bothriolepis (Joseph Smit, 1905).*

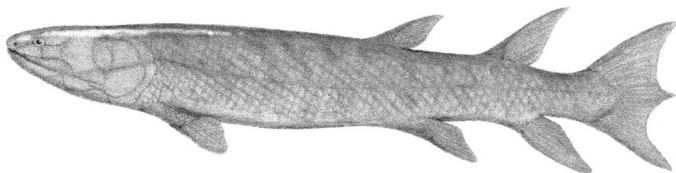

Figure 185: *Eusthenopteron, ∼385 Ma*

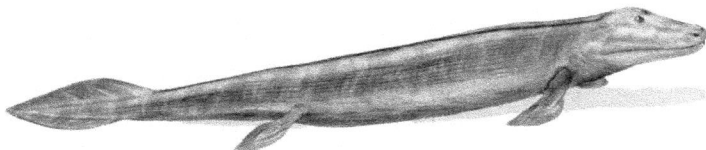

Figure 186: *Tiktaalik, ∼375 Ma*

Figure 187: *Acanthostega,* ~365 Ma

were shared with many early lobed-finned fishes. As early tetrapods are found on two Devonian continents, Laurussia (Euramerica) and Gondwana, as well as the island of North China, it is widely supposed that early tetrapods were capable of swimming across the shallow (and relatively narrow) continental-shelf seas that separated these landmasses.

Since the early 20th century, several families of tetrapodomorph fishes have been proposed as the nearest relatives of tetrapods, among them the rhizodonts (notably Sauripterus), the osteolepidids, the tristichopterids (notably Eusthenopteron), and more recently the elpistostegalians (also known as Panderichthyida) notably the genus Tiktaalik.

A notable feature of Tiktaalik is the absence of bones covering the gills. These bones would otherwise connect the shoulder girdle with skull, making the shoulder girdle part of the skull. With the loss of the gill-covering bones, the shoulder girdle is separated from the skull, connected to the torso by muscle and other soft-tissue connections. The result is the appearance of the neck. This feature appears only in tetrapods and Tiktaalik, not other tetrapodomorph fishes. Tiktaalik also had a pattern of bones in the skull roof (upper half of the skull) that is similar to the end-Devonian tetrapod Ichthyostega. The two also shared a semi-rigid ribcage of overlapping ribs, which may have substituted for a rigid spine. In conjunction with robust forelimbs and shoulder girdle, both Tiktaalik and Ichthyostega may have had the ability to locomote on land in the manner of a seal, with the forward portion of the torso elevated, the hind part dragging behind. Finally, Tiktaalik fin bones are somewhat similar to the limb bones of tetrapods.

However, there are issues with supposing that Tiktaalik is a tetrapod ancestor. For example, Tiktaalik had a long spine with far more vertebrae than any known tetrapod or other tetrapodomorph fish. Also the oldest tetrapod trace fossils (tracks and trackways) predate Tiktaalik by a considerable margin. Several hypotheses have been proposed to explain this date discrepancy: 1) The nearest common ancestor of tetrapods and Tiktaalik dates to the Early Devonian. By this hypothesis, the Tiktaalik lineage is the closest to tetrapods, but Tiktaalik itself was a late-surviving relic. 2) Tiktaalik represents a case of parallel evolution. 3) Tetrapods evolved more than once.

<templatestyles src="Template:Clade/styles.css" />

Euteleostomi

<templatestyles src="Template:Clade/-styles.css" />

Actinopterygii

(ray-finned fishes)

Sarcopterygii

<templatestyles src="Template:Clade/styles.css" />

Crossopterygii　　(Actinistia) Coelacanths

Rhipidistia

<templatestyles src="Template:Clade/styles.css" />

Dipnomorpha　　Dipnoi (Lungfish)

Tetrapodomorpha

<templatestyles src="Template:Clade/styles.css" />

†Tetrapodomorph fishes

Tetrapoda

(fleshy-limbed vertebrates)

(bony vertebrates)

Devonian fossils

The oldest evidence for the existence of tetrapods comes from trace fossils, tracks (footprints) and trackways found in Zacheɫmie, Poland, dated to the Eifelian stage of the Middle Devonian, 390[518] million years ago, although these traces have also been interpreted as the ichnogenus Piscichnus (fish nests/feeding traces).[519] The adult tetrapods had an estimated length of 2.5 m (8 feet), and lived in a lagoon with an average depth of 1–2 m, although it is not known at what depth the underwater tracks were made. The lagoon was inhabited by a variety of marine organisms and was apparently salt water. The average water temperature was 30 degrees C (86 F). The second oldest evidence for tetrapods, also tracks and trackways, date from ca. 385 Mya (Valentia Island, Ireland).[520,521]

The oldest partial fossils of tetrapods date from the Frasnian beginning ~380 mya. These include *Elginerpeton* and *Obruchevichthys*. Some paleontologists dispute their status as true (digit-bearing) tetrapods.

All known forms of Frasnian tetrapods became extinct in the Late Devonian extinction, also known as the end-Frasnian extinction. This marked the beginning of a gap in the tetrapod fossil record known as the Famennian gap, occupying roughly the first half of the Famennian stage.

The oldest near-complete tetrapod fossils, Acanthostega and Ichthyostega, date from the second half of the Fammennian. Although both were essentially four-footed fish, Ichthyostega is the earliest known tetrapod that may have had the ability to pull itself onto land and drag itself forward with its forelimbs. There is no evidence that it did so, only that it may have been anatomically capable of doing so.

The end-Fammenian marked another extinction, known as the end-Fammenian extinction or the Hangenberg event, which is followed by another gap in the tetrapod fossil record, Romer's gap, also known as the Tournaisian gap. This gap, which was initially 30 million years, but has been gradually reduced over time, currently occupies much of the 13.9-million year Tournaisian, the first stage of the Carboniferous period.

Figure 188: *Ichthyostega, 374–359 Ma*

Palaeozoic tetrapods

Devonian stem-tetrapods

Tetrapod-like vertebrates first appeared in the early Devonian period. These early "stem-tetrapods" would have been animals similar to *Ichthyostega*, with legs and lungs as well as gills, but still primarily aquatic and unsuited to life on land. The Devonian stem-tetrapods went through two major bottlenecks during the Late Devonian extinctions, also known as the end-Frasnian and end-Fammenian extinctions. These extinction events led to the disappearance of stem-tetrapods with fish-like features. When stem-tetrapods reappear in the fossil record in early Carboniferous deposits, some 10 million years later, the adult forms of some are somewhat adapted to a terrestrial existence.[522] Why they went to land in the first place is still debated.

Carboniferous tetrapods

During the early Carboniferous, the number of digits on hands and feet of stem-tetropods became standardized at no more than five, as lineages with more digits died out. By mid-Carboniferous times, the stem-tetrapods had radiated into two branches of true ("crown group") tetrapods. Modern amphibians are derived from either the temnospondyls or the lepospondyls (or possibly both), whereas the anthracosaurs were the relatives and ancestors of the amniotes (reptiles, mammals, and kin). The first amniotes are known from the early part of the Late Carboniferous. Amphibians must return to water to lay eggs; in contrast, amniote eggs have a membrane ensuring gas exchange out of water and can therefore be laid on land.

Amphibians and amniotes were affected by the Carboniferous Rainforest Collapse (CRC), an extinction event that occurred ~300 million years ago. The sudden collapse of a vital ecosystem shifted the diversity and abundance of major groups. Amniotes were more suited to the new conditions. They invaded

Figure 189: *Edops, 323-299 Ma*

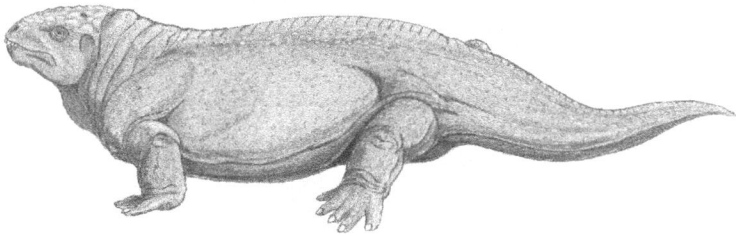

Figure 190: *Diadectes, 290–272 Ma*

new ecological niches and began diversifying their diets to include plants and other tetrapods, previously having been limited to insects and fish.

Permian tetrapods

In the Permian period, in addition to temnospondyl and anthracosaur clades, there were two important clades of amniote tetrapods, the sauropsids and the synapsids. The latter were the most important and successful Permian animals.

The end of the Permian saw a major turnover in fauna during the Permian–Triassic extinction event. There was a protracted loss of species, due

to multiple extinction pulses. Many of the once large and diverse groups died
out or were greatly reduced.

Mesozoic tetrapods

The diapsids (a subgroup of the sauropsids) began to diversify during the Triassic, giving rise to the turtles, crocodiles, and dinosaurs. In the Jurassic,
lizards developed from other diapsids. In the Cretaceous, snakes developed
from lizards and modern birds branched from a group of theropod dinosaurs.
By the late Mesozoic, the groups of large, primitive tetrapod that first appeared during the Paleozoic such as temnospondyls and amniote-like tetrapods
had gone extinct. Many groups of synapsids, such as anomodonts and therocephalians, that once comprised the dominant terrestrial fauna of the Permian,
also became extinct during the Mesozoic; however, during the Jurassic, one
synapsid group (Cynodontia) gave rise to the modern mammals, which survived through the Mesozoic to later diversify during the Cenozoic.

Cenozoic tetrapods

Following the great faunal turnover at the end of the Mesozoic, representatives of seven major groups of tetrapods persisted into the Cenozoic era. One
of them, the Choristodera, became extinct 20 million years ago for unknown
reasons. The surviving six are:

- Lissamphibia: frogs and toads, newts and salamanders, and caecilians
- Lepidosauria: tuataras, lizards, amphisbaenians and snakes
- Testudines: turtles and tortoises
- Crocodilia: crocodiles, alligators, caimans and gharials
- Dinosauria: avian dinosaurs (i.e. modern birds)
- Mammalia: mammals

Classification

The classification of tetrapods has a long history. Traditionally, tetrapods are
divided into four classes based on gross anatomical and physiological traits.[523]
Snakes and other legless reptiles are considered tetrapods because they are
sufficiently like other reptiles that have a full complement of limbs. Similar
considerations apply to caecilians and aquatic mammals. Newer taxonomy
is frequently based on cladistics instead, giving a variable number of major
"branches" (clades) of the tetrapod family tree.

As is the case throughout evolutionary biology today, there is debate over how
to properly classify the groups within Tetrapoda. Traditional biological classification sometimes fails to recognize evolutionary transitions between older

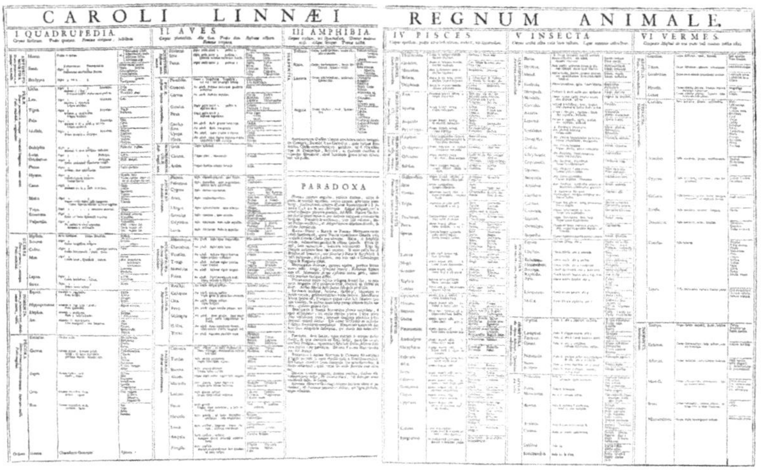

Figure 191: *Linnaeus' 1735 classification of animals, with tetrapods occupying the first three classes*

groups and descendant groups with markedly different characteristics. For example, the birds, which evolved from the dinosaurs, are defined as a separate group from them, because they represent a distinct new type of physical form and functionality. In phylogenetic nomenclature, in contrast, the newer group is always included in the old. For this school of taxonomy, dinosaurs and birds are not groups in contrast to each other, but rather birds are a sub-type *of* dinosaurs.

History of classification

The tetrapods, including all large- and medium-sized land animals, have been among the best understood animals since earliest times. By Aristotle's time, the basic division between mammals, birds and egg-laying tetrapods (the "herptiles") was well known, and the inclusion of the legless snakes into this group was likewise recognized. With the birth of modern biological classification in the 18th century, Linnaeus used the same division, with the tetrapods occupying the first three of his six classes of animals. While reptiles and amphibians can be quite similar externally, the French zoologist Pierre André Latreille recognized the large physiological differences at the beginning of the 19th century and split the herptiles into two classes, giving the four familiar classes of tetrapods: amphibians, reptiles, birds and mammals.[524]

Modern classification

With the basic classification of tetrapods settled, a half a century followed
where the classification of living and fossil groups was predominately done
by experts working within classes. In the early 1930s, American vertebrate
palaeontologist Alfred Romer (1894–1973) produced an overview, drawing
together taxonomic work from the various subfields to create an orderly taxonomy in his *Vertebrate Paleontology*.[525] This classical scheme with minor
variations is still used in works where systematic overview is essential, e.g.
Benton (1998) and Knobill and Neill (2006).[526,527] While mostly seen in general works, it is also still used in some specialist works like Fortuny & al.
(2011). The taxonomy down to subclass level shown here is from Hildebrand
and Goslow (2001):

- **Superclass Tetrapoda** - four-limbed vertebrates
  - **Class Amphibia** - amphibians
    - **Subclass Ichthyostegalia** - early fish-like amphibians-now outside
      tetrapoda
    - **Subclass Anthracosauria** - reptile-like amphibians (once thought to
      be the ancestors of the amniotes)
    - **Subclass Temnospondyli** - large-headed Paleozoic and Mesozoic
      amphibians
    - **Subclass Lissamphibia** - modern amphibians
  - **Class Sauropsida** - reptiles (includes birds)
    - **Subclass Diapsida** - diapsids, including crocodiles, dinosaurs (in-
      cludes birds), lizards, snakes and turtles
  - **Class Synapsida** - mammals and their ancestors
    - **Subclass Prototheria** - monotremes
    - **Subclass Allotheria** - multituberculates
    - **Subclass Theria** - live bearing mammals

This classification is the one most commonly encountered in school textbooks
and popular works. While orderly and easy to use, it has come under critique
from cladistics. The earliest tetrapods are grouped under Class Amphibia, although several of the groups are more closely related to amniotes than to modern day amphibians. Traditionally, birds are not considered a type of reptile,
but crocodiles are more closely related to birds than they are to other reptiles,
such as lizards. Birds themselves are thought to be descendents of theropod
dinosaurs. Basal non-mammalian synapsids ("mammal-like reptiles") traditionally also sort under Class Reptilia as a separate subclass, but they are more
closely related to mammals than to living reptiles. Considerations like these
have led some authors to argue for a new classification based purely on phylogeny, disregarding the anatomy and physiology.

Relationships

Stem-tetrapods

Stem tetrapods are all animals more closely related to tetrapods than to lung-fish, but excluding the tetrapod crown group. The cladogram below illustrates the relationships of stem-tetrapods, from Swartz, 2012:

<templatestyles src="Template:Clade/styles.css" />

Rhipidistia

<templatestyles src="Template:Clade/styles.css" />

TetrapodomorphaDipnomorpha (lungfishes and relatives)

<templatestyles src="Template:Clade/styles.css" />

Kenichthys

<templatestyles src="Template:Clade/styles.css" />

Rhizodontidae

<templatestyles src="Template:Clade/styles.css" />

Canowindridae

<templatestyles src="Template:Clade/styles.css" />

Marsdenichthys

<templatestyles src="Template:Clade/styles.css" />
Canowindra

<templatestyles src="Template:Clade/styles.css" />
Koharalepis

Beelarongia

<templatestyles src="Template:Clade/styles.css" />

MegalichthyiformesEotetrapodiformes

<templatestyles src="Template:Clade/styles.css" />

Gogonasus

<templatestyles src=”Template:Clade/styles.css” />
  Gyroptychius

  <templatestyles src=”Template:Clade/styles.css” />
    Osteolepis

    <templatestyles src=”Template:Clade/styles.css” />
      Medoevia

      Megalichthyidae

<templatestyles src=”Template:Clade/styles.css” />

Tristichopteridae
<templatestyles src=”Template:Clade/styles.css” />

  Spodichthys

  <templatestyles src=”Template:Clade/styles.css” />
    Tristichopterus

    <templatestyles src=”Template:Clade/styles.css” />
      Eusthenopteron

      <templatestyles src=”Template:Clade/styles.css” />
        Jarvikina

        <templatestyles src=”Template:Clade/styles.css” />
          Cabbonichthys

          <templatestyles src=”Template:Clade/styles.css” />
            Mandageria

            Eusthenodon

<templatestyles src=”Template:Clade/styles.css” />

Tinirau
<templatestyles src=”Template:Clade/styles.css” />

Elpistostegalia*Platycephalichthys*

<templatestyles src="Template:Clade/styles.css" />

Stegocephalia*Panderichthys*

<templatestyles src="Template:Clade/styles.css" />

  <templatestyles src="Template:Clade/styles.css" />
   Tiktaalik

   Elpistostege

  <templatestyles src="Template:Clade/styles.css" />
   Elginerpeton

   <templatestyles src="Template:Clade/styles.css" />
    Ventastega

    <templatestyles src="Template:Clade/styles.css" />
     Acanthostega

     <templatestyles src="Template:Clade/styles.css" />
      Ichthyostega

      <templatestyles src="Template:Clade/styles.css" />
       Whatcheeriidae

       <templatestyles src="Template:Clade/styles.css" />
        Colosteidae

        <templatestyles src="Template:Clade/styles.css" />
         Crassigyrinus

         <templatestyles src="Template:Clade/styles.css" />
          Baphetidae

          Tetrapoda

Crown group

Crown tetrapods are defined as the nearest common ancestor of all living tetrapods (amphibians, reptiles, birds, and mammals) along with all of the descendants of that ancestor.

The inclusion of certain extinct groups in the crown Tetrapoda depends on the relationships of modern amphibians, or lissamphibians. There are currently three major hypotheses on the origins of lissamphibians. In the temnospondyl hypothesis (TH), lissamphibians are most closely related to dissorophoid temnospondyls, which would make temnospondyls tetrapods. In the lepospondyl hypothesis (LH), lissamphibians are the sister taxon of lysorophian lepospondyls, making lepospondyls tetrapods and temnospondyls stem-tetrapods. In the polyphyletic hypothesis (PH), frogs and salamanders evolved from dissorophoid temnospondyls while caecilians come out of microsaur lepospondyls, making both lepospondyls and temnospondyls true tetrapods.

Temnospondyl hypothesis (TH)

This hypothesis comes in a number of variants, most of which have lissamphibians coming out of the dissorophoid temnospondyls, usually with the focus on amphibamids and branchiosaurids.

The Temnospondyl Hypothesis is the currently favored or majority view, supported by Ruta *et al* (2003a,b), Ruta and Coates (2007), Coates *et al* (2008), Sigurdsen and Green (2011), and Schoch (2013,2014).

Cladogram modified after Coates, Ruta and Friedman (2008).

<templatestyles src="Template:Clade/styles.css" />

Crown-group Tetrapoda

<templatestyles src="Template:Clade/styles.css" />

<templatestyles src="Template:Clade/styles.css" />

Tem-   <templatestyles src="Template:Clade/styles.css" />
nospondyli   **crown group Lissamphibia**

<templatestyles src="Template:Clade/styles.css" />

<templatestyles src="Template:Clade/styles.css" />

Embolomeri

<templatestyles src="Template:Clade/styles.css" />
Gephyrostegidae

<templatestyles src="Template:Clade/styles.css" />
Seymouriamorpha

<templatestyles src="Template:Clade/styles.css" />
<templatestyles src="Template:Clade/styles.css" />
Diadectomorpha

Crown-group Amniota

<templatestyles src="Template:Clade/styles.css" />
Microsauria

<templatestyles src="Template:Clade/styles.css" />
Lysorophia

<templatestyles src="Template:Clade/styles.css" />
Adelospondyli

<templatestyles src="Template:Clade/styles.css" />
Nectridea

<templatestyles src="Template:Clade/styles.css" />
Aistopoda

Lepospondyl hypothesis (LH)

Cladogram modified after Laurin, *How Vertebrates Left the Water* (2010).
<templatestyles src="Template:Clade/styles.css" />

Stegocephalia (''Tetrapoda'')
<templatestyles src="Template:Clade/styles.css" />

Acanthostega  *Ichthyostega* 
<templatestyles src="Template:Clade/styles.css" />

Temnospondyli

<templatestyles src="Template:Clade/styles.css" />
Embolomeri

<templatestyles src="Template:Clade/styles.css" />
Seymouriamorpha

<templatestyles src="Template:Clade/styles.css" />
Reptiliomor-pha <templatestyles src="Template:Clade/styles.css" />
Amniota

Diadectomorpha

Amphibia <templatestyles src="Template:Clade/styles.css" />
<templatestyles src="Template:Clade/styles.css" />
Aistopoda

Adelogyrinidae

<templatestyles src="Template:Clade/styles.css" />
Nectridea

<templatestyles src="Template:Clade/styles.css" />
Lysorophia

<templatestyles src="Template:Clade/styles.css" />
Lissamphibia

Polyphyly hypothesis (PH)

This hypothesis has batrachians (frogs and salamander) coming out of dissorophoid temnospondyls, with caecilians out of microsaur lepospondyls. There are two variants, one developed by Carroll, the other by Anderson.

Cladogram modified after Schoch, Frobisch, (2009). <templatestyles src="Template:Clade/styles.css" />

Tetrapoda

<templatestyles src="Template:Clade/styles.css" />

stem tetrapods

<templatestyles src="Template:Clade/styles.css" />

<templatestyles src="Template:Clade/styles.css" />

Tem-nospondyli <templatestyles src="Template:Clade/styles.css" />

basal temnospondyls

Dissorophoidea <templatestyles src="Template:Clade/styles.css" />

<templatestyles src="Template:Clade/styles.css" />

Amphibamidae

Frogs

<templatestyles src="Template:Clade/styles.css" />

Branchiosauridae

Salamanders

<templatestyles src="Template:Clade/styles.css" />

Lep-ospondyli <templatestyles src="Template:Clade/styles.css" />

Lysorophia

<templatestyles src="Template:Clade/styles.css" />

Caecilians

Microsauria

```
<templatestyles
src="Template:Clade/styles.css"
/>
    <templatestyles
    src="Template:Clade/styles.css"
    />
```

Seymouriamorpha

```
    <templatestyles
    src="Template:Clade/styles.css" />
```

Diadectomorpha

Amniota

Anatomy and physiology of early tetrapods

The tetrapod's ancestral fish, tetrapodomorph, possessed similar traits to those inherited by the early tetrapods, including internal nostrils and a large fleshy fin built on bones that could give rise to the tetrapod limb. Their palatal and jaw structures were similar to those of early tetrapods, and their dentition was similar too, with labyrinthine teeth fitting in a pit-and-tooth arrangement on the palate.

A major difference between early tetrapodomorph fishes and early tetrapods was in the relative development of the front and back skull portions; the snout is much less developed than in most early tetrapods and the post-orbital skull is exceptionally longer than an amphibian's.

To propagate in the terrestrial environment, animals had to overcome certain challenges. Their bodies needed additional support, because buoyancy was no longer a factor. Water retention was now important, since it was no longer the living matrix, and could be lost easily to the environment. Finally, animals needed new sensory input systems to have any ability to function reasonably on land.

Skull

A notable characteristic that make a tetrapod's skull different from a fish's are the relative frontal and rear portion lengths. The fish had a long rear portion while the front was short; the orbital vacuities were thus located towards the anterior end. In the tetrapod, the front of the skull lengthened, positioning the orbits farther back on the skull.

Neck

In tetrapodomorph fishes such as Eusthenopteron, the part of the body that would later become the neck was covered by a number of gill-covering bones known as the opercular series. These bones functioned as part of pump mechanism for forcing water through the mouth and past the gills. When the mouth opened to take in water, the gill flaps closed (including the gill-covering bones), thus ensuring that water entered only through the mouth. When the mouth closed, the gill flaps opened and water was forced through the gills. In Acanthostega, a basal tetrapod, the gill-covering bones have disappeared, although the underlying gill arches are still present. Besides the opercular series, Acanthostega also lost the throat-covering bones (gular series). The opercular series and gular series combined are sometimes known as the operculo-gular or operculogular series. Other bones in the neck region lost in Acanthostega (and later tetrapods) include the extrascapular series and the supracleithral series. Both sets of bones connect the shoulder girdle to the skull. With the loss of these bones, tetrapods acquired a neck, allowing the head to rotate somewhat independently of the torso. This, in turn, required stronger soft-tissue connections between head and torso, including muscles and ligaments connecting the skull with the spine and shoulder girdle. Bones and groups of bones were also consolidated and strengthened.

In Carboniferous tetrapods, the neck joint (occiput) provided a pivot point for the spine against the back of the skull. In tetrapodomorph fishes such as Eusthenopteron, no such neck joint existed. Instead, the notochord (a sort of spine made of cartilage) entered a hole in the back of the braincase and continued to the middle of the braincase. Acanthostega had the same arrangement as Eusthenopteron, and thus no neck joint. The neck joint evolved independently in different lineages of early tetrapods.

Dentition

Tetrapods had a tooth structure known as "plicidentine" characterized by infolding of the enamel as seen in cross-section. The more extreme version found in early tetrapods is known as "labyrinthodont" or "labyrinthodont plicidentine." This type of tooth structure has evolved independently in several types of bony fishes, both ray-finned and lobe finned, some modern lizards, and in a number of tetrapodomorph fishes. The infolding appears to evolve when a fang or large tooth grows in a small jaw, erupting when it still weak and immature. The infolding provides added strength to the young tooth, but offers little advantage when the tooth is mature. Such teeth are associated with feeding on soft prey in juveniles.

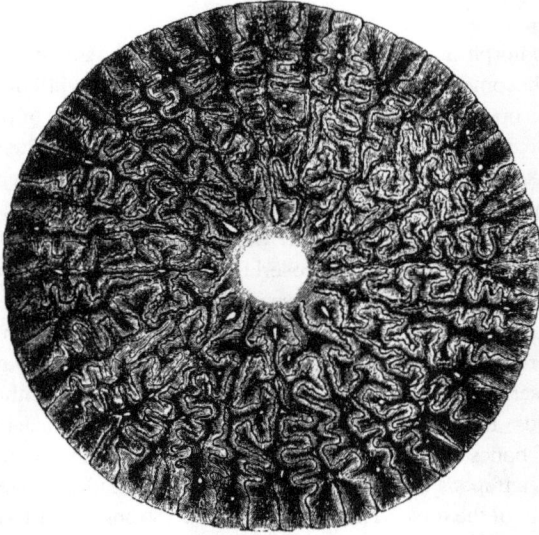

Figure 192: *Cross-section of a labyrinthodont tooth*

Axial skeleton

With the move from water to land, the spine had to resist the bending caused by body weight and had to provide mobility where needed. Previously, it could bend along its entire length. Likewise, the paired appendages had not been formerly connected to the spine, but the slowly strengthening limbs now transmitted their support to the axis of the body.

Girdles

The shoulder girdle was disconnected from the skull, resulting in improved terrestrial locomotion. The early sarcopterygians cleithrum was retained as the clavicle, and the interclavicle was well-developed, lying on the underside of the chest. In primitive forms, the two clavicles and the interclavical could have grown ventrally in such a way as to form a broad chest plate. The upper portion of the girdle had a flat, scapular blade, with the glenoid cavity situated below performing as the articulation surface for the humerus, while ventrally there was a large, flat coracoid plate turning in toward the midline.

The pelvic girdle also was much larger than the simple plate found in fishes, accommodating more muscles. It extended far dorsally and was joined to the backbone by one or more specialized sacral ribs. The hind legs were some-what specialized in that they not only supported weight, but also provided propulsion. The dorsal extension of the pelvis was the ilium, while the broad ventral plate was composed of the pubis in front and the ischium in behind. The three bones met at a single point in the center of the pelvic triangle called the acetabulum, providing a surface of articulation for the femur.

Limbs

Fleshy lobe-fins supported on bones seem to have been an ancestral trait of all bony fishes (Osteichthyes). The ancestors of the ray-finned fishes (Actinopterygii) evolved their fins in a different direction. The Tetrapodomorph ancestors of the Tetrapods further developed their lobe fins. The paired fins had bones distinctly homologous to the humerus, ulna, and radius in the fore-fins and to the femur, tibia, and fibula in the pelvic fins.

The paired fins of the early sarcopterygians were smaller than tetrapod limbs, but the skeletal structure was very similar in that the early sarcopterygians had a single proximal bone (analogous to the humerus or femur), two bones in the next segment (forearm or lower leg), and an irregular subdivision of the fin, roughly comparable to the structure of the carpus / tarsus and phalanges of a hand.

Locomotion

In typical early tetrapod posture, the upper arm and upper leg extended nearly straight horizontal from its body, and the forearm and the lower leg extended downward from the upper segment at a near right angle. The body weight was not centered over the limbs, but was rather transferred 90 degrees outward and down through the lower limbs, which touched the ground. Most of the animal's strength was used to just lift its body off the ground for walking, which was probably slow and difficult. With this sort of posture, it could only make short broad strides. This has been confirmed by fossilized footprints found in Carboniferous rocks.

Feeding

Early tetrapods had a wide gaping jaw with weak muscles to open and close it. In the jaw were moderate-sized palatal and vomerine (upper) and coronoid (lower) fangs, as well rows of smaller teeth. This was in contrast to the larger fangs and small marginal teeth of earlier tetrapodomorph fishes such as Eusthenopteron. Although this indicates a change in feeding habits, the exact nature of the change in unknown. Some scholars have suggested a change to bottom-feeding or feeding in shallower waters (Ahlberg and Milner 1994). Others have suggesting a mode of feeding comparable to that of the Japanese giant salamander, which uses both suction feeding and direct biting to eat small crustaceans and fish. A study of these jaws shows that they were used for feeding underwater, not on land.

In later terrestrial tetrapods, two methods of jaw closure emerge: static and kinetic inertial (also known as snapping). In the static system, the jaw muscles are arranged in such a way that the jaws have maximum force when shut or nearly shut. In the kinetic inertial system, maximum force is applied when the jaws are wide open, resulting in the jaws snapping shut with great velocity and momentum. Although the kinetic inertial system is occasionally found in fish, it requires special adaptations (such as very narrow jaws) to deal with the high viscosity and density of water, which would otherwise impede rapid jaw closure.

The tetrapod tongue is built from muscles that once controlled gill openings. The tongue is anchored to the hyoid bone, which was once the lower half of a pair of gill bars (the second pair after the ones that evolved into jaws). The tongue did not evolve until the gills began to disappear. Acanthostega still had gills, so this would have been a later development. In an aquatically feeding animals, the food supported by water and can literally float (or get sucked in) to the mouth. On land, the tongue becomes important.

Respiration

The evolution of early tetrapod respiration was influenced by an event known as the "charcoal gap," a period of more than 20 million years, in the middle and late Devonian, when atmospheric oxygen levels were too low to sustain wildfires. During this time, fish inhabiting anoxic waters (very low in oxygen) would have been under evolutionary pressure to develop their air-breathing ability.

Early tetrapods probably relied on four methods of respiration: with lungs, with gills, cutaneous respiration (skin breathing), and breathing through the lining of the digestive tract, especially the mouth.

Gills

The early tetrapod Acanthostega had at least three and probably four pairs of gill bars, each containing deep grooves in the place where one would expect to find the afferent branchial artery. This strongly suggests that functional gills were present. Some aquatic temnospondyls retained internal gills at least into the early Jurassic. Evidence of clear fish-like internal gills is present in *Archegosaurus*.

Lungs

Lungs originated as an extra pair of pouches in the throat, behind the gill pouches. They were probably present in the last common ancestor of bony fishes. In some fishes they evolved into swim bladders for maintaining buoyancy. Lungs and swim bladders are homologous (descended from a common ancestral form) as is the case for the pulmonary artery (which delivers deoxygenated blood from the heart to the lungs) and the arteries that supply swim bladders. Air was introduced into the lungs by a process known as buccal pumping.

In the earliest tetrapods, exhalation was probably accomplished with the aid of the muscles of the torso (the thoracoabdominal region). Inhaling with the ribs was either primitive for amniotes, or evolved independently in at least two different lineages of amniotes. It is not found in amphibians. The muscularized diaphragm is unique to mammals.

Recoil aspiration

Although tetrapods are widely thought to have inhaled through buccal pumping (mouth pumping), according to an alternative hypothesis, aspiration (inhalation) occurred through passive recoil of the exoskeleton in a manner similar to the contemporary primitive ray-finned fish polypterus. This fish inhales through its spiracle (blowhole), an anatomical feature present in early tetrapods. Exhalation is powered by muscles in the torso. During exhalation, the bony scales in the upper chest region become indented. When the muscles are relaxed, the bony scales spring back into position, generating considerable negative pressure within the torso, resulting in a very rapid intake of air through the spiracle.

Cutaneous respiration

Skin breathing, known as cutaneous respiration, is common in fish and amphibians, and occur both in and out of water. In some animals waterproof barriers impede the exchange of gases through the skin. For example, keratin in human skin, the scales of reptiles, and modern proteinaceous fish scales impede the exchange of gases. However, early tetrapods had scales made of highly vascularized bone covered with skin. For this reason, it is thought that early tetrapods could engage some significant amount of skin breathing.

Carbon dioxide metabolism

Although air-breathing fish can absorb oxygen through their lungs, the lungs tend to be ineffective for discharging carbon dioxide. In tetrapods, the ability of lungs to discharge CO_2 came about gradually, and was not fully attained until the evolution of amniotes. The same limitation applies to gut air breathing (GUT), i.e., breathing with the lining of the digestive tract. Tetrapod skin would have been effective for both absorbing oxygen and discharging CO_2, but only up to a point. For this reason, early tetrapods may have experienced chronic hypercapnia (high levels of blood CO_2). This is not uncommon in fish that inhabit waters high in CO_2. According to one hypothesis, the "sculpted" or "ornamented" dermal skull roof bones found in early tetrapods may have been related to a mechanism for relieving respiratory acidosis (acidic blood caused by excess CO_2) through compensatory metabolic alkalosis.

Circulation

Early tetrapods probably had a three-chambered heart, as do modern amphibians and reptiles, in which oxygenated blood from the lungs and de-oxygenated blood from the respiring tissues enters by separate atria, and is directed via a spiral valve to the appropriate vessel — aorta for oxygenated blood and pulmonary vein for deoxygenated blood. The spiral valve is essential to keeping the mixing of the two types of blood to a minimum, enabling the animal to have higher metabolic rates, and be more active than otherwise.

Senses

Olfaction

The difference in density between air and water causes smells (certain chemical compounds detectable by chemoreceptors) to behave differently. An animal first venturing out onto land would have difficulty in locating such chemical signals if its sensory apparatus had evolved in the context of aquatic detection.

Lateral line system

Fish have a lateral line system that detects pressure fluctuations in the water. Such pressure is non-detectable in air, but grooves for the lateral line sense organs were found on the skull of early tetrapods, suggesting either an aquatic or largely aquatic habitat. Modern amphibians, which are semi-aquatic, exhibit this feature whereas it has been retired by the higher vertebrates.

Vision

Changes in the eye came about because the behavior of light at the surface of the eye differs between an air and water environment due to the difference in refractive index, so the focal length of the lens altered to function in air. The eye was now exposed to a relatively dry environment rather than being bathed by water, so eyelids developed and tear ducts evolved to produce a liquid to moisten the eyeball.

Early tetrapods inherited a set of five rod and cone opsins known as the vertebrate opsins.

Four cone opsins were present in the first vertebrate, inherited from invertebrate ancestors:

- LWS/MWS (long- to medium-wave sensitive) - green, yellow, or red
- SWS1 (short-wave sensitive) - ultraviolet or violet - lost in monotremes (platypus, echidna)
- SWS2 (short-wave sensitive) - violet or blue - lost in therians (placental mammals and marsupials)
- RH2 (rhodopsin-like cone opsin) - green - lost separately in amphibians and mammals, retained in reptiles and birds

A single rod opsin, rhodopsin, was present in the first jawed vertebrate, inherited from a jawless vertebrate ancestor:

- RH1 (rhodposin) - blue-green - used night vision and color correction in low-light environments

Balance

Tetrapods retained the balancing function of the middle ear from fish ancestry.

Hearing

Air vibrations could not set up pulsations through the skull as in a proper auditory organ. The spiracle was retained as the otic notch, eventually closed in by the tympanum, a thin, tight membrane.

The hyomandibula of fish migrated upwards from its jaw supporting position, and was reduced in size to form the stapes. Situated between the tympanum and braincase in an air-filled cavity, the stapes was now capable of transmitting vibrations from the exterior of the head to the interior. Thus the stapes became an important element in an impedance matching system, coupling airborne sound waves to the receptor system of the inner ear. This system had evolved independently within several different amphibian lineages.

The impedance matching ear had to meet certain conditions to work. The stapes had to be perpendicular to the tympanum, small and light enough to reduce its inertia, and suspended in an air-filled cavity. In modern species that are sensitive to over 1 kHz frequencies, the footplate of the stapes is 1/20th the area of the tympanum. However, in early amphibians the stapes was too large, making the footplate area oversized, preventing the hearing of high frequencies. So it appears they could only hear high intensity, low frequency sounds—and the stapes more probably just supported the brain case against the cheek.

Only in the early Triassic, about hundred million years after they conquered land, did the tympanic middle ear evolve (independently) in all the tetrapod lineages.[528]

References

Literature

* Benton, Michael (5 February 2009). *Vertebrate Palaeontology*[529] (3 ed.). John Wiley & Sons. p. 1. ISBN 978-1-4051-4449-0. Retrieved 10 June 2015.
* Clack, J.A. (2012). *Gaining ground: the origin and evolution of tetrapods*[530] (2nd ed.). Bloomington, Indiana, USA.: Indiana University Press. ISBN 9780253356758.
* Laurin, Michel (2010). *How Vertebrates Left the Water*[531]. University of California Press. ISBN 978-0-520-26647-6. Retrieved 26 May 2015.
* McGhee, George R., Jr. (2013). *When the Invasion of Land Failed: The Legacy of the Devonian Extinctions*[532]. Columbia University Press. ISBN 978-0-231-16057-5. Retrieved 2 May 2015.
* Steyer, Sebastien (2012). *Earth Before the Dinosaurs*[533]. Indiana University Press. p. 59. ISBN 0-253-22380-6. Retrieved 1 June 2015.

Further reading

Wikimedia Commons has media related to *Tetrapoda*.

* Clack, Jennifer A. (2009). "The Fin to Limb Transition: New Data, Interpretations, and Hypotheses from Paleontology and Developmental Biology". *Annual Review of Earth and Planetary Sciences*. **37** (1): 163–179. Bibcode: 2009AREPS..37..163C[534]. doi: 10.1146/annurev.earth.36.031207.124146[535].
* Hall, Brian K., ed. (2007). *Fins Into Limbs: Evolution, Development, and Transformation*[536]. Chicago: University of Chicago Press. ISBN 978-0-226-31340-5.

- Long JA, Young GC, Holland T, Senden TJ, Fitzgerald EM (November 2006). "An exceptional Devonian fish from Australia sheds light on tetrapod origins". *Nature*. **444** (7116): 199–202. Bibcode: 2006Natur.444..199L[537]. doi: 10.1038/nature05243[538]. PMID 17051154[539].
- Benton, Michael (2005). *Vertebrate Palaeontology* (3rd ed.). Blackwell Publishing.

Evolution of tetrapods

<table>
<tr><td align="center">Part of a series on
Paleontology</td></tr>
<tr><td align="center"></td></tr>
<tr><td></td></tr>
<tr><td></td></tr>
<tr><td></td></tr>
<tr><td></td></tr>
<tr><td align="center">**Paleontology Portal**
Category</td></tr>
<tr><td>
• v̲
• t̲
• e̲[540]
</td></tr>
</table>

The **evolution of tetrapods** began about 400 million years ago in the Devonian Period with the earliest tetrapods evolved from lobe-finned fishes. Tetrapods are categorized as a biological superclass, Tetrapoda, which includes all living and extinct amphibians, reptiles, birds, and mammals. While most species today are terrestrial, little evidence supports the idea that any of the earliest tetrapods could move about on land, as their limbs could not have held their midsections off the ground and the known trackways do not indicate they dragged their bellies around. Presumably, the tracks were made by animals walking along the bottoms of shallow bodies of water. The specific aquatic

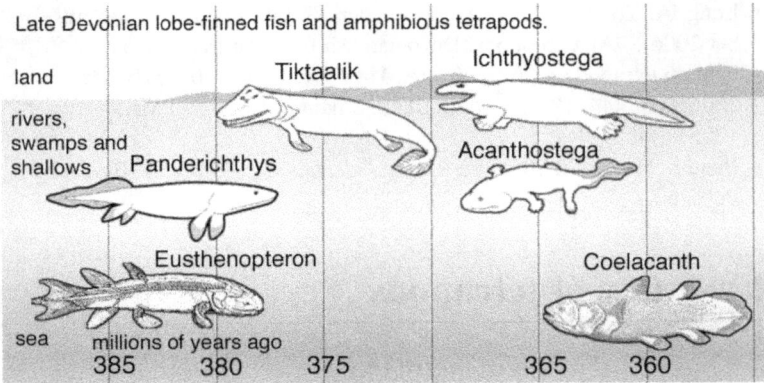

Figure 193: *In Late Devonian vertebrate speciation, descendants of pelagic lobe-finned fish — like Eusthenopteron — exhibited a sequence of adaptations:*
* *Panderichthys, suited to muddy shallows*
* *Tiktaalik with limb-like fins that could take it onto land*
* *Early tetrapods in weed-filled swamps, such as:*
  * *Acanthostega, which had feet with eight digits*
  * *Ichthyostega with limbs*
  Descendants also included pelagic lobe-finned fish such as coelacanth species.

ancestors of the tetrapods, and the process by which land colonization occurred, remain unclear, and are areas of active research and debate among palaeontologists at present.

Most amphibians today remain semiaquatic, living the first stage of their lives as fish-like tadpoles. Several groups of tetrapods, such as the snakes and cetaceans, have lost some or all of their limbs. In addition, many tetrapods have returned to partially aquatic or fully aquatic lives throughout the history of the group (modern examples of fully aquatic tetrapods include cetaceans and sirenians). The first returns to an aquatic lifestyle may have occurred as early as the Carboniferous Period whereas other returns occurred as recently as the Cenozoic, as in cetaceans, pinnipeds, and several modern amphibians.

The change from a body plan for breathing and navigating in water to a body plan enabling the animal to move on land is one of the most profound evolutionary changes known.[541] It is also one of the best understood, largely thanks to a number of significant transitional fossil finds in the late 20th century combined with improved phylogenetic analysis.

Origin

Evolution of fish

The Devonian period is traditionally known as the "Age of Fish", marking the diversification of numerous extinct and modern major fish groups. Among them were the early bony fishes, who diversified and spread in freshwater and brackish environments at the beginning of the period. The early types resembled their cartilaginous ancestors in many features of their anatomy, including a shark-like tailfin, spiral gut, large pectoral fins stiffened in front by skeletal elements and a largely unossified axial skeleton.

They did, however, have certain traits separating them from cartilaginous fishes, traits that would become pivotal in the evolution of terrestrial forms. With the exception of a pair of spiracles, the gills did not open singly to the exterior as they do in sharks; rather, they were encased in a gill chamber stiffened by membrane bones and covered by a bony operculum, with a single opening to the exterior. The cleithrum bone, forming the posterior margin of the gill chamber, also functioned as anchoring for the pectoral fins. The cartilaginous fishes do not have such an anchoring for the pectoral fins. This allowed for a movable joint at the base of the fins in the early bony fishes, and would later function in a weight bearing structure in tetrapods. As part of the overall armour of rhomboid cosmin scales, the skull had a full cover of dermal bone, constituting a skull roof over the otherwise shark-like cartilaginous inner cranium. Importantly, they also had a pair of ventral paired lungs, a feature lacking in sharks and rays.

Lungs before land

The lung/swim bladder originated as an outgrowth of the gut, forming a gas-filled bladder above the digestive system. In its primitive form, the air bladder was open to the alimentary canal, a condition called physostome and still found in many fish. The primary function is not entirely certain. One consideration is buoyancy. The heavy scale armour of the early bony fishes would certainly weigh the animals down. In cartilaginous fishes, lacking a swim bladder, the open sea sharks need to swim constantly to avoid sinking into the depths, the pectoral fins providing lift. Another factor is oxygen consumption. Ambient oxygen was relatively low in the early Devonian, possibly about half of modern values. Per unit volume, there is much more oxygen in air than in water, and vertebrates are active animals with a high energy requirement compared to invertebrates of similar sizes. The Devonian saw increasing oxygen levels which opened up new ecological niches by allowing groups able to exploit the additional oxygen to develop into active, large-bodied animals. Particularly in tropical swampland habitats, atmospheric oxygen is much more stable, and may have prompted a reliance of lungs rather than gills for primary oxygen uptake. In the end, both buoyancy and breathing may have been important, and some modern physostome fishes do indeed use their bladders for both.

To function in gas exchange, lungs require a blood supply. In cartilaginous fishes and teleosts, the heart lies low in the body and pumps blood forward through the ventral aorta, which splits up in a series of paired aortic arches, each corresponding to a gill arch.[542] The aortic arches then merge above the gills to form a dorsal aorta supplying the body with oxygenated blood. In lungfishes, bowfin and bichirs, the swim bladder is supplied with blood by paired pulmonary arteries branching off from the hindmost (6th) aortic arch. The same basic pattern is found in the lungfish *Protopterus* and in terrestrial salamanders, and was probably the pattern found in the tetrapods' immediate ancestors as well as the first tetrapods. In most other bony fishes the swim bladder is supplied with blood by the dorsal aorta.

The breath

In order for the lungs to allow gas exchange, the lungs first need to have gas in them. In modern tetrapods, three important breathing mechanisms are conserved from early ancestors, the first being a $CO_2/H+$ detection system. In modern tetrapod breathing, the impulse to take a breath is triggered by a buildup of CO_2 in the bloodstream and not a lack of O_2. A similar $CO_2/H+$ detection system is found in all Osteichthyes, which implies that the LCA of all Osteichthyes had a need of this sort of detection system. The second mechanism for a breath is a surfactant system in the lungs to facilitate gas exchange. This is also found in all Osteichthyes, even those that are almost entirely aquatic. The highly conserved nature of this system suggests that even aquatic Osteichthyes have some need for a surfactant system, which may seem strange as there is no gas underwater. The third mechanism for a breath is the actual motion of the breath. This mechanism predates the LCA of Osteichthyes, as it can be observed in *Lampetra camtshatica,* the sister clade to Osteichthyes. In Lampreys, this mechanism takes the form of a "cough", where the lamprey shakes its body to allow water flow across its gills. When CO_2 levels in the lamprey's blood climb too high, a signal is sent to a central pattern generator that causes the lamprey to "cough" and allow CO_2 to leave its body. This linkage between the CO_2 detection system and the central pattern generator is extremely similar to the linkage between these two systems in tetrapods, which implies homology.

External and internal nares

The nostrils in most bony fish differ from those of tetrapods. Normally, bony fish have four nares (nasal openings), one naris behind the other on each side. As the fish swims, water flows into the forward pair, across the olfactory tissue, and out through the posterior openings. This is true not only of ray-finned fish but also of the coelacanth, a fish included in the Sarcopterygii, the group that also includes the tetrapods. In contrast, the tetrapods have only one pair of nares externally but also sport a pair of internal nares, called choanae, allowing them to draw air through the nose. Lungfish are also sarcopterygians with internal nostrils, but these are sufficiently different from tetrapod choanae that they have long been recognized as an independent development.

The evolution of the tetrapods' internal nares was hotly debated in the 20th century. The internal nares could be one set of the external ones (usually presumed to be the posterior pair) that have migrated into the mouth, or the internal pair could be a newly evolved structure. To make way for a migration, however, the two tooth-bearing bones of the upper jaw, the maxilla and the premaxilla, would have to separate to let the nostril through and then rejoin; until recently, there was no evidence for a transitional stage, with the two bones disconnected. Such evidence is now available: a small lobe-finned fish called *Kenichthys*, found in China and dated at around 395 million years old, represents evolution "caught in mid-act", with the maxilla and premaxilla separated and an aperture—the incipient choana—on the lip in between the two bones. *Kenichthys* is more closely related to tetrapods than is the coelacanth, which has only external nares; it thus represents an intermediate stage in the evolution of the tetrapod condition. The reason for the evolutionary movement of the posterior nostril from the nose to lip, however, is not well understood.

Into the shallows

The relatives of *Kenichthys* soon established themselves in the waterways and brackish estuaries and became the most numerous of the bony fishes throughout the Devonian and most of the Carboniferous. The basic anatomy of group is well known thanks to the very detailed work on *Eusthenopteron* by Erik Jarvik in the second half of the 20th century. The bones of the skull roof were broadly similar to those of early tetrapods and the teeth had an infolding of the enamel similar to that of labyrinthodonts. The paired fins had a build with bones distinctly homologous to the humerus, ulna, and radius in the fore-fins and to the femur, tibia, and fibula in the pelvic fins.

There were a number of families: Rhizodontida, Canowindridae, Elpistostegidae, Megalichthyidae, Osteolepidae and Tristichopteridae. Most were openwater fishes, and some grew to very large sizes; adult specimens are several

Figure 194: *Devonian fishes, including an early shark Cla-*
doselache, Eusthenopteron and other lobe-finned fishes,
and the placoderm Bothriolepis (Joseph Smit, 1905).

meters in length. The Rhizodontid *Rhizodus* is estimated to have grown to 7
meters (23 feet), making it the largest freshwater fish known.

While most of these were open-water fishes, one group, the Elpistostegalians,
adapted to life in the shallows. They evolved flat bodies for movement in very
shallow water, and the pectoral and pelvic fins took over as the main propulsion
organs. Most median fins disappeared, leaving only a protocercal tailfin. Since
the shallows were subject to occasional oxygen deficiency, the ability to breathe
atmospheric air with the swim bladder became increasingly important. The
spiracle became large and prominent, enabling these fishes to draw air.

Skull morphology

The tetrapods have their root in the early Devonian tetrapodomorph fish.
Primitive tetrapods developed from an osteolepid tetrapodomorph lobe-finned
fish (sarcopterygian-crossopterygian), with a two-lobed brain in a flattened
skull. The coelacanth group represents marine sarcopterygians that never acquired these shallow-water adaptations. The sarcopterygians apparently took
two different lines of descent and are accordingly separated into two major
groups: the Actinistia (including the coelacanths) and the Rhipidistia (which
include extinct lines of lobe-finned fishes that evolved into the lungfish and the
tetrapodomorphs).

Figure 195: *Stalked fins like those of the bichirs can be used for terrestrial movement*

From fins to feet

The oldest known tetrapodomorph is *Kenichthys* from China, dated at around 395 million years old. Two of the earliest tetrapodomorphs, dating from 380 Ma, were *Gogonasus* and *Panderichthys*.[543] They had choanae and used their fins to move through tidal channels and shallow waters choked with dead branches and rotting plants. Their fins could have been used to attach themselves to plants or similar while they were lying in ambush for prey. The universal tetrapod characteristics of front limbs that bend forward from the elbow and hind limbs that bend backward from the knee can plausibly be traced to early tetrapods living in shallow water. Pelvic bone fossils from *Tiktaalik* shows, if representative for early tetrapods in general, that hind appendages and pelvic-propelled locomotion originated in water before terrestrial adaptations.[544]

Another indication that feet and other tetrapod traits evolved while the animals were still aquatic is how they were feeding. They did not have the modifications of the skull and jaw that allowed them to swallow prey on land. Prey could be caught in the shallows, at the water's edge or on land, but had to be eaten in water where hydrodynamic forces from the expansion of their buccal cavity would force the food into their esophagus.[545]

It has been suggested that the evolution of the tetrapod limb from fins in lobe-finned fishes is related to expression of the HOXD13 gene or the loss of the proteins actinodin 1 and actinodin 2, which are involved in fish fin development. Robot simulations suggest that the necessary nervous circuitry for walking evolved from the nerves governing swimming, utilizing the sideways oscillation of the body with the limbs primarily functioning as anchoring points and providing limited thrust. This type of movement, as well as changes to the pectoral girdle are similar to those seen in the fossil record can be induced in bichirs by raising then out of water.

A 2012 study using 3D reconstructions of *Ichthyostega* concluded that it was incapable of typical quadrupedal gaits. The limbs could not move alternately as they lacked the necessary rotary motion range. In addition, the hind limbs lacked the necessary pelvic musculature for hindlimb-driven land movement. Their most likely method of terrestrial locomotion is that of synchronous "crutching motions", similar to modern mudskippers. *(Viewing several videos of mudskipper "walking" shows that they move by pulling themselves forward with both pectoral fins at the same time (left & right pectoral fins move simultaneously, not alternatively). The fins are brought forward and planted; the shoulders then rotate rearward, advancing the body & dragging the tail as a third point of contact. There are no rear "limbs"/fins, and there is no significant flexure of the spine involved.)*

Denizens of the swamp

The first tetrapods probably evolved in coastal and brackish marine environments, and in shallow and swampy freshwater habitats. Formerly, researchers thought the timing was towards the end of the Devonian. In 2010, this belief was challenged by the discovery of the oldest known tetrapod tracks, preserved in marine sediments of the southern coast of Laurasia, now Świętokrzyskie (Holy Cross) Mountains of Poland. They were made during the Eifelian stage at the end of the Middle Devonian. The tracks, some of which show digits, date to about 395 million years ago—18 million years earlier than the oldest known tetrapod body fossils. Additionally, the tracks show that the animal was capable of thrusting its arms and legs forward, a type of motion that would have been impossible in tetrapodomorph fish like *Tiktaalik*. The animal that produced the tracks is estimated to have been up to 2.5 metres (8.2 ft) long with footpads up to 26 centimetres (10 in) wide, although most tracks are only 15 centimetres (5.9 in) wide. The new finds suggest that the first tetrapods may have lived as opportunists on the tidal flats, feeding on marine animals that were washed up or stranded by the tide. Currently, however, fish are stranded in significant numbers only at certain times of year, as in alewife spawning season; such strandings could not provide a significant supply of food for predators. There is no reason to suppose that Devonian fish were less prudent than those of today. According to Melina Hale of University of Chicago, not all ancient trackways are necessarily made by early tetrapods, but could also be created by relatives of the tetrapods who used their fleshy appendages in a similar substrate-based locomotion.[546]

Palaeozoic tetrapods

Devonian tetrapods

Research by Jennifer A. Clack and her colleagues showed that the very earliest tetrapods, animals similar to *Acanthostega*, were wholly aquatic and quite unsuited to life on land. This is in contrast to the earlier view that fish had first invaded the land — either in search of prey (like modern mudskippers) or to find water when the pond they lived in dried out — and later evolved legs, lungs, etc.

By the late Devonian, land plants had stabilized freshwater habitats, allowing the first wetland ecosystems to develop, with increasingly complex food webs that afforded new opportunities. Freshwater habitats were not the only places to find water filled with organic matter and dense vegetation near the water's edge. Swampy habitats like shallow wetlands, coastal lagoons and large brackish river deltas also existed at this time, and there is much to suggest that this is the kind of environment in which the tetrapods evolved. Early fossil tetrapods have been found in marine sediments, and because fossils of primitive tetrapods in general are found scattered all around the world, they must have spread by following the coastal lines — they could not have lived in freshwater only.

One analysis from the University of Oregon suggests no evidence for the "shrinking waterhole" theory - transitional fossils are not associated with evidence of shrinking puddles or ponds - and indicates that such animals would probably not have survived short treks between depleted waterholes. The new theory suggests instead that proto-lungs and proto-limbs were useful adaptations to negotiate the environment in humid, wooded floodplains.

The Devonian tetrapods went through two major bottlenecks during what is known as the Late Devonian extinction; one at the end of the Frasnian stage, and one twice as large at the end of the following Famennian stage. These events of extinctions led to the disappearance of primitive tetrapods with fishlike features like Ichthyostega and their primary more aquatic relatives.[547] When tetrapods reappear in the fossil record after the Devonian extinctions, the adult forms are all fully adapted to a terrestrial existence, with later species secondary adapted to an aquatic lifestyle.[548]

Excretion in tetrapods

The common ancestor of all present gnathostomes (jawed-vertebrates) lived in freshwater, and later migrated back to the seaWikipedia:Citation needed. To deal with the much higher salinity in sea water, they evolved the ability to turn the nitrogen waste product ammonia into harmless urea, storing it in the body to give the blood the same osmolarity as the sea water without poisoning the organism. This is the system currently found in cartilaginous fishes. Ray-finned fishes (Actinopterygii) later returned to freshwater and lost this ability, while the fleshy-finned fishes (Sarcopterygii) retained it. Since the blood of ray-finned fishes contains more salt than freshwater, they could simply get rid of ammonia through their gills. When they finally returned to the sea again, they did not recover their old trick of turning ammonia to urea, and they had to evolve salt excreting glands instead. Lungfishes do the same when they are living in water, making ammonia and no urea, but when the water dries up and they are forced to burrow down in the mud, they switch to urea production. Like cartilaginous fishes, the coelacanth can store urea in its blood, as can the only known amphibians that can live for long periods of time in salt water (the toad *Bufo marinus* and the frog *Rana cancrivora*). These are traits they have inherited from their ancestors.

If early tetrapods lived in freshwater, and if they lost the ability to produce urea and used ammonia only, they would have to evolve it from scratch again later. Not a single species of all the ray-finned fishes living today has been able to do that, so it is not likely the tetrapods would have done so either. Terrestrial animals that can only produce ammonia would have to drink constantly, making a life on land impossible (a few exceptions exist, as some terrestrial woodlice can excrete their nitrogenous waste as ammonia gas). This probably also was a problem at the start when the tetrapods started to spend time out of water, but eventually the urea system would dominate completely. Because of this it is not likely they emerged in freshwater (unless they first migrated into freshwater habitats and then migrated onto land so shortly after that they still retained the ability to make urea), although some species never left, or returned to, the water could of course have adapted to freshwater lakes and rivers.

Lungs

It is now clear that the common ancestor of the bony fishes (Osteichthyes) had a primitive air-breathing lung—later evolved into a swim bladder in most actinopterygians (ray-finned fishes). This suggests that crossopterygians evolved in warm shallow waters, using their simple lung when the oxygen level in the water became too low.

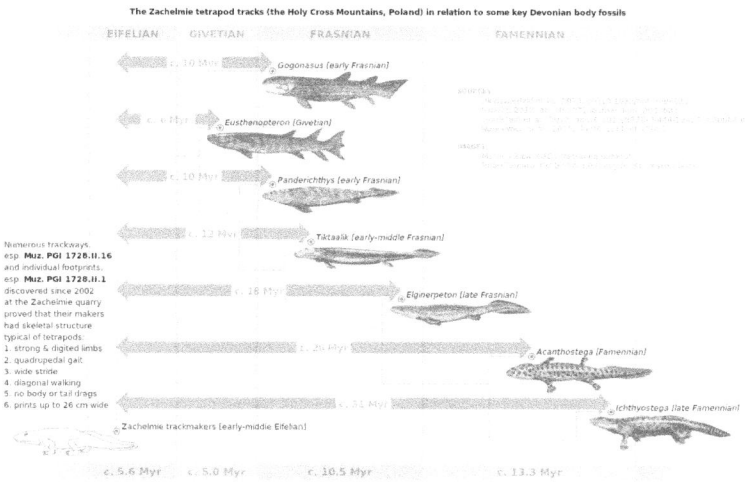

Figure 196: *Oldest tetrapod tracks from Zachelmie in relation to key Devonian tetrapodomorph body fossils*

Fleshy lobe-fins supported on bones rather than ray-stiffened fins seem to have been an ancestral trait of all bony fishes (Osteichthyes). The lobe-finned ancestors of the tetrapods evolved them further, while the ancestors of the ray-finned fishes (Actinopterygii) evolved their fins in a different direction. The most primitive group of actinopterygians, the bichirs, still have fleshy frontal fins.

Fossils of early tetrapods

Nine genera of Devonian tetrapods have been described, several known mainly or entirely from lower jaw material. All but one were from the Laurasian supercontinent, which comprised Europe, North America and Greenland. The only exception is a single Gondwanan genus, *Metaxygnathus*, which has been found in Australia.

The first Devonian tetrapod identified from Asia was recognized from a fossil jawbone reported in 2002. The Chinese tetrapod *Sinostega pani* was discovered among fossilized tropical plants and lobe-finned fish in the red sandstone sediments of the Ningxia Hui Autonomous Region of northwest China. This finding substantially extended the geographical range of these animals and has raised new questions about the worldwide distribution and great taxonomic diversity they achieved within a relatively short time.

Figure 197: *Eusthenopteron*

Figure 198: *Panderichthys*

These earliest tetrapods were not terrestrial. The earliest confirmed terrestrial forms are known from the early Carboniferous deposits, some 20 million years later. Still, they may have spent very brief periods out of water and would have used their legs to paw their way through the mud.

Why they went to land in the first place is still debated. One reason could be that the small **juveniles** who had completed their metamorphosis **had what it took** to make use of what land had to offer. Already adapted to breathe air and move around in shallow waters near land as a protection (just as modern fish

Figure 199: *Tiktaalik*

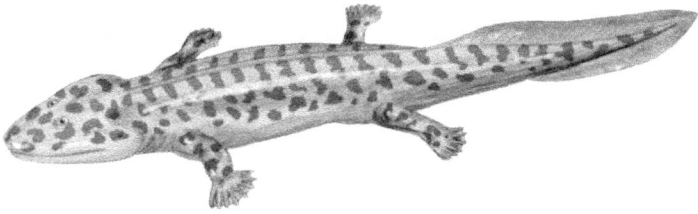

Figure 200: *Acanthostega*

Figure 201: *Ichthyostega*

Figure 202: *Hynerpeton*

Figure 203: *Tulerpeton*

Figure 204: *Crassigyrinus*

Figure 205: *Diadectes*

and amphibians often spend the first part of their life in the comparative safety of shallow waters like mangrove forests), two very different niches partially overlapped each other, with the young **juveniles** in the diffuse line between. One of them was overcrowded and dangerous while the other was much safer and much less crowded, offering less competition over resources. The terrestrial niche was also a much more challenging place for primary aquatic animals, but because of the way evolution and selection pressure work, those **juveniles** who could take advantage of this would be rewarded. Once they gained a small foothold on land, thanks to their preadaptations and **being at the right place at the right time**, favourable variations in their descendants would gradually result in continuing evolution and diversification.

At this time the abundance of invertebrates crawling around on land and near water, in moist soil and wet litter, offered a food supply. Some were even big enough to eat small tetrapods, but the land was free from dangers common in the water.

From water to land

Initially making only tentative forays onto land, tetrapods adapted to terrestrial environments over time and spent longer periods away from the water. It is also possible that the adults started to spend some time on land (as the skeletal modifications in early tetrapods such as *Ichthyostega* suggests) to bask in the sun close to the water's edgeWikipedia:Citation needed, while otherwise being mostly aquatic.

Carboniferous tetrapods

Until the 1990s, there was a 30 million year gap in the fossil record between the late Devonian tetrapods and the reappearance of tetrapod fossils in recognizable mid-Carboniferous amphibian lineages. It was referred to as "Romer's Gap", which now covers the period from about 360 to 345 million years ago (the Devonian-Carboniferous transition and the early Mississippian), after the palaeontologist who recognized it.

During the "gap", tetrapod backbones developed, as did limbs with digits and other adaptations for terrestrial life. Ears, skulls and vertebral columns all underwent changes too. The number of digits on hands and feet became standardized at five, as lineages with more digits died out. Thus, those very few tetrapod fossils found in this "gap" are all the more prized by palaeontologists because they document these significant changes and clarify their history.

The transition from an aquatic, lobe-finned fish to an air-breathing amphibian was a significant and fundamental one in the evolutionary history of the vertebrates. For an organism to live in a gravity-neutral aqueous environment, then colonize one that requires an organism to support its entire weight and possess a mechanism to mitigate dehydration, required significant adaptations or exaptations within the overall body plan, both in form and in function. *Eryops*, an example of an animal that made such adaptations, refined many of the traits found in its fish ancestors. Sturdy limbs supported and transported its body while out of water. A thicker, stronger backbone prevented its body from sagging under its own weight. Also, through the reshaping of vestigial fish jaw bones, a rudimentary middle ear began developing to connect to the piscine inner ear, allowing *Eryops* to amplify, and so better sense, airborne sound.

By the Visean (mid-Carboniferous) stage, the early tetrapods had radiated into at least three or four main branches. Recognizable basal-group tetrapods are representative of the temnospondyls (e.g. *Eryops*) lepospondyls (e.g. *Diplocaulus*), anthracosaurs, which were the relatives and ancestors of the Amniota, and possibly the baphetids, which are thought to be related to temnospondyls and whose status as a main branch is yet unresolved. Depending on which authorities one follows, modern amphibians (frogs, salamanders and caecilians) are most probably derived from either temnospondyls or lepospondyls (or possibly both, although this is now a minority position).

The first amniotes (clade of vertebrates that today includes reptiles, mammals, and birds) are known from the early part of the Late Carboniferous. By the Triassic, this group had already radiated into the earliest mammals, turtles, and crocodiles (lizards and birds appeared in the Jurassic, and snakes in the Cretaceous). This contrasts sharply with the (possibly fourth) Carboniferous group, the baphetids, which have left no extant surviving lineages.

Carboniferous rainforest collapse

Amphibians and reptiles were strongly affected by the Carboniferous rainforest collapse (CRC), an extinction event that occurred ~307 million years ago. The Carboniferous period has long been associated with thick, steamy swamps and humid rainforests. Since plants form the base of almost all of Earth's ecosystems, any changes in plant distribution have always affected animal life to some degree. The sudden collapse of the vital rainforest ecosystem profoundly affected the diversity and abundance of the major tetrapod groups that relied on it. The CRC, which was a part of one of the top two most devastating plant extinctions in Earth's history, was a self-reinforcing and very rapid change of environment wherein the worldwide climate became much drier and cooler overall (although much new work is being done to better understand the fine-grained historical climate changes in the Carboniferous-Permian transition and how they arose).

The ensuing worldwide plant reduction resulting from the difficulties plants encountered in adjusting to the new climate caused a progressive fragmentation and collapse of rainforest ecosystems. This reinforced and so further accelerated the collapse by sharply reducing the amount of animal life which could be supported by the shrinking ecosystems at that time. The outcome of this animal reduction was a crash in global carbon dioxide levels, which impacted the plants even more. The aridity and temperature drop which resulted from this runaway plant reduction and decrease in a primary greenhouse gas caused the Earth to rapidly enter a series of intense Ice Ages.

This impacted amphibians in particular in a number of ways. The enormous drop in sealevel due to greater quantities of the world's water being locked into glaciers profoundly affected the distribution and size of the semiaquatic ecosystems which amphibians favored, and the significant cooling of the climate further narrowed the amount of new territory favorable to amphibians. Given that among the hallmarks of amphibians are an obligatory return to a body of water to lay eggs, a delicate skin prone to desiccation (thereby often requiring the amphibian to be relatively close to water throughout its life), and a reputation of being a bellwether species for disrupted ecosystems due to the resulting low resilience to ecological change, amphibians were particularly devastated, with the Labyrinthodonts among the groups faring worst. In contrast, reptiles - whose amniotic eggs have a membrane that enables gas exchange out of water, and which thereby can be laid on land - were better adapted to the new conditions. Reptiles invaded new niches at a faster rate and began diversifying their diets, becoming herbivorous and carnivorous, rather than feeding exclusively on insects and fish. Meanwhile, the severely impacted amphibians simply could not out-compete reptiles in mastering the new ecological niches, and so were obligated to pass the tetrapod evolutionary torch to the increasingly successful and swiftly radiating reptiles.

Permian tetrapods

In the Permian period: early "amphibia" (labyrinthodonts) clades included temnospondyl and anthracosaur; while amniote clades included the Sauropsida and the Synapsida. Sauropsida would eventually evolve into today's reptiles and birds; whereas Synapsida would evolve into today's mammals. During the Permian, however, the distinction was less clear—amniote fauna being typically described as either *reptile* or as *mammal-like reptile*. The latter (synapsida) were the most important and successful Permian animals.

The end of the Permian saw a major turnover in fauna during the Permian–Triassic extinction event: probably the most severe mass extinction event of the phanerozoic. There was a protracted loss of species, due to multiple extinction pulses. Many of the once large and diverse groups died out or were greatly reduced.

Mesozoic tetrapods

Life on Earth seemed to recover quickly after the Permian extinctions, though this was mostly in the form of disaster taxa such as the hardy *Lystrosaurus*. Specialized animals that formed complex ecosystems with high biodiversity, complex food webs, and a variety of niches, took much longer to recover. Current research indicates that this long recovery was due to successive waves of extinction, which inhibited recovery, and to prolonged environmental stress to organisms that continued into the Early Triassic. Recent research indicates that recovery did not begin until the start of the mid-Triassic, 4M to 6M years after the extinction; and some writers estimate that the recovery was not complete until 30M years after the P-Tr extinction, i.e. in the late Triassic.

A small group of reptiles, the diapsids, began to diversify during the Triassic, notably the dinosaurs. By the late Mesozoic, the large labyrinthodont groups that first appeared during the Paleozoic such as temnospondyls and reptile-like amphibians had gone extinct. All current major groups of sauropsids evolved during the Mesozoic, with birds first appearing in the Jurassic as a derived clade of theropod dinosaurs. Many groups of synapsids such as anomodonts and therocephalians that once comprised the dominant terrestrial fauna of the Permian also became extinct during the Mesozoic; during the Triassic, however, one group (Cynodontia) gave rise to the descendant taxon Mammalia, which survived through the Mesozoic to later diversify during the Cenozoic.

Cenozoic tetrapods

The Cenozoic era began with the end of the Mesozoic era and the Cretaceous epoch; and continues to this day. The beginning of the Cenozoic was marked by the Cretaceous-Paleogene extinction event during which all non-avian dinosaurs became extinct. The Cenozoic is sometimes called the "Age of Mammals".

During the Mesozoic, the prototypical mammal was a small nocturnal insectivore something like a tree shrew. Due to their nocturnal habits, most mammals lost their color vision, and greatly improved their sense of hearing. All mammals of today are shaped by this origin. Primates later re-evolved color-vision.

During the Paleocene and Eocene, most mammals remained small (under 20 kg). Cooling climate in the Oligocene and Miocene, and the expansion of grasslands favored the evolution of larger mammalian species.

Ratites run, and penguins swim and waddle: but the majority of birds are rather small, and can fly. Some birds use their ability to fly to complete epic globe-crossing migrations, while others such as frigate birds fly over the oceans for months on end.

Bats have also taken flight, and along with cetaceans have developed echolocation or sonar.

Whales, seals, manatees, and sea otters have returned to the ocean and an aquatic lifestyle.

Vast herds of ruminant ungulates populate the grasslands and forests. Carnivores have evolved to keep the herd-animal populations in check.

Extant (living) tetrapods

Following the great faunal turnover at the end of the Mesozoic, only six major groups of tetrapods were left, all of which also include many extinct groups:

- Lissamphibia: frogs and toads, newts and salamanders, and caecilians
- Testudines: turtles and tortoises
- Lepidosauria: tuataras, lizards, amphisbaenians and snakes
- Crocodilia: crocodiles, alligators, caimans and gharials
- Neornithes: modern birds
- Mammalia: mammals

Dinosaurs, birds and mammals

Evolution of dinosaurs

This article gives an outline and examples of dinosaur evolution. For a detailed list of interrelationships see Dinosaur classification.

Dinosaurs evolved within a single lineage of archosaurs 232-234 Ma (million years ago) in the Ladinian age, the latter part of the middle Triassic. Dinosauria is a well-supported clade, present in 98% of bootstraps. It is diagnosed by many features including loss of the postfrontal on the skull and an elongate deltopectoral crest on the humerus.[549]

In March 2017, scientists reported a new way of classifying the dinosaur family tree, based on newer and more evidence than available earlier. According to the new classification, the original dinosaurs, arising 200 million years ago, were small, two-footed omnivorous animals with large grasping hands. Descendants (for the non-avian dinosaurs) lasted until 66 million years ago.

Origins amongst archosaurs

The process leading up to the Dinosauromorpha and the first true dinosaurs can be followed through fossils of the early Archosaurs such as the Proterosuchidae, *Erythrosuchidae* and *Euparkeria* which have fossils dating back to 250 Ma, through mid-Triassic archosaurs such as *Ticinosuchus* 232-236 Ma. Crocodiles are also descendants of mid-Triassic archosaurs.

Dinosaurs can be defined as the last common ancestor of birds (Saurischia) and *Triceratops* (Ornithischia) and all the descendants of that ancestor. With that definition, the pterosaurs and several species of archosaurs narrowly miss out on being classified as dinosaurs. The pterosaurs are famous for flying through the Mesozoic skies on leathery wings and reaching the largest sizes of any flying

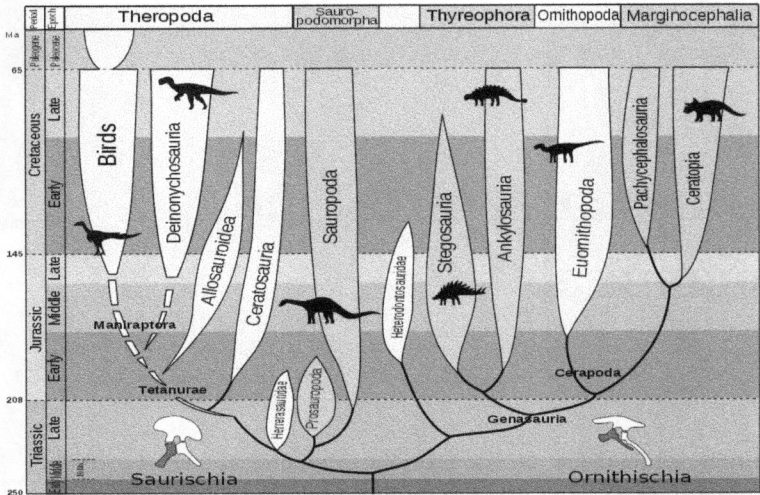

Figure 206: *Evolution of dinosaurs*

animal that ever existed. Archosaur genera that also narrowly miss out on being classified as dinosaurs include *Schleromochlus* 220-225 Ma, *Lagerpeton* 230-232 Ma and *Marasuchus* 230-232 Ma.

Earliest dinosaurs

The first known dinosaurs were bipedal predators that were 1-2 metres (3.3-6.5 ft) long.

Spondylosoma may or may not be a dinosaur; the fossils (all postcranial) are tentatively dated at 235-242 Ma.

The earliest confirmed dinosaur fossils include saurischian ('lizard-hipped') dinosaurs *Nyasasaurus* 243 Ma, *Saturnalia* 225-232 Ma, *Herrerasaurus* 220-230 Ma, *Staurikosaurus* possibly 225-230 Ma, *Eoraptor* 220-230 Ma and *Alwalkeria* 220-230 Ma. *Saturnalia* may be a basal saurischian or a prosauropod. The others are basal saurischians.

Among the earliest ornithischian ('bird-hipped') dinosaurs is *Pisanosaurus* 220-230 Ma. Although *Lesothosaurus* comes from 195-206 Ma, skeletal features suggest that it branched from the main Ornithischia line at least as early as *Pisanosaurus*.

    A. *Eoraptor*, an early saurischian, **B** *Lesothosaurus*, a primitive ornithischian, **C** *Staurikosaurus* (Saurischia) pelvis, **D** *Lesothosaurus* pelvis

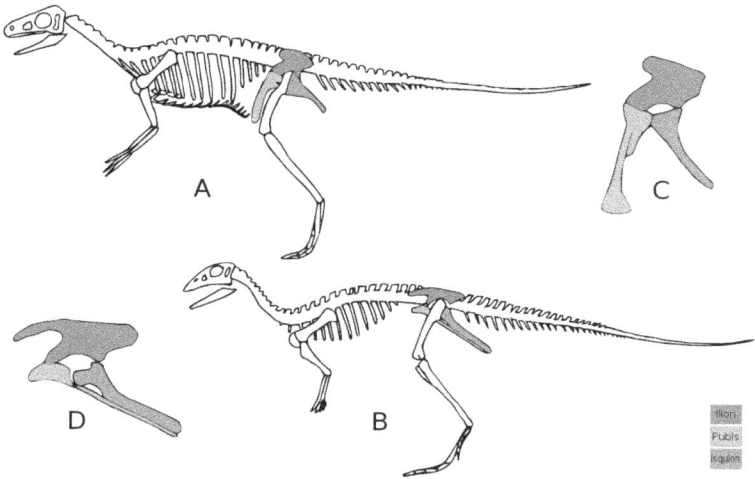

It is clear from this figure that early saurischians resembled early ornithischians, but not modern crocodiles. Saurischians are distinguished from the ornithischians by retaining the ancestral configuration of bones in the pelvis. Another difference is in the skull, the upper skull of the Ornithischia is more solid and the joint connecting the lower jaw is more flexible; both are adaptations to herbivory and both can already be seen in *Lesothosaurus*.

Saurischia

Setting aside the basal Saurischia, the rest of the Saurischia are split into the Sauropodomorpha and Theropoda. The Sauropodomorpha is split into Prosauropoda and Sauropoda. The evolutionary paths taken by the Theropoda are very complicated. *The Dinosauria* (2004), a major reference work on dinosaurs, splits the Theropoda into groups Ceratosauria, Basal Tetanurae, Tyrannosauroidea, Ornithomimosauria, Therizinosauroidea, Oviraptorosauria, Troodontidae, Dromaeosauridae and Basal Avialae in turn. Each group branches off the main trunk at a later date. See Dinosaur classification for the detailed interrelationships between these.

Sauropodomorpha

The first sauropodomorphs were prosauropods. Prosauropod fossils are known from the late Triassic to early Jurassic 227-180 Ma. They could be bipedal or quadrupedal and had developed long necks and tails and relatively small heads. They had lengths of 2.5 (8.2 ft) to 10 m (33 ft) and were primarily herbivorous.

The earliest prosauropods, such as *Thecodontosaurus* from 205-220 Ma, still retained the ancestral bipedal stance and large head to body ratio.

These evolved into the sauropods which became gigantic quadrupedal herbivores, some of which reached lengths of at least 26 m (85 ft). Features defining this clade include a ratio of forelimb length to hindlimb length greater than 0.6. Most sauropods still had hindlimbs larger than forelimbs; one notable exception is *Brachiosaurus* whose long forelimbs suggest that it had evolved to feed from tall trees like a modern-day giraffe.

Sauropod fossils are found from the times of the earliest dinosaurs right up to the Cretaceous–Paleogene extinction event, from 227 to 66 Ma. Most sauropods are known from the Jurassic, to be more precise between 227 and 121 Ma.

The Cretaceous sauropods form two groups. The Diplodocoidea lived from 121 to 66 Ma. The Titanosauriformes lived from 132 to 66 Ma. The latter clade consists of series of nested subgroups, the Titanosauria, the Titanosauridae and Saltasauridae. Both the Diplodocoidea and Titanosauriformes are descended from the Neosauropoda, the earliest of which lived in about 169 Ma.

The sauropods are famous for being the largest land animals that ever lived, and for having relatively small skulls. The enlargement of prosauropod and sauropod dinosaurs into these giants and the change in skull length is illustrated in the following charts.

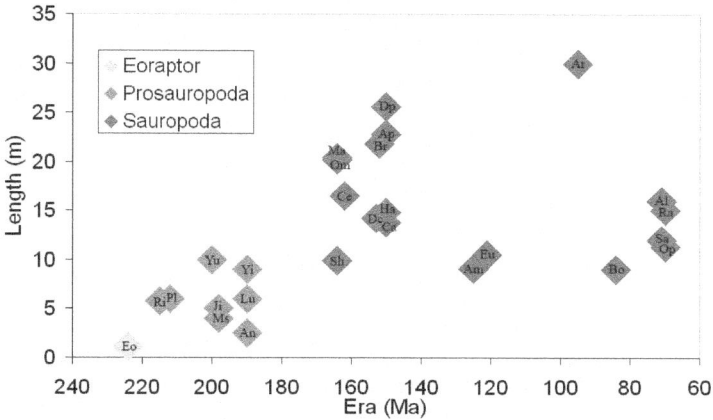

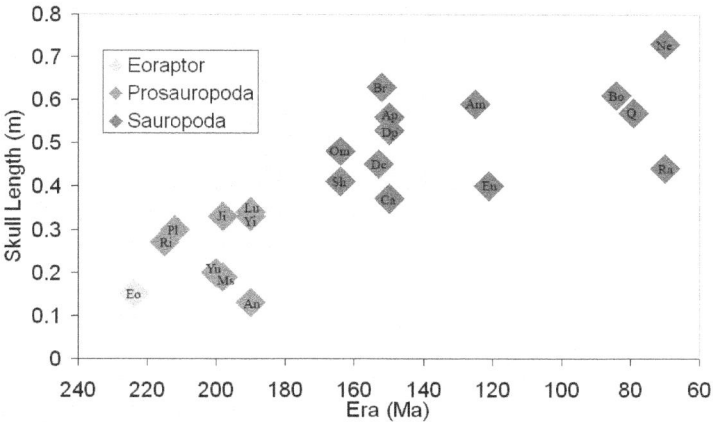

Dinosaurs used in creating these charts are (in date order): Eo *Eoraptor*; Prosauropods Ri *Riojasaurus*, Pl *Plateosaurus*, Yu *Yunnanosaurus*, Ms *Massospondylus*, Ji *Jingshanosaurus*, An *Anchisaurus*, Lu *Lufengosaurus*, Yi *Yimenosaurus*, ; and Sauropods Sh *Shunosaurus*, Om *Omeisaurus*, Mm *Mamenchisaurus*, Ce *Cetiosaurus*, Dc *Dicraeosaurus*, Br *Brachiosaurus*, Eu *Euhelopus*, Ap *Apatosaurus*, Ca *Camarasaurus*, Dp *Diplodocus*, Ha *Haplocanthosaurus*, Am *Amargasaurus*, Ar *Argentinosaurus (approx)*, Bo *Bonitasaura*, Q *Quaesitosaurus*, Al *Alamosaurus*, Sa *Saltasaurus*, Ra *Rapetosaurus*, Op *Opisthocoelicaudia*, Ne *Nemegtosaurus*.

With the exception of *Argentinosaurus* (included to fill a gap in time), these graphs show only the length of sauropods for whom near-complete fossil skeletons are known. It doesn't show other very large sauropods (see Dinosaur size#Sauropods) because these are only known from very incomplete skeletons. The ratio of skull length to body length is much higher in *Eoraptor* than in sauropods. The longest skull graphed is of *Nemegtosaurus*, which is not thought be a particularly large sauropod. The skull of *Nemegtosaurus* was found near the headless skeleton of 11 metre (36 ft) long *Opisthocoelicaudia*, and it has been suggested that they may be the same species, but see Nemegtosauridae.

The relationship between the evolution of large herbivores and large plants remains uncertain. About 50% of the plants over the time of the dinosaurs were conifers, they increased in number in the Triassic until stabilising in about 190 Ma. Cycads formed the second largest group until about 120 Ma. Ferns were present in roughly constant numbers the whole time. Flowering plants began about 120 Ma and by the end of the period had taken over from the cycads. All dinosaur herbivores appear to have been adversely affected by the extinction event at the end of the Jurassic.

Theropoda

By far the earliest fossils of Theropoda (not counting the basal saurischians) are of the Coelophysoidea, including *Coelophysis* and others, from late Triassic and early Jurassic 227-180 Ma. Cladistic analysis sometimes connects these to the group called Ceratosauria. Principal features of both include changes in the pelvic girdle and hind limb that differ between the sexes. Other ceratosauria first appear in the late Jurassic of western North America.

These are followed by the basal Tetanurae, of whom fossils have been found from the mid Jurassic to past the end of the early Cretaceous 180 Ma to 94 Ma. They have a relatively short maxillary tooth row. They did not all branch off the evolutionary line leading to coelurosaurs at the same time. Basal tetanurans include Megalosauridae, spinosaurids, a diverse clade of allosaurs, and several genera of less certain affinities, including *Compsognathus*. With the exception of *Compsognathus* they are large-bodied. Allosaurs form a distinct long-lived clade that share some cranial characters. They include the well known *Allosaurus* and *Sinraptor* among others.

The great radiation of Theropoda into many different clades of Coelurosauria must have happened in the mid to late Jurassic, because *Archaeopteryx* was around in about 152-154 Ma, and cladistic analysis has shown that many other groups of Coelurosauria branched off before that.[550] Fossil evidence from China suggests that the earliest feathers were found on the primitive Coelurosauria. The most primitive of these, e.g. on the tyrannosauroid *Dilong*, were simply hollow-cored fibres that would have been useful for insulation but useless for flying.

Occasional bones and cladistic analyses point to the Tyrannosauroidea branching off from the other Theropoda early, in the middle Jurassic, although nearly complete skeletons haven't yet appeared before *Eotyrannus* from 121-127 Ma, and the many close relatives of *Tyrannosaurus* itself don't appear before 84 Ma, near the end of the late Cretaceous.

Ornithomimosauria fossils are known from 127 to 65 Ma. The earliest branch from the main line of Ornithomimosauria is believed to be *Harpymimus*.

The Therizinosauroidea are unusual theropods in being almost all vegetarian. Fossil Therizinosauroidea are known from 127 to 65 Ma.

Maniraptorans include Oviraptorosauria, Deinonychosaurs and birds. They are characterized by an ulna with a curved shaft.

Oviraptorosaurian fossils are known from 127 to 65 Ma. They have a toothless skull that is extremely modified. The skeleton has an unusually short tail.

Deinonychosaurs, named after the enlarged sickle-shaped second digit of the foot, are closely related to birds. They have two distinct families, Troodontidae and Dromaeosauridae. Troodontid fossils are known from 127 to 65 Ma. They have a more slender build and longer limbs. The earliest named troodontid fossil known is *Sinornithoides*. Dromaeosaurid fossils are known from about 127 to 65 Ma with the exception of *Utahraptor*. The skeletal remains of *Utahraptor* are about 127-144 Ma. This is interesting because according to a recent cladistic analysis, *Utahraptor* is about as far from the ancestral Theropoda as it is possible to get, further than *Archaeopteryx*. Dromaeosaurids have a larger second digit; this family includes the well known dinosaurs *Dromaeosaurus*, *Deinonychus* and *Velociraptor*.

Ancient birds (Avialae) include both the Aves, which are defined as descendants of the common ancestor of modern birds and *Archaeopteryx*, and the more primitive *Epidendrosaurus*. Fossil birds stretch down from 154 Ma through the Cretaceous–Paleogene extinction event at 65 Ma to the present day. Scores of complete skeletons have now been found of the more recent *Confuciusornis*, which is an early representative of the Ornithurae. Ornithurans all have a bony pygostyle, to which tail feathers are anchored. For more details on the evolution of birds, see Evolution of birds.

Ornithischia

Ornithischia, as the name indicates, was coined for the birdlike pelvic girdle, although they are not the ancestors of birds.

The ornithischian skull and dentition was modified very early by a herbivorous diet.[551] *Lesothosaurus* separated early, but the skull of *Lesothosaurus* already shows such adaptations, with broad proportions, a less flexible upper jaw, and a more mobile connection for the lower jaw.

Heterodontosauridae has been shown to be the basalmost group within Ornithischia.[552] Heterodontosaurids are very small (body length < 1 m) and lived from the Late Triassic to Early Cretaceous. Apart from *Abrictosaurus* all have a short upper canine and longer lower canine. The forelimbs in known fossils are relatively long.

The major clades were already established by the early Jurassic. The ornithischians divided into armoured thyreophorans and unarmoured ornithopods and marginocephalians.

Thyreophorans

Surface body armour (scutes) is the most striking feature of the thyreophorans. *Scutellosaurus* has these but otherwise differs little from *Lesothosaurus*. It has a long tail and combined bipedal-quadrupedal posture that separates it from all later thyreophorans including Stegosauria and Ankylosauria. These two clades, although quite different in overall appearance, share many unusual features in the skull and skeleton.

Stegosaurs are easily recognised by the prominent row of plates above the spine and long spines on the tail. Most stegosaurs, but not *Stegosaurus*, also have a spine over each shoulder. These spines and plates have evolved from the earlier surface scutes. *Huayangosaurus* is the oldest and most primitive known stegosaur.

Ankylosaurs are easily recognised by their extensive body armour. The skull is heavily ossified. Early in their evolution, ankylosaurs split into the Nodosauridae and Ankylosauridae, distinguished by features of the skull.

Ornithopoda

Ornithopods fall into distinct clades - Hypsilophodontidae, and Iguanodontia.

Hypsilophodontids more closely resemble their ancestors than the heterodontosaurids do. The most distinctive features are short scapula and rod-shaped pre-pubic process. The earliest is *Agilisaurus* from the middle Jurassic of China.

Iguanodontians are a diverse but morphologically tight knit array of genera known from fossils of the late Cretaceous. Significant modifications include the evolution of tooth batteries, a ligament-bound metacarpus and a digitigrade hand posture. *Tenontosaurus* is the most basal iguanodontian. Others include *Iguanodon*, *Camptosaurus* and *Muttaburrasaurus*.

Marginocephalia

Marginocephalia are named for a shelf that projects over the back of the skull. They include the pachycephalosaurians and ceratopsians.

Pachycephalosaurs are best known for their thick upper fronts to their skull. The oldest known is *Stenopelix*, from the early Cretaceous of Europe.

Ceratopsians, famous for *Protoceratops*, *Triceratops* and *Styracosaurus* illustrate the evolution of frilled and horned skulls. The frills evolved from the shelf common to all Marginocephalia. Ceratopsians are separated into basal ceratopsians, including the parrot-beaked *Psittacosaurus*, and neoceratopsians.

Diversity of ceratopsian skulls. A) Skeleton of *Protoceratops*. B) to I) Skulls. B) & C) *Psittacosaurus* side & top. D) & E) *Protoceratops* side & top. F) & G) *Triceratops* side & top. H) & I) *Styracosaurus* side (without lower jaw) & top.

The evolution of ceratopsid dinosaurs shares characteristics with the evolution of some mammal groups, both were "geologically brief" events precipitating the simultaneous evolution of large body size, derived feeding structures, and "varied hornlike organs."

The sequence of ceratopsian evolution in the Cretaceous is roughly from *Psittacosaurus* (121 -99 Ma) to *Protoceratops* (83 Ma) to (*Triceratops* 67 Ma and

Styracosaurus 72 Ma). In side view the skull of *Psittacosaurus* bears very little resemblance to that of *Styracosaurus* but in top view a similar pentagonal arrangement can be seen.

Fossil record

The first few lines of primitive dinosaurs diversified rapidly through the Triassic period; dinosaur species quickly evolved the specialised features and range of sizes needed to exploit nearly every terrestrial ecological niche. During the period of dinosaur predominance, which encompassed the ensuing Jurassic and Cretaceous periods, nearly every known land animal larger than 1 meter in length was a dinosaur.

One measure of the quality of the fossil record is obtained by comparing the date of first appearance with the order of branching of a cladogram based on the shape of fossil elements. Close correspondence exists for ornithiscians, saurischians and subgroups. The cladogram link between coelophysids and ceratosaurs is an exception, it would place the origin of coelophysids much too late. The simplest explanation is convergent evolution - ceratosaur bones evolved independently into a shape that resembles that of the earlier coelophysids. The other possibility is that ceratosaurs evolved much earlier than the fossil record suggests.

Most dinosaur fossils have been found in the Norian-Sinemurian, Kimmeridgian-Tithonian, and Campanian-Maastrichtian periods. Continuity of lineages across the intervening gaps shows that those gaps are artifacts of preservation rather than any reduction in diversity or abundance.

In many instances, cladistic analysis shows that ancestral lineages of varying durations fall in those gaps. The length of missing ancestral lineages in 1997 range from 25 Ma (*Lesothosaurus*, Genasauria, Hadrosauroidea, Sauropoda, Neoceratopsia, Coelurosauria) to 85 Ma (Carcharodontosauridae). Because the dinosaurian radiation began at small body size, the unrecorded early history may be due to less reliable fossilization of smaller species. However, some missing lineages, notably of Carcharodontosauridae and Abelisauridae, require alternative explanations because the missing range extends across stages rich in fossil materials.

Evolutionary trends

Body size

Body size is important because of its correlation with metabolism, diet, life history, geographic range and extinction rate. The modal body mass of dinosaurs lies between 1 and 10 tons throughout the Mesozoic and across all major continental regions. There was a trend towards increasing body size within many dinosaur clades, including the Thyreophora, Ornithopoda, Pachycephalosauria, Ceratopsia, Sauropomorpha, and basal Theropoda. Marked decreases in body size have also occurred in some lineages, but are more sporadic. The best known example is the decrease in body size leading up to the first birds; *Archaeopteryx* was below 10 kg in weight, and later birds *Confuciusornis* and *Sinornis* are starling- to pigeon-sized. This occurred for easier flight.

Mobility

The ancestral dinosaur was a biped. The evolution of a quadrupedal posture occurred four times, among the ancestors of Euornithopoda, Thyreophora, Ceratopsia and Sauropodomorpha. In all four cases this was associated with an increase in body size, and in all four cases the trend is unidirectional without reversal.

Dinosaurs exhibit a pattern of the reduction and loss of fingers on the lateral side of the hand (digits III, IV and V). The primitive function of the dinosaur hand is grasping with a partially opposable thumb, rather than weight-bearing. The reduction of digits is one of the defining features of tyrannosaurids, only having two functional digits on very short forelimbs.

Effect of food sources

The ancestral dinosaur was a carnivore. Herbivory among dinosaurs arose three times, at the origin of the ornithischian, sauropodomorph, and therizinosaurid clades. Individual therizinosaurids are herbivorous or omnivorous. Herbivory among the ornithischians and sauropodomorphs was never reversed.

The potential co-evolution of plants and herbivorous dinosaurs has been subject to extensive speculation. The appearance of prosauropods in the late Triassic has been tentatively linked either to the demise or diversification of types of flora at that time. The rise of ceratopsids and iguanodont and hadrosaurid ornithopods in the Cretaceous has been tentatively linked to the angiosperm radiation. Unfortunately, there are still no hard data on dietary preferences of herbivorous dinosaurs, apart from data on chewing technique and gastroliths.

Biogeography

Dinosaurian faunas, which were relatively uniform in character when Pangaea began to break up, became markedly differentiated by the close of the Cretaceous. Biogeography is based on the splitting of an ancestral species by the emplacement of a geographic barrier. Interpretation is limited by a lack of fossil evidence for eastern North America, Madagascar, India, Antarctica and Australia. No unequivocal proof of the biogeographical action on dinosaur species has been obtained, but some authors have outlined centres of origin for many dinosaur groups, multiple dispersal routes, and intervals of geographic isolation.

Dinosaurs that have been given as evidence of biogeography include abelisaurid theropods from South America and possibly elsewhere on Gondwana.

Relationships between dinosaurs show abundant evidence of dispersal from one region of the globe to another. Tetanuran theropods travelled widely through western North America, Asia, South America, Africa and Antarctica. Pachycephalosaurs and ceratopsians show clear evidence of multiple bidirectional dispersion events across Beringa.

Extinction

The Cretaceous–Paleogene extinction event, which occurred 66 million years ago at the end of the Cretaceous period, caused the extinction of all dinosaurs except for the line that had already given rise to the first birds.

References

- Sampson, S. D., 2001, Speculations on the socioecology of Ceratopsid dinosaurs (Orinthischia: Neoceratopsia): In: Mesozoic Vertebrate Life, edited by Tanke, D. H., and Carpenter, K., Indiana University Press, pp. 263-276.
- Paul C. Sereno (1999) *The evolution of dinosaurs*, Science, Vol 284, pp. 2137–2146 http://www.sciencemag.org/cgi/content/abstract/284/5423/2137

Origin of birds

The scientific question of within which larger group of animals birds evolved, has traditionally been called the **origin of birds**. The present scientific consensus is that birds are a group of theropod dinosaurs that originated during the Mesozoic Era.

A close relationship between birds and dinosaurs was first proposed in the nineteenth century after the discovery of the primitive bird *Archaeopteryx* in Germany. Birds and extinct non-avian dinosaurs share many unique skeletal traits. Moreover, fossils of more than thirty species of non-avian dinosaur have been collected with preserved feathers. There are even very small dinosaurs, such as *Microraptor* and *Anchiornis*, which have long, vaned, arm and leg feathers forming wings. The Jurassic basal avialan *Pedopenna* also shows these long foot feathers. Witmer in 2009 concluded that this evidence is sufficient to demonstrate that avian evolution went through a four-winged stage. Fossil evidence also demonstrates that birds and dinosaurs shared features such as hollow, pneumatized bones, gastroliths in the digestive system, nest-building and brooding behaviors.

Although the origin of birds has historically been a contentious topic within evolutionary biology, only a few scientists still debate the dinosaurian origin of birds, suggesting descent from other types of archosaurian reptiles. Within the consensus that supports dinosaurian ancestry, the exact sequence of evolutionary events that gave rise to the early birds within maniraptoran theropods is disputed. The origin of bird flight is a separate but related question for which there are also several proposed answers.

Research history

Huxley, *Archaeopteryx* and early research

Scientific investigation into the origin of birds began shortly after the 1859 publication of Charles Darwin's *On the Origin of Species*. In 1860, a fossilized feather was discovered in Germany's Late Jurassic Solnhofen limestone. Christian Erich Hermann von Meyer described this feather as *Archaeopteryx lithographica* the next year. Richard Owen described a nearly complete skeleton in 1863, recognizing it as a bird despite many features reminiscent of reptiles, including clawed forelimbs and a long, bony tail.

Biologist Thomas Henry Huxley, known as "Darwin's Bulldog" for his tenacious support of the new theory of evolution by means of natural selection, almost immediately seized upon *Archaeopteryx* as a transitional fossil between birds and reptiles. Starting in 1868, and following earlier suggestions by Karl

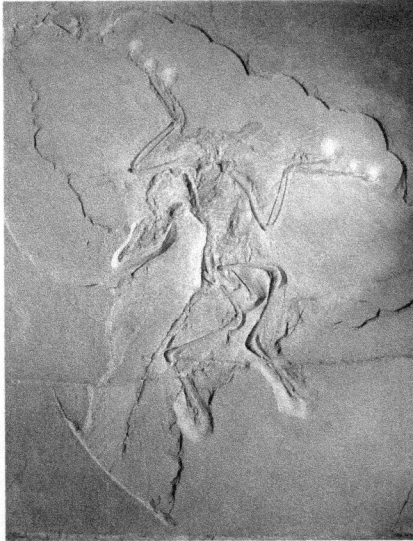

Figure 207: *The Berlin specimen of Archaeopteryx lithographica*

Figure 208: *Thomas Henry Huxley (1825–1895).*

Gegenbaur, and Edward Drinker Cope,[553] Huxley made detailed comparisons of *Archaeopteryx* with various prehistoric reptiles and found that it was most similar to dinosaurs like *Hypsilophodon* and *Compsognathus*. The discovery in the late 1870s of the iconic "Berlin specimen" of *Archaeopteryx*, complete with a set of reptilian teeth, provided further evidence. Huxley was the first to propose an evolutionary relationship between birds and dinosaurs. Although Huxley was opposed by the very influential Owen, his conclusions were accepted by many biologists, including Baron Franz Nopcsa, while others, notably Harry Seeley, argued that the similarities were due to convergent evolution.

Heilmann and the thecodont hypothesis

A turning point came in the early twentieth century with the writings of Gerhard Heilmann of Denmark. An artist by trade, Heilmann had a scholarly interest in birds and from 1913 to 1916, expanding on earlier work by Othenio Abel, published the results of his research in several parts, dealing with the anatomy, embryology, behavior, paleontology, and evolution of birds. His work, originally written in Danish as *Vor Nuvaerende Viden om Fuglenes Afstamning*, was compiled, translated into English, and published in 1926 as *The Origin of Birds*.

Like Huxley, Heilmann compared *Archaeopteryx* and other birds to an exhaustive list of prehistoric reptiles, and also came to the conclusion that theropod dinosaurs like *Compsognathus* were the most similar. However, Heilmann noted that birds had clavicles (collar bones) fused to form a bone called the furcula ("wishbone"), and while clavicles were known in more primitive reptiles, they had not yet been recognized in dinosaurs. Since he was a firm believer in Dollo's law, which states that evolution is not reversible, Heilmann could not accept that clavicles were lost in dinosaurs and re-evolved in birds. He was therefore forced to rule out dinosaurs as bird ancestors and ascribe all of their similarities to convergent evolution. Heilmann stated that bird ancestors would instead be found among the more primitive "thecodont" grade of reptiles. Heilmann's extremely thorough approach ensured that his book became a classic in the field, and its conclusions on bird origins, as with most other topics, were accepted by nearly all evolutionary biologists for the next four decades.

Clavicles are relatively delicate bones and therefore in danger of being destroyed or at least damaged beyond recognition. Nevertheless, some fossil theropod clavicles had actually been excavated before Heilmann wrote his book but these had been misidentified.[554] The absence of clavicles in dinosaurs became the orthodox view despite the discovery of clavicles in the primitive

Figure 209: *Heilmann's hypothetical illustration of a pair of fighting 'Proaves' from 1916*

theropod *Segisaurus* in 1936. The next report of clavicles in a dinosaur was in a Russian article in 1983.[555]

Contrary to what Heilmann believed, paleontologists now accept that clavicles and in most cases furculae are a standard feature not just of theropods but of saurischian dinosaurs. Up to late 2007 ossified furculae (i.e. made of bone rather than cartilage) have been found in all types of theropods except the most basal ones, *Eoraptor* and *Herrerasaurus*.[556] The original report of a furcula in the primitive theropod *Segisaurus* (1936) was confirmed by a re-examination in 2005. Joined, furcula-like clavicles have also been found in *Massospondylus*, an Early Jurassic sauropodomorph.

Ostrom, *Deinonychus* and the dinosaur renaissance

The tide began to turn against the 'thecodont' hypothesis after the 1964 discovery of a new theropod dinosaur in Montana. In 1969, this dinosaur was described and named *Deinonychus* by John Ostrom of Yale University. The next year, Ostrom redescribed a specimen of *Pterodactylus* in the Dutch Teyler Museum as another skeleton of *Archaeopteryx*. The specimen consisted mainly of a single wing and its description made Ostrom aware of the similarities between the wrists of *Archaeopteryx* and *Deinonychus*.

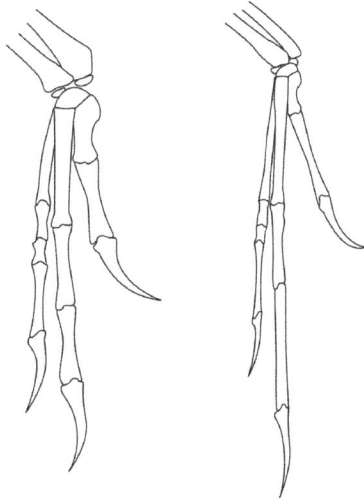

Figure 210: *The similarity of the forelimbs of Deinonychus (left) and Archaeopteryx (right) led John Ostrom to revive the link between dinosaurs and birds.*

In 1972, British paleontologist Alick Walker hypothesized that birds arose not from 'thecodonts' but from crocodile ancestors like *Sphenosuchus*. Ostrom's work with both theropods and early birds led him to respond with a series of publications in the mid-1970s in which he laid out the many similarities between birds and theropod dinosaurs, resurrecting the ideas first put forth by Huxley over a century before. Ostrom's recognition of the dinosaurian ancestry of birds, along with other new ideas about dinosaur metabolism, activity levels, and parental care, began what is known as the dinosaur renaissance, which began in the 1970s and continues to this day.

Ostrom's revelations also coincided with the increasing adoption of phylogenetic systematics (cladistics), which began in the 1960s with the work of Willi Hennig. Cladistics is a method of arranging species based strictly on their evolutionary relationships, using a statistical analysis of their anatomical characteristics. In the 1980s, cladistic methodology was applied to dinosaur phylogeny for the first time by Jacques Gauthier and others, showing unequivocally that birds were a derived group of theropod dinosaurs. Early analyses suggested that dromaeosaurid theropods like *Deinonychus* were particularly closely related to birds, a result that has been corroborated many times since.

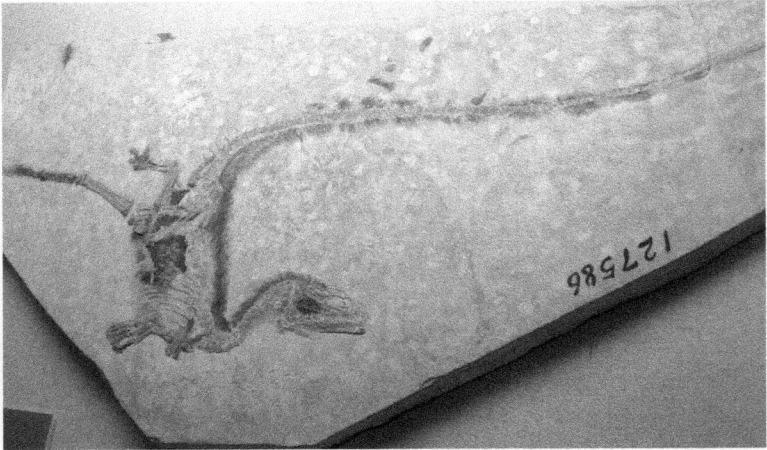

Figure 211: *Fossil of Sinosauropteryx prima*

Feathered dinosaurs in China

The early 1990s saw the discovery of spectacularly preserved bird fossils in several Early Cretaceous geological formations in the northeastern Chinese province of Liaoning. In 1996, Chinese paleontologists described *Sinosauropteryx* as a new genus of bird from the Yixian Formation, but this animal was quickly recognized as a theropod dinosaur closely related to *Compsognathus*. Surprisingly, its body was covered by long filamentous structures. These were dubbed 'protofeathers' and considered homologous with the more advanced feathers of birds, although some scientists disagree with this assessment. Chinese and North American scientists described *Caudipteryx* and *Protarchaeopteryx* soon after. Based on skeletal features, these animals were non-avian dinosaurs, but their remains bore fully formed feathers closely resembling those of birds. "Archaeoraptor", described without peer review in a 1999 issue of *National Geographic*, turned out to be a smuggled forgery, but legitimate remains continue to pour out of the Yixian, both legally and illegally. Feathers or "protofeathers" have been found on a wide variety of theropods in the Yixian, and the discoveries of extremely bird-like dinosaurs, as well as dinosaur-like primitive birds, have almost entirely closed the morphological gap between theropods and birds.

Digit homology

There is a debate between embryologists and paleontologists whether the hands of theropod dinosaurs and birds are essentially different, based on phalangeal counts, a count of the number of phalanges (fingers) in the hand. This is an important and fiercely debated area of research because its results may challenge the consensus that birds are descendants of dinosaurs.

Embryologists and some paleontologists who oppose the bird-dinosaur link have long numbered the digits of birds II-III-IV on the basis of multiple studies of the development in the egg. This is based on the fact that in most amniotes, the first digit to form in a 5-fingered hand is digit IV, which develops a primary axis. Therefore, embryologists have identified the primary axis in birds as digit IV, and the surviving digits as II-III-IV. The fossils of advanced theropod (Tetanurae) hands appear to have the digits I-II-III (some genera within Avetheropoda also have a reduced digit IV[557]). If this is true, then the II-III-IV development of digits in birds is an indication against theropod (dinosaur) ancestry. However, with no ontogenical (developmental) basis to definitively state which digits are which on a theropod hand (because no non-avian theropods can be observed growing and developing today), the labelling of the theropod hand is not absolutely conclusive.

Paleontologists have traditionally identified avian digits as I-II-III. They argue that the digits of birds number I-II-III, just as those of theropod dinosaurs do, by the conserved phalangeal formula. The phalangeal count for archosaurs is 2-3-4-5-3; many archosaur lineages have a reduced number of digits, but have the same phalangeal formula in the digits that remain. In other words, paleontologists assert that archosaurs of different lineages tend to lose the same digits when digit loss occurs, from the outside to the inside. The three digits of dromaeosaurs, and *Archaeopteryx* have the same phalangeal formula of I-II-III as digits I-II-III of basal archosaurs. Therefore, the lost digits would be V and IV. If this is true, then modern birds would also possess digits I-II-III. Also, one 1999 publication proposed a frame-shift in the digits of the theropod line leading to birds (thus making digit I into digit II, II to III, and so forth).[558] However, such frame shifts are rare in amniotes and—to be consistent with the theropod origin of birds—would have had to occur solely in the bird-theropod lineage forelimbs and not the hindlimbs (a condition unknown in any animal).[559] This is called *Lateral Digit Reduction* (LDR) versus *Bilateral Digit Reduction* (BDR) (see also *Limusaurus*[560])

A small minority, including ornithologists Alan Feduccia and Larry Martin, continues to assert that birds are instead the descendants of earlier archosaurs, such as *Longisquama* or *Euparkeria*. Embryological studies of bird developmental biology have raised questions about digit homology in bird and dinosaur forelimbs. However, due to the cogent evidence provided by comparative anatomy and phylogenetics, as well as the dramatic feathered dinosaur fossils from China, the idea that birds are derived dinosaurs, first championed by Huxley and later by Nopcsa and Ostrom, enjoys near-unanimous support among today's paleontologists.

Thermogenic muscle hypothesis

A 2011 publication suggested that selection for the expansion of skeletal muscle, rather than the evolution of flight, was the driving force for the emergence of this clade. Muscles became larger in prospectively endothermic saurians, according to this hypothesis, as a response to the loss of the vertebrate mitochondrial uncoupling protein, UCP1, which is thermogenic. In mammals, UCP1 functions within brown adipose tissue to protect newborns against hypothermia. In modern birds, skeletal muscle serves a similar function and is presumed to have done so in their ancestors. In this view, bipedality and other avian skeletal alterations were side effects of muscle hyperplasia, with further evolutionary modifications of the forelimbs, including adaptations for flight or swimming, and vestigiality, being secondary consequences of two-leggedness.

Phylogeny

Archaeopteryx has historically been considered the first bird, or *Urvogel*. Although newer fossil discoveries filled the gap between theropods and *Archaeopteryx*, as well as the gap between *Archaeopteryx* and modern birds, phylogenetic taxonomists, in keeping with tradition, almost always use *Archaeopteryx* as a specifier to help define Aves. Aves has more rarely been defined as a crown group consisting only of modern birds. Nearly all palaeontologists regard birds as coelurosaurian theropod dinosaurs. Within Coelurosauria, multiple cladistic analyses have found support for a clade named Maniraptora, consisting of therizinosauroids, oviraptorosaurs, troodontids, dromaeosaurids, and birds. Of these, dromaeosaurids and troodontids are usually united in the clade Deinonychosauria, which is a sister group to birds (together forming the node-clade Eumaniraptora) within the stem-clade Paraves.

Other studies have proposed alternative phylogenies, in which certain groups of dinosaurs usually considered non-avian may have evolved from avian ancestors. For example, a 2002 analysis found that oviraptorosaurs were basal avians. Alvarezsaurids, known from Asia and the Americas, have been variously classified as basal maniraptorans, paravians, the sister taxon of ornithomimosaurs, as well as specialized early birds. The genus *Rahonavis*, originally described as an early bird, has been identified as a non-avian dromaeosaurid in several studies. Dromaeosaurids and troodontids themselves have also been suggested to lie within Aves rather than just outside it.

Features linking birds and dinosaurs

Many anatomical features are shared by birds and theropod dinosaurs.

Feathers

Archaeopteryx, the first good example of a "feathered dinosaur", was discovered in 1861. The first specimen was found in the Solnhofen limestone in southern Germany, which is a *lagerstätte*, a rare and remarkable geological formation known for its superbly detailed fossils. *Archaeopteryx* is a transitional fossil, with features clearly intermediate between those of non-avian theropod dinosaurs and birds. Discovered just two years after Darwin's seminal *Origin of Species*, its discovery spurred the nascent debate between proponents of evolutionary biology and creationism. This early bird is so dinosaurlike that, without a clear impression of feathers in the surrounding rock, at least one specimen was mistaken for *Compsognathus*.

File:Parts of feather modified.jpg

Parts of a feather

Since the 1990s, a number of additional feathered dinosaurs have been found, providing even stronger evidence of the close relationship between dinosaurs and modern birds. The first of these were initially described as simple filamentous *protofeathers*, which were reported in dinosaur lineages as primitive as compsognathids and tyrannosauroids. However, feathers indistinguishable from those of modern birds were soon after found in non-avialan dinosaurs as well.

A small minority of researchers have claimed that the simple filamentous "protofeather" structures are simply the result of the decomposition of collagen fiber under the dinosaurs' skin or in fins along their backs, and that species with unquestionable feathers, such as oviraptorosaurs and dromaeosaurs are not dinosaurs, but true birds unrelated to dinosaurs.[561] However, a majority of studies have concluded that feathered dinosaurs are in fact dinosaurs, and that the simpler filaments of unquestionable theropods represent simple feathers. Some researchers have demonstrated the presence of color-bearing melanin in the structures—which would be expected in feathers but not collagen fibers. Others have demonstrated, using studies of modern bird decomposition, that even advanced feathers appear filamentous when subjected to the crushing forces experienced during fossilization, and that the supposed "protofeathers"

Figure 212: *Fossil cast of NGMC 91, a probable specimen of Sinornithosaurus*

may have been more complex than previously thought.[562] Detailed examination of the "protofeathers" of *Sinosauropteryx prima* showed that individual feathers consisted of a central quill (*rachis*) with thinner *barbs* branching off from it, similar to but more primitive in structure than modern bird feathers.

Skeleton

Because feathers are often associated with birds, feathered dinosaurs are often touted as the missing link between birds and dinosaurs. However, the multiple skeletal features also shared by the two groups represent the more important link for paleontologists. Furthermore, it is increasingly clear that the relationship between birds and dinosaurs, and the evolution of flight, are more complex topics than previously realized. For example, while it was once believed that birds evolved from dinosaurs in one linear progression, some scientists, most notably Gregory S. Paul, conclude that dinosaurs such as the dromaeosaurs may have evolved from birds, losing the power of flight while keeping their feathers in a manner similar to the modern ostrich and other ratites.

Comparisons of bird and dinosaur skeletons, as well as cladistic analysis, strengthens the case for the link, particularly for a branch of theropods called maniraptors. Skeletal similarities include the neck, pubis, wrist (semi-lunate carpal), arm and pectoral girdle, shoulder blade, clavicle, and breast bone.

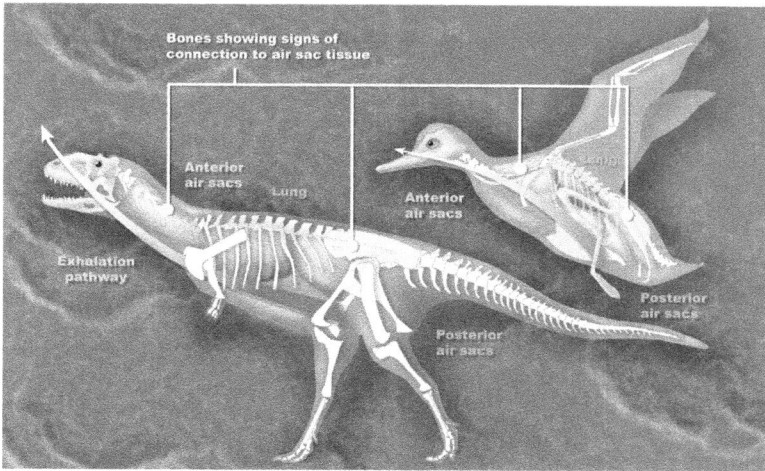

Figure 213: *Comparison between the air sacs of Majungasaurus and a bird (duck)*

A study comparing embryonic, juvenile and adult archosaur skulls concluded that bird skulls are derived from those of theropod dinosaurs by progenesis, a type of paedomorphic heterochrony, which resulted in retention of juvenile characteristics of their ancestors.

Lungs

Large meat-eating dinosaurs had a complex system of air sacs similar to those found in modern birds, according to an investigation led by Patrick M. O'Connor of Ohio University. In theropod dinosaurs (carnivores that walked on two legs and had birdlike feet) flexible soft tissue air sacs likely pumped air through the stiff lungs, as is the case in birds. "What was once formally considered unique to birds was present in some form in the ancestors of birds", O'Connor said.

Heart

Computed tomography (CT) scans conducted in 2000 of the chest cavity of a specimen of the ornithopod *Thescelosaurus* found the apparent remnants of complex four-chambered hearts, much like those found in today's mammals and birds. The idea is controversial within the scientific community, coming under fire for bad anatomical science or simply wishful thinking.

A study published in 2011 applied multiple lines of inquiry to the question of the object's identity, including more advanced CT scanning, histology, X-ray diffraction, X-ray photoelectron spectroscopy, and scanning electron microscopy. From these methods, the authors found that: the object's internal structure does not include chambers but is made up of three unconnected areas of lower density material, and is not comparable to the structure of an ostrich's heart; the "walls" are composed of sedimentary minerals not known to be produced in biological systems, such as goethite, feldspar minerals, quartz, and gypsum, as well as some plant fragments; carbon, nitrogen, and phosphorus, chemical elements important to life, were lacking in their samples; and cardiac cellular structures were absent. There was one possible patch with animal cellular structures. The authors found their data supported identification as a concretion of sand from the burial environment, not the heart, with the possibility that isolated areas of tissues were preserved.

The question of how this find reflects metabolic rate and dinosaur internal anatomy is moot, though, regardless of the object's identity. Both modern crocodilians and birds, the closest living relatives of dinosaurs, have four-chambered hearts (albeit modified in crocodilians), so dinosaurs probably had them as well; the structure is not necessarily tied to metabolic rate.[563]

Sleeping posture

Fossils of the troodonts *Mei* and *Sinornithoides* demonstrate that the dinosaurs slept like certain modern birds, with their heads tucked under their arms.[564] This behavior, which may have helped to keep the head warm, is also characteristic of modern birds.

Reproductive biology

When laying eggs, female birds grow a special type of bone in their limbs. This medullary bone forms as a calcium-rich layer inside the hard outer bone, and is used as a calcium source to make eggshells. The presence of endosteally derived bone tissues lining the interior marrow cavities of portions of a *Tyrannosaurus rex* specimen's hind limb suggested that *T. rex* used similar reproductive strategies, and revealed that the specimen is female. Further research has found medullary bone in the theropod *Allosaurus* and ornithopod *Tenontosaurus*. Because the line of dinosaurs that includes *Allosaurus* and *Tyrannosaurus* diverged from the line that led to *Tenontosaurus* very early in the evolution of dinosaurs, this suggests that dinosaurs in general produced medullary tissue.

Figure 214: *A nesting Citipati osmolskae specimen, at the American Museum of Natural History in New York City*

Brooding and care of young

Several *Citipati* specimens have been found resting over the eggs in its nest in a position most reminiscent of brooding.

Numerous dinosaur species, for example *Maiasaura*, have been found in herds mixing both very young and adult individuals, suggesting rich interactions between them.

A dinosaur embryo was found without teeth, which suggests some parental care was required to feed the young dinosaur, possibly the adult dinosaur regurgitated food into the young dinosaur's mouth (*see* altricial). This behaviour is seen in numerous bird species; parent birds regurgitate food into the hatchling's mouth.

Gizzard stones

Both birds and dinosaurs use gizzard stones. These stones are swallowed by animals to aid digestion and break down food and hard fibres once they enter the stomach. When found in association with fossils, gizzard stones are called gastroliths. Gizzard stones are also found in some fish (mullets, mud shad, and the gillaroo, a type of trout) and in crocodiles.

Molecular evidence

On several occasions, the extraction of DNA and proteins from Mesozoic dinosaurs fossils has been claimed, allowing for a comparison with birds. Several proteins have putatively been detected in dinosaur fossils, including hemoglobin.

In the March 2005 issue of *Science*, Dr. Mary Higby Schweitzer and her team announced the discovery of flexible material resembling actual soft tissue inside a 68-million-year-old *Tyrannosaurus rex* leg bone of specimen MOR 1125 from the Hell Creek Formation in Montana. The seven collagen types obtained from the bone fragments, compared to collagen data from living birds (specifically, a chicken), suggest that older theropods and birds are closely related.[565] The soft tissue allowed a molecular comparison of cellular anatomy and protein sequencing of collagen tissue published in 2007, both of which indicated that *T. rex* and birds are more closely related to each other than either is to *Alligator*.[566] A second molecular study robustly supported the relationship of birds to dinosaurs, though it did not place birds within Theropoda, as expected. This study utilized eight additional collagen sequences extracted from a femur of the "mummified" *Brachylophosaurus canadensis* specimen MOR 2598, a hadrosaur. However, these results have been very controversial. No other peptides of a Mesozoic age have been reported. In 2008, it was suggested that the presumed soft tissue was in fact a bacterial microfilm.[567] In response, it was argued that these very microfilms protected the soft tissue.[568] Another objection was that the results could have been caused by contamination.[569] In 2015, under more controlled conditions safeguarding against contamination, the peptides were still identified.[570] In 2017, a study found that a peptide was present in the bone of the modern ostrich that was identical to that found in the *Tyrannosaurus* and *Brachylophosaurus* specimens, highlighting the danger of a cross-contamination.[571]

The successful extraction of ancient DNA from dinosaur fossils has been reported on two separate occasions, but upon further inspection and peer review, neither of these reports could be confirmed.

Origin of bird flight

Debates about the origin of bird flight are almost as old as the idea that birds evolved from dinosaurs, which arose soon after the discovery of *Archaeopteryx* in 1862. Two theories have dominated most of the discussion since then: the cursorial ("from the ground up") theory proposes that birds evolved from small, fast predators that ran on the ground; the arboreal ("from the trees down") theory proposes that powered flight evolved from unpowered gliding

Figure 215: *Reconstruction of Rahonavis, a ground-dwelling feathered dinosaur that some researchers think was well equipped for flight*

by arboreal (tree-climbing) animals. A more recent theory, "wing-assisted incline running" (WAIR), is a variant of the cursorial theory and proposes that wings developed their aerodynamic functions as a result of the need to run quickly up very steep slopes such as trees, which would help small feathered dinosaurs escape from predators.

In March 2018, scientists reported that *Archaeopteryx* was likely capable of flight, but in a manner substantially different from that of modern birds.

Cursorial ("from the ground up") theory

The cursorial theory of the origin of flight was first proposed by Samuel Wendell Williston, and elaborated upon by Baron Nopcsa. This hypothesis proposes that some fast-running animals with long tails used their arms to keep their balance while running. Modern versions of this theory differ in many details from the Williston-Nopcsa version, mainly as a result of discoveries since Nopcsa's time.

Nopcsa theorized that increasing the surface area of the outstretched arms could have helped small cursorial predators keep their balance, and that the scales of the forearms elongated, evolving into feathers. The feathers could also have been used to trap insects or other prey. Progressively, the animals leapt for longer distances, helped by their evolving wings. Nopcsa also proposed three stages in the evolution of flight. First, animals developed passive flight, in which developing wing structures served as a sort of parachute. Second, they achieved active flight by flapping the wings. He used *Archaeopteryx* as an example of this second stage. Finally, birds gained the ability to soar.

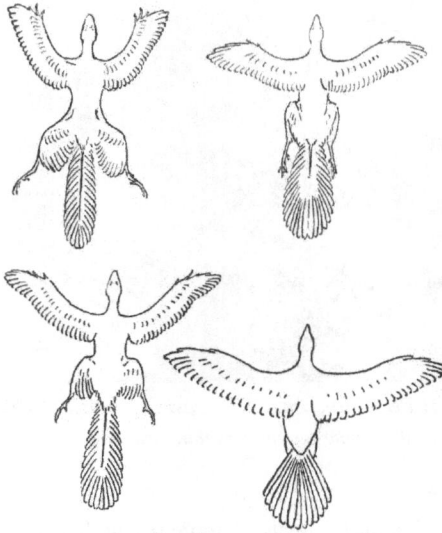

Figure 216: *Proposed development of flight in a book from 1922:*
Tetrapteryx, Archaeopteryx, Hypothetical Stage, Modern Bird

Current thought is that feathers did not evolve from scales, as feathers are made of different proteins. More seriously, Nopcsa's theory assumes that feathers evolved as part of the evolution of flight, and recent discoveries prove that assumption is false.

Feathers are very common in coelurosaurian dinosaurs (including the early tyrannosauroid *Dilong*). Modern birds are classified as coelurosaurs by nearly all palaeontologists, though not by a few ornithologists.[572,573] The modern version of the "from the ground up" hypothesis argues that birds' ancestors were small, *feathered*, ground-running predatory dinosaurs (rather like roadrunners in their hunting style[574]) that used their forelimbs for balance while pursuing prey, and that the forelimbs and feathers later evolved in ways that provided gliding and then powered flight. The most widely suggested original functions of feathers include thermal insulation and competitive displays, as in modern birds.

All of the *Archaeopteryx* fossils come from marine sediments, and it has been suggested that wings may have helped the birds run over water in the manner of the *Jesus Christ Lizard* (common basilisk).[575]

Most recent refutations of the "from the ground up" hypothesis attempt to refute the modern version's assumption that birds are modified coelurosaurian dinosaurs. The strongest attacks are based on embryological analyses that conclude that birds' wings are formed from digits 2, 3, and 4, (corresponding to the index, middle, and ring fingers in humans. The first of a bird's three digits forms the alula, which they use to avoid stalling in low-speed flight—for example, when landing). The hands of coelurosaurs, however, are formed by digits 1, 2, and 3 (thumb and first two fingers in humans).[576] However, these embryological analyses were immediately challenged on the embryological grounds that the "hand" often develops differently in clades that have lost some digits in the course of their evolution, and that birds' "hands" do develop from digits 1, 2, and 3. This debate is complex and not yet resolved - see "Digit homology".

Wing-assisted incline running

The wing-assisted incline running (WAIR) hypothesis was prompted by observation of young chukar chicks, and proposes that wings developed their aerodynamic functions as a result of the need to run quickly up very steep slopes such as tree trunks, for example to escape from predators.[577] This makes it a specialized type of cursorial ("from the ground up") theory. Note that in this scenario birds need *downforce* to give their feet increased grip.[578] But early birds, including *Archaeopteryx*, lacked the shoulder mechanism by which modern birds' wings produce swift, powerful upstrokes. Since the downforce WAIR depends on is generated by upstrokes, it seems that early birds were incapable of WAIR. Because WAIR is a behavioural trait without osteological specializations, the phylogenetic placement of the flight stroke before the divergence of Neornithes makes it impossible to determine if WAIR is ancestral to the avian flight stroke or derived from it.

Arboreal ("from the trees down") theory

Most versions of the arboreal hypothesis state that the ancestors of birds were very small dinosaurs that lived in trees, springing from branch to branch. This small dinosaur already had feathers, which were co-opted by evolution to produce longer, stiffer forms that were useful in aerodynamics, eventually producing wings. Wings would have then evolved and become increasingly refined as devices to give the leaper more control, to parachute, to glide, and to fly in stepwise fashion. The arboreal hypothesis also notes that, for arboreal animals, aerodynamics are far more energy efficient, since such animals simply fall to achieve minimum gliding speeds.[579,580]

Several small dinosaurs from the Jurassic or Early Cretaceous, all with feathers, have been interpreted as possibly having arboreal and/or aerodynamic adaptations. These include *Scansoriopteryx*, *Epidexipteryx*, *Microraptor*, *Pedopenna*, and *Anchiornis*. *Anchiornis* is particularly important to this subject, as it lived at the beginning of the Late Jurassic, long before *Archaeopteryx*.

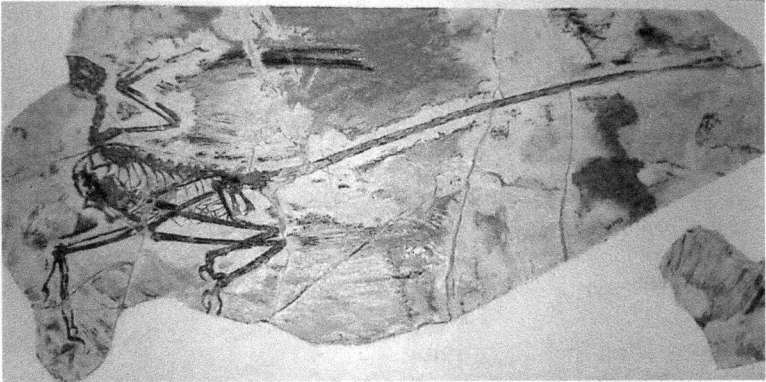

Figure 217: *The four-winged Microraptor, a member of the Dro-maeosauridae, a group of dinosaurs closely related to birds*

Analysis of the proportions of the toe bones of the most primitive birds *Archaeopteryx* and *Confuciusornis*, compared to those of living species, suggest that the early species may have lived both on the ground and in trees.[581]

One study suggested that the earliest birds and their immediate ancestors did not climb trees. This study determined that the amount of toe claw curvature of early birds was more like that seen in modern ground-foraging birds than in perching birds.

Diminished significance of *Archaeopteryx*

Archaeopteryx was the first and for a long time the only known feathered Mesozoic animal. As a result, discussion of the evolution of birds and of bird flight centered on *Archaeopteryx* at least until the mid-1990s.

There has been debate about whether *Archaeopteryx* could really fly. It appears that *Archaeopteryx* had the brain structures and inner-ear balance sensors that birds use to control their flight. *Archaeopteryx* also had a wing feather arrangement like that of modern birds and similarly asymmetrical flight feathers on its wings and tail. But *Archaeopteryx* lacked the shoulder mechanism by which modern birds' wings produce swift, powerful upstrokes (see diagram above of supracoracoideus pulley); this may mean that it and other early birds were incapable of flapping flight and could only glide.

But the discovery since the early 1990s of many feathered dinosaurs means that *Archaeopteryx* is no longer the key figure in the evolution of bird flight. Other small feathered coelurosaurs from the Cretaceous and Late Jurassic show possible precursors of avian flight. These include *Rahonavis*, a ground-runner

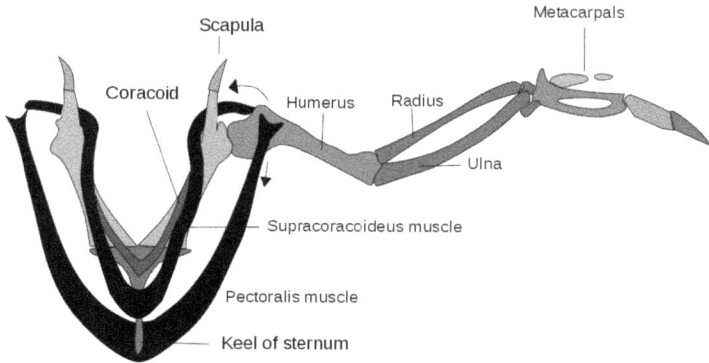

Figure 218: *The supracoracoideus works using a pulley-like system to lift the wing while the pectorals provide the powerful downstroke.*

with a *Velociraptor*-like raised sickle claw on the second toe, that some pale-ontologists assume to have been better adapted for flight than *Archaeopteryx*, *Scansoriopteryx*, an arboreal dinosaur that may support the "from the trees down" theory, and *Microraptor*, an arboreal dinosaur possibly capable of pow-ered flight but, if so, more like a biplane, as it had well-developed feathers on its legs. As early as 1915, some scientists argued that the evolution of bird flight may have gone through a four-winged (or *tetrapteryx*) stage.

Secondary flightlessness in dinosaurs

<templatestyles src="Template:Clade/styles.css" />

Coelurosaurs <templatestyles src="Template:Clade/styles.css" />

Tyrannosauroids

<templatestyles src="Template:Clade/styles.css" />
Ornithomimosaurs

<templatestyles src="Template:Clade/styles.css" />
Alvarezsaurids

<templatestyles src="Template:Clade/styles.css" />
<templatestyles src="Template:Clade/styles.css" />

Therizinosauroids

Oviraptosaurs

<templatestyles src="Template:Clade/styles.css" />
Archaeopteryx and *Rahonavis*
(birds)

Deinony-
chosaurs

<templatestyles src="Template:Clade/styles.css" />
Troodontids

<templatestyles src="Template:Clade/styles.css" />
<templatestyles src="Template:Clade/styles.css" />

Confuciusornis
(bird)

Microraptor
(dromaeosaur)

Dromaeosaurs

Simplified cladogram from Mayr *et al.* (2005)
Groups usually regarded as birds are in bold type.

A hypothesis, credited to Gregory Paul and propounded in his books *Predatory Dinosaurs of the World* (1988) and *Dinosaurs of the Air* (2002), suggests that some groups of non-flying carnivorous dinosaurs - especially deinonychosaurs, but perhaps others such as oviraptorosaurs, therizinosaurs, alvarezsaurids and ornithomimosaurs - actually descend from birds. Paul also proposed that the bird ancestor of these groups was more advanced in its flight adaptations than *Archaeopteryx*. The hypothesis would mean that *Archaeopteryx* is less closely related to extant birds than these dinosaurs are.[582]

Paul's hypothesis received additional support when Mayr *et al.* (2005) analyzed a new, tenth specimen of *Archaeopteryx*, and concluded that *Archaeopteryx* was the sister clade to the Deinonychosauria, but that the more advanced bird *Confuciusornis* was within the Dromaeosauridae. This result supports Paul's hypothesis, suggesting that the Deinonychosauria and the Troodontidae are part of Aves, the bird lineage proper, and secondarily flightless. This paper, however, excluded all other birds and thus did not sample their character distributions. The paper was criticized by Corfe and Butler

(2006) who found the authors could not support their conclusions statistically. Mayr *et al.* agreed that the statistical support was weak, but added that it is also weak for the alternative scenarios.

Current cladistic analyses do not support Paul's hypothesis about the position of *Archaeopteryx*. Instead, they indicate that *Archaeopteryx* is closer to birds, within the clade *Avialae*, than it is to deinonychosaurs or oviraptorosaurs. However, some fossils support the version of this theory that holds that some non-flying carnivorous dinosaurs may have had flying ancestors. In particular, *Microraptor*, *Pedopenna*, and *Anchiornis* all have winged feet, share many features, and lie close to the base of the clade Paraves. This suggests that the ancestral paravian was a four-winged glider, and that larger Deinonychosaurs secondarily lost the ability to glide, while the bird lineage increased in aerodynamic ability as it progressed. *Deinonychus* may also display partial volancy, with the young being capable of flight or gliding and the adults being flightless.[583] In 2018, a study concluded that the last common ancestor of the Pennaraptora had joint surfaces on the fingers, and between the metatarsus and the wrist, that were optimised to stabilise the hand in flight. This was seen as an indication for secondary flightlessness in heavy basal members of that group[584]

References

<templatestyles src="Template:Refbegin/styles.css" />

- Barsbold, Rinchen (1983): O ptich'ikh chertakh v stroyenii khishchnykh dinozavrov. ["Avian" features in the morphology of predatory dinosaurs]. *Transactions of the Joint Soviet Mongolian Paleontological Expedition* **24**: 96-103. [Original article in Russian.] Translated by W. Robert Welsh, copy provided by Kenneth Carpenter and converted by Matthew Carrano. PDF fulltext[585]
- Borenstein, Seth (July 31, 2014). "Study traces dinosaur evolution into early birds"[586]. *AP News*. Retrieved August 3, 2014.
- Bostwick, Kimberly S. (2003): Bird origins and evolution: data accumulates, scientists integrate, and yet the "debate" still rages. *Cladistics* **19**: 369–371. doi: 10.1016/S0748-3007(03)00069-0[587] PDF fulltext[588]
- Dingus, Lowell & Rowe, Timothy (1997): *The Mistaken Extinction: Dinosaur Evolution and the Origin of Birds*. W. H. Freeman and Company, New York. ISBN 0-7167-2944-X
- Dinosauria On-Line[589] (1995): Archaeopteryx's Relationship With Modern Birds[590]. Retrieved 2006-09-30.
- Dinosauria On-Line[589] (1996): Dinosaurian Synapomorphies Found In *Archaeopteryx*[591]. Retrieved 2006-09-30.

- Heilmann, G. (1926): *The Origin of Birds*. Witherby, London. ISBN 0-486-22784-7 (1972 Dover reprint)
- Mayr, Gerald; Pohl, B. & Peters, D. S. (2005): A Well-Preserved *Archaeopteryx* Specimen with Theropod Features. *Science* **310**(5753): 1483-1486. doi: 10.1126/science.1120331[592]
- Olson, Storrs L. (1985): The fossil record of birds. *In:* Farner, D.S.; King, J.R. & Parkes, Kenneth C. (eds.): *Avian Biology* **8**: 79-238. Academic Press, New York.

External links

- DinoBuzz[593] A popular-level discussion of the dinosaur-bird hypothesis
- *Archaeopteryx* - FAQs[594] from the Usenet newsgroup talk.origins.
- Dinosaurs among us[595] Article and Video[596] American Museum of Natural History exhibit of dinosaur evolution leading to birds

Wikimedia Commons has media related to *Aves fossils*.

Evolution of mammals

Part of a series on
Evolutionary biology

- ꞏ Evolutionary biology portal
- Category
- Book
- Related topics

- v
- t
- e[597]

The **evolution of mammals** has passed through many stages since the first appearance of their synapsid ancestors in the late Carboniferous period. The most ancestral forms in the class Mammalia are the egg-laying mammals in the subclass Prototheria. This class first started out as something close to the platypus and evolved to modern day mammals.[598] By the mid-Triassic, there were many synapsid species that looked like mammals. The lineage leading to today's mammals split up in the Jurassic; synapsids from this period include *Dryolestes*, more closely related to extant placentals and marsupials than to monotremes, as well as *Ambondro*, more closely related to monotremes. Later on, the eutherian and metatherian lineages separated; the metatherians are the animals more closely related to the marsupials, while the eutherians are those more closely related to the placentals. Since *Juramaia*, the earliest known eutherian, lived 160 million years ago in the Jurassic, this divergence must have occurred in the same period.

After the Cretaceous–Paleogene extinction event wiped out the non-avian dinosaurs (birds being the only surviving dinosaurs) and several mammalian groups, placental and marsupial mammals diversified into many new forms and ecological niches throughout the Paleogene and Neogene, by the end of which all modern orders had appeared.

Mammals are the only living synapsids. The synapsid lineage became distinct from the sauropsid lineage in the late Carboniferous period, between 320 and 315 million years ago. The sauropsids are today's reptiles and birds along with all the extinct animals more closely related to them than to mammals. This

Figure 219: *Restoration of Procynosuchus, a member of the cynodont group, which includes the ancestors of mammals*

does not include the mammal-like reptiles, a group more closely related to the mammals.

Throughout the Permian period, the synapsids included the dominant carnivores and several important herbivores. In the subsequent Triassic period, however, a previously obscure group of sauropsids, the archosaurs, became the dominant vertebrates. The mammaliaforms appeared during this period; their superior sense of smell, backed up by a large brain, facilitated entry into nocturnal niches with less exposure to archosaur predation. The nocturnal lifestyle may have contributed greatly to the development of mammalian traits such as endothermy and hair. Later in the Mesozoic, after theropod dinosaurs replaced rauisuchians as the dominant carnivores, mammals spread into other ecological niches. For example, some became aquatic, some were gliders, and some even fed on juvenile dinosaurs.

Most of the evidence consists of fossils. For many years, fossils of Mesozoic mammals and their immediate ancestors were very rare and fragmentary; but, since the mid-1990s, there have been many important new finds, especially in China. The relatively new techniques of molecular phylogenetics have also shed light on some aspects of mammalian evolution by estimating the timing of important divergence points for modern species. When used carefully, these techniques often, but not always, agree with the fossil record.

Although mammary glands are a signature feature of modern mammals, little is known about the evolution of lactation as these soft tissues are not often preserved in the fossil record. Most research concerning the evolution of mammals centers on the shapes of the teeth, the hardest parts of the tetrapod body. Other important research characteristics include the evolution of the middle ear bones, erect limb posture, a bony secondary palate, fur, hair, and warm-bloodedness.

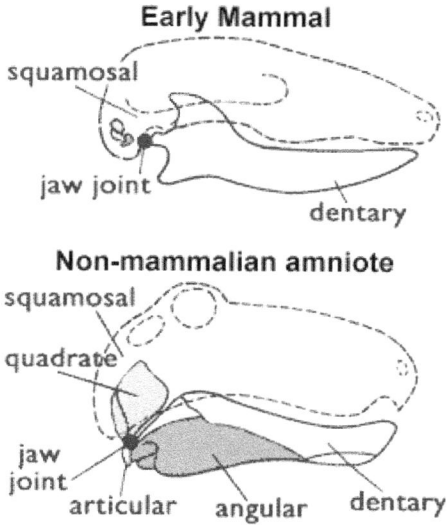

Figure 220: *Figure 1:In mammals, the quadrate and articular bones are small and part of the middle ear; the lower jaw consists only of dentary bone.*

Definition of "mammal"

While living mammal species can be identified by the presence of milk-producing mammary glands in the females, other features are required when classifying fossils, because mammary glands and other soft-tissue features are not visible in fossils.

One such feature available for paleontology, shared by all living mammals (including monotremes), but not present in any of the early Triassic therapsids, is shown in Figure 1 (on the right), namely: mammals use two bones for hearing that all other amniotes use for eating. The earliest amniotes had a jaw joint composed of the articular (a small bone at the back of the lower jaw) and the quadrate (a small bone at the back of the upper jaw). All non-mammalian tetrapods use this system including amphibians, turtles, lizards, snakes, crocodilians, dinosaurs (and their descendants the birds), ichthyosaurs, pterosaurs and therapsids. But mammals have a different jaw joint, composed only of the dentary (the lower jaw bone, which carries the teeth) and the squamosal (another small skull bone). In the Jurassic, their quadrate and articular bones evolved into the incus and malleus bones in the middle ear.[599] Mammals also have a double occipital condyle; they have two knobs at the base of the skull that fit into the topmost neck vertebra, while other tetrapods have a single occipital condyle.

In a 1981 article, Kenneth A. Kermack and his co-authors argued for drawing the line between mammals and earlier synapsids at the point where the mammalian pattern of molar occlusion was being acquired and the dentary-squamosal joint had appeared. The criterion chosen, they noted, is merely a matter of convenience; their choice was based on the fact that "the lower jaw is the most likely skeletal element of a Mesozoic mammal to be preserved." Today, most paleontologists consider that animals are mammals if they satisfy this criterion.

The ancestry of mammals

<templatestyles src="Template:Clade/styles.css" />

| Tetra-pods | <templatestyles src="Template:Clade/styles.css" /> |
| | Amphibians |

| | Am-niotes | <templatestyles src="Template:Clade/styles.css" /> |
| | | Sauropsids (including dinosaurs) |

| | | Synapsids | <templatestyles src="Template:Clade/styles.css" /> |
| | | | Caseids | *Cotylorhynchus* |

| | | | Eupely-cosaurs | <templatestyles src="Template:Clade/styles.css" /> |
| | | | | Edaphosaurids | *Edaphosaurus* |

				Sphenacodontians	<templatestyles src="Template:Clade/styles.css" />
					<templatestyles src="Template:Clade/styles.css" />
					Sphenacodontids	*Dimetrodon*

| | | | | | <templatestyles src="Template:Clade/styles.css" /> |
| | | | | | Therapsids | Mammals |

Amniotes

Life timeline

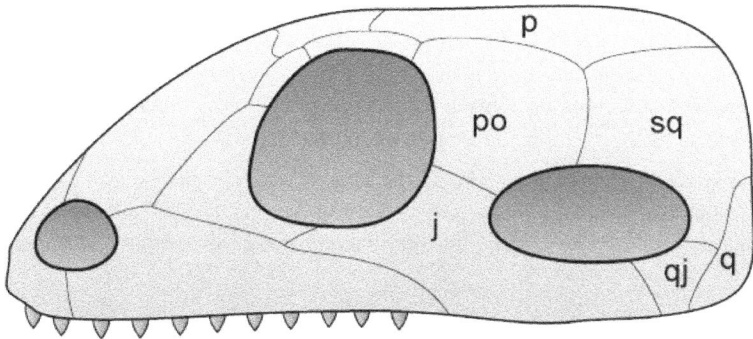

Figure 221: *The original synapsid skull structure has one hole behind each eye, in a fairly low position on the skull (lower right in this image).*

Axis scale: million years

Also see: *Human timeline* and *Nature timeline*

The first fully terrestrial vertebrates were amniotes — their eggs had internal membranes that allowed the developing embryo to breathe but kept water in. This allowed amniotes to lay eggs on dry land, while amphibians generally need to lay their eggs in water (a few amphibians, such as the Surinam toad, have evolved other ways of getting around this limitation). The first amniotes apparently arose in the middle Carboniferous from the ancestral reptiliomorphs.[600]

Within a few million years, two important amniote lineages became distinct: mammals' synapsid ancestors and the sauropsids, from which lizards, snakes, crocodilians, dinosaurs, and birds are descended. The earliest known fossils of synapsids and sauropsids (such as *Archaeothyris* and *Hylonomus*, respectively) date from about 320 to 315 million years ago. The times of origin are difficult to know, because vertebrate fossils from the late Carboniferous are very rare, and therefore the actual first occurrences of each of these types of animal might have been considerably earlier than the first fossil.

Synapsids

Synapsid skulls are identified by the distinctive pattern of the holes behind each eye, which served the following purposes:

• made the skull lighter without sacrificing strength.
• saved energy by using less bone.

- probably provided attachment points for jaw muscles. Having attachment points further away from the jaw made it possible for the muscles to be longer and therefore to exert a strong pull over a wide range of jaw movement without being stretched or contracted beyond their optimum range.

The synapsid pelycosaurs included the largest land vertebrates of the Early Permian, such as the 6 m (20 ft) long *Cotylorhynchus hancocki*. Among the other large pelycosaurs were *Dimetrodon grandis* and *Edaphosaurus cruciger*.

Therapsids

Therapsids descended from pelycosaurs in the middle Permian and took over their position as the dominant land vertebrates. They differ from pelycosaurs in several features of the skull and jaws, including larger temporal fenestrae and incisors that are equal in size.

The therapsid lineage that led to mammals went through a series of stages, beginning with animals that were very like their pelycosaur ancestors and ending with some that could easily be mistaken for mammals:

- gradual development of a bony secondary palate. Most books and articles interpret this as a prerequisite for the evolution of mammals' high metabolic rate, because it enabled these animals to eat and breathe at the same time. But some scientists point out that some modern ectotherms use a fleshy secondary palate to separate the mouth from the airway, and that a *bony* palate provides a surface on which the tongue can manipulate food, facilitating chewing rather than breathing. The interpretation of the bony secondary palate as an aid to chewing also suggests the development of a faster metabolism, because chewing reduces the size of food particles delivered to the stomach and can therefore speed their digestion. In mammals, the palate is formed by two specific bones, but various Permian therapsids had other combinations of bones in the right places to function as a palate.
- the dentary gradually becomes the main bone of the lower jaw.
- progress towards an erect limb posture, which would increase the animals' stamina by avoiding Carrier's constraint. But this process was erratic and very slow — for example: all herbivorous therapsids retained sprawling limbs (some late forms may have had semi-erect hind limbs); Permian carnivorous therapsids had sprawling forelimbs, and some late Permian ones also had semi-sprawling hindlimbs. In fact, modern monotremes still have semi-sprawling limbs.

Therapsid family tree

(simplified from; only those that are most relevant to the evolution of mammals are described below)

<templatestyles src="Template:Clade/styles.css" />

Ther-   <templatestyles src="Template:Clade/styles.css" />
ap-
sids

        Biarmosuchia

     Eu-   <templatestyles src="Template:Clade/styles.css" />
     ther-   <templatestyles src="Template:Clade/styles.css" />
     ap-   <templatestyles src="Template:Clade/styles.css" />
     sida   Dinocephalia

          Neother-   <templatestyles src="Template:Clade/styles.css" />
          apsida   Anomodonts   <templatestyles src="Template:Clade/styles.css" />
                       Dicynodonts

              Theriodontia   <templatestyles src="Template:Clade/styles.css" />
                       Gorgonopsia

                  Eutheri-   <templatestyles src="Template:Clade/styles.css" />
                  odontia   Therocephalia

                     Cyn-   <templatestyles src="Template:Clade/styles.css" />
                     odontia   styles.css" />
                         (Mammals, eventually)

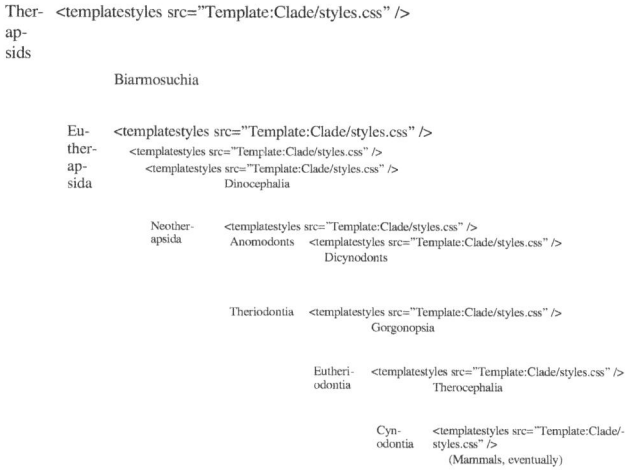

Only the dicynodonts, therocephalians, and cynodonts survived into the Triassic.

Biarmosuchia

The Biarmosuchia were the most primitive and pelycosaur-like of the therapsids.

Dinocephalians

Dinocephalians ("terrible heads") included both carnivores and herbivores. They were large; *Anteosaurus* was up to 6 m (20 ft) long. Some of the carnivores had semi-erect hindlimbs, but all dinocephalians had sprawling forelimbs. In many ways they were very primitive therapsids; for example, they had no secondary palate and their jaws were rather "reptilian".

Figure 222: *Lystrosaurus, one of the few genera of dicyn-odonts that survived the Permian-Triassic extinction event*

Anomodonts

The anomodonts ("anomalous teeth") were among the most successful of the herbivorous therapsids — one sub-group, the dicynodonts, survived almost to the end of the Triassic. But anomodonts were very different from modern herbivorous mammals, as their only teeth were a pair of fangs in the upper jaw and it is generally agreed that they had beaks like those of birds or ceratopsians.

Theriodonts

The theriodonts ("beast teeth") and their descendants had jaw joints in which the lower jaw's articular bone tightly gripped the skull's very small quadrate bone. This allowed a much wider gape, and one group, the carnivorous gorgonopsians ("gorgon faces"), took advantage of this to develop "sabre teeth". But the theriodont's jaw hinge had a longer term significance — the much reduced size of the quadrate bone was an important step in the development of the mammalian jaw joint and middle ear.

The gorgonopsians still had some primitive features: no bony secondary palate (but other bones in the right places to perform the same functions); sprawling forelimbs; hindlimbs that could operate in both sprawling and erect postures. But the therocephalians ("beast heads"), which appear to have arisen at about the same time as the gorgonopsians, had additional mammal-like features, e.g.

Figure 223: *Artist's conception of the cynodont Trirachodon within a burrow*

their finger and toe bones had the same number of phalanges (segments) as in early mammals (and the same number that primates have, including humans).

Cynodonts

The cynodonts, a theriodont group that also arose in the late Permian, include the ancestors of all mammals. Cynodonts' mammal-like features include further reduction in the number of bones in the lower jaw, a secondary bony palate, cheek teeth with a complex pattern in the crowns, and a brain which filled the endocranial cavity.

Multi-chambered burrows have been found, containing as many as 20 skeletons of the Early Triassic cynodont *Trirachodon*; the animals are thought to have been drowned by a flash flood. The extensive shared burrows indicate that these animals were capable of complex social behaviors.

Triassic takeover

The catastrophic Permian-Triassic mass extinction slightly more than 250 million years ago killed off about 70 percent of terrestrial vertebrate species and the majority of land plants.

As a result, ecosystems and food chains collapsed, and the establishment of new stable ecosystems took about 30 million years. With the disappearance of the gorgonopsians, which were dominant predators in the late Permian, the cynodonts' principal competitors for dominance of the carnivorous niches were a previously obscure sauropsid group, the archosaurs, which includes the ancestors of crocodilians and dinosaurs.

The archosaurs quickly became the dominant carnivores, a development often called the "Triassic takeover". Their success may have been due to the fact that the early Triassic was predominantly arid and therefore archosaurs' superior water conservation gave them a decisive advantage. All known archosaurs have glandless skins and eliminate nitrogenous waste in a uric acid paste containing little water, while the cynodonts probably excreted most such waste in a solution of urea, as mammals do today; considerable water is required to keep urea dissolved.

However, this theory has been questioned, since it implies synapsids were necessarily less advantaged in water retention, that synapsid decline coincides with climate changes or archosaur diversity (neither of which has been tested) and the fact that desert-dwelling mammals are as well adapted in this department as archosaurs,[601] and some cynodonts like *Trucidocynodon* were large-sized predators.

The Triassic takeover was probably a vital factor in the evolution of the mammals. Two groups stemming from the early cynodonts were successful in niches that had minimal competition from the archosaurs: the tritylodonts, which were herbivores, and the mammals, most of which were small nocturnal insectivores (although some, like *Sinoconodon*, were carnivores that fed on vertebrate prey, while still others were herbivores or omnivores).[602] As a result:

- The therapsid trend towards differentiated teeth with precise occlusion accelerated, because of the need to hold captured arthropods and crush their exoskeletons.
- As the body length of the mammals' ancestors fell below 50 mm (2 inches), advances in thermal insulation and temperature regulation would have become necessary for nocturnal life.
- Acute senses of hearing and smell became vital.
  - This accelerated the development of the mammalian middle ear.
  - The increase in the size of the olfactory lobes of the brain increased brain weight as a percentage of total body weight. Brain tissue requires a disproportionate amount of energy. The need for more food to support the enlarged brains increased the pressures for improvements in insulation, temperature regulation and feeding.
- Probably as a side-effect of the nocturnal life, mammals lost two of the four cone opsins, photoreceptors in the retina, present in the eyes of the

earliest amniotes. Paradoxically, this might have improved their ability to discriminate colors in dim light.

This retreat to a nocturnal role is called a nocturnal bottleneck, and is thought to explain many of the features of mammals.

From cynodonts to crown mammals

Fossil record

Mesozoic synapsids that had evolved to the point of having a jaw joint composed of the dentary and squamosal bones are preserved in few good fossils, mainly because they were mostly smaller than rats:

- They were largely restricted to environments that are less likely to provide good fossils. Floodplains as the best terrestrial environments for fossilization provide few mammal fossils, because they are dominated by medium to large animals, and the mammals could not compete with archosaurs in the medium to large size range. Tracks from the Early Cretaceous of Angola show the existence of raccoon-size mammals 118 Million years ago.
- Their delicate bones were vulnerable to being destroyed before they could be fossilized — by scavengers (including fungi and bacteria) and by being trodden on.
- Small fossils are harder to spot and more vulnerable to being destroyed by weathering and other natural stresses before they are discovered.

In the past 50 years, however, the number of Mesozoic fossil mammals has increased decisively; only 116 genera were known in 1979, for example, but about 310 in 2007, with an increase in quality such that "at least 18 Mesozoic mammals are represented by nearly complete skeletons".

Mammals or mammaliaforms

Some writers restrict the term "mammal" to the crown group mammals, the group consisting of the most recent common ancestor of the monotremes, marsupials, and placentals, together with all the descendants of that ancestor. In an influential 1988 paper, Timothy Rowe advocated this restriction, arguing that "ancestry... provides the only means of properly defining taxa" and, in particular, that the divergence of the monotremes from the animals more closely related to marsupials and placentals "is of central interest to any study of Mammalia as a whole." To accommodate some related taxa falling outside the crown group, he defined the Mammaliaformes as comprising "the last common ancestor of Morganucodontidae and Mammalia [as he had defined the latter term] and all its descendants." Besides Morganucodontidae, the newly defined taxon includes Docodonta and Kuehneotheriidae. Though haramiyids have been referred to the mammals since the 1860s, Rowe excluded them from the Mammaliaformes as falling outside his definition, putting them in a larger clade, the Mammaliamorpha.

Some writers have adopted this terminology noting, to avoid misunderstanding, that they have done so. Most paleontologists, however, still think that animals with the dentary-squamosal jaw joint and the sort of molars characteristic of modern mammals should formally be members of Mammalia.

Where the ambiguity in the term "mammal" may be confusing, this article uses "mammaliaform" and "crown mammal".

Family tree – cynodonts to crown group mammals

(based on Cynodontia:Dendrogram – Palaeos[603])

<templatestyles src="Template:Clade/styles.css" />

Cyn-   <templatestyles src="Template:Clade/styles.css" />
odon-
tia

<templatestyles src="Template:Clade/styles.css" />

Dvinia

Procynosuchidae

Epicyn-
odontia <templatestyles src="Template:Clade/styles.css" />
Thrinaxodon

Eucyn-
odontia <templatestyles src="Template:Clade/styles.css" />
<templatestyles src="Template:Clade/styles.css" />
Cynognathus

<templatestyles src="Template:Clade/styles.css" />

Tritylodontidae

Traversodontidae

Probainog-
nathia <templatestyles src="Template:Clade/styles.css" />
<templatestyles src="Template:Clade/styles.css" />

Tritheledontidae

Chiniquodontidae

<templatestyles src="Template:Clade/styles.css" />
Prozostrodon

Mammali-
aformes <templatestyles src="Template:Clade/styles.css" />
Morganucodontidae

<templatestyles src="Template:Clade/styles.css" />
Docodonta

<templatestyles src="Template:Clade/styles.css"
/>
Hadrocodium

<templatestyles src="Template:Clade/-
styles.css" />

Kuehneotheriidae

crown group Mammals

Morganucodontidae

The Morganucodontidae first appeared in the late Triassic, about 205M years ago. They are an excellent example of transitional fossils, since they have both the dentary-squamosal and articular-quadrate jaw joints. They were also one of the first discovered and most thoroughly studied of the mammaliaforms outside of the crown-group mammals, since an unusually large number of morganucodont fossils have been found.

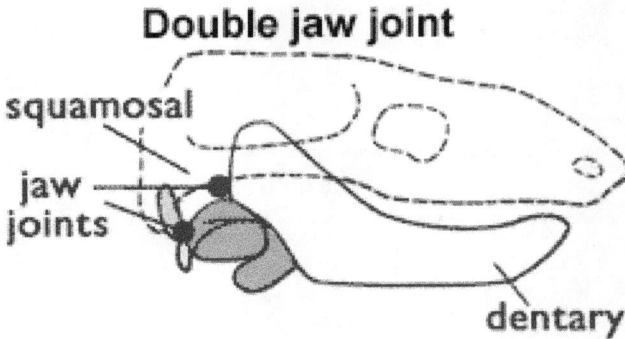

Figure 224: *Morganucodontidae and other transitional forms had both types of jaw joint: dentary-squamosal (front) and articular-quadrate (rear).*

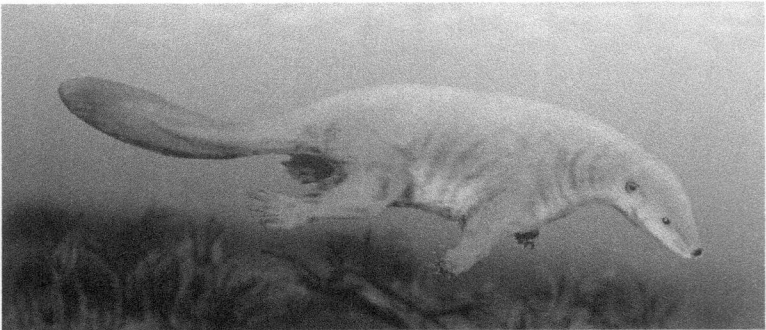

Figure 225: *Reconstruction of Castorocauda. Note the fur and the adaptations for swimming (broad, flat tail; webbed feet) and for digging (robust limbs and claws).*

Docodonts

Docodonts, among the most common Jurassic mammaliaforms, are noted for the sophistication of their molars. They are thought to have had general semi-aquatic tendencies, with the fish-eating *Castorocauda* ("beaver tail"), which lived in the mid-Jurassic about 164M years ago and was first discovered in 2004 and described in 2006, being the most well-understood example. *Castorocauda* was not a crown group mammal, but it is extremely important in the study of the evolution of mammals because the first find was an almost complete skeleton (a real luxury in paleontology) and it breaks the "small nocturnal insectivore" stereotype:[604]

- It was noticeably larger than most Mesozoic mammaliaform fossils — about 17 in (43 cm) from its nose to the tip of its 5-inch (130 mm) tail, and may have weighed 500–800 g (18–28 oz).
- It provides the earliest absolutely certain evidence of hair and fur. Previously the earliest was *Eomaia*, a crown group mammal from about 125M years ago.
- It had aquatic adaptations including flattened tail bones and remnants of soft tissue between the toes of the back feet, suggesting that they were webbed. Previously the earliest known semi-aquatic mammaliaforms were from the Eocene, about 110M years later.
- *Castorocauda*'s powerful forelimbs look adapted for digging. This feature and the spurs on its ankles make it resemble the platypus, which also swims and digs.
- Its teeth look adapted for eating fish: the first two molars had cusps in a straight row, which made them more suitable for gripping and slicing than for grinding; and these molars are curved backwards, to help in grasping slippery prey.

Hadrocodium

The family tree above shows *Hadrocodium* as an "aunt" of crown mammals. This mammaliaform, dated about 195M years ago in the very early Jurassic, exhibits some important features:

- The jaw joint consists only of the squamosal and dentary bones, and the jaw contains no smaller bones to the rear of the dentary, unlike the therapsid design.
- In therapsids and early mammaliaforms the eardrum may have stretched over a trough at the rear of the lower jaw. But *Hadrocodium* had no such trough, which suggests its ear was part of the cranium, as it is in crown-group mammals — and hence that the former articular and quadrate had migrated to the middle ear and become the malleus and incus. On the other hand, the dentary has a "bay" at the rear that mammals lack. This suggests that *Hadrocodium*'s dentary bone retained the same shape that it would have had if the articular and quadrate had remained part of the jaw joint, and therefore that *Hadrocodium* or a very close ancestor may have been the first to have a fully mammalian middle ear.
- Therapsids and earlier mammaliaforms had their jaw joints very far back in the skull, partly because the ear was at the rear end of the jaw but also had to be close to the brain. This arrangement limited the size of the braincase, because it forced the jaw muscles to run round and over it. *Hadrocodium*'s braincase and jaws were no longer bound to each other by the need to support the ear, and its jaw joint was further forward. In its

descendants or those of animals with a similar arrangement, the brain case was free to expand without being constrained by the jaw and the jaw was free to change without being constrained by the need to keep the ear near the brain — in other words it now became possible for mammaliaforms both to develop large brains and to adapt their jaws and teeth in ways that were purely specialized for eating.

Earliest crown mammals

The crown group mammals, sometimes called 'true mammals', are the extant mammals and their relatives back to their last common ancestor. Since this group has living members, DNA analysis can be applied in an attempt to explain the evolution of features that do not appear in fossils. This endeavor often involves molecular phylogenetics, a technique that has become popular since the mid-1980s.

Family tree of early crown mammals

Cladogram after Z.-X Luo. († marks extinct groups)

<templatestyles src="Template:Clade/styles.css" />

Crown group mammals <templatestyles src="Template:Clade/styles.css" />

Aus- <templatestyles src="Template:Clade/styles.css" />
tralosphenida
Ausktribosphenidae †

Monotremes

<templatestyles src="Template:Clade/styles.css" />
Eutriconodonta †

<templatestyles src="Template:Clade/styles.css" />
Allothe- Multituberculates †
ria

Trech- <templatestyles src="Template:Clade/styles.css" />
notheria
Spalacotheroidea †

Cladotheria <templatestyles src="Template:Clade/-
styles.css" />
Dryolestoidea †

The- <templatestyles
ria src="Template:Clade/styles.css"
/>

Metatheria Marsupials

Eutheria Placentals

Color vision

Early amniotes had four opsins in the cones of their retinas to use for distinguishing colours: one sensitive to red, one to green, and two corresponding to different shades of blue. The green opsin was not inherited by any crown mammals, but all normal individuals did inherit the red one. Early crown mammals thus had three cone opsins, the red one and both of the blues. All their extant descendants have lost one of the blue-sensitive opsins but not always the same one: marsupials and placentals (except for cetaceans, which later lost the other blue opsin as well) retain one blue-sensitive opsin while monotremes retain the other. Some placentals and marsupials, including humans, subsequently evolved green-sensitive opsins; like early crown mammals, therefore, their vision is trichromatic.

Australosphenida and Ausktribosphenidae

Ausktribosphenidae is a group name that has been given to some rather puzzling finds that:

- appear to have tribosphenic molars, a type of tooth that is otherwise known only in placentals and marsupials.
- come from mid-Cretaceous deposits in Australia — but Australia was connected only to Antarctica, and placentals originated in the Northern Hemisphere and were confined to it until continental drift formed land connections from North America to South America, from Asia to Africa and from Asia to India (the late Cretaceous map here[605] shows how the southern continents are separated).
- are represented only by teeth and jaw fragments, which is not very helpful.

Australosphenida is a group that has been defined in order to include the Ausktribosphenidae and monotremes. *Asfaltomylos* (mid- to late Jurassic, from Patagonia) has been interpreted as a basal australosphenid (animal that has features shared with both Ausktribosphenidae and monotremes; lacks features that are peculiar to Ausktribosphenidae or monotremes; also lacks features that are absent in Ausktribosphenidae and monotremes) and as showing that australosphenids were widespread throughout Gondwanaland (the old Southern Hemisphere super-continent).

Recent analysis of *Teinolophos*, which lived somewhere between 121 and 112.5 million years ago, suggests that it was a "crown group" (advanced and relatively specialised) monotreme. This was taken as evidence that the basal (most primitive) monotremes must have appeared considerably earlier, but this has been disputed (see the following section). The study also indicated that some alleged Australosphenids were also "crown group" monotremes (e.g.

Steropodon) and that other alleged Australosphenids (e.g. *Ausktribosphenos*, *Bishops*, *Ambondro*, *Asfaltomylos*) are more closely related to and possibly members of the Therian mammals (group that includes marsupials and placentals, see below).

Monotremes

Teinolophos, from Australia, is the earliest known monotreme. A 2007 study (published 2008) suggests that it was not a basal (primitive, ancestral) monotreme but a full-fledged platypus, and therefore that the platypus and echidna lineages diverged considerably earlier. A more recent study (2009), however, has suggested that, while *Teinolophos* was a type of platypus, it was also a basal monotreme and predated the radiation of modern monotremes. The semi-aquatic lifestyle of platypuses prevented them from being outcompeted by the marsupials that migrated to Australia millions of years ago, since joeys need to remain attached to their mothers and would drown if their mothers ventured into water (though there are exceptions like the water opossum and the lutrine opossum; however, they both live in South America and thus don't come into contact with monotremes). Genetic evidence has determined that echidnas diverged from the platypus lineage as recently as 19-48M, when they made their transition from semi-aquatic to terrestrial lifestyle.

Monotremes have some features that may be inherited from the cynodont ancestors:

- like lizards and birds, they use the same orifice to urinate, defecate and reproduce ("monotreme" means "one hole").
- they lay eggs that are leathery and uncalcified, like those of lizards, turtles and crocodilians.

Unlike other mammals, female monotremes do not have nipples and feed their young by "sweating" milk from patches on their bellies.

Of course these features are not visible in fossils, and the main characteristics from paleontologists' point of view are:

- a slender dentary bone in which the coronoid process is small or non-existent.
- the external opening of the ear lies at the posterior base of the jaw.
- the jugal bone is small or non-existent.
- a primitive pectoral girdle with strong ventral elements: coracoids, clavicles and interclavicle. Note: therian mammals have no interclavicle.
- sprawling or semi-sprawling forelimbs.

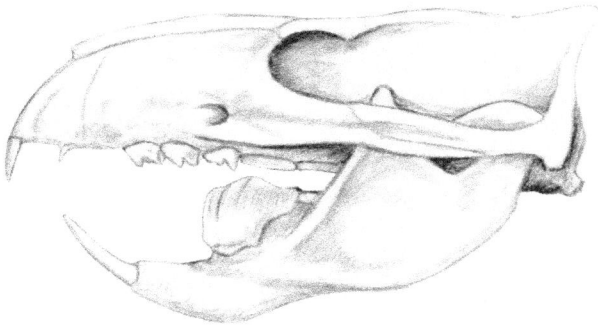

Figure 226: *Skull of the multituberculate Ptilodus*

Multituberculates

Multituberculates (named for the multiple tubercles on their "molars") are often called the "rodents of the Mesozoic", but this is an example of convergent evolution rather than meaning that they are closely related to the Rodentia. They existed for approximately 120 million years—the longest fossil history of any mammal lineage—but were eventually outcompeted by rodents, becoming extinct during the early Oligocene.

Some authors have challenged the phylogeny represented by the cladogram above. They exclude the multituberculates from the mammalian crown group, holding that multituberculates are more distantly related to extant mammals than even the Morganucodontidae. Multituberculates are like undisputed crown mammals in that their jaw joints consist of only the dentary and squamosal bones-whereas the quadrate and articular bones are part of the middle ear; their teeth are differentiated, occlude, and have mammal-like cusps; they have a zygomatic arch; and the structure of the pelvis suggests that they gave birth to tiny helpless young, like modern marsupials.[606] On the other hand, they differ from modern mammals:

• Their "molars" have two parallel rows of tubercles, unlike the tribosphenic (three-peaked) molars of uncontested early crown mammals.

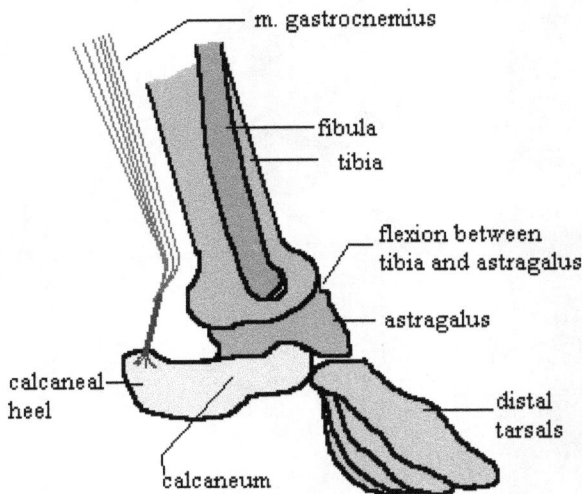

Figure 227: *Therian form of crurotarsal ankle. Adapted with permission from Palaeos*[607]

- The chewing action differs in that undisputed crown mammals chew with a side-to-side grinding action, which means that the molars usually occlude on only one side at a time, while multituberculates' jaws were incapable of side-to-side movement—they chewed, rather, by dragging the lower teeth backwards against the upper ones as the jaw closed.
- The anterior (forward) part of the zygomatic arch mostly consists of the maxilla (upper jawbone) rather than the jugal, a small bone in a little slot in the maxillary process (extension).
- The squamosal does not form part of the braincase.
- The rostrum (snout) is unlike that of undisputed crown mammals; in fact it looks more like that of a pelycosaur, such as *Dimetrodon*. The multituberculate rostrum is box-like, with the large flat maxillae forming the sides, the nasal the top, and the tall premaxilla at the front.

Theria

Theria ("beasts") is the clade originating with the last common ancestor of the Eutheria (including placentals) and Metatheria (including marsupials). Common features include:

- no interclavicle.

- coracoid bones non-existent or fused with the shoulder blades to form coracoid processes.
- a type of crurotarsal ankle joint in which: the main joint is between the tibia and astragalus; the calcaneum has no contact with the tibia but forms a heel to which muscles can attach. (The other well-known type of cruro-tarsal ankle is seen in crocodilians and works differently — most of the bending at the ankle is between the calcaneum and astragalus).
- tribosphenic molars.

Metatheria

The living Metatheria are all marsupials (animals with pouches). A few fossil genera, such as the Mongolian late Cretaceous *Asiatherium*, may be marsupials or members of some other metatherian group(s).

The oldest known metatherian is *Sinodelphys*, found in 125M-year-old early Cretaceous shale in China's northeastern Liaoning Province. The fossil is nearly complete and includes tufts of fur and imprints of soft tissues.

Didelphimorphia (common opossums of the Western Hemisphere) first appeared in the late Cretaceous and still have living representatives, probably because they are mostly semi-arboreal unspecialized omnivores.

The best-known feature of marsupials is their method of reproduction:

- The mother develops a kind of yolk sack in her womb that delivers nutrients to the embryo. Embryos of bandicoots, koalas and wombats additionally form placenta-like organs that connect them to the uterine wall, although the placenta-like organs are smaller than in placental mammals and it is not certain that they transfer nutrients from the mother to the embryo.
- Pregnancy is very short, typically four to five weeks. The embryo is born at a very early stage of development, and is usually less than 2 in (5.1 cm) long at birth. It has been suggested that the short pregnancy is necessary to reduce the risk that the mother's immune system will attack the embryo.
- The newborn marsupial uses its forelimbs (with relatively strong hands) to climb to a nipple, which is usually in a pouch on the mother's belly. The mother feeds the baby by contracting muscles over her mammary glands, as the baby is too weak to suck. The newborn marsupial's need to use its forelimbs in climbing to the nipple was historically thought to have restricted metatherian evolution, as it was assumed that the forelimb couldn't become specialised intro structures like wings, hooves or flippers. However, several bandicoots, most notably the pig-footed bandicoot, have true hooves similar to those of placental ungulates, and several marsupial gliders have evolved.

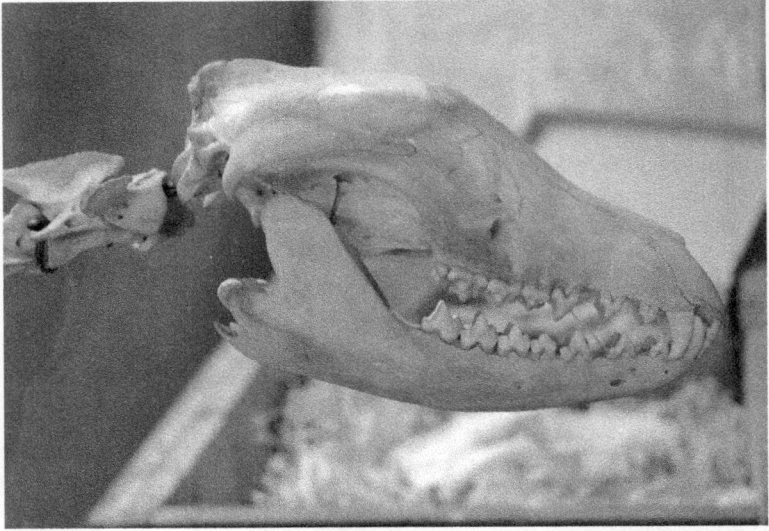

Figure 228: *Skull of thylacine, showing marsupial pattern of molars*

Although some marsupials look very like some placentals (the thylacine or "marsupial wolf" is a good example), marsupial skeletons have some features that distinguish them from placentals:

- Some, including the thylacine, have four molars; whereas no known placental has more than three.
- All have a pair of palatal fenestrae, window-like openings on the bottom of the skull (in addition to the smaller nostril openings).

Marsupials also have a pair of marsupial bones (sometimes called "epubic bones"), which support the pouch in females. But these are not unique to marsupials, since they have been found in fossils of multituberculates, monotremes, and even eutherians — so they are probably a common ancestral feature that disappeared at some point after the ancestry of living placental mammals diverged from that of marsupials. Some researchers think the epubic bones' original function was to assist locomotion by supporting some of the muscles that pull the thigh forwards.

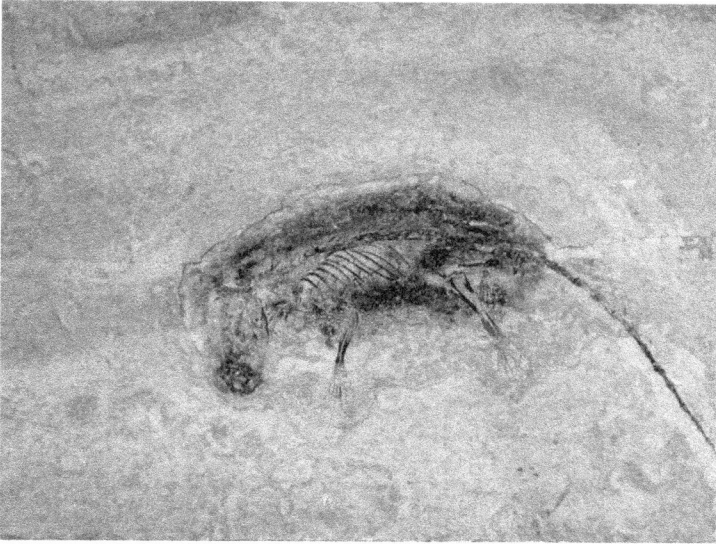

Figure 229: *Fossil of Eomaia in the Hong Kong Science Museum.*

Eutheria

The time of appearance of the earliest eutherians has been a matter of controversy. On one hand, recently discovered fossils of *Juramaia* have been dated to 160 million years ago and classified as eutherian. Fossils of *Eomaia* from 125 million years ago in the Early Cretaceous have also been classified as eutherian. A recent analysis of phenomic characters, however, classified *Eomaia* as pre-eutherian and reported that the earliest clearly eutherian specimens came from *Maelestes*, dated to 91 million years ago. That study also reported that eutherians did not significantly diversify until after the catastrophic extinction at the Cretaceous–Paleogene boundary, about 66 million years ago.

Eomaia was found to have some features that are more like those of marsupials and earlier metatherians:

- Epipubic bones extending forwards from the pelvis, which are not found in any modern placental, but are found in all other mammals — early mammaliaforms, non-placental eutherians, marsupials, and monotremes — as well as in the cynodont therapsids that are closest to mammals. Their function is to stiffen the body during locomotion. This stiffening would be harmful in pregnant placentals, whose abdomens need to expand.

- A narrow pelvic outlet, which indicates that the young were very small at birth and therefore pregnancy was short, as in modern marsupials. This suggests that the placenta was a later development.
- Five incisors in each side of the upper jaw. This number is typical of metatherians, and the maximum number in modern placentals is three, except for homodonts, such as the armadillo. But *Eomaia's* molar to pre-molar ratio (it has more pre-molars than molars) is typical of eutherians, including placentals, and not normal in marsupials.

Eomaia also has a Meckelian groove, a primitive feature of the lower jaw that is not found in modern placental mammals.

These intermediate features are consistent with molecular phylogenetics estimates that the placentals diversified about 110M years ago, 15M years after the date of the *Eomaia* fossil.

Eomaia also has many features that strongly suggest it was a climber, including several features of the feet and toes; well-developed attachment points for muscles that are used a lot in climbing; and a tail that is twice as long as the rest of the spine.

Placentals' best-known feature is their method of reproduction:

- The embryo attaches itself to the uterus via a large placenta via which the mother supplies food and oxygen and removes waste products.
- Pregnancy is relatively long and the young are fairly well-developed at birth. In some species (especially herbivores living on plains) the young can walk and even run within an hour of birth.

It has been suggested that the evolution of placental reproduction was made possible by retroviruses that:

- make the interface between the placenta and uterus into a syncytium, i.e. a thin layer of cells with a shared external membrane. This allows the passage of oxygen, nutrients and waste products, but prevents the passage of blood and other cells that would cause the mother's immune system to attack the fetus.
- reduce the aggressiveness of the mother's immune system, which is good for the foetus but makes the mother more vulnerable to infections.

From a paleontologist's point of view, eutherians are mainly distinguished by various features of their teeth, ankles and feet.

Figure 230: *Restoration of Volaticotherium, a Middle Jurassic eutriconodont and the earliest known gliding mammal.*

Expansion of ecological niches in the Mesozoic

There is still some truth in the "small, nocturnal insectivores" stereotype, but
recent finds, mainly in China, show that some mammaliaforms and crown
group mammals were larger and had a variety of lifestyles. For example:

- *Castorocauda*, a member of Docodonta which lived in the middle Jurassic about 164 million years, was about 42.5 cm (16.7 in) long, weighed
  500–800 g (18–28 oz), had a beaver-like tail that was adapted for swimming, limbs adapted for swimming and digging, and teeth adapted for
  eating fish. Another docodont, *Haldanodon*, also had semi-aquatic habits,
  and indeed aquatic tendencies were probably common among docodonts
  based on their prevalence in wetland environments.[608] The eutriconodonts
  Liaoconodon and *Yanoconodon* have more recently also have been suggested to be freshwater swimmers, lacking *Castorocauda*'s powerful tail
  but possessing paddle-like limbs;[609] the eutriconodont *Astroconodon* has
  similarly been suggested as being semi-aquatic in the past, albeit to less
  convincing evidence.
- Multituberculates are allotherians that survived for over 125 million
  years (from mid-Jurassic, about 160M years ago, to late Eocene, about
  35M years ago) are often called the "rodents of the Mesozoic". As noted

Figure 231: *Skull cast of Late Cretaceous Didelphodon, showing its robust teeth adapted to a durophagous diet.*

above, they may have given birth to tiny live neonates rather than laying eggs.

- *Fruitafossor*, from the late Jurassic period about 150 million years ago, was about the size of a chipmunk and its teeth, forelimbs and back suggest that it broke open the nest of social insects to prey on them (probably termites, as ants had not yet appeared).
- Similarly, the gobiconodontid *Spinolestes* possessed adaptations for fossoriality and convergent traits with placental xenarthrans like scutes and xenarthrous vertebrae, so it too might have had anteater like habits. It is also notable for the presence of quills akin to those of modern spiny mice.
- *Volaticotherium*, from the boundary the early Cretaceous about 125M years ago, is the earliest-known gliding mammal and had a gliding membrane that stretched out between its limbs, rather like that of a modern flying squirrel. This also suggests it was active mainly during the day. The closely related *Argentoconodon* also shows similar adaptations that may also suggest aerial locomotion.
- *Repenomamus*, a eutriconodont from the early Cretaceous 130 million years ago, was a stocky, badger-like predator that sometimes preyed on young dinosaurs. Two species have been recognized, one more than 1 m (39 in) long and weighing about 12–14 kg (26–31 lb), the other less than 0.5 m (20 in) long and weighing 4–6 kg (8.8–13.2 lb).[610]

- *Schowalteria* is a Late Cretaceous species almost as large if not larger than *R. giganticus* that shows speciations towards herbivory, comparable to those of modern ungulates.
- Zhelestidae is a lineage of Late Cretaceous herbivorous eutherians, to the point of being mistaken for stem-ungulates.[611]
- Similarly, mesungulatids are also fairly large sized herbivorous mammals from the Late Cretaceous
- Deltatheroidans were metatherians that were specialised towards carnivorous habits,[612] and possible forms like *Oxlestes* and *Khudulestes* might have been among the largest Mesozoic mammals, though their status as deltatheroidans is questionable.
- *Ichthyoconodon*, a eutriconodont from the Berriasian of Morocco, is currently known from molariforms found in marine deposits. These teeth are sharp-cusped and similar in shape to those of piscivorous mammals, and unlike the teeth of contemporary mammals they do not show degradation, so rather than being carried down by river deposits the animal died *in situ* or close. This has been taken to mean that it was a marine mammal, likely one of the few examples known from the Mesozoic. Alternatively, its close relations to *Volaticotherium* and *Argentoconodon* might suggest that it was a flying mammal.
- *Didelphodon* is a Late Cretaceous riverine species of stagodontid marsupialiform with a durophagous dentition, robust jaws similar to a modern Tasmanian devil, and a postcranial skeleton very similar in size and shape to an otter. This animal has been lauded as the strongest bite of all Mesozoic mammals. It possibly specialized on eating freshwater crabs and molluscs.

Evolution of major groups of living mammals

There are currently vigorous debates between traditional paleontologists and molecular phylogeneticists about how and when the modern groups of mammals diversified, especially the placentals. Generally, the traditional paleontologists date the appearance of a particular group by the earliest known fossil whose features make it likely to be a member of that group, while the molecular phylogeneticists suggest that each lineage diverged earlier (usually in the Cretaceous) and that the earliest members of each group were anatomically very similar to early members of other groups and differed only in their genetics. These debates extend to the definition of and relationships between the major groups of placentals — the controversy about Afrotheria is a good example.

Fossil-based family tree of placental mammals

Here is a very simplified version of a typical family tree based on fossils, based on Cladogram of Mammalia – Palaeos[613]. It tries to show the nearest thing there is at present to a consensus view, but some paleontologists have very different views, for example:

- The most common view is that placentals originated in the Southern Hemisphere, but some paleontologists argue that they first appeared in Laurasia (old supercontinent containing modern Asia, N. America and Europe).
- Paleontologists differ as to when the first placentals appeared, with estimates ranging from 20M years before the end of the Cretaceous to just after the end of the Cretaceous. Molecular biologists argue for a much earlier origin, even suggesting appearance in the Middle Jurassic.
- Molecular data suggest that either Xenarthra, Afrotheria, or Atlantogenata (Xenarthra + Afrotheria), was the earliest-diverging group from the rest of the placental mammals.

For the sake of brevity and simplicity, the diagram omits some extinct groups in order to focus on the ancestry of well-known modern groups of placentals — † marks extinct groups. The diagram also shows the following:

- the age of the oldest known fossils in many groups, since one of the major debates between traditional paleontologists and molecular phylogeneticists is about when various groups first became distinct.
- well-known modern members of most groups.

<templatestyles src="Template:Clade/styles.css" />

Eu-   <templatestyles src="Template:Clade/styles.css" />
the-
ria

Xenarthra (late cretaceous)
(armadillos, anteaters, sloths)

<templatestyles src="Template:Clade/styles.css" />
Pholidota (late cretaceous)
(pangolins)

Epithe-
ria (latest Cretaceous)   <templatestyles src="Template:Clade/styles.css" />
(some extinct groups) †

<templatestyles src="Template:Clade/styles.css" />
Insectivora (latest Cretaceous)
(hedgehogs, shrews, moles, tenrecs)

<templatestyles src="Template:Clade/styles.css" />
<templatestyles src="Template:Clade/styles.css" />
Ana-   <templatestyles src="Template:Clade/styles.css" />
galida   Zalambdalestidae † (late Cretaceous)

<templatestyles src="Template:Clade/styles.css" />
Macroscelidea (late Eocene)
(elephant shrews)

<templatestyles src="Template:Clade/styles.css" />
Anagaloidea †

Glires (early Paleocene)   <templatestyles src="Template:Clade/styles.css" />

Lagomorpha (Eocene)
(rabbits, hares, pikas)

Rodentia (late Paleocene)
(mice & rats, squirrels, porcupines)

Ar-   <templatestyles src="Template:Clade/styles.css" />
chonta   <templatestyles src="Template:Clade/styles.css" />
Scandentia (mid-Eocene)
(tree shrews)

Primato-   <templatestyles src="Template:Clade/styles.css" />
morpha
Plesiadapiformes †

Primates (early Paleocene)
(tarsiers, lemurs, monkeys, apes including humans)

<templatestyles src="Template:Clade/styles.css" />
Dermoptera (late Eocene)
(colugos)

Chiroptera (late Paleocene)
(bats)

<templatestyles src="Template:Clade/styles.css" />
Carnivora (early Paleocene)
(cats, dogs, bears, seals)

Ungulatomor-   <templatestyles src="Template:Clade/styles.css" />
pha (late Cretaceous)   Eparctocy-   <templatestyles src="Template:Clade/styles.css" />
ona (late Cretaceous)   (some extinct groups) †

<templatestyles src="Template:Clade/styles.css" />
Arctostylopida † (late Paleocene)

<templatestyles src="Template:Clade/-
styles.css" />
Mesonychia † (mid-Paleocene)
(predators / scavengers, but
not closely related to modern
carnivores)

Cetar-   <templatestyles

This family tree contains some surprises and puzzles. For example:

- The closest living relatives of cetaceans (whales, dolphins, porpoises) are artiodactyls, hoofed animals, which are almost all pure herbivores.
- Bats are fairly close relatives of primates.
- The closest living relatives of elephants are the aquatic sirenians, while their next relatives are hyraxes, which look more like well-fed guinea pigs.
- There is little correspondence between the structure of the family (what was descended from what) and the dates of the earliest fossils of each group. For example, the earliest fossils of perissodactyls (the living members of which are horses, rhinos and tapirs) date from the late Paleocene, but the earliest fossils of their "sister group", the Tubulidentata, date from the early Miocene, nearly 50M years later. Paleontologists are fairly confident about the family relationships, which are based on cladistic analyses, and believe that fossils of the ancestors of modern aardvarks have simply not been found yet.

Molecular phylogenetics based family tree of placental mammals

Molecular phylogenetics uses features of organisms' genes to work out family trees in much the same way as paleontologists do with features of fossils — if two organisms' genes are more similar to each other than to those of a third organism, the two organisms are more closely related to each other than to the third.

Molecular phylogeneticists have proposed a family tree that is very different from the one with which paleontologists are familiar. Like paleontologists, molecular phylogeneticists have different ideas about various details, but here is a typical family tree according to molecular phylogenetics:[614] Note that the diagram shown here omits extinct groups, as one cannot extract DNA from fossils.

<templatestyles src="Template:Clade/styles.css" />

Eu- <templatestyles src="Template:Clade/styles.css" />
the-
ria

Atlantogenata ("born round the Atlantic ocean") <templatestyles src="Template:Clade/styles.css" />

Xenarthra (armadillos, anteaters, sloths)

Afrotheria <templatestyles src="Template:Clade/styles.css" />

Afrosoricida (golden moles, tenrecs, otter shrews)

<templatestyles src="Template:Clade/styles.css" />

<templatestyles src="Template:Clade/styles.css" />

Macroscelidea (elephant shrews)

Tubulidentata (aardvarks)

Paenungu- <templatestyles
lata ("not quite ungulates") src="Template:Clade/styles.css" />

Hyracoidea (hyraxes)

Proboscidea (elephants)

Sirenia (manatees, dugongs)

Boreoeutheria ("northern true /- placental mammals") <templatestyles src="Template:Clade/styles.css" />

Laurasiatheria <templatestyles src="Template:Clade/styles.css" />

Erinaceomorpha (hedgehogs, gymnures)

Soricomorpha (moles, shrews, solenodons)

Cetartiodactyla (camels and llamas, pigs and peccaries, ruminants, whales and hippos)

Pega- <templatestyles src="Template:Clade/
soferae styles.css" />

Pholidota (pangolins)

Chiroptera (bats)

Carnivora (cats, dogs, bears, seals)

Perissodactyla (horses, rhinos, tapirs).

Euarchon- <templatestyles src="Template:Clade/styles.css" />
toglires Glires <templatestyles src="Template:Clade/styles.css" />

Lagomorpha (rabbits, hares, pikas)

Rodentia (late Paleocene) (mice and rats, squirrels, porcupines)

Euar- <templatestyles src="Template:Clade/
chonta styles.css" />

Scandentia (tree shrews)

Dermoptera (colugos)

Primates (tarsiers, lemurs, monkeys, apes including humans)

Here are the most significant of the many differences between this family tree and the one familiar to paleontologists:

- The top-level division is between Atlantogenata and Boreoeutheria, instead of between Xenarthra and the rest. However, analysis of transposable element insertions supports a three-way top-level split between Xenarthra, Afrotheria and Boreoeutheria and the Atlantogenata clade does not receive significant support in recent distance-based molecular phylogenetics.
- Afrotheria contains several groups that are only distantly related according to the paleontologists' version: Afroinsectiphilia ("African insectivores"), Tubulidentata (aardvarks, which paleontologists regard as much closer to odd-toed ungulates than to other members of Afrotheria), Macroscelidea (elephant shrews, usually regarded as close to rabbits and rodents). The only members of Afrotheria that paleontologists would regard as closely related are Hyracoidea (hyraxes), Proboscidea (elephants) and Sirenia (manatees, dugongs).
- Insectivores are split into three groups: one is part of Afrotheria and the other two are distinct sub-groups within Boreoeutheria.
- Bats are closer to Carnivora and odd-toed ungulates than to Primates and Dermoptera (colugos).
- Perissodactyla (odd-toed ungulates) are closer to Carnivora and bats than to Artiodactyla (even-toed ungulates).

The grouping together of the Afrotheria has some geological justification. All surviving members of the Afrotheria originate from South American or (mainly) African lineages — even the Indian elephant, which diverged from an African lineage about 7.6[615] million years ago. As Pangaea broke up, Africa and South America separated from the other continents less than 150M years ago, and from each other between 100M and 80M years ago.[616,617] So it would not be surprising if the earliest eutherian immigrants into Africa and South America were isolated there and radiated into all the available ecological niches.

Nevertheless, these proposals have been controversial. Paleontologists naturally insist that fossil evidence must take priority over deductions from samples of the DNA of modern animals. More surprisingly, these new family trees have been criticised by other molecular phylogeneticists, sometimes quite harshly:[618]

- Mitochondrial DNA's mutation rate in mammals varies from region to region — some parts hardly ever change and some change extremely quickly and even show large variations between individuals within the same species.

- Mammalian mitochondrial DNA mutates so fast that it causes a problem called "saturation", where random noise drowns out any information that may be present. If a particular piece of mitochondrial DNA mutates randomly every few million years, it will have changed several times in the 60 to 75M years since the major groups of placental mammals diverged.

Timing of placental evolution

Recent molecular phylogenetic studies suggest that most placental orders diverged late in the Cretaceous period, about 100 to 85 million years ago, but that modern families first appeared later, in the late Eocene and early Miocene epochs of the Cenozoic period. Fossil-based analyses, on the contrary, limit the placentals to the Cenozoic. Many Cretaceous fossil sites contain well-preserved lizards, salamanders, birds, and mammals, but not the modern forms of mammals. It is likely that they simply did not exist, and that the molecular clock runs fast during major evolutionary radiations. On the other hand, there is fossil evidence from 85^{619} million years ago of hoofed mammals that may be ancestors of modern ungulates.

Fossils of the earliest members of most modern groups date from the Paleocene, a few date from later and very few from the Cretaceous, before the extinction of the dinosaurs. But some paleontologists, influenced by molecular phylogenetic studies, have used statistical methods to extrapolate *backwards* from fossils of members of modern groups and concluded that primates arose in the late Cretaceous.[620] However, statistical studies of the fossil record confirm that mammals were restricted in size and diversity right to the end of the Cretaceous, and rapidly grew in size and diversity during the Early Paleocene.

Evolution of mammalian features

Jaws and middle ears

Hadrocodium, whose fossils date from the early Jurassic, provides the first clear evidence of fully mammalian jaw joints and middle ears, in which the jaw joint is formed by the dentary and squamosal bones while the articular and quadrate move to the middle ear, where they are known as the incus and malleus.

One analysis of the monotreme *Teinolophos* suggested that this animal had a pre-mammalian jaw joint formed by the angular and quadrate bones and that the definitive mammalian middle ear evolved twice independently, in monotremes and in therian mammals, but this idea has been disputed.[621] In fact, two of the suggestion's authors co-authored a later paper that reinterpreted the same features as evidence that *Teinolophos* was a full-fledged platypus, which means it would have had a mammalian jaw joint and middle ear.

Lactation

It has been suggested that lactation's original function was to keep eggs moist. Much of the argument is based on monotremes (egg-laying mammals):[622]

- While the amniote egg is usually described as able to evolve away from water, most reptile eggs actually need moisture if they are not to dry out.
- Monotremes do not have nipples, but secrete milk from a hairy patch on their bellies.
- During incubation, monotreme eggs are covered in a sticky substance whose origin is not known. Before the eggs are laid, their shells have only three layers. Afterwards, a fourth layer appears with a composition different from that of the original three. The sticky substance and the fourth layer may be produced by the mammary glands.
- If so, that may explain why the patches from which monotremes secrete milk are hairy. It is easier to spread moisture and other substances over the egg from a broad, hairy area than from a small, bare nipple.

Later research demonstrated that caseins already appeared in the common mammalian ancestor approximately 200–310 million years ago. The question of whether secretion of a substance to keep egg moist translated into actual lactation in therapsids is open. A small mammaliomorph called *Sinocodon*, generally assumed to be the sister group of all later mammals, had front teeth in even the smallest individuals. Combined with a poorly ossified jaw, they very probably did not suckle.[623] Thus suckling may have evolved right at the pre-mammal/mammal transition. However, tritylodontids, generally assumed to be more basal, show evidence of suckling. Morganucodontans, also assumed to be basal Mammaliaformes, also show evidence of lactation.[624]

Hair and fur

The first clear evidence of hair or fur is in fossils of *Castorocauda* and *Megaconus*, from 164M years ago in the mid-Jurassic. As both mammals *Megaconus* and *Castorocauda* have a double coat of hair, with both guard hairs and an undercoat, it may be assumed that their last common ancestor did as well. This animal must have been Triassic as it was an ancestor of the Triassic *Tikitherium*. More recently, the discovery of hair remnants in Permian coprolites pushes back the origin of mammalian hair much further back in the synapsid line to Paleozoic therapsids.[625]

In the mid-1950s, some scientists interpreted the foramina (passages) in the maxillae (upper jaws) and premaxillae (small bones in front of the maxillae) of cynodonts as channels that supplied blood vessels and nerves to vibrissae (whiskers) and suggested that this was evidence of hair or fur. It was soon pointed out, however, that foramina do not necessarily show that an animal had vibrissae; the modern lizard *Tupinambis* has foramina that are almost identical to those found in the non-mammalian cynodont *Thrinaxodon*. Popular sources, nevertheless, continue to attribute whiskers to *Thrinaxodon*. A trace fossil from the Lower Triassic had been erroneously regarded as a cynodont footprint showing hair, but this interpretation has been refuted.[626] A study of cranial openings for facial nerves connected whiskers in extant mammals indicate the Prozostrodontia, small immediate ancestors of mammals, presented whiskers similar to mammals, but that less advanced therapsids would either have immobile whiskers or no whisker at all. Fur may have evolved from whiskers. Whiskers themselves may have evolved as a response to nocturnal and/or burrowing lifestyle.

Ruben & Jones (2000) note that the Harderian glands, which secrete lipids for coating the fur, were present in the earliest mammals like *Morganucodon*, but were absent in near-mammalian therapsids like *Thrinaxodon*. The Msx2 gene associated with hair follicle maintenance is also linked to the closure of the parietal eye in mammals, indicating that fur and lack of pineal eye is linked. The pineal eye is present in *Thrinaxodon*, but absent in more advanced cynognaths (the Probainognathia).

Insulation is the "cheapest" way to maintain a fairly constant body temperature, without consuming energy to produce more body heat. Therefore, the possession of hair or fur would be good evidence of homeothermy, but would not be such strong evidence of a high metabolic rate.

Erect limbs

Understanding of the evolution of erect limbs in mammals is incomplete — living and fossil monotremes have sprawling limbs. Some scientists think that the parasagittal (non-sprawling) limb posture is limited to the Boreosphenida, a group that contains the therians but not, for example, the multituberculates. In particular, they attribute a parasagittal stance to the therians *Sinodelphys* and *Eomaia*, which means that the stance had arisen by 125 million years ago, in the Early Cretaceous. However, they also discuss that earlier mammals had more erect forelimbs as opposed to the more sprawling hindlimbs, a trend still continued to some extent in modern placentals and marsupials.

Warm-bloodedness

"Warm-bloodedness" is a complex and rather ambiguous term, because it includes some or all of the following:

- **Endothermy**, the ability to generate heat internally rather than via behaviors such as basking or muscular activity.

- **Homeothermy**, maintaining a fairly constant body temperature. Most enzymes have an optimum operating temperature; efficiency drops rapidly outside the preferred range. A homeothermic organism needs only to possess enzymes that function well in a small range of temperatures.
- **Tachymetabolism**, maintaining a high metabolic rate, particularly when at rest. This requires a fairly high and stable body temperature because of the Q_{10} effect: biochemical processes run about half as fast if an animal's temperature drops by 10 °C.

Since scientists cannot know much about the internal mechanisms of extinct creatures, most discussion focuses on homeothermy and tachymetabolism. However, it is generally agreed that endothermy first evolved in non-mammalian synapsids such as dicynodonts, which possess body proportions associated with heat retention,[627] high vascularised bones with Haversian canals,[628] and possibly hair.[629] More recently, it has been suggested that endothermy evolved as far back as *Ophiacodon*.

Modern monotremes have a low body temperature compared to marsupials and placental mammals, around 32 °C (90 °F). Phylogenetic bracketing suggests that the body temperatures of early crown-group mammals were not less than that of extant monotremes. There is cytological evidence that the low metabolism of monotremes is a secondarily evolved trait.

Respiratory turbinates

Modern mammals have respiratory turbinates, convoluted structures of thin bone in the nasal cavity. These are lined with mucous membranes that warm and moisten inhaled air and extract heat and moisture from exhaled air. An animal with respiratory turbinates can maintain a high rate of breathing without the danger of drying its lungs out, and therefore may have a fast metabolism. Unfortunately these bones are very delicate and therefore have not yet been found in fossils. But rudimentary ridges like those that support respiratory turbinates have been found in advanced Triassic cynodonts, such as *Thrinaxodon* and *Diademodon*, which suggests that they may have had fairly high metabolic rates.

Bony secondary palate

Mammals have a secondary bony palate, which separates the respiratory passage from the mouth, allowing them to eat and breathe at the same time. Secondary bony palates have been found in the more advanced cynodonts and have been used as evidence of high metabolic rates. But some cold-blooded vertebrates have secondary bony palates (crocodilians and some lizards), while birds, which are warm-blooded, do not.

Diaphragm

A muscular diaphragm helps mammals to breathe, especially during strenuous activity. For a diaphragm to work, the ribs must not restrict the abdomen, so that expansion of the chest can be compensated for by reduction in the volume of the abdomen and *vice versa*. Diaphragms are known in caseid pelycosaurs, indicating an early origin within synapsids, though they were still fairly inefficient and likely required support from other muscle groups and limb motion.[630]

The advanced cynodonts have very mammal-like rib cages, with greatly reduced lumbar ribs. This suggests that these animals had more developed diaphragms, were capable of strenuous activity for fairly long periods and therefore had high metabolic rates. On the other hand, these mammal-like rib cages may have evolved to increase agility. However, the movement of even advanced therapsids was "like a wheelbarrow", with the hindlimbs providing all the thrust while the forelimbs only steered the animal, in other words advanced therapsids were not as agile as either modern mammals or the early dinosaurs. So the idea that the main function of these mammal-like rib cages was to increase agility is doubtful.

Limb posture

The therapsids had sprawling forelimbs and semi-erect hindlimbs. This suggests that Carrier's constraint would have made it rather difficult for them to move and breathe at the same time, but not as difficult as it is for animals such as lizards, which have completely sprawling limbs. Advanced therapsids may therefore have been significantly less active than modern mammals of similar size and so may have had slower metabolisms overall or else been bradymetabolic (lower metabolism when at rest).

Brain

Mammals are noted for their large brain size relative to body size, compared to other animal groups. Recent findings suggest that the first brain area to expand was that involved in smell. Scientists scanned the skulls of early mammal species dating back to 190–200 million years ago and compared the brain case shapes to earlier pre-mammal species; they found that the brain area involved in the sense of smell was the first to enlarge. This change may have allowed these early mammals to hunt insects at night when dinosaurs were not active.

References

- Robert L. Carroll, *Vertebrate Paleontology and Evolution*, W. H. Freeman and Company, New York, 1988 ISBN 0-7167-1822-7. Chapters XVII through XXI
- Nicholas Hotton III, Paul D. MacLean, Jan J. Roth, and E. Carol Roth, editors, *The Ecology and Biology of Mammal-like Reptiles*, Smithsonian Institution Press, Washington and London, 1986 ISBN 0-87474-524-1
- T. S. Kemp, *The Origin and Evolution of Mammals*, Oxford University Press, New York, 2005 ISBN 0-19-850760-7
- Zofia Kielan-Jaworowska, Richard L. Cifelli, and Zhe-Xi Luo, *Mammals from the Age of Dinosaurs: Origins, Evolution, and Structure*, Columbia University Press, New York, 2004 ISBN 0-231-11918-6. Comprehensive coverage from the first mammals up to the time of the Cretaceous–Paleogene extinction event.
- Luo, Zhe-Xi (13 December 2007). "Transformation and diversification in early mammal evolution"[631] (PDF). *Nature*. **450** (7172): 1011–1019. Bibcode: 2007Natur.450.1011L[632]. doi: 10.1038/nature06277[633]. PMID 18075580[634]. Archived from the original[635] (PDF) on 2012-11-24. A survey article with 98 references to the scientific literature.

External links

- The Cynodontia[636] covers several aspects of the evolution of cynodonts into mammals, with plenty of references.
- Mammals[637], BBC Radio 4 discussion with Richard Corfield, Steve Jones & Jane Francis (*In Our Time*, Oct. 13, 2005)

Flowering plants

Flowering plant

Flowering plants	
Temporal range: Early Cretaceous - present, 130–0 Ma	
PreꞒ Ꞓ O S D C P T J K Pg N	
Magnolia virginiana sweet bay	
Scientific classification	
Kingdom:	Plantae
Subkingdom:	Embryophyta
(unranked):	Spermatophyta
(unranked):	**Angiosperms**
Groups (APG IV)[638]	

Basal angiosperms
- Amborellales
- Nymphaeales
- Austrobaileyales

Core angiosperms
- magnoliids
- Chloranthales
- monocots
- Ceratophyllales
- eudicots

Synonyms

- Anthophyta Cronquist[639]
- Angiospermae Lindl.
- Magnoliophyta Cronquist, Takht. & W.Zimm.
- Magnolicae Takht.[640]

The **flowering plants**, also known as **angiosperms**, **Angiospermae** or **Magnoliophyta**,[641] are the most diverse group of land plants, with 416 families, approximately 13,164 known genera and c. 295,383 known species. Like gymnosperms, angiosperms are seed-producing plants. However, they are distinguished from gymnosperms by characteristics including flowers, endosperm within the seeds, and the production of fruits that contain the seeds. Etymologically, *angiosperm* means a plant that produces seeds within an enclosure; in other words, a fruiting plant. The term comes from the Greek words *angeion* ("case" or "casing") and *sperma* ("seed").

The ancestors of flowering plants diverged from gymnosperms in the Triassic Period, 245 to 202 million years ago (mya), and the first flowering plants are known from 160 mya. They diversified extensively during the Lower Cretaceous, became widespread by 120 mya, and replaced conifers as the dominant trees from 100 to 60 mya.

Description

Angiosperm derived characteristics

Angiosperms differ from other seed plants in several ways, described in the table below. These distinguishing characteristics taken together have made the angiosperms the most diverse and numerous land plants and the most commercially important group to humans.[642]

Distinctive features of angiosperms

Feature	Description
Flowering organs	Flowers, the reproductive organs of flowering plants, are the most remarkable feature distinguishing them from the other seed plants. Flowers provided angiosperms with the means to have a more species-specific breeding system, and hence a way to evolve more readily into different species without the risk of crossing back with related species. Faster speciation enabled the Angiosperms to adapt to a wider range of ecological niches. This has allowed flowering plants to largely dominate terrestrial ecosystems.Wikipedia:Citation needed
Stamens with two pairs of pollen sacs	Stamens are much lighter than the corresponding organs of gymnosperms and have contributed to the diversification of angiosperms through time with adaptations to specialized pollination syndromes, such as particular pollinators. Stamens have also become modified through time to prevent self-fertilization, which has permitted further diversification, allowing angiosperms eventually to fill more niches.
Reduced male parts, three cells	The male gametophyte in angiosperms is significantly reduced in size compared to those of gymnosperm seed plants. The smaller size of the pollen reduces the amount of time between pollination — the pollen grain reaching the female plant — and fertilization. In gymnosperms, fertilization can occur up to a year after pollination, whereas in angiosperms, fertilization begins very soon after pollination. The shorter amount of time between pollination and fertilization allows angiosperms to produce seeds earlier after pollination than gymnosperms, providing angiosperms a distinct evolutionary advantage.
Closed carpel enclosing the ovules (carpel or carpels and accessory parts may become the fruit)	The closed carpel of angiosperms also allows adaptations to specialized pollination syndromes and controls. This helps to prevent self-fertilization, thereby maintaining increased diversity. Once the ovary is fertilized, the carpel and some surrounding tissues develop into a fruit. This fruit often serves as an attractant to seed-dispersing animals. The resulting cooperative relationship presents another advantage to angiosperms in the process of dispersal.
Reduced female gametophyte, seven cells with eight nuclei	The reduced female gametophyte, like the reduced male gametophyte, may be an adaptation allowing for more rapid seed set, eventually leading to such flowering plant adaptations as annual herbaceous life-cycles, allowing the flowering plants to fill even more niches.
Endosperm	In general, endosperm formation begins after fertilization and before the first division of the zygote. Endosperm is a highly nutritive tissue that can provide food for the developing embryo, the cotyledons, and sometimes the seedling when it first appears.

Vascular anatomy

The amount and complexity of tissue-formation in flowering plants exceeds that of gymnosperms. The vascular bundles of the stem are arranged such that the xylem and phloem form concentric rings.

In the dicotyledons, the bundles in the very young stem are arranged in an open ring, separating a central pith from an outer cortex. In each bundle, separating the xylem and phloem, is a layer of meristem or active formative tissue known

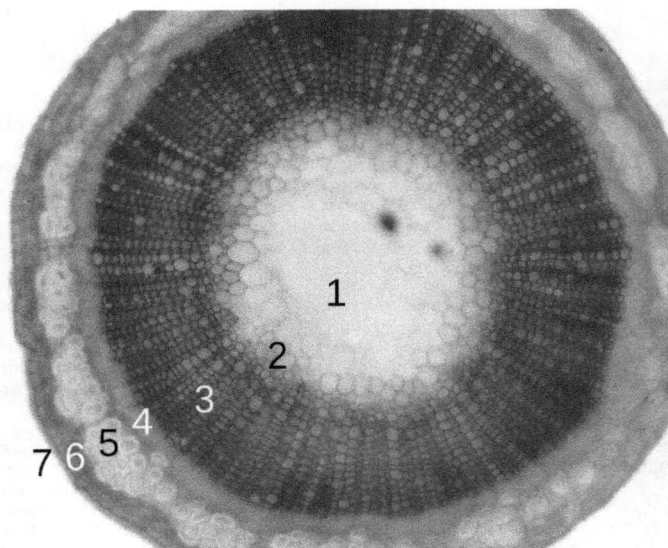

Figure 232: *Cross-section of a stem of the angiosperm flax:*
1. Pith, 2. Protoxylem, 3. Xylem I, 4. Phloem I, 5.
Sclerenchyma (bast fibre), 6. Cortex, 7. Epidermis

as cambium. By the formation of a layer of cambium between the bundles (interfascicular cambium), a complete ring is formed, and a regular periodical increase in thickness results from the development of xylem on the inside and phloem on the outside. The soft phloem becomes crushed, but the hard wood persists and forms the bulk of the stem and branches of the woody perennial. Owing to differences in the character of the elements produced at the beginning and end of the season, the wood is marked out in transverse section into concentric rings, one for each season of growth, called annual rings.

Among the monocotyledons, the bundles are more numerous in the young stem and are scattered through the ground tissue. They contain no cambium and once formed the stem increases in diameter only in exceptional cases.

Reproductive anatomy

The characteristic feature of angiosperms is the flower. Flowers show remarkable variation in form and elaboration, and provide the most trustworthy external characteristics for establishing relationships among angiosperm species. The function of the flower is to ensure fertilization of the ovule and development of fruit containing seeds. The floral apparatus may arise terminally

Figure 233: *A collection of flowers forming an inflorescence.*

on a shoot or from the axil of a leaf (where the petiole attaches to the stem). Occasionally, as in violets, a flower arises singly in the axil of an ordinary foliage-leaf. More typically, the flower-bearing portion of the plant is sharply distinguished from the foliage-bearing or vegetative portion, and forms a more or less elaborate branch-system called an inflorescence.

There are two kinds of reproductive cells produced by flowers. Microspores, which will divide to become pollen grains, are the "male" cells and are borne in the stamens (or microsporophylls). The "female" cells called megaspores, which will divide to become the egg cell (megagametogenesis), are contained in the ovule and enclosed in the carpel (or megasporophyll).

The flower may consist only of these parts, as in willow, where each flower comprises only a few stamens or two carpels. Usually, other structures are present and serve to protect the sporophylls and to form an envelope attractive to pollinators. The individual members of these surrounding structures are known as sepals and petals (or tepals in flowers such as *Magnolia* where sepals and petals are not distinguishable from each other). The outer series (calyx of sepals) is usually green and leaf-like, and functions to protect the rest of the flower, especially the bud. The inner series (corolla of petals) is, in general, white or brightly colored, and is more delicate in structure. It functions to attract insect or bird pollinators. Attraction is effected by color, scent,

Clariſ **LINN ÆI.M.D.**
METHODUS plantarum SEXUALIS
in SISTEMATE NATURÆ
deſcripta

Monandria.
Diandria.
Triandria.
Tetrandria.
Pentandria.
Hexandria.
Heptandria.
Octandria.
Enneandria.
Decandria
Dodecandria.
Icoſandria.
Polyandria.
Didynamia.
Tetradinamia.
Monadelphia.
Diadelphia.
Polyadelphia.
Syngeneſia.
Gynandria.
Monoecia.
Dioecia.
Polygamia.
Cryptogamia.

Lugd. bat: 1736

G.D.EHRET. Palatheidelb:
fecit & edidit

Figure 234: *From 1736, an illustration of Linnaean classification*

and nectar, which may be secreted in some part of the flower. The characteristics that attract pollinators account for the popularity of flowers and flowering plants among humans.

While the majority of flowers are perfect or hermaphrodite (having both pollen and ovule producing parts in the same flower structure), flowering plants have developed numerous morphological and physiological mechanisms to reduce or prevent self-fertilization. Heteromorphic flowers have short carpels and long stamens, or vice versa, so animal pollinators cannot easily transfer pollen to the pistil (receptive part of the carpel). Homomorphic flowers may employ a biochemical (physiological) mechanism called self-incompatibility to discriminate between self and non-self pollen grains. In other species, the male and female parts are morphologically separated, developing on different flowers.

Taxonomy

History of classification

The botanical term "Angiosperm", from the Ancient Greek αγγείον, *angeíon* (bottle, vessel) and σπέρμα, (seed), was coined in the form Angiospermae by Paul Hermann in 1690, as the name of one of his primary divisions of the plant kingdom. This included flowering plants possessing seeds enclosed in capsules,

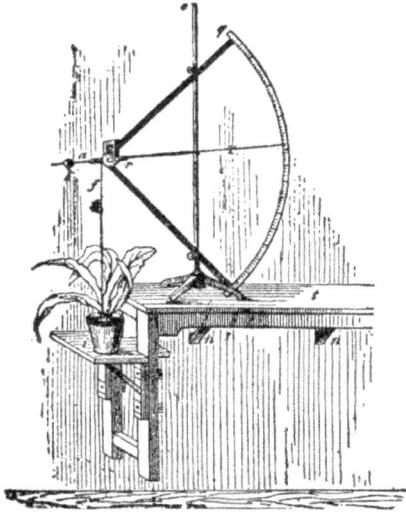

Figure 235: *An auxanometer, a device for measuring increase or rate of growth in plants*

distinguished from his Gymnospermae, or flowering plants with achenial or schizo-carpic fruits, the whole fruit or each of its pieces being here regarded as a seed and naked. The term and its antonym were maintained by Carl Linnaeus with the same sense, but with restricted application, in the names of the orders of his class Didynamia. Its use with any approach to its modern scope became possible only after 1827, when Robert Brown established the existence of truly naked ovules in the Cycadeae and Coniferae,[643] and applied to them the name Gymnosperms.Wikipedia:Citation needed From that time onward, as long as these Gymnosperms were, as was usual, reckoned as dicotyledonous flowering plants, the term Angiosperm was used antithetically by botanical writers, with varying scope, as a group-name for other dicotyledonous plants.

In 1851, Hofmeister discovered the changes occurring in the embryo-sac of flowering plants, and determined the correct relationships of these to the Cryptogamia. This fixed the position of Gymnosperms as a class distinct from Dicotyledons, and the term Angiosperm then gradually came to be accepted as the suitable designation for the whole of the flowering plants other than Gymnosperms, including the classes of Dicotyledons and Monocotyledons. This is the sense in which the term is used today.

In most taxonomies, the flowering plants are treated as a coherent group. The most popular descriptive name has been Angiospermae (Angiosperms), with

Anthophyta ("flowering plants") a second choice. These names are not linked to any rank. The Wettstein system and the Engler system use the name Angiospermae, at the assigned rank of subdivision. The Reveal system treated flowering plants as subdivision Magnoliophytina (Frohne & U. Jensen ex Reveal, Phytologia 79: 70 1996), but later split it to Magnoliopsida, Liliopsida, and Rosopsida. The Takhtajan system and Cronquist system treat this group at the rank of division, leading to the name Magnoliophyta (from the family name Magnoliaceae). The Dahlgren system and Thorne system (1992) treat this group at the rank of class, leading to the name Magnoliopsida. The APG system of 1998, and the later 2003[644] and 2009[645] revisions, treat the flowering plants as a clade called angiosperms without a formal botanical name. However, a formal classification was published alongside the 2009 revision in which the flowering plants form the Subclass Magnoliidae.[646]

The internal classification of this group has undergone considerable revision. The Cronquist system, proposed by Arthur Cronquist in 1968 and published in its full form in 1981, is still widely used but is no longer believed to accurately reflect phylogeny. A consensus about how the flowering plants should be arranged has recently begun to emerge through the work of the Angiosperm Phylogeny Group (APG), which published an influential reclassification of the angiosperms in 1998. Updates incorporating more recent research were published as the APG II system in 2003,[644] the APG III system in 2009,[645] and the APG IV system in 2016.

Traditionally, the flowering plants are divided into two groups,

• Dicotyledoneae or Magnoliopsida
• Monocotyledoneae or Liliopsida

which in the Cronquist system are called Magnoliopsida (at the rank of class, formed from the family name Magnoliaceae) and Liliopsida (at the rank of class, formed from the family name Liliaceae). Other descriptive names allowed by Article 16 of the ICBN include Dicotyledones or Dicotyledoneae, and Monocotyledones or Monocotyledoneae, which have a long history of use. In English a member of either group may be called a dicotyledon (plural dicotyledons) and monocotyledon (plural monocotyledons), or abbreviated, as dicot (plural dicots) and monocot (plural monocots). These names derive from the observation that the dicots most often have two cotyledons, or embryonic leaves, within each seed. The monocots usually have only one, but the rule is not absolute either way. From a broad diagnostic point of view, the number of cotyledons is neither a particularly handy nor a reliable character.

Recent studies, as by the APG, show that the monocots form a monophyletic group (clade) but that the dicots do not (they are paraphyletic). Nevertheless, the majority of dicot species do form a monophyletic group, called the eudicots

Figure 236: *Monocot (left) and dicot seedlings*

or tricolpates. Of the remaining dicot species, most belong to a third major clade known as the magnoliids, containing about 9,000 species. The rest include a paraphyletic grouping of early branching taxa known collectively as the basal angiosperms, plus the families Ceratophyllaceae and Chloranthaceae.

Modern classification

There are eight groups of living angiosperms:

- Basal angiosperms (ANA: *Amborella*, Nymphaeales, Austrobaileyales)
  - *Amborella*, a single species of shrub from New Caledonia;
  - Nymphaeales, about 80 species,[647] water lilies and Hydatellaceae;
  - Austrobaileyales, about 100 species of woody plants from various parts of the world
- Core angiosperms (Mesangiospermae)[646]
  - Chloranthales, several dozen species of aromatic plants with toothed leaves;
  - Magnoliids, about 9,000 species, characterized by trimerous flowers, pollen with one pore, and usually branching-veined leaves—for example magnolias, bay laurel, and black pepper;
  - Monocots, about 70,000 species, characterized by trimerous flowers, a single cotyledon, pollen with one pore, and usually parallel-veined leaves—for example grasses, orchids, and palms;

- *Ceratophyllum*, about 6 species of aquatic plants, perhaps most familiar as aquarium plants;
- Eudicots, about 175,000 species, characterized by 4- or 5-merous flowers, pollen with three pores, and usually branching-veined leaves—for example sunflowers, petunia, buttercup, apples, and oaks.

The exact relationship between these eight groups is not yet clear, although there is agreement that the first three groups to diverge from the ancestral angiosperm were Amborellales, Nymphaeales, and Austrobaileyales. The term basal angiosperms refers to these three groups. Among the remaining five groups (core angiosperms), the relationship between the three broadest of these groups (magnoliids, monocots, and eudicots) remains unclear. Zeng and colleagues (Fig. 1) describe four competing schemes.[648] Of these, eudicots and monocots are the largest and most diversified, with $\sim$ 75% and 20% of angiosperm species, respectively. Some analyses make the magnoliids the first to diverge, others the monocots.[649] *Ceratophyllum* seems to group with the eudicots rather than with the monocots. The 2016 Angiosperm Phylogeny Group revision (APG IV) retained the overall higher order relationship described in APG III.[645]

<templatestyles src="Template:Clade/styles.css" />

an-   <templatestyles src="Template:Clade/styles.css" />
giosperms

    Amborella

    <templatestyles src="Template:Clade/styles.css" />
    Nymphaeales

      <templatestyles src="Template:Clade/styles.css" />
      Austrobaileyales

        <templatestyles src="Template:Clade/styles.css" />
          <templatestyles src="Template:Clade/styles.css" />
            <templatestyles src="Template:Clade/styles.css" />

        magnoliids

        Chloranthales

          <templatestyles src="Template:Clade/styles.css" />
          monocots

            <templatestyles src="Template:Clade/styles.css" />
            Ceratophyllum

          eudicots

1. Phylogeny of the flowering plants, as of APG III (2009).[645]

<templatestyles src="Template:Clade/styles.css" />

an-   <templatestyles src="Template:Clade/styles.css" />
giosperms

      <templatestyles src="Template:Clade/styles.css" />
      <templatestyles src="Template:Clade/styles.css" />

       Amborella

       Nymphaeales

      <templatestyles src="Template:Clade/styles.css" />
      Austrobaileyales

       <templatestyles src="Template:Clade/styles.css" />
       monocots

        <templatestyles src="Template:Clade/styles.css" />
        Chloranthales

         <templatestyles src="Template:Clade/styles.css" />
         magnoliids

          <templatestyles src="Template:Clade/styles.css" />
          Ceratophyllum

          eudicots

2. Example of alternative phylogeny (2010)[649]

<templatestyles src="Template:Clade/styles.css" />

an-   <templatestyles src="Template:Clade/styles.css" />
giosperms   Amborellales

           <templatestyles src="Template:Clade/styles.css" />
           Nymphaeales

           <templatestyles src="Template:Clade/styles.css" />
           Austrobaileyales

           <templatestyles src="Template:Clade/styles.css" />
           <templatestyles src="Template:Clade/styles.css" />
           <templatestyles src="Template:Clade/styles.css" />

           magnoliids

           Chloranthales

           <templatestyles src="Template:Clade/styles.css" />
           monocots

           <templatestyles src="Template:Clade/styles.css" />

           Ceratophyllales

           eudicots

3. APG IV (2016)[638]

Detailed Cladogram of the Angiosperm Phylogeny Group (APG) IV classification.[638]

<templatestyles src="Template:Clade/styles.css" />

<templatestyles src="Template:Clade/styles.css" />

Amborellales Melikyan, Bobrov & Zaytzeva 1999

<templatestyles src="Template:Clade/styles.css" />

Nymphaeales Salisbury ex von Berchtold & Presl 1820

<templatestyles src="Template:Clade/styles.css" />

MesangiospermsAustrobaileyales Takhtajan ex Reveal 1992

<templatestyles src="Template:Clade/styles.css" />

Chloranthales Mart. 1835

<templatestyles src="Template:Clade/styles.css" />

<templatestyles src="Template:Clade/styles.css" />

Magnoliids

<templatestyles src="Template:Clade/styles.css" />

<templatestyles src="Template:Clade/styles.css" />

Canellales Cronquist 1957

Piperales von Berchtold & Presl 1820

<templatestyles src="Template:Clade/styles.css" />

Magnoliales de Jussieu ex von Berchtold & Presl 1820

Laurales de Jussieu ex von Berchtold & Presl 1820

`<templatestyles src="Template:Clade/styles.css" />`

Monocots `<templatestyles src="Template:Clade/styles.css" />`

Acorales Link 1835

`<templatestyles src="Template:Clade/styles.css" />`

Alismatales Brown ex von Berchtold & Presl 1820

`<templatestyles src="Template:Clade/styles.css" />`

Petrosaviales Takhtajan 1997

`<templatestyles src="Template:Clade/styles.css" />`
`<templatestyles src="Template:Clade/styles.css" />`

Dioscoreales Brown 1835

Pandanales Brown ex von Berchtold & Presl 1820

`<templatestyles src="Template:Clade/styles.css" />`

Liliales Perleb 1826

`<templatestyles src="Template:Clade/styles.css" />`

Asparagales Link 1829

Commelin-
ids `<templatestyles src="Template:Clade/styles.css" />`

Arecales Bromhead 1840

`<templatestyles src="Template:Clade/styles.css" />`

Poales Small 1903

`<templatestyles src="Template:Clade/styles.css" />`

Zingiberales Grisebach 1854

Commelinales de Mirbel ex von Berchtold & Presl 1820

`<templatestyles src="Template:Clade/styles.css" />`

Evolution

Fossilized spores suggest that higher plants (embryophytes) have lived on land for at least 475 million years. Early land plants reproduced sexually with flagellated, swimming sperm, like the green algae from which they evolved. An adaptation to terrestrialization was the development of upright meiosporangia for dispersal by spores to new habitats. This feature is lacking in the descendants of their nearest algal relatives, the Charophycean green algae. A later terrestrial adaptation took place with retention of the delicate, avascular sexual stage, the gametophyte, within the tissues of the vascular sporophyte. This occurred by spore germination within sporangia rather than spore release, as in non-seed plants. A current example of how this might have happened can be seen in the precocious spore germination in *Selaginella*, the spike-moss. The result for the ancestors of angiosperms was enclosing them in a case, the seed. The first seed bearing plants, like the ginkgo, and conifers (such as pines and firs), did not produce flowers. The pollen grains (male gametophytes) of *Ginkgo* and cycads produce a pair of flagellated, mobile sperm cells that "swim" down the developing pollen tube to the female and her eggs.

The apparently sudden appearance of nearly modern flowers in the fossil record initially posed such a problem for the theory of evolution that Charles Darwin called it an "*abominable mystery*". However, the fossil record has considerably grown since the time of Darwin, and recently discovered angiosperm fossils such as *Archaefructus*, along with further discoveries of fossil gymnosperms, suggest how angiosperm characteristics may have been acquired in a series of steps. Several groups of extinct gymnosperms, in particular seed ferns, have been proposed as the ancestors of flowering plants, but there is no continuous fossil evidence showing exactly how flowers evolved. Some older fossils, such as the upper Triassic *Sanmiguelia*, have been suggested. Based on current evidence, some propose that the ancestors of the angiosperms diverged from an unknown group of gymnosperms in the Triassic period (245–202 million years ago). Fossil angiosperm-like pollen from the Middle Triassic (247.2–242.0 Ma) suggests an older date for their origin. A close relationship between angiosperms and gnetophytes, proposed on the basis of morphological evidence, has more recently been disputed on the basis of molecular evidence that suggest gnetophytes are instead more closely related to other gymnosperms.Wikipedia:Citation needed

The evolution of seed plants and later angiosperms appears to be the result of two distinct rounds of whole genome duplication events. These occurred at 319[650] million years ago and 192[651] million years ago. Another possible whole genome duplication event at 160[652] million years ago perhaps created the ancestral line that led to all modern flowering plants. That event was studied

Figure 237: *Flowers of Malus sylvestris (crab apple)*

by sequencing the genome of an ancient flowering plant, *Amborella trichopoda*, and directly addresses Darwin's *"abominable mystery."*

The earliest known macrofossil confidently identified as an angiosperm, *Archaefructus liaoningensis*, is dated to about 125 million years BP (the Cretaceous period), whereas pollen considered to be of angiosperm origin takes the fossil record back to about 130 million years BP. However, one study has suggested that the early-middle Jurassic plant *Schmeissneria*, traditionally considered a type of ginkgo, may be the earliest known angiosperm, or at least a close relative. In addition, circumstantial chemical evidence has been found for the existence of angiosperms as early as 250 million years ago. Oleanane, a secondary metabolite produced by many flowering plants, has been found in Permian deposits of that age together with fossils of gigantopterids.[653] Gigantopterids are a group of extinct seed plants that share many morphological traits with flowering plants, although they are not known to have been flowering plants themselves.Wikipedia:Citation needed

In 2013 flowers encased in amber were found and dated 100 million years before present. The amber had frozen the act of sexual reproduction in the process of taking place. Microscopic images showed tubes growing out of pollen and penetrating the flower's stigma. The pollen was sticky, suggesting it was carried by insects.

Figure 238: *Flowers and leaves of Oxalis pes-caprae (Bermuda buttercup)*

Recent DNA analysis based on molecular systematics[654] showed that *Amborella trichopoda*, found on the Pacific island of New Caledonia, belongs to a sister group of the other flowering plants, and morphological studies[655] suggest that it has features that may have been characteristic of the earliest flowering plants.

The orders Amborellales, Nymphaeales, and Austrobaileyales diverged as separate lineages from the remaining angiosperm clade at a very early stage in flowering plant evolution.

The great angiosperm radiation, when a great diversity of angiosperms appears in the fossil record, occurred in the mid-Cretaceous (approximately 100 million years ago). However, a study in 2007 estimated that the division of the five most recent (the genus *Ceratophyllum*, the family Chloranthaceae, the eudicots, the magnoliids, and the monocots) of the eight main groups occurred around 140 million years ago. By the late Cretaceous, angiosperms appear to have dominated environments formerly occupied by ferns and cycadophytes, but large canopy-forming trees replaced conifers as the dominant trees only close to the end of the Cretaceous 66 million years ago or even later, at the beginning of the Tertiary. The radiation of herbaceous angiosperms occurred

Figure 239: *Two bees on a flower head of Creeping Thistle, Cirsium arvense*

much later. Yet, many fossil plants recognizable as belonging to modern families (including beech, oak, maple, and magnolia) had already appeared by the late Cretaceous.

It has been proposed that the swift rise of angiosperms to dominance was facilitated by a reduction in their genome size. During the early Cretaceous period, only angiosperms underwent rapid genome downsizing, while genome sizes of ferns and gymnosperms remained unchanged. Smaller genomes–and smaller nuclei–allow for faster rates of cell division and smaller cells. Thus, species with smaller genomes can pack more, smaller cells–in particular veins and stomata–into a given leaf volume. Genome downsizing therefore facilitated higher rates of leaf gas exchange (transpiration and photosynthesis) and faster rates of growth. This would have countered some of the negative physiological effects of genome duplications, facilitated increased uptake of carbon dioxide despite concurrent declines in atmospheric CO_2 concentrations, and allowed the flowering plants to outcompete other land plants.

It is generally assumed that the function of flowers, from the start, was to involve mobile animals in their reproduction processes. That is, pollen can be scattered even if the flower is not brightly colored or oddly shaped in a way that attracts animals; however, by expending the energy required to create such traits, angiosperms can enlist the aid of animals and, thus, reproduce more efficiently.

Island genetics provides one proposed explanation for the sudden, fully developed appearance of flowering plants. Island genetics is believed to be a common source of speciation in general, especially when it comes to radical adaptations that seem to have required inferior transitional forms. Flowering plants may have evolved in an isolated setting like an island or island chain, where the plants bearing them were able to develop a highly specialized relationship with some specific animal (a wasp, for example). Such a relationship, with a hypothetical wasp carrying pollen from one plant to another much the way fig wasps do today, could result in the development of a high degree of specialization in both the plant(s) and their partners. Note that the wasp example is not incidental; bees, which, it is postulated, evolved specifically due to mutualistic plant relationships, are descended from wasps.

Animals are also involved in the distribution of seeds. Fruit, which is formed by the enlargement of flower parts, is frequently a seed-dispersal tool that attracts animals to eat or otherwise disturb it, incidentally scattering the seeds it contains (see frugivory). Although many such mutualistic relationships remain too fragile to survive competition and to spread widely, flowering proved to be an unusually effective means of reproduction, spreading (whatever its origin) to become the dominant form of land plant life.

Flower ontogeny uses a combination of genes normally responsible for forming new shoots.[656] The most primitive flowers probably had a variable number of flower parts, often separate from (but in contact with) each other. The flowers tended to grow in a spiral pattern, to be bisexual (in plants, this means both male and female parts on the same flower), and to be dominated by the ovary (female part). As flowers evolved, some variations developed parts fused together, with a much more specific number and design, and with either specific sexes per flower or plant or at least "ovary-inferior".

Flower evolution continues to the present day; modern flowers have been so profoundly influenced by humans that some of them cannot be pollinated in nature. Many modern domesticated flower species were formerly simple weeds, which sprouted only when the ground was disturbed. Some of them tended to grow with human crops, perhaps already having symbiotic companion plant relationships with them, and the prettiest did not get plucked because of their beauty, developing a dependence upon and special adaptation to human affection.[657]

A few paleontologists have also proposed that flowering plants, or angiosperms, might have evolved due to interactions with dinosaurs. One of the idea's strongest proponents is Robert T. Bakker. He proposes that herbivorous dinosaurs, with their eating habits, provided a selective pressure on plants, for which adaptations either succeeded in deterring or coping with predation by herbivores.

Figure 240: *A poster of twelve different species of flowers of the Asteraceae family*

In August 2017, scientists presented a detailed description and 3D model image of what the first flower possibly looked like, and presented the hypothesis that it may have lived about 140 million years ago.

A Bayesian analysis of 52 angiosperm taxa suggested that the crown group of angiosperms evolved between 178[658] million years ago and 198[659] million years ago.[660]

Diversity

The number of species of flowering plants is estimated to be in the range of 250,000 to 400,000. This compares to around 12,000 species of moss or 11,000 species of pteridophytes,[661] showing that the flowering plants are much more diverse. The number of families in APG (1998) was 462. In APG II[644] (2003) it is not settled; at maximum it is 457, but within this number there are 55 optional segregates, so that the minimum number of families in this system is 402. In APG III (2009) there are 415 families.[645]

The diversity of flowering plants is not evenly distributed. Nearly all species belong to the eudicot (75%), monocot (23%), and magnoliid (2%) clades. The remaining 5 clades contain a little over 250 species in total; i.e. less than 0.1% of flowering plant diversity, divided among 9 families. The 43 most-diverse of 443 families of flowering plants by species, in their APG circumscriptions, are

Figure 241: *Lupinus pilosus*

Figure 242: *Bud of a pink rose*

1. Asteraceae or Compositae (daisy family): 22,750 species;
2. Orchidaceae (orchid family): 21,950;
3. Fabaceae or Leguminosae (bean family): 19,400;
4. Rubiaceae (madder family): 13,150;
5. Poaceae or Gramineae (grass family): 10,035;
6. Lamiaceae or Labiatae (mint family): 7,175;
7. Euphorbiaceae (spurge family): 5,735;
8. Melastomataceae or Melastomaceae (melastome family): 5,005;
9. Myrtaceae (myrtle family): 4,625;
10. Apocynaceae (dogbane family): 4,555;
11. Cyperaceae (sedge family): 4,350;
12. Malvaceae (mallow family): 4,225;
13. Araceae (arum family): 4,025;
14. Ericaceae (heath family): 3,995;
15. Gesneriaceae (gesneriad family): 3,870;
16. Apiaceae or Umbelliferae (parsley family): 3,780;
17. Brassicaceae or Cruciferae (cabbage family): 3,710:
18. Piperaceae (pepper family): 3,600;
19. Bromeliaceae (bromeliad family): 3,540;
20. Acanthaceae (acanthus family): 3,500;
21. Rosaceae (rose family): 2,830;
22. Boraginaceae (borage family): 2,740;
23. Urticaceae (nettle family): 2,625;
24. Ranunculaceae (buttercup family): 2,525;
25. Lauraceae (laurel family): 2,500;
26. Solanaceae (nightshade family): 2,460;
27. Campanulaceae (bellflower family): 2,380;
28. Arecaceae (palm family): 2,361;
29. Annonaceae (custard apple family): 2,220;
30. Caryophyllaceae (pink family): 2,200;
31. Orobanchaceae (broomrape family): 2,060;
32. Amaranthaceae (amaranth family): 2,050;
33. Iridaceae (iris family): 2,025;
34. Aizoaceae or Ficoidaceae (ice plant family): 2,020;
35. Rutaceae (rue family): 1,815;
36. Phyllanthaceae (phyllanthus family): 1,745;
37. Scrophulariaceae (figwort family): 1,700;
38. Gentianaceae (gentian family): 1,650;
39. Convolvulaceae (bindweed family): 1,600;
40. Proteaceae (protea family): 1,600;
41. Sapindaceae (soapberry family): 1,580;
42. Cactaceae (cactus family): 1,500;

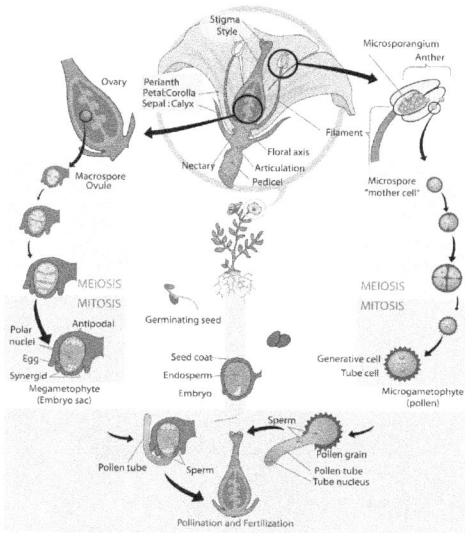

Figure 243: *Angiosperm life cycle*

43. Araliaceae (*Aralia* or ivy family): 1,450.

Of these, the Orchidaceae, Poaceae, Cyperaceae, Araceae, Bromeliaceae, Arecaceae, and Iridaceae are monocot families; Piperaceae, Lauraceae, and Annonaceae are magnoliid dicots; the rest of the families are eudicots.

Reproduction

Fertilization and embryogenesis

Double fertilization refers to a process in which two sperm cells fertilize cells in the ovule. This process begins when a pollen grain adheres to the stigma of the pistil (female reproductive structure), germinates, and grows a long pollen tube. While this pollen tube is growing, a haploid generative cell travels down the tube behind the tube nucleus. The generative cell divides by mitosis to produce two haploid (*n*) sperm cells. As the pollen tube grows, it makes its way from the stigma, down the style and into the ovary. Here the pollen tube reaches the micropyle of the ovule and digests its way into one of the synergids, releasing its contents (which include the sperm cells). The synergid that the cells were released into degenerates and one sperm makes its way to fertilize the egg cell, producing a diploid (2*n*) zygote. The second sperm cell fuses with both central cell nuclei, producing a triploid (3*n*) cell. As the zygote develops

Figure 244: *The fruit of the Aesculus or Horse Chestnut tree*

into an embryo, the triploid cell develops into the endosperm, which serves as the embryo's food supply. The ovary will now develop into a fruit and the ovule will develop into a seed.

Fruit and seed

As the development of embryo and endosperm proceeds within the embryo sac, the sac wall enlarges and combines with the nucellus (which is likewise enlarging) and the integument to form the *seed coat*. The ovary wall develops to form the fruit or pericarp, whose form is closely associated with type of seed dispersal system.

Frequently, the influence of fertilization is felt beyond the ovary, and other parts of the flower take part in the formation of the fruit, e.g., the floral receptacle in the apple, strawberry, and others.Wikipedia:Citation needed

The character of the seed coat bears a definite relation to that of the fruit. They protect the embryo and aid in dissemination; they may also directly promote germination. Among plants with indehiscent fruits, in general, the fruit provides protection for the embryo and secures dissemination. In this case, the seed coat is only slightly developed. If the fruit is dehiscent and the seed is exposed, in general, the seed-coat is well developed, and must discharge the functions otherwise executed by the fruit.Wikipedia:Citation needed

Meiosis

Flowering plants generate gametes using a specialized cell division called meiosis. Meiosis takes place in the ovule (a structure within the ovary that is located within the pistil at the center of the flower) (see diagram labeled "Angiosperm lifecycle"). A diploid cell (megaspore mother cell) in the ovule undergoes meiosis (involving two successive cell divisions) to produce four cells (megaspores) with haploid nuclei.[662] One of these four cells (megaspore) then undergoes three successive mitotic divisions to produce an immature embryo sac (megagametophyte) with eight haploid nuclei. Next, these nuclei are segregated into separate cells by cytokinesis to producing 3 antipodal cells, 2 synergid cells and an egg cell. Two polar nuclei are left in the central cell of the embryo sac.Wikipedia:Citation needed

Pollen is also produced by meiosis in the male anther (microsporangium). During meiosis, a diploid microspore mother cell undergoes two successive meiotic divisions to produce 4 haploid cells (microspores or male gametes). Each of these microspores, after further mitoses, becomes a pollen grain (microgametophyte) containing two haploid generative (sperm) cells and a tube nucleus. When a pollen grain makes contact with the female stigma, the pollen grain forms a pollen tube that grows down the style into the ovary. In the act of fertilization, a male sperm nucleus fuses with the female egg nucleus to form a diploid zygote that can then develop into an embryo within the newly forming seed. Upon germination of the seed, a new plant can grow and mature.Wikipedia:Citation needed

The adaptive function of meiosis is currently a matter of debate. A key event during meiosis in a diploid cell is the pairing of homologous chromosomes and homologous recombination (the exchange of genetic information) between homologous chromosomes. This process promotes the production of increased genetic diversity among progeny and the recombinational repair of damages in the DNA to be passed on to progeny. To explain the adaptive function of meiosis in flowering plants, some authors emphasize diversity and others emphasize DNA repair.

Apomixis

Apomixis (reproduction via asexually formed seeds) is found naturally in about 2.2% of angiosperm genera. One type of apomixis, gametophytic apomixis found in a dandelion species involves formation of an unreduced embryo sac due to incomplete meiosis (apomeiosis) and development of an embryo from the unreduced egg inside the embryo sac, without fertilization (parthenogenesis).Wikipedia:Citation needed

Uses

Agriculture is almost entirely dependent on angiosperms, which provide virtually all plant-based food, and also provide a significant amount of livestock feed. Of all the families of plants, the Poaceae, or grass family (providing grains), is by far the most important, providing the bulk of all feedstocks (rice, maize, wheat, barley, rye, oats, pearl millet, sugar cane, sorghum). The Fabaceae, or legume family, comes in second place. Also of high importance are the Solanaceae, or nightshade family (potatoes, tomatoes, and peppers, among others); the Cucurbitaceae, or gourd family (including pumpkins and melons); the Brassicaceae, or mustard plant family (including rapeseed and the innumerable varieties of the cabbage species *Brassica oleracea*); and the Apiaceae, or parsley family. Many of our fruits come from the Rutaceae, or rue family (including oranges, lemons, grapefruits, etc.), and the Rosaceae, or rose family (including apples, pears, cherries, apricots, plums, etc.).Wikipedia:Citation needed

In some parts of the world, certain single species assume paramount importance because of their variety of uses, for example the coconut (*Cocos nucifera*) on Pacific atolls, and the olive (*Olea europaea*) in the Mediterranean region.

Flowering plants also provide economic resources in the form of wood, paper, fiber (cotton, flax, and hemp, among others), medicines (digitalis, camphor), decorative and landscaping plants, and many other uses. The main area in which they are surpassed by other plants—namely, coniferous trees (Pinales), which are non-flowering (gymnosperms)—is timber and paper production.[663]

Bibliography

<templatestyles src="Template:Refbegin/styles.css" />

Articles, books and chapters

- ⊕ This article incorporates text from a publication now in the public domain: Chisholm, Hugh, ed. (1911). "Angiosperms". *Encyclopædia Britannica* (11th ed.). Cambridge University Press.
- APG (2003). "An update of the Angiosperm Phylogeny Group classification for the orders and families of flowering plants: APG II"[664]. *Botanical Journal of the Linnean Society*. **141** (4): 399–436. doi: 10.1046/j.1095-8339.2003.t01-1-00158.x[665].

- APG (2009). "An update of the Angiosperm Phylogeny Group classification for the orders and families of flowering plants: APG III"[666]. *Botanical Journal of the Linnean Society*. **161** (2): 105–121. doi: 10.1111/j.1095-8339.2009.00996.x[667]. Retrieved 2010-12-10.
- APG (2016). "An update of the Angiosperm Phylogeny Group classification for the orders and families of flowering plants: APG IV"[668]. *Botanical Journal of the Linnean Society*. **181** (1): 1–20. doi: 10.1111/boj.12385[669]. Retrieved 2016-05-20.
- Becker, Kenneth M. (February 1973). "A Comparison of Angiosperm Classification Systems". *Taxon*. **22** (1): 19–50. doi: 10.2307/1218032[670].
- Bell, Adrian D. (2008) [1991]. *Plant Form. An Illustrated Guide to Flowering Plant Morphology*[671]. Portland, Oregon: Timber Press. ISBN 978-0-88192-850-1.
  - 1st edition published by Oxford University Press in 1991[672] ISBN 978-0-19854-219-3
- Bell, C.D.; Soltis, D.E.; Soltis, P.S. (2010). "The Age and Diversification of the Angiosperms Revisited". *American Journal of Botany*. **97** (8): 1296–1303. doi: 10.3732/ajb.0900346[673]. PMID 21616882[674].
- Chase, Mark W. & Reveal, James L. (2009). "A phylogenetic classification of the land plants to accompany APG III". *Botanical Journal of the Linnean Society*. **161** (2): 122–127. doi: 10.1111/j.1095-8339.2009.01002.x[675].
- Cromie, William J. (December 16, 1999). "Oldest Known Flowering Plants Identified By Genes"[676]. Harvard University Gazette.
- Cronquist, Arthur (October 1960). "The divisions and classes of plants". *The Botanical Review*. **26** (4): 425–482. doi: 10.1007/BF02940572[677].
- Cronquist, Arthur (1981). *An Integrated System of Classification of Flowering Plants*. New York: Columbia Univ. Press. ISBN 0-231-03880-1.
- Dahlgren, R. M. T. (February 1980). "A revised system of classification of the angiosperms". *Botanical Journal of the Linnean Society*. **80** (2): 91–124. doi: 10.1111/j.1095-8339.1980.tb01661.x[678].
- Dahlgren, Rolf (February 1983). "General aspects of angiosperm evolution and macrosystematics". *Nordic Journal of Botany*. **3** (1): 119–149. doi: 10.1111/j.1756-1051.1983.tb01448.x[679].
- Dilcher, D. (2000). "Toward a new synthesis: Major evolutionary trends in the angiosperm fossil record"[680]. *Proceedings of the National Academy of Sciences*. **97** (13): 7030. Bibcode: 2000PNAS...97.7030D[681]. doi: 10.1073/pnas.97.13.7030[682]. PMC 34380[680] ⓐ .
- Dilcher, David L; Cronquist, Arthur; Zimmermann, Martin Huldrych; Stevens, Peter; Stevenson, Dennis William; Berry, Paul E. (8 March 2016). "Angiosperm"[683]. *Encyclopedia Britannica*. Retrieved 31 January 2017.

- Heywood, V. H., Brummitt, R. K., Culham, A. & Seberg, O. (2007). *Flowering Plant Families of the World*. Richmond Hill, Ontario, Canada: Firefly Books. ISBN 1-55407-206-9.
- Hill, Christopher; Crane, Peter (January 1982). "Evolutionary Cladistics and the origin of Angiosperms"[684]. In Joysey, Kenneth Alan; Friday, A.E. *Problems of Phylogenetic Reconstruction*. Special Volumes. **21**. London: Systematics Association. pp. 269–361. ISBN 978-0-12-391250-3.
- Lersten, Nels R. (2004). *Flowering plant embryology with emphasis on economic species*[685]. Ames, Iowa: Blackwell Pub. ISBN 9780470752678.
- Pooja (2004). *Angiosperms*[686]. New Delhi: Discovery. ISBN 9788171417889. Retrieved 7 January 2016.
- Raven, P.H., R.F. Evert, S.E. Eichhorn. *Biology of Plants*, 7th Edition. W.H. Freeman. 2004
- Sattler, R. 1973. *Organogenesis of Flowers. A Photographic Text-Atlas*. University of Toronto Press.
- Simpson, Michael G. (2010). *Plant Systematics*[687] (2nd ed.). Academic Press. ISBN 9780080922089.
- Soltis, Pamela S; Soltis, Douglas E (April 2016). "Ancient WGD events as drivers of key innovations in angiosperms". *Current Opinion in Plant Biology*. **30**: 159–165. doi: 10.1016/j.pbi.2016.03.015[688].
- Takhtajan, A. (June 1964). "The Taxa of the Higher Plants above the Rank of Order". *Taxon*. **13** (5): 160–164. doi: 10.2307/1216134[689]. JSTOR 10.2307/1216134[690].
- Takhtajan, A. (July–September 1980). "Outline of the Classification of Flowering Plants (Magnoliophyta)". *Botanical Review*. **46** (3): 225–359. doi: 10.1007/bf02861558[691]. JSTOR 10.2307/4353970[692].
- Zeng, Liping; Zhang, Qiang; Sun, Renran; Kong, Hongzhi; Zhang, Ning; Ma, Hong (24 September 2014). "Resolution of deep angiosperm phylogeny using conserved nuclear genes and estimates of early divergence times". *Nature Communications*. **5** (4956). Bibcode: 2014NatCo...5E4956Z[693]. doi: 10.1038/ncomms5956[694].

Websites

- Cole, Theodor C.H.; Hilger, Harmut H.; Stevens, Peter F. (2017). "Angiosperm Phylogeny Poster – Flowering Plant Systematics"[695] (PDF).
- Watson, L.; Dallwitz, M.J. (1992). "The Families of Flowering Plants: Descriptions, Illustrations, Identification, and Information Retrieval"[696]. 14 December 2000. Archived from the original[697] on 2014-08-02.
- *Flowering plant*[698] at the Encyclopedia of Life

External links

Wikimedia Commons has media related to *Magnoliophyta*.

Wikispecieshas information related to *Magnoliophyta*

The Wikibook *Dichotomous Key*has a page on the topic of: *Magnoliophyta*

Gymnosperm

Gymnospermae	
Temporal range: Carboniferous - Present Pre꞊Є OSD C P T J K PgN	

Various gymnosperms.	
Scientific classification	
Kingdom:	Plantae
Subkingdom:	Embryophyta
(unranked):	**Gymnospermae** (paraphyletic)
Divisions	
Pinophyta (or Coniferophyta) - Conifers Ginkgophyta - *Ginkgo* Cycadophyta - Cycads Gnetophyta - *Gnetum, Ephedra, Welwitschia*	

Figure 245: *Encephalartos sclavoi cone, about 30 cm long*

The **gymnosperms** are a group of seed-producing plants that includes conifers, cycads, *Ginkgo*, and gnetophytes. The term "gymnosperm" comes from the Greek composite word γυμνόσπερμος (γυμνός gymnos, "naked" and σπέρμα sperma, "seed"), meaning "naked seeds". The name is based on the unenclosed condition of their seeds (called ovules in their unfertilized state). The non-encased condition of their seeds stands in contrast to the seeds and ovules of flowering plants (angiosperms), which are enclosed within an ovary. Gymnosperm seeds develop either on the surface of scales or leaves, which are often modified to form cones, or solitary as in Yew, *Torreya*, *Ginkgo*.

The gymnosperms and angiosperms together compose the spermatophytes or seed plants. The gymnosperms are divided into six phyla. Organisms that belong to the Cycadophyta, Ginkgophyta, Gnetophyta, and Pinophyta (also known as Coniferophyta) phyla are still in existence while those in the Pteridospermales and Cordaitales phyla are now extinct.

By far the largest group of living gymnosperms are the conifers (pines, cypresses, and relatives), followed by cycads, gnetophytes (*Gnetum*, *Ephedra* and *Welwitschia*), and *Ginkgo biloba* (a single living species).

Roots in some genera have fungal association with roots in the form of mycorrhiza (*Pinus*), while in some others (*Cycas*) small specialised roots called coralloid roots are associated with nitrogen-fixing cyanobacteria.

Classification

In early classification schemes, the gymnosperms (Gymnospermae) were regarded as a "natural" group. There is conflicting evidence on the question of whether the living gymnosperms form a clade. The fossil record of gymnosperms includes many distinctive taxa that do not belong to the four modern groups, including seed-bearing trees that have a somewhat fern-like vegetative morphology (the so-called "seed ferns" or pteridosperms.)[699] When fossil gymnosperms such as Bennettitales, *Caytonia* and the glossopterids are considered, it is clear that angiosperms are nested within a larger gymnosperm clade, although which group of gymnosperms is their closest relative remains unclear.

For the most recent classification on extant gymnosperms see Christenhusz *et al.* (2011). There are 12 families, 83 known genera with a total of ca 1080 known species (Christenhusz & Byng 2016).

Subclass **Cycadidae**

- Order **Cycadales**
  - Family **Cycadaceae**: *Cycas*
  - Family **Zamiaceae**: *Dioon, Bowenia, Macrozamia, Lepidozamia, Encephalartos, Stangeria, Ceratozamia, Microcycas, Zamia.*

Subclass **Ginkgoidae**

- Order **Ginkgoales**
  - Family **Ginkgoaceae**: *Ginkgo*

Subclass **Gnetidae**

- Order **Welwitschiales**
  - Family **Welwitschiaceae**: *Welwitschia*
- Order **Gnetales**
  - Family **Gnetaceae**: *Gnetum*
- Order **Ephedrales**
  - Family **Ephedraceae**: *Ephedra*

Subclass **Pinidae**

- Order **Pinales**
  - Family **Pinaceae**: *Cedrus, Pinus, Cathaya, Picea, Pseudotsuga, Larix, Pseudolarix, Tsuga, Nothotsuga, Keteleeria, Abies*
- Order **Araucariales**
  - Family **Araucariaceae**: *Araucaria, Wollemia, Agathis*

- Family **Podocarpaceae**: *Phyllocladus, Lepidothamnus, Prumnopitys, Sundacarpus, Halocarpus, Parasitaxus, Lagarostrobos, Manoao, Saxegothaea, Microcachrys, Pherosphaera, Acmopyle, Dacrycarpus, Dacrydium, Falcatifolium, Retrophyllum, Nageia, Afrocarpus, Podocarpus*
- Order **Cupressales**
- Family **Sciadopityaceae**: *Sciadopitys*
- Family **Cupressaceae**: *Cunninghamia, Taiwania, Athrotaxis, Metasequoia, Sequoia, Sequoiadendron, Cryptomeria, Glyptostrobus, Taxodium, Papuacedrus, Austrocedrus, Libocedrus, Pilgerodendron, Widdringtonia, Diselma, Fitzroya, Callitris* (incl. *Actinostrobus* and *Neocallitropsis*), *Thujopsis, Thuja, Fokienia, Chamaecyparis, Callitropsis, Cupressus, Juniperus, Xanthocyparis, Calocedrus, Tetraclinis, Platycladus, Microbiota*
- Family **Taxaceae**: *Austrotaxus, Pseudotaxus, Taxus, Cephalotaxus, Amentotaxus, Torreya*

Diversity and origin

There are more than 1000 extant or currently living species of gymnosperms in 88 plant genera belonging to 14 plant families.

It is widely accepted that the gymnosperms originated in the late Carboniferous period, replacing the lycopsid rainforests of the tropical region.[700] This appears to have been the result of a whole genome duplication event around 319[701] million years ago.[702] Early characteristics of seed plants were evident in fossil progymnosperms of the late Devonian period around 383 million years ago. It has been suggested that during the mid-Mesozoic era, pollination of some extinct groups of gymnosperms was by extinct species of scorpionflies that had specialized proboscis for feeding on pollination drops. The scorpionflies likely engaged in pollination mutualisms with gymnosperms, long before the similar and independent coevolution of nectar-feeding insects on angiosperms. Evidence has also been found that mid-Mesozoic gymnosperms were pollinated by Kalligrammatid lacewings, a now-extinct genus with members which (in an example of convergent evolution) resembled the modern butterflies that arose far later.

Conifers are by far the most abundant extant group of gymnosperms with six to eight families, with a total of 65-70 genera and 600-630 species (696 accepted names).[703] Conifers are woody plants and most are evergreens.[704] The leaves of many conifers are long, thin and needle-like, other species, including most Cupressaceae and some Podocarpaceae, have flat, triangular scale-like leaves. *Agathis* in Araucariaceae and *Nageia* in Podocarpaceae have broad, flat strap-shaped leaves.

Figure 246: *Zamia integrifolia, a cycad native to Florida*

Cycads are the next most abundant group of gymnosperms, with two or three families, 11 genera, and approximately 338 species. A majority of cycads are native to tropical climates and are most abundantly found in regions near the equator. The other extant groups are the 95-100 species of Gnetales and one species of Ginkgo.

Uses

Gymnosperms have major economic uses. Pine, fir, spruce, and cedar are all examples of conifers that are used for lumber, paper production, and resin. Some other common uses for gymnosperms are soap, varnish, nail polish, food, gum, and perfumes.

Life cycle

Gymnosperms, like all vascular plants, have a sporophyte-dominant life cycle, which means they spend most of their life cycle with diploid cells, while the gametophyte (gamete-bearing phase) is relatively short-lived. Two spore types, microspores and megaspores, are typically produced in pollen cones or ovulate cones, respectively. Gametophytes, as with all heterosporous plants, develop within the spore wall. Pollen grains (microgametophytes) mature from

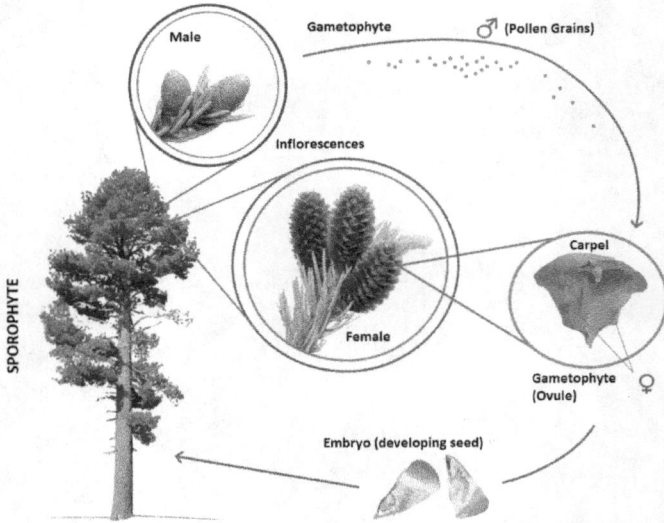

Figure 247: *Example of gymnosperm lifecycle*

microspores, and ultimately produce sperm cells. Megagametophytes develop from megaspores and are retained within the ovule. Gymnosperms produce multiple archegonia, which produce the female gamete. During pollination, pollen grains are physically transferred between plants from the pollen cone to the ovule. Pollen is usually moved by wind or insects. Whole grains enter each ovule through a microscopic gap in the ovule coat (integument) called the micropyle. The pollen grains mature further inside the ovule and produce sperm cells. Two main modes of fertilization are found in gymnosperms. Cycads and *Ginkgo* have motile sperm that swim directly to the egg inside the ovule, whereas conifers and gnetophytes have sperm with no flagella that are moved along a pollen tube to the egg. After syngamy (joining of the sperm and egg cell), the zygote develops into an embryo (young sporophyte). More than one embryo is usually initiated in each gymnosperm seed. The mature seed comprises the embryo and the remains of the female gametophyte, which serves as a food supply, and the seed coat.

Genetics

The first published sequenced genome for any gymnosperm was the genome of *Picea abies* in 2013.

External links

> Wikimedia Commons has media related to *Gymnosperms*.

- Gymnosperm Database[705]
- Gymnosperms on the Tree of Life[706]
- Albert Seward (1911). "Gymnosperms". *Encyclopædia Britannica* (11th ed.).

Social insects

Eusociality

Eusociality (from Greek εὖ *eu* "good" and social), the highest level of organization of animal sociality, is defined by the following characteristics: cooperative brood care (including care of offspring from other individuals), overlapping generations within a colony of adults, and a division of labor into reproductive and non-reproductive groups. The division of labor creates specialized behavioral groups within an animal society which are sometimes called castes. Eusociality is distinguished from all other social systems because individuals of at least one caste usually lose the ability to perform at least one behavior characteristic of individuals in another caste.

Eusociality exists in certain insects, crustaceans and mammals. It is mostly observed and studied in the Hymenoptera (ants, bees, and wasps) and in the termites. A colony has caste differences: queens and reproductive males take the roles of the sole reproducers, while soldiers and workers work together to create a living situation favorable for the brood. In addition to Hymenoptera and Isoptera, there are two known eusocial vertebrates among rodents: the naked mole-rat and the Damaraland mole-rat. Some shrimps, such as *Synalpheus regalis*, are also eusocial. E. O. Wilson has claimed that humans are eusocial, but his arguments have been disputed by a large number of evolutionary biologists, who note that humans do not have division of reproductive labor.

Several other levels of animal sociality have been distinguished. These include presocial (solitary but social), subsocial, and parasocial (including communal, quasisocial, and semisocial).

Figure 248: *Termite mound: termites developed euso-*
ciality in the Jurassic period, over 145 million years ago.

History

The term "eusocial" was introduced in 1966 by Suzanne Batra who used it to describe nesting behavior in Halictine bees. Batra observed the cooperative behavior of the bees, males and females alike, as they took responsibility for at least one duty (i.e. burrowing, cell construction, oviposition) within the colony. The cooperativeness was essential as the activity of one labor division greatly influenced the activity of another.

For example, the size of pollen balls, a source of food, depended on when the egg-laying females oviposited. If the provisioning by pollen collectors was incomplete by the time the egg-laying female occupied a cell and oviposited, the size of the pollen balls would be small, leading to small offspring. Batra applied this term to species in which a colony is started by a single individual. Batra described other species, where the founder is accompanied by numerous helpers—as in a swarm of bees or ants—as "hypersocial".

In 1969, Charles D. Michener further expanded Batra's classification with his comparative study of social behavior in bees. He observed multiple species of bees (Apoidea) in order to investigate the different levels of animal sociality, all of which are different stages that a colony may pass through. Eusociality,

Figure 249: *Co-operative brood rearing in honeybees*

which is the highest level of animal sociality a species can attain, specifically had three characteristics that distinguished it from the other levels:

1. "Egg-layers and worker-like individuals among adult females" (division of labor)
2. The overlap of generations (mother and adult offspring)
3. Cooperative work on the cells of the bees' honeycomb

E. O. Wilson then extended the terminology to include other social insects; such as ants, wasps, and termites. Originally, it was defined to include organisms (only invertebrates) that had the following three features:

1. Reproductive division of labor (with or without sterile castes)
2. Overlapping generations
3. Cooperative care of young

As eusociality became a recognized widespread phenomenon, however, it was also discovered in a group of chordates, the mole-rats. Further research also distinguished another possibly important criterion for eusociality, known as "the point of no return". This is characterized by eusocial individuals that become fixed into one behavioral group, which usually occurs before reproductive maturity. This prevents them from transitioning between behavioral groups and creates an animal society that is truly dependent on each other for survival and reproductive success. For many insects, this irreversibility has

Figure 250: *A swarming meat-eater ant colony*

changed the anatomy of the worker caste, which is sterile and provides support for the reproductive caste.

Taxonomic range

Most eusocial societies exist in arthropods, while a few are found in mammals.

In insects

The order Hymenoptera contains the largest group of eusocial insects, including ants, bees, and wasps—those with reproductive "queens" and more or less sterile "workers" and/or "soldiers" that perform specialized tasks. For example, in the well-studied social wasp *Polistes versicolor*, dominant females perform tasks such as building new cells and ovipositing, while subordinate females tend to perform tasks like feeding the larvae and foraging. The task differentiation between castes can be seen in the fact that subordinates complete 81.4% of the total foraging activity, while dominants only complete 18.6% of the total foraging.

While only a moderate percentage of species in bees (families Apidae and Halictidae) and wasps (Crabronidae and Vespidae) are eusocial, nearly all species of ants (Formicidae) are eusocial. Some major lineages within these groups are mostly or entirely eusocial, as well, such as the bee tribes Apini, Bombini, Euglossini, and Meliponini, and the wasp subfamilies Polistinae and Vespinae.

Eusociality in these families is sometimes managed by a set of pheromones that alter the behavior of specific castes in the colony. These pheromones may act across different species, as observed in *Apis andreniformis* (black dwarf honey bee), where worker bees responded to queen pheromone from the related *Apis florea* (red dwarf honey bee). Pheromones are sometimes used in these castes to assist with foraging. Workers of the Australian stingless bee *Tetragonula carbonaria*, for instance, mark food sources with a pheromone, helping their nest mates to find the food.

Reproductive specialization generally involves the production of sterile members of the species, which carry out specialized tasks to care for the reproductive members. It can manifest in the appearance of individuals within a group whose behavior or morphology is modified for group defense, including self-sacrificing behavior ("altruism"). An example of a species whose sterile caste displays this altruistic behavior is *Myrmecocystus mexicanus*, one of the species of honey ant. Select sterile workers fill their abdomens with liquid food until they become immobile and hang from the ceilings of the underground nests, acting as food storage for the rest of the colony.[707] Not all social species of insects have distinct morphological differences between castes. For example, in the Neotropical social wasp *Synoeca surinama*, social displays determine the caste ranks of individuals in the developing brood. These castes are sometimes further specialized in their behavior based on age. For example, *Scaptotrigona postica* workers assume different roles in the nest based on their age. Between approximately 0–40 days old, the workers perform tasks within the nest such as provisioning cell broods, colony cleaning, and nectar reception and dehydration. Once older than 40 days, *Scaptotrigona postica* workers move outside of the nest to practice colony defense and foraging.

In *Lasioglossum aeneiventre*, a halictid bee from Central America, nests may be headed by more than one female; such nests have more cells, and the number of active cells per female is correlated with the number of females in the nest, implying that having more females leads to more efficient building and provisioning of cells. In similar species with only one queen, such as *Lasioglossum malachurum* in Europe, the degree of eusociality depends on the clime in which the species is found.

Termites (order Blattodea, infraorder Isoptera) make up another large portion of highly advanced eusocial animals. The colony is differentiated into various castes: the queen and king are the sole reproducing individuals; workers forage and maintain food and resources;[708] and soldiers defend the colony against ant attacks. The latter two castes, which are sterile and perform highly specialized, complex social behaviors, are derived from different stages of pluripotent larvae produced by the reproductive caste. Some soldiers have jaws so enlarged (specialized for defense and attack) that they are unable to feed themselves and must be fed by workers.

Austroplatypus incompertus is a species of ambrosia beetle native to Australia, and is the first beetle (order Coleoptera) to be recognized as eusocial. This species forms colonies in which a single female is fertilized, and is protected by many unfertilized females, which also serve as workers excavating tunnels in trees. This species also participates in cooperative brood care, in which individuals care for juveniles that are not their own.

Some species of gall-inducing insects, including the gall-forming aphid, *Pemphigus spyrothecae* (order Hemiptera), and thrips (order Thysanoptera), were also described as eusocial. These species have very high relatedness among individuals due to their partially asexual mode of reproduction (sterile soldier castes being clones of the reproducing female), but the gall-inhabiting behavior gives these species a defensible resource that sets them apart from related species with similar genetics. They produce soldier castes capable of fortress defense and protection of their colony against both predators and competitors. In these groups, therefore, high relatedness alone does not lead to the evolution of social behavior, but requires that groups occur in a restricted, shared area. These species have morphologically distinct soldier castes that defend against kleptoparasites (parasitism by theft) and are able to reproduce parthenogenetically (without fertilization).

In crustaceans

Eusociality has also arisen three different times among some crustaceans that live in separate colonies. *Synalpheus regalis*, *Synalpheus filitigitus*, and *Synalpheus chacei*, three species of parasitic shrimp that rely on fortress defense and live in groups of closely related individuals in tropical reefs and sponges, live eusocially with a single breeding female and a large number of male defenders, armed with enlarged snapping claws. As with other eusocial societies, there is a single shared living space for the colony members, and the non-breeding members act to defend it.

The fortress defense hypothesis additionally points out that because sponges provide both food and shelter, there is an aggregation of relatives (because the shrimp do not have to disperse to find food), and much competition for those nesting sites. Being the target of attack promotes a good defense system (soldier caste); soldiers therefore promote the fitness of the whole nest by ensuring safety and reproduction of the queen.

Eusociality offers a competitive advantage in shrimp populations. Eusocial species were found to be more abundant, occupy more of the habitat, and use more of the available resources than non-eusocial species. Other studies add

Figure 251: *Naked mole-rat, one of two eusocial species in the Bathyergidae*

to these findings by pointing out that cohabitation was more rare than expected by chance, and that most sponges were dominated by one species, which was frequently eusocial.

In mammals

Among mammals, eusociality is known in two species in the Bathyergidae, the naked mole-rat (*Heterocephalus glaber*) and the Damaraland mole-rat (*Fukomys damarensis*), both of which are diploid and highly inbred. Usually living in harsh or limiting environments, these mole-rats aid in raising siblings and relatives born to a single reproductive queen. However, this classification is controversial owing to disputed definitions of 'eusociality'. To avoid inbreeding, mole rats sometimes outbreed and establish new colonies when resources are sufficient. Most of the individuals cooperatively care for the brood of a single reproductive female (the queen) to which they are most likely related. Thus, it is uncertain whether mole rats classify as true eusocial organisms, since their social behavior depends largely on their resources and environment.

Some mammals in the Carnivora and Primates exhibit eusocial tendencies. Perhaps most notable are meerkats (*Suricata suricatta*) and dwarf mongooses

(*Helogale parvula*). These show cooperative breeding and marked reproductive skews. In the dwarf mongoose, the breeding pair receives food priority and protection from subordinates and rarely has to defend against predators.

In humans

An early 21st century debate focused on whether humans are prosocial versus eusocial. Edward O. Wilson, in his controversial 2012 book, *The Social Conquest of the Earth*, referred to humans as a species of eusocial ape. He supported his reasoning by stating our eusocial similarities to ants, and by observing that early hominins cooperated to rear their children while other members of the same group hunted and foraged; he noted that all eusocial species went through just such a stage of collective rearing. Humans also fall under some of Wilson's original criteria of eusociality (some kind of division of labor, overlapping generations, and cooperative care of young including ones that are not their own). Wilson argued that through cooperation and teamwork, ants and humans gain a type of "superpower" that is unavailable to other social animals that have failed to make the leap from social to eusocial. In his view, eusociality creates the superorganism.

Wilson's claims created a vigorous debate. Summarizing the positions taken, Herbert Gintis noted that Wilson explicitly diverged from the general view that human sociality is accounted for by W. D. Hamilton's inclusive fitness theory, formulating instead a controversial theory of group selection. Wilson's claims, especially as formulated in his 2010 paper with Nowak and Tarnita, were "vigorously rejected" by up to 134 scientists at a time, writing in the same journal, *Nature*, as well as by well-known individuals such as Richard Dawkins and Steven Pinker. Dawkins noted that human groups involve many unrelated families who may cooperate to a high level, but do not have reproductive division of labor. Gintis further observed that biological altruism cannot exist in "advanced eusocial species", because altruism by definition reduces fitness: but non-reproductive workers already have a fitness of zero. Gintis thus agreed with Dawkins that humans are not eusocial, though he suggested some changes to Hamilton's rule to make it more complete.

Evolution

Phylogenetic distribution

Eusociality is a rare but widespread phenomenon in species in at least seven orders in the animal kingdom, as shown in the phylogenetic tree (non-eusocial groups not shown). All species of termites are eusocial, and it is believed that they were the first eusocial animals to evolve, sometime in the upper Jurassic period ($\sim$150 million years ago). All other orders also contain non-eusocial species, including many lineages where eusociality was inferred to be the ancestral state. Thus the number of independent evolutions of eusociality is still under investigation.

<templatestyles src="Template:Clade/styles.css" />

Animalia

<templatestyles src="Template:Clade/styles.css" />

ChordataArthropodaMole-rats

<templatestyles src="Template:Clade/styles.css" />

Synalpheus spp.

In-   <templatestyles src="Template:Clade/styles.css" />
secta

| Rhipi-neoptera | all Termites | |

Eu-metabola	<templatestyles src="Template:Clade/styles.css" />	
	Parane-optera	<templatestyles src="Template:Clade/styles.css" />
		Thysanoptera	*Kladothrips* spp.

| | | Hemiptera | various Aphids |

| | Metabola | <templatestyles src="Template:Clade/styles.css" /> |
| | | Coleoptera | *Austroplatypus incompertus* |

| | | Hymenoptera | <templatestyles src="Template:Clade/styles.css" /> |
| | | | many Vespidae (wasps) |

| | | Apoidea | <templatestyles src="Template:Clade/styles.css" /> |
| | | | all Ants |

many Bees

Paradox

Prior to the gene-centered view of evolution, eusociality was seen as an apparent evolutionary paradox: if adaptive evolution unfolds by differential reproduction of individual organisms, how can individuals incapable of passing on their genes evolve and persist? In *On the Origin of Species*, Darwin referred to the existence of sterile castes as the "one special difficulty, which at first appeared to me insuperable, and actually fatal to my theory".[709] Darwin anticipated that a possible resolution to the paradox might lie in the close family relationship, which W.D. Hamilton quantified a century later with his 1964 inclusive fitness theory. After the gene-centered view of evolution was developed in the mid 1970s, non-reproductive individuals were seen as an extended phenotype of the genes, which are the primary beneficiaries of natural selection.Wikipedia:Citation needed

Inclusive fitness and haplodiploidy

According to inclusive fitness theory, organisms can gain fitness not just through increasing their own reproductive output, but also via increasing the reproductive output of other individuals that share their genes, especially their close relatives. Individuals are selected to help their relatives when the cost of helping is less than the benefit gained by their relative multiplied by the fraction of genes that they share, i.e. when *Cost < relatedness * Benefit*. Under inclusive fitness theory, the necessary conditions for eusociality to evolve are more easily fulfilled by haplodiploid species because of their unusual relatedness structure.

In haplodiploid species, females develop from fertilized eggs and males develop from unfertilized eggs. Because a male is haploid, his daughters share 100% of his genes and 50% of their mother's. Therefore, they share 75% of their genes with each other. This mechanism of sex determination gives rise to what W. D. Hamilton first termed "supersisters" which are more related to their sisters than they would be to their own offspring. Even though workers often do not reproduce, they can potentially pass on more of their genes by helping to raise their sisters than they would by having their own offspring (each of which would only have 50% of their genes). This unusual situation, where females may have greater fitness when they help rear siblings rather than producing offspring, is often invoked to explain the multiple independent evolutions of eusociality (arising at least nine separate times) within the haplodiploid group Hymenoptera.

However, not all eusocial species are haplodiploid (termites, some snapping shrimps, and mole rats are not). Conversely, many bees are haplodiploid yet are not eusocial, and among eusocial species many queens mate with multiple males, resulting in a hive of half-sisters that share only 25% of their genes. The association between haplodiploidy and eusociality is below statistical significance. Haplodiploidy alone is thus neither necessary nor sufficient for eusociality to emerge. However relatedness does still play a part, as monogamy (queens mating singly) has been shown to be the ancestral state for all eusocial species so far investigated.

Ecology

Many scientists citing the close phylogenetic relationships between eusocial and non-eusocial species are making the case that environmental factors are especially important in the evolution of eusociality. The relevant factors primarily involve the distribution of food and predators.

Increased parasitism and predation rates are the primary ecological drivers of social organization. Group living affords colony members defense against enemies, specifically predators, parasites, and competitors, and allows them to gain advantage from superior foraging methods.

With the exception of some aphids, all eusocial species live in a communal nest which provides both shelter and access to food resources. Mole rats and ants live in underground burrows; wasps, bees, and some termites build above-ground hives; thrips and aphids inhabit galls (neoplastic outgrowths) induced on plants; ambrosia beetles and some termites nest together in dead wood; and snapping shrimp inhabit crevices in marine sponges. For many species the habitat outside the nest is often extremely arid or barren, creating such a high cost to dispersal that the chance to take over the colony following parental death is greater than the chance of dispersing to form a new colony. Defense of such fortresses from both predators and competitors often favors the evolution of non-reproductive soldier castes, while the high costs of nest construction and expansion favor non-reproductive worker castes.

The importance of ecology is supported by evidence such as experimentally induced reproductive division of labor, for example when normally solitary queens are forced together. Conversely, female Damaraland mole-rats undergo hormonal changes that promote dispersal after periods of high rainfall, supporting the plasticity of eusocial traits in response to environmental cues.

Climate also appears to be a selective agent driving social complexity; across bee lineages and Hymenoptera in general, higher forms of sociality are more likely to occur in tropical than temperate environments. Similarly, social transitions within Halictidae bees, where eusociality has been gained and lost multiple times, are correlated with periods of climatic warming. Social behavior in facultative social bees is often reliably predicted by ecological conditions, and switches in behavioral type have been experimentally induced by translocating offspring of solitary or social populations to warm and cool climates. In *H. rubicundus*, solitary females produce a single brood in cooler regions and social females produce two broods in warmer regions. In another species of sweat bees, *L. calceatum*, social phenotype has been predicted by altitude and micro-habitat composition, with social nests found in warmer, sunnier sites, and solitary nests found in adjacent, cooler, shaded locations.

Multilevel selection

Once pre-adaptations such as group formation, nest building, high cost of dispersal, and morphological variation are present, between-group competition has been cited as a quintessential force in the transition to advanced eusociality. Because the hallmarks of eusociality will produce an extremely altruistic society, such groups will out-reproduce their less cooperative competitors, eventually eliminating all non-eusocial groups from a species. Multilevel selection has however been heavily criticized by some for its conflict with the kin selection theory.

Reversal to solitarity

A reversal to solitarity is an evolutionary phenomenon in which descendants of a eusocial group evolve solitary behavior once again. Bees have been model organisms for the study of reversal to solitarity, because of the diversity of their social systems. Each of the four origins of eusociality in bees was followed by at least one reversal to solitarity, giving a total of at least nine reversals. This suggests that eusociality is costly to maintain, and can only persist when ecological variables favor it. Disadvantages of eusociality include the cost of investing in non-reproductive offspring, and an increased risk of disease.

All reversals to solitarity have occurred among primitively eusocial groups; none have followed the emergence of advanced eusociality. The "point of no return" hypothesis posits that the morphological differentiation of reproductive and non-reproductive castes prevents highly eusocial species such as the honeybee from reverting to the solitary state.

Physiological and developmental mechanisms

An understanding of the physiological causes and consequences of the eusocial condition has been somewhat slow; nonetheless, major advancements have been made in learning more about the mechanistic and developmental processes that lead to eusociality.

Involvement of pheromones

Pheromones are thought to play an important role in the physiological mechanisms underlying the development and maintenance of eusociality. In fact the evolution of enzymes involved both in the production and perception of pheromones has been shown to be important for the emergence of eusociality both within termites and in Hymenoptera. The most well-studied queen pheromone system in social insects is that of the honey bee *Apis mellifera*. Queen mandibular glands were found to produce a mixture of five compounds, three aliphatic and two aromatic, which have been found to control workers. Mandibular gland extracts inhibit workers from constructing queen cells in which new queens are reared which can delay the hormonally based behavioral development of workers and can suppress ovarian development in workers. Both behavioral effects mediated by the nervous system often leading to recognition of queens (releaser) and physiological effects on the reproductive and endocrine system (primer) are attributed to the same pheromones. These pheromones volatilize or are deactivated within thirty minutes, allowing workers to respond rapidly to the loss of their queen.

The levels of two of the aliphatic compounds increase rapidly in virgin queens within the first week after eclosion (emergence from the pupal case), which is consistent with their roles as sex attractants during the mating flight. It is only after a queen is mated and begins laying eggs, however, that the full blend of compounds is made. The physiological factors regulating reproductive development and pheromone production are unknown.

In several ant species, reproductive activity has also been associated with pheromone production by queens. In general, mated egg laying queens are attractive to workers whereas young winged virgin queens, which are not yet mated, elicit little or no response. However, very little is known about when pheromone production begins during the initiation of reproductive activity or about the physiological factors regulating either reproductive development or queen pheromone production in ants.

Among ants, the queen pheromone system of the fire ant *Solenopsis invicta* is particularly well studied. Both releaser and primer pheromones have been demonstrated in this species. A queen recognition (releaser) hormone is stored in the poison sac along with three other compounds. These compounds were reported to elicit a behavioral response from workers. Several primer effects have also been demonstrated. Pheromones initiate reproductive development in new winged females, called female sexuals. These chemicals also inhibit workers from rearing male and female sexuals, suppress egg production in other queens of multiple queen colonies and cause workers to execute excess queens. The action of these pheromones together maintains the eusocial phenotype which includes one queen supported by sterile workers and sexually active males (drones). In queenless colonies that lack such pheromones, winged females will quickly shed their wings, develop ovaries and lay eggs. These virgin replacement queens assume the role of the queen and even start to produce queen pheromones. There is also evidence that queen weaver ants *Oecophylla longinoda* have a variety of exocrine glands that produce pheromones, which prevent workers from laying reproductive eggs.

Similar mechanisms are used for the eusocial wasp species *Vespula vulgaris*. In order for a *Vespula vulgaris* queen to dominate all the workers, usually numbering more than 3000 in a colony, she exerts pheromone to signal her dominance. The workers were discovered to regularly lick the queen while feeding her, and the air-borne pheromone from the queen's body alerts those workers of her dominance.

The mode of action of inhibitory pheromones which prevent the development of eggs in workers has been convincingly demonstrated in the bumble bee *Bombus terrestris*. In this species, pheromones suppress activity of the corpora allata and juvenile hormone (JH) secretion. The corpora allata is an endocrine gland that produces JH, a group of hormones that regulate many aspects of insect physiology. With low JH, eggs do not mature. Similar inhibitory effects of lowering JH were seen in halictine bees and polistine wasps, but not in honey bees.

Other strategies

A variety of strategies in addition to the use of pheromones have evolved that give the queens of different species of social insects a measure of reproductive control over their nest mates. In many Polistes wasp colonies, monogamy is established soon after colony formation by physical dominance interactions among foundresses of the colony including biting, chasing and food soliciting. Such interactions created a dominance hierarchy headed by individuals with the greatest ovarian development. Larger, older individuals often have an advantage during the establishment of dominance hierarchies. The rank of subordinates is positively correlated with the degree of ovarian development and the highest ranking individual usually becomes queen if the established queen disappears. Workers do not oviposit when queens are present because of a variety of reasons: colonies tend to be small enough that queens can effectively dominate workers, queens practice selective oophagy or egg eating, or the flow of nutrients favors queen over workers and queens rapidly lay eggs in new or vacated cells. However, it is also possible that morphological differences favor the worker. In certain species of wasps, such as *Apoica flavissima* queens are smaller than their worker counterparts. This can lead to interesting worker-queen dynamics, often with the worker policing queen behaviors. Other wasps, like *Polistes instabilis* have workers with the potential to develop into reproductives, but only in cases where there are no queens to suppress them.

In primitively eusocial bees (where castes are morphologically similar and colonies usually small and short-lived), queens frequently nudge their nest mates and then burrow back down into the nest. This behavior draws workers into the lower part of the nest where they may respond to stimuli for cell construction and maintenance. Being nudged by the queen may play a role in inhibiting ovarian development and this form of queen control is supplemented by oophagy of worker laid eggs. Furthermore, temporally discrete production of workers and gynes (actual or potential queens) can cause size dimorphisms between different castes as size is strongly influenced by the season during which the individual is reared. In many wasp species worker caste determination is characterized by a temporal pattern in which workers precede non-workers of the same generation. In some cases, for example in the bumble bee, queen control weakens late in the season and the ovaries of workers develop to an increasing extent. The queen attempts to maintain her dominance by aggressive behavior and by eating worker laid eggs; her aggression is often directed towards the worker with the greatest ovarian development.

In highly eusocial wasps (where castes are morphologically dissimilar), both the quantity and quality of food seem to be important for caste differentiation. Recent studies in wasps suggest that differential larval nourishment may be the environmental trigger for larval divergence into one of two developmental classes destined to become either a worker or a gyne. All honey bee larvae are initially fed with royal jelly, which is secreted by workers, but normally they are switched over to a diet of pollen and honey as they mature; if their diet is exclusively royal jelly, however, they grow larger than normal and differentiate into queens. This jelly seems to contain a specific protein, designated as royalactin, which increases body size, promotes ovary development and shortens the developmental time period. Furthermore, the differential expression in *Polistes* of larval genes and proteins (also differentially expressed during queen versus caste development in honey bees) indicate that regulatory mechanisms may occur very early in development.

Definition debates

Subsequent to Edward O. Wilson's original definition, other authors have sought to narrow or expand the definition of eusociality by focusing on the nature and degree of the division of labor, which was not originally specified. A narrower and more widely accepted definition specifies the requirement for irreversibly distinct behavioral groups or castes (with respect to sterility and/or other features). Such a definition, however, excludes social vertebrates, like the mole rats, and some Hymenoptera species, like the weaver ants.

For example, depending on the availability of resources and the condition of the environment, mole rats can change between different types of social behaviors. In 2005, according to Wilson and Hölldobler, these types of animals would be considered primitively eusocial, which is different from eusocial, since the labor division is not permanent. A broader definition, on the other hand, would

Figure 252: *Weaver ants, here collaborating to pull nest leaves together, can be considered primitively eusocial, as they do not have permanent division of labor.*

include mole rats because it allows for any temporary division of labor or non-random distribution of reproductive success to constitute eusociality.

In 2010, Nowak, Tarnita and Wilson challenged the theoretical explanation of the evolution of eusociality. Based on the concept of inclusive fitness, the kin selection theory considers the relatedness of individuals to be one of the most important factors that lead to eusociality. This brought up issues like the irrelevance of haplodiploidy and maternal control. Nowak et al. argued that the kin selection theory is inadequate because it can explain only a subset of eusocial populations due to its assumptions (i.e. "all interactions must be additive and pairwise" which excludes any interaction that involves more than two players). To them, the standard natural selection theory, which is a more general approach than the current explanation of eusociality, is the appropriate theory to use since it explains the same phenomenon and it would work for a larger number of eusocial cases. It also requires simpler mathematical calculations when explaining the evolution of eusociality. This paper led to a large influx of publications that refuted Nowak et al.'s ideas and supported the validity and specificity of the kin selection theory. Among these, Trivers and Hare studied the haplodiploid Hymenoptera, finding that the workers were able to win the parent-offspring conflict, countering the parents' best interest and selecting the outcome that has the greater benefit for the workers (i.e. reproductive success with higher relatedness to workers than the queen). This refuted Nowak et al.'s defense that future offspring development could be based solely on the fitness of parents.

External links

- International Union for the Study of Social Insects[710]
- Eusociality in naked mole-rats[711]

Humans

Human evolution

Human evolution is the evolutionary process that led to the emergence of anatomically modern humans, beginning with the evolutionary history of primates – in particular genus *Homo* – and leading to the emergence of *Homo sapiens* as a distinct species of the hominid family, the great apes. This process involved the gradual development of traits such as human bipedalism and language.

The study of human evolution involves many scientific disciplines, including physical anthropology, primatology, archaeology, paleontology, neurobiology, ethology, linguistics, evolutionary psychology, embryology and genetics. Genetic studies show that primates diverged from other mammals about 85[712] million years ago, in the Late Cretaceous period, and the earliest fossils appear in the Paleocene, around 55[713] million years ago.

Within the Hominoidea (apes) superfamily, the Hominidae family diverged from the Hylobatidae (gibbon) family some 15–20 million years ago; African great apes (subfamily Homininae) diverged from orangutans (Ponginae) about 14[714] million years ago; the Hominini tribe (humans, *Australopithecines* and other extinct biped genera, and chimpanzee) parted from the Gorillini tribe (gorillas) between 8-9 million years ago; and, in turn, the subtribes Hominina (humans and biped ancestors) and Panina (chimps) separated 4-7.5 million years ago.

Anatomical changes

Human evolution from its first separation from the last common ancestor of humans and chimpanzees is characterized by a number of morphological, developmental, physiological, and behavioral changes. The most significant of these adaptations are bipedalism, increased brain size, lengthened ontogeny

Figure 253: *Homo sapiens*

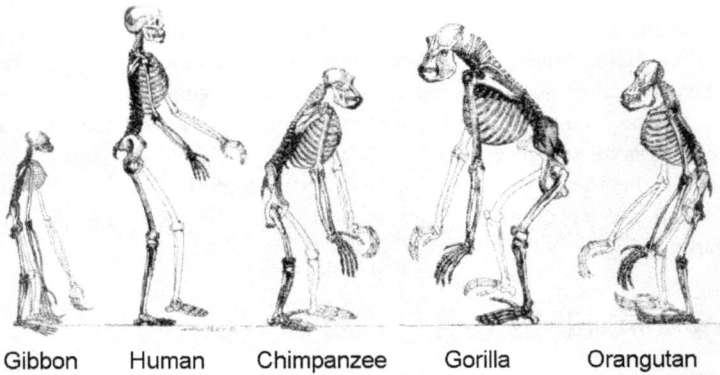

Gibbon   Human   Chimpanzee   Gorilla   Orangutan

Figure 254: *The hominoids are descendants of a common ancestor*

(gestation and infancy), and decreased sexual dimorphism. The relationship between these changes is the subject of ongoing debate.Wikipedia:Citing sources Other significant morphological changes included the evolution of a power and precision grip, a change first occurring in *H. erectus*.

Bipedalism

Bipedalism is the basic adaptation of the hominid and is considered the main cause behind a suite of skeletal changes shared by all bipedal hominids. The earliest hominin, of presumably primitive bipedalism, is considered to be either *Sahelanthropus* or *Orrorin*, both of which arose some 6 to 7 million years ago. The non-bipedal knuckle-walkers, the gorilla and chimpanzee, diverged from the hominin line over a period covering the same time, so either of *Sahelanthropus* or *Orrorin* may be our last shared ancestor. *Ardipithecus*, a full biped, arose somewhat later. Wikipedia:Citation needed

The early bipeds eventually evolved into the australopithecines and still later into the genus *Homo*. There are several theories of the adaptation value of bipedalism. It is possible that bipedalism was favored because it freed the hands for reaching and carrying food, saved energy during locomotion, enabled long distance running and hunting, provided an enhanced field of vision, and helped avoid hyperthermia by reducing the surface area exposed to direct sun; features all advantageous for thriving in the new savanna and woodland environment created as a result of the East African Rift Valley uplift versus the previous closed forest habitat. A new study provides support for the hypothesis that walking on two legs, or bipedalism, evolved because it used less energy than quadrupedal knuckle-walking. However, recent studies suggest that bipedality without the ability to use fire would not have allowed global dispersal. This change in gait saw a lengthening of the legs proportionately when compared to the length of the arms, which were shortened through the removal of the need for brachiation. Another change is the shape of the big toe. Recent studies suggest that Australopithecines still lived part of the time in trees as a result of maintaining a grasping big toe. This was progressively lost in Habilines.

Anatomically, the evolution of bipedalism has been accompanied by a large number of skeletal changes, not just to the legs and pelvis, but also to the vertebral column, feet and ankles, and skull. The femur evolved into a slightly more angular position to move the center of gravity toward the geometric center of the body. The knee and ankle joints became increasingly robust to better support increased weight. To support the increased weight on each vertebra in the upright position, the human vertebral column became S-shaped and the lumbar vertebrae became shorter and wider. In the feet the big toe moved into alignment with the other toes to help in forward locomotion. The arms and forearms shortened relative to the legs making it easier to run. The foramen magnum migrated under the skull and more anterior.

The most significant changes occurred in the pelvic region, where the long downward facing iliac blade was shortened and widened as a requirement for keeping the center of gravity stable while walking; bipedal hominids have a shorter but broader, bowl-like pelvis due to this. A drawback is that the birth canal of bipedal apes is smaller than in knuckle-walking apes, though there has been a widening of it in comparison to that of australopithecine and modern humans, permitting the passage of newborns due to the increase in cranial size but this is limited to the upper portion, since further increase can hinder normal bipedal movement.

The shortening of the pelvis and smaller birth canal evolved as a requirement for bipedalism and had significant effects on the process of human birth which is much more difficult in modern humans than in other primates. During human birth, because of the variation in size of the pelvic region, the fetal head must be in a transverse position (compared to the mother) during entry into the birth canal and rotate about 90 degrees upon exit. The smaller birth canal became a limiting factor to brain size increases in early humans and prompted a shorter gestation period leading to the relative immaturity of human offspring, who are unable to walk much before 12 months and have greater neoteny, compared to other primates, who are mobile at a much earlier age. The increased brain growth after birth and the increased dependency of children on mothers had a big effect upon the female reproductive cycle,[715] and the more frequent appearance of alloparenting in humans when compared with other hominids.[716] Delayed human sexual maturity also led to the evolution of menopause with one explanation providing that elderly women could better pass on their genes by taking care of their daughter's offspring, as compared to having more children of their own.

Encephalization

The human species eventually developed a much larger brain than that of other primates—typically 1,330 cm^3 (81 cu in) in modern humans, nearly three times the size of a chimpanzee or gorilla brain. After a period of stasis with *Australopithecus anamensis* and *Ardipithecus*, species which had smaller brains as a result of their bipedal locomotion, the pattern of encephalization started with *Homo habilis*, whose 600 cm^3 (37 cu in) brain was slightly larger than that of chimpanzees. This evolution continued in *Homo erectus* with 800–1,100 cm^3 (49–67 cu in), and reached a maximum in Neanderthals with 1,200–1,900 cm^3 (73–116 cu in), larger even than modern *Homo sapiens*. This brain increase manifested during postnatal brain growth, far exceeding that of other apes (heterochrony). It also allowed for extended periods

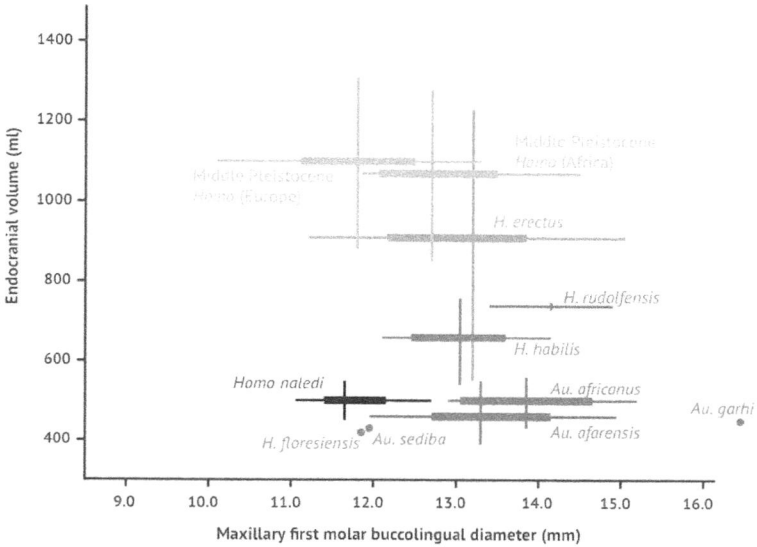

Figure 255: *Brain size and tooth size in hominins.*

of social learning and language acquisition in juvenile humans, beginning as much as 2 million years ago.

Furthermore, the changes in the structure of human brains may be even more significant than the increase in size.

The temporal lobes, which contain centers for language processing, have increased disproportionately, as has the prefrontal cortex, which has been related to complex decision-making and moderating social behavior. Encephalization has been tied to increased meat and starches in the diet, and the development of cooking, and it has been proposed that intelligence increased as a response to an increased necessity for solving social problems as human society became more complex. Changes in skull morphology, such as smaller mandibles and mandible muscle attachments, allowed more room for the brain to grow.

The increase in volume of the neocortex also included a rapid increase in size of the cerebellum. Its function has traditionally been associated with balance and fine motor control, but more recently with speech and cognition. The great apes, including hominids, had a more pronounced cerebellum relative to the neocortex than other primates. It has been suggested that because of its function of sensory-motor control and learning complex muscular actions, the cerebellum may have underpinned human technological adaptations, including the preconditions of speech.

Figure 256: *The size and shape of the skull changed over time.*
The leftmost, and largest, is a replica of a modern human skull.

The immediate survival advantage of encephalization is difficult to discern, as the major brain changes from *Homo erectus* to *Homo heidelbergensis* were not accompanied by major changes in technology. It has been suggested that the changes were mainly social and behavioural, including increased empathic abilities, increases in size of social groups, and increased behavioural plasticity

Sexual dimorphism

The reduced degree of sexual dimorphism in humans is visible primarily in the reduction of the male canine tooth relative to other ape species (except gibbons) and reduced brow ridges and general robustness of males. Another important physiological change related to sexuality in humans was the evolution of hidden estrus. Humans are the only hominoids in which the female is fertile year round and in which no special signals of fertility are produced by the body (such as genital swelling or overt changes in proceptivity during estrus).Wikipedia:Citation needed

Nonetheless, humans retain a degree of sexual dimorphism in the distribution of body hair and subcutaneous fat, and in the overall size, males being around 15% larger than females. These changes taken together have been interpreted as a result of an increased emphasis on pair bonding as a possible solution to the requirement for increased parental investment due to the prolonged infancy of offspring.

Ulnar opposition

The ulnar opposition – the contact between the thumb and the tip of the little finger of the same hand – is unique to anatomically modern humans. In other primates the thumb is short and unable to touch the little finger. The ulnar opposition facilitates the precision grip and power grip of the human hand, underlying all the skilled manipulations.

Other changes

A number of other changes have also characterized the evolution of humans, among them an increased importance on vision rather than smell; a longer juvenile developmental period and higher infant dependency; a smaller gut; faster basal metabolism; loss of body hair; evolution of sweat glands; a change in the shape of the dental arcade from being u-shaped to being parabolic; development of a chin (found in *Homo sapiens* alone); development of styloid processes; and the development of a descended larynx.

History of study

Hominin timeline

Axis scale: million years

🖑

Also see: *Life timeline* and *Nature timeline*

Before Darwin

The word *homo*, the name of the biological genus to which humans belong, is Latin for "human". It was chosen originally by Carl Linnaeus in his classification system. The word "human" is from the Latin *humanus*, the adjectival form of *homo*. The Latin "homo" derives from the Indo-European root *dhghem, or "earth". Linnaeus and other scientists of his time also considered the great apes to be the closest relatives of humans based on morphological and anatomical similarities.

Darwin

The possibility of linking humans with earlier apes by descent became clear only after 1859 with the publication of Charles Darwin's *On the Origin of Species*, in which he argued for the idea of the evolution of new species from earlier ones. Darwin's book did not address the question of human evolution, saying only that "Light will be thrown on the origin of man and his history."

The first debates about the nature of human evolution arose between Thomas Henry Huxley and Richard Owen. Huxley argued for human evolution from apes by illustrating many of the similarities and differences between humans and apes, and did so particularly in his 1863 book *Evidence as to Man's Place in Nature*. However, many of Darwin's early supporters (such as Alfred Russel Wallace and Charles Lyell) did not initially agree that the origin of the mental capacities and the moral sensibilities of humans could be explained by natural selection, though this later changed. Darwin applied the theory of evolution and sexual selection to humans when he published *The Descent of Man* in 1871.

First fossils

A major problem in the 19th century was the lack of fossil intermediaries. Neanderthal remains were discovered in a limestone quarry in 1856, three years before the publication of *On the Origin of Species*, and Neanderthal fossils had been discovered in Gibraltar even earlier, but it was originally claimed that these were human remains of a creature suffering some kind of illness. Despite the 1891 discovery by Eugène Dubois of what is now called *Homo erectus* at Trinil, Java, it was only in the 1920s when such fossils were discovered in Africa, that intermediate species began to accumulate.Wikipedia:Citation

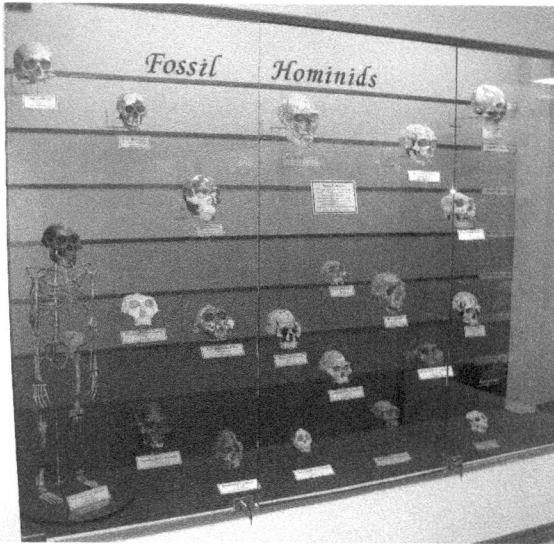

Figure 257: *Fossil hominid evolution display at The Museum of Osteology, Oklahoma City, Oklahoma, U.S..*

needed In 1925, Raymond Dart described *Australopithecus africanus*. The type specimen was the Taung Child, an australopithecine infant which was discovered in a cave. The child's remains were a remarkably well-preserved tiny skull and an endocast of the brain.

Although the brain was small (410 cm^3), its shape was rounded, unlike that of chimpanzees and gorillas, and more like a modern human brain. Also, the specimen showed short canine teeth, and the position of the foramen magnum (the hole in the skull where the spine enters) was evidence of bipedal locomotion. All of these traits convinced Dart that the Taung Child was a bipedal human ancestor, a transitional form between apes and humans.

The East African fossils—and *Homo naledi* in South Africa

During the 1960s and 1970s, hundreds of fossils were found in East Africa in the regions of the Olduvai Gorge and Lake Turkana. The driving force of these searches was the Leakey family, with Louis Leakey and his wife Mary Leakey, and later their son Richard and daughter-in-law Meave—all successful and world-renowned fossil hunters and paleoanthropologists. From the fossil beds of Olduvai and Lake Turkana they amassed specimens of the early hominins: the australopithecines and *Homo* species, and even *Homo erectus*.

These finds cemented Africa as the cradle of humankind. In the late 1970s and the 1980s, Ethiopia emerged as the new hot spot of paleoanthropology after "Lucy", the most complete fossil member of the species *Australopithecus afarensis*, was found in 1974 by Donald Johanson near Hadar in the desertic Afar Triangle region of northern Ethiopia. Although the specimen had a small brain, the pelvis and leg bones were almost identical in function to those of modern humans, showing with certainty that these hominins had walked erect. Lucy was classified as a new species, *Australopithecus afarensis*, which is thought to be more closely related to the genus *Homo* as a direct ancestor, or as a close relative of an unknown ancestor, than any other known hominid or hominin from this early time range; *see* terms "hominid" and "hominin". (The specimen was nicknamed "Lucy" after the Beatles' song "Lucy in the Sky with Diamonds", which was played loudly and repeatedly in the camp during the excavations.) The Afar Triangle area would later yield discovery of many more hominin fossils, particularly those uncovered or described by teams headed by Tim D. White in the 1990s, including *Ardipithecus ramidus* and *Ardipithecus kadabba*.

In 2013, fossil skeletons of *Homo naledi*, an extinct species of hominin assigned (provisionally) to the genus *Homo*, were found in the Rising Star Cave system, a site in South Africa's Cradle of Humankind region in Gauteng province near Johannesburg. As of September 2015[717], fossils of at least fifteen individuals, amounting to 1550 specimens, have been excavated from the cave. The species is characterized by a body mass and stature similar to small-bodied human populations, a smaller endocranial volume similar to *Australopithecus*, and a cranial morphology (skull shape) similar to early *Homo* species. The skeletal anatomy combines primitive features known from australopithecines with features known from early hominins. The individuals show signs of having been deliberately disposed of within the cave near the time of death. The fossils have not yet been dated.

The genetic revolution

The genetic revolution in studies of human evolution started when Vincent Sarich and Allan Wilson measured the strength of immunological cross-reactions of blood serum albumin between pairs of creatures, including humans and African apes (chimpanzees and gorillas). The strength of the reaction could be expressed numerically as an immunological distance, which was in turn proportional to the number of amino acid differences between homologous proteins in different species. By constructing a calibration curve of the ID of species' pairs with known divergence times in the fossil record, the data could be used as a molecular clock to estimate the times of divergence of pairs with poorer or unknown fossil records.

Figure 258: *Louis Leakey examining skulls from Olduvai Gorge, Tanzania*

In their seminal 1967 paper in *Science*, Sarich and Wilson estimated the divergence time of humans and apes as four to five million years ago, at a time when standard interpretations of the fossil record gave this divergence as at least 10 to as much as 30 million years. Subsequent fossil discoveries, notably "Lucy", and reinterpretation of older fossil materials, notably *Ramapithecus*, showed the younger estimates to be correct and validated the albumin method.

Progress in DNA sequencing, specifically mitochondrial DNA (mtDNA) and then Y-chromosome DNA (Y-DNA) advanced the understanding of human origins. Application of the molecular clock principle revolutionized the study of molecular evolution.

On the basis of a separation from the orangutan between 10 and 20 million years ago, earlier studies of the molecular clock suggested that there were about 76 mutations per generation that were not inherited by human children from their parents; this evidence supported the divergence time between hominins and chimps noted above. However, a 2012 study in Iceland of 78 children and their parents suggests a mutation rate of only 36 mutations per generation; this datum extends the separation between humans and chimps to an earlier period greater than 7 million years ago (Ma). Additional research with 226 offspring of wild chimp populations in 8 locations suggests that chimps reproduce at age 26.5 years, on average; which suggests the human divergence from chimps occurred between 7 and 13 million years ago. And these data suggest that *Ardipithecus* (4.5 Ma), *Orrorin* (6 Ma) and *Sahelanthropus* (7 Ma) all may be on the hominid lineage, and even that the separation may have occurred outside the East African Rift region.

Furthermore, analysis of the two species' genes in 2006 provides evidence that after human ancestors had started to diverge from chimpanzees, interspecies mating between "proto-human" and "proto-chimps" nonetheless occurred regularly enough to change certain genes in the new gene pool:

> A new comparison of the human and chimp genomes suggests that after the two lineages separated, they may have begun interbreeding... A principal finding is that the X chromosomes of humans and chimps appear to have diverged about 1.2 million years more recently than the other chromosomes.

The research suggests:

> There were in fact two splits between the human and chimp lineages, with the first being followed by interbreeding between the two populations and then a second split. The suggestion of a hybridization has startled paleoanthropologists, who nonetheless are treating the new genetic data seriously.

The quest for the earliest hominin

In the 1990s, several teams of paleoanthropologists were working throughout Africa looking for evidence of the earliest divergence of the hominin lineage from the great apes. In 1994, Meave Leakey discovered *Australopithecus anamensis*. The find was overshadowed by Tim D. White's 1995 discovery of *Ardipithecus ramidus*, which pushed back the fossil record to 4.2[718] million years ago.

In 2000, Martin Pickford and Brigitte Senut discovered, in the Tugen Hills of Kenya, a 6-million-year-old bipedal hominin which they named *Orrorin tugenensis*. And in 2001, a team led by Michel Brunet discovered the skull of *Sahelanthropus tchadensis* which was dated as 7.2[719] million years ago, and which Brunet argued was a bipedal, and therefore a hominid—that is, a hominin (*cf* Hominidae; terms "hominids" and hominins).

Human dispersal

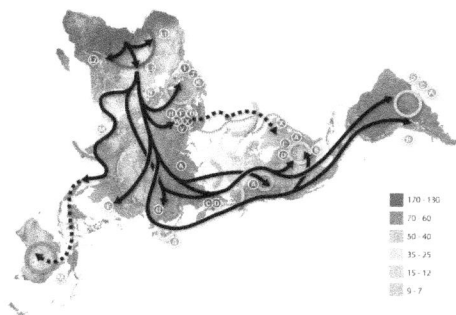

A global mapping model of human migration, based from divergence of the mitochondrial DNA (which indicates the matrilineage).[720] Timescale (ka) indicated by colours.

A "trellis" (as Milford H. Wolpoff called it) that emphasizes back-and-forth gene flow among geographic regions.

Different models for the beginning of the present human species.

Anthropologists in the 1980s were divided regarding some details of reproductive barriers and migratory dispersals of the genus *Homo*. Subsequently, genetics has been used to investigate and resolve these issues. According to the Sahara pump theory evidence suggests that genus *Homo* have migrated out of Africa at least three and possibly four times (e.g. *Homo erectus*, *Homo heidelbergensis* and two or three times for *Homo sapiens*). Recent evidence suggests these dispersals are closely related to fluctuating periods of climate change.[721]

Recent evidence suggests that humans may have left Africa half a million years earlier than previously thought. A joint Franco-Indian team has found human artifacts in the Siwalk Hills north of New Delhi dating back at least 2.6 million years. This is earlier than the previous earliest finding of genus *Homo* at Dmanisi, in Georgia, dating to 1.85 million years. Although controversial, tools found at a Chinese cave strengthen the case that humans used tools as far back as 2.48 million years ago.[722] This suggests that the Asian "Chopper" tool tradition, found in Java and northern China may have left Africa before the appearance of the Acheulian hand axe.

Dispersal of modern *Homo sapiens*

Up until the genetic evidence became available there were two dominant models for the dispersal of modern humans. The multiregional hypothesis proposed that the genus *Homo* contained only a single interconnected population as it does today (not separate species), and that its evolution took place worldwide continuously over the last couple of million years. This model was proposed in 1988 by Milford H. Wolpoff. In contrast the "out of Africa" model proposed that modern *H. sapiens* speciated in Africa recently (that is, approximately 200,000 years ago) and the subsequent migration through Eurasia resulted in nearly complete replacement of other *Homo* species. This model has been developed by Chris B. Stringer and Peter Andrews.

Sequencing mtDNA and Y-DNA sampled from a wide range of indigenous populations revealed ancestral information relating to both male and female genetic heritage, and strengthened the Out of Africa theory and weakened the views of Multiregional Evolutionism. Aligned in genetic tree differences were interpreted as supportive of a recent single origin. Analyses have shown a greater diversity of DNA patterns throughout Africa, consistent with the idea that Africa is the ancestral home of mitochondrial Eve and Y-chromosomal Adam, and that modern human dispersal out of Africa has only occurred over the last 55,000 years.

"Out of Africa" has thus gained much support from research using female mitochondrial DNA and the male Y chromosome. After analysing genealogy trees constructed using 133 types of mtDNA, researchers concluded that all were descended from a female African progenitor, dubbed Mitochondrial Eve. "Out of Africa" is also supported by the fact that mitochondrial genetic diversity is highest among African populations.

A broad study of African genetic diversity, headed by Sarah Tishkoff, found the San people had the greatest genetic diversity among the 113 distinct populations sampled, making them one of 14 "ancestral population clusters". The research also located a possible origin of modern human migration in southwestern Africa, near the coastal border of Namibia and Angola.[723] The fossil evidence was insufficient for archaeologist Richard Leakey to resolve the debate about exactly where in Africa modern humans first appeared. Studies of haplogroups in Y-chromosomal DNA and mitochondrial DNA have largely supported a recent African origin. All the evidence from autosomal DNA also predominantly supports a Recent African origin. However, evidence for archaic admixture in modern humans, both in Africa and later, throughout Eurasia has recently been suggested by a number of studies.

Recent sequencing of Neanderthal and Denisovan genomes shows that some admixture with these populations has occurred. Modern humans outside

Africa have 2–4% Neanderthal alleles in their genome, and some Melane-sians have an additional 4–6% of Denisovan alleles. These new results do not contradict the "out of Africa" model, except in its strictest interpretation, although they make the situation more complex. After recovery from a genetic bottleneck that could possibly be due to the Toba supervolcano catastrophe, a fairly small group left Africa and later briefly interbred on three separate occa-sions with Neanderthals, probably in the middle-east, on the Eurasian steppe or even in North Africa before their departure. Their still predominantly African descendants spread to populate the world. A fraction in turn interbred with Denisovans, probably in south-east Asia, before populating Melanesia. HLA haplotypes of Neanderthal and Denisova origin have been identified in mod-ern Eurasian and Oceanian populations. The Denisovan EPAS1 gene has also been found in Tibetan populations.[724]

There are still differing theories on whether there was a single exodus from Africa or several. A multiple dispersal model involves the Southern Dispersal theory, which has gained support in recent years from genetic, linguistic and archaeological evidence. In this theory, there was a coastal dispersal of modern humans from the Horn of Africa crossing the Bab el Mandib to Yemen at a lower sea level around 70,000 years ago. This group helped to populate Southeast Asia and Oceania, explaining the discovery of early human sites in these areas much earlier than those in the Levant. This group seems to have been dependent upon marine resources for their survival.

Stephen Oppenheimer has proposed a second wave of humans may have later dispersed through the Persian Gulf oases, and the Zagros mountains into the Middle East. Alternatively it may have come across the Sinai Peninsula into Asia, from shortly after 50,000 yrs BP, resulting in the bulk of the human populations of Eurasia. It has been suggested that this second group possi-bly possessed a more sophisticated "big game hunting" tool technology and was less dependent on coastal food sources than the original group. Much of the evidence for the first group's expansion would have been destroyed by the rising sea levels at the end of each glacial maximum. The multiple disper-sal model is contradicted by studies indicating that the populations of Eurasia and the populations of Southeast Asia and Oceania are all descended from the same mitochondrial DNA L3 lineages, which support a single migration out of Africa that gave rise to all non-African populations.

Stephen Oppenheimer, on the basis of the early date of Badoshan Iranian Au-rignacian, suggests that this second dispersal, may have occurred with a pluvial period about 50,000 years before the present, with modern human big-game hunting cultures spreading up the Zagros Mountains, carrying modern human genomes from Oman, throughout the Persian Gulf, northward into Armenia and Anatolia, with a variant travelling south into Israel and to Cyrenicia.[725]

```
                        Hominoidea                    Superfamily

            Hominidae              Hylobatidae         Family

        Homininae      Ponginae                        Subfamily

    Hominini    Gorillini                              Tribe

  Homo    Pan      Gorilla     Pongo    Hylobates      Genus
```

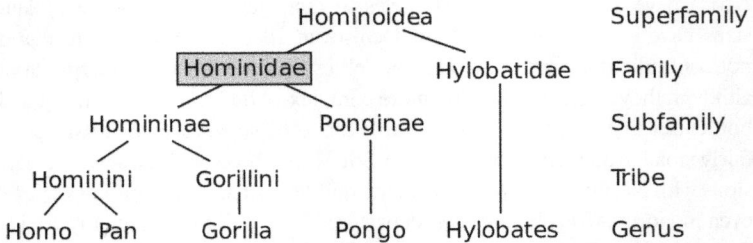

Figure 259: *Family tree showing the extant hominoids: humans (genus Homo),*
chimpanzees and bonobos (genus Pan), gorillas (genus Gorilla), orangutans
(genus Pongo), and gibbons (four genera of the family Hylobatidae: Hylobates,
Hoolock, Nomascus, and Symphalangus). All except gibbons are hominids.

Evidence

The evidence on which scientific accounts of human evolution are based comes
from many fields of natural science. The main source of knowledge about the
evolutionary process has traditionally been the fossil record, but since the development of genetics beginning in the 1970s, DNA analysis has come to occupy a place of comparable importance. The studies of ontogeny, phylogeny
and especially evolutionary developmental biology of both vertebrates and invertebrates offer considerable insight into the evolution of all life, including
how humans evolved. The specific study of the origin and life of humans is
anthropology, particularly paleoanthropology which focuses on the study of
human prehistory.

Evidence from molecular biology

The closest living relatives of humans are bonobos and chimpanzees (both
genus *Pan*) and gorillas (genus *Gorilla*). With the sequencing of both the human and chimpanzee genome, as of 2012 estimates of the similarity between
their DNA sequences range between 95% and 99%. By using the technique
called the molecular clock which estimates the time required for the number
of divergent mutations to accumulate between two lineages, the approximate
date for the split between lineages can be calculated.

The gibbons (family Hylobatidae) and then orangutans (genus *Pongo*) were
the first groups to split from the line leading to the hominins, including humans—followed by gorillas, and, ultimately, by the chimpanzees (genus *Pan*).
The splitting date between hominin and chimpanzee lineages is placed by some
between 4 to 8[726] million years ago, that is, during the Late Miocene. Speciation, however, appears to have been unusually drawn-out. Initial divergence

occurred sometime between 7 to 13[727] million years ago, but ongoing hybridization blurred the separation and delayed complete separation during several millions of years. Patterson (2006) dated the final divergence at 5 to 6[728] million years ago.

Genetic evidence has also been employed to resolve the question of whether there was any gene flow between early modern humans and Neanderthals, and to enhance our understanding of the early human migration patterns and splitting dates. By comparing the parts of the genome that are not under natural selection and which therefore accumulate mutations at a fairly steady rate, it is possible to reconstruct a genetic tree incorporating the entire human species since the last shared ancestor.

Each time a certain mutation (single-nucleotide polymorphism) appears in an individual and is passed on to his or her descendants a haplogroup is formed including all of the descendants of the individual who will also carry that mutation. By comparing mitochondrial DNA which is inherited only from the mother, geneticists have concluded that the last female common ancestor whose genetic marker is found in all modern humans, the so-called mitochondrial Eve, must have lived around 200,000 years ago.

Genetics

Human evolutionary genetics studies how one human genome differs from the other, the evolutionary past that gave rise to it, and its current effects. Differences between genomes have anthropological, medical and forensic implications and applications. Genetic data can provide important insight into human evolution.

Evidence from the fossil record

There is little fossil evidence for the divergence of the gorilla, chimpanzee and hominin lineages. The earliest fossils that have been proposed as members of the hominin lineage are *Sahelanthropus tchadensis* dating from 7[729] million years ago, *Orrorin tugenensis* dating from 5.7[730] million years ago, and *Ardipithecus kadabba* dating to 5.6[731] million years ago. Each of these have been argued to be a bipedal ancestor of later hominins but, in each case, the claims have been contested. It is also possible that one or more of these species are ancestors of another branch of African apes, or that they represent a shared ancestor between hominins and other apes.

The question then of the relationship between these early fossil species and the hominin lineage is still to be resolved. From these early species, the australopithecines arose around 4[732] million years ago and diverged into robust (also called *Paranthropus*) and gracile branches, one of which (possibly *A. garhi*)

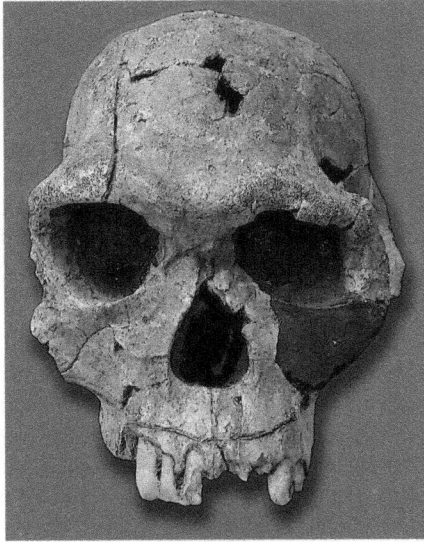

Figure 260: *Replica of fossil skull of Homo habilis. Fossil number KNM ER 1813, found at Koobi Fora, Kenya.*

Figure 261: *Replica of fossil skull of Homo ergaster (African Homo erectus). Fossil number Khm-Heu 3733 discovered in 1975 in Kenya.*

probably went on to become ancestors of the genus *Homo*. The australopithecine species that is best represented in the fossil record is *Australopithecus afarensis* with more than one hundred fossil individuals represented, found from Northern Ethiopia (such as the famous "Lucy"), to Kenya, and South Africa. Fossils of robust australopithecines such as *Au. robustus* (or alternatively *Paranthropus robustus*) and *Au./P. boisei* are particularly abundant in South Africa at sites such as Kromdraai and Swartkrans, and around Lake Turkana in Kenya.

The earliest member of the genus *Homo* is *Homo habilis* which evolved around 2.8[733] million years ago. *Homo habilis* is the first species for which we have positive evidence of the use of stone tools. They developed the Oldowan lithic technology, named after the Olduvai Gorge in which the first specimens were found. Some scientists consider *Homo rudolfensis*, a larger bodied group of fossils with similar morphology to the original *H. habilis* fossils, to be a separate species while others consider them to be part of *H. habilis*—simply representing intraspecies variation, or perhaps even sexual dimorphism. The brains of these early hominins were about the same size as that of a chimpanzee, and their main adaptation was bipedalism as an adaptation to terrestrial living.

During the next million years, a process of encephalization began and, by the arrival (about 1.9[734] million years ago) of *Homo erectus* in the fossil record, cranial capacity had doubled. *Homo erectus* were the first of the hominins to emigrate from Africa, and, from 1.8 to 1.3[735] million years ago, this species spread through Africa, Asia, and Europe. One population of *H. erectus*, also sometimes classified as a separate species *Homo ergaster*, remained in Africa and evolved into *Homo sapiens*. It is believed that these species, *H. erectus* and *H. ergaster*, were the first to use fire and complex tools.

The earliest transitional fossils between *H. ergaster/erectus* and archaic *H. sapiens* are from Africa, such as *Homo rhodesiensis*, but seemingly transitional forms were also found at Dmanisi, Georgia. These descendants of African *H. erectus* spread through Eurasia from ca. 500,000 years ago evolving into *H. antecessor*, *H. heidelbergensis* and *H. neanderthalensis*. The earliest fossils of anatomically modern humans are from the Middle Paleolithic, about 200,000 years ago such as the Omo remains of Ethiopia; later fossils from Es Skhul cave in Israel and Southern Europe begin around 90,000 years ago (0.09[736] million years ago).

As modern humans spread out from Africa, they encountered other hominins such as *Homo neanderthalensis* and the so-called Denisovans, who may have evolved from populations of *Homo erectus* that had left Africa around 2[737] million years ago. The nature of interaction between early humans and these sister species has been a long-standing source of controversy, the question being whether humans replaced these earlier species or whether they were

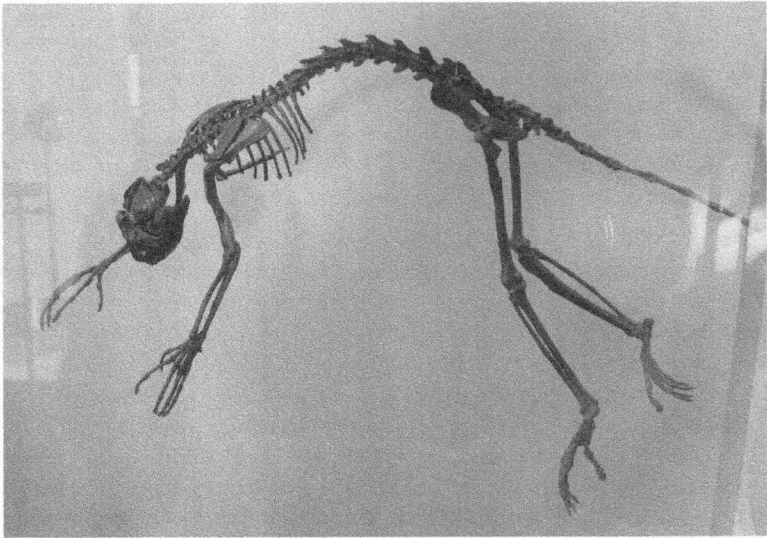

Figure 262: *Notharctus tenebrosus, American Museum of Natural History, New York City, New York, US.*

in fact similar enough to interbreed, in which case these earlier populations may have contributed genetic material to modern humans.

This migration out of Africa is estimated to have begun about 70,000 years BP and modern humans subsequently spread globally, replacing earlier hominins either through competition or hybridization. They inhabited Eurasia and Oceania by 40,000 years BP, and the Americas by at least 14,500 years BP.

Before *Homo*

Early evolution of primates

Evolutionary history of the primates can be traced back 65 million years. One of the oldest known primate-like mammal species, the *Plesiadapis*, came from North America; another, *Archicebus*, came from China. Other similar basal primates were widespread in Eurasia and Africa during the tropical conditions of the Paleocene and Eocene.

David R. Begun concluded that early primates flourished in Eurasia and that a lineage leading to the African apes and humans, including to *Dryopithecus*, migrated south from Europe or Western Asia into Africa. The surviving

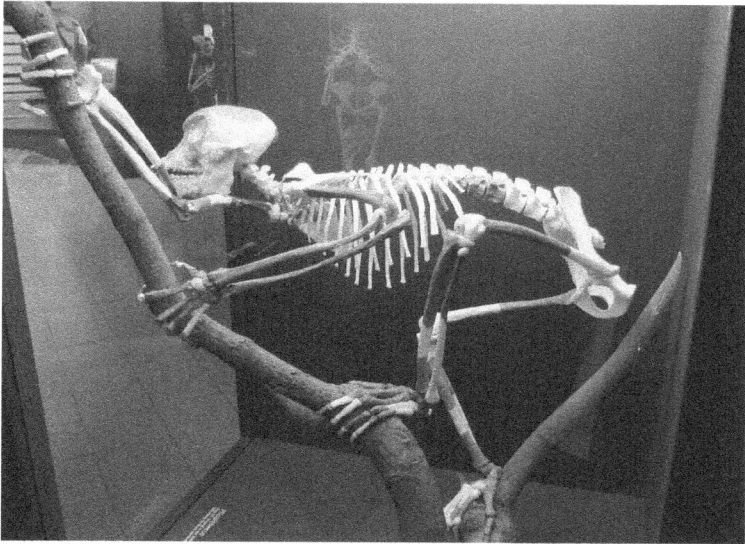

Figure 263: *Reconstructed tailless Proconsul skeleton*

tropical population of primates—which is seen most completely in the Upper Eocene and lowermost Oligocene fossil beds of the Faiyum depression southwest of Cairo—gave rise to all extant primate species, including the lemurs of Madagascar, lorises of Southeast Asia, galagos or "bush babies" of Africa, and to the anthropoids, which are the Platyrrhines or New World monkeys, the Catarrhines or Old World monkeys, and the great apes, including humans and other hominids.

The earliest known catarrhine is *Kamoyapithecus* from uppermost Oligocene at Eragaleit in the northern Great Rift Valley in Kenya, dated to 24 million years ago. Its ancestry is thought to be species related to *Aegyptopithecus*, *Propliopithecus*, and *Parapithecus* from the Faiyum, at around 35 million years ago. In 2010, *Saadanius* was described as a close relative of the last common ancestor of the crown catarrhines, and tentatively dated to 29–28 million years ago, helping to fill an 11-million-year gap in the fossil record.

In the Early Miocene, about 22 million years ago, the many kinds of arboreally adapted primitive catarrhines from East Africa suggest a long history of prior diversification. Fossils at 20 million years ago include fragments attributed to *Victoriapithecus*, the earliest Old World monkey. Among the genera thought to be in the ape lineage leading up to 13 million years ago are

Proconsul, Rangwapithecus, Dendropithecus, Limnopithecus, Nacholapithecus, Equatorius, Nyanzapithecus, Afropithecus, Heliopithecus, and *Kenyapithecus,* all from East Africa.

The presence of other generalized non-cercopithecids of Middle Miocene from sites far distant—*Otavipithecus* from cave deposits in Namibia, and *Pierolapithecus* and *Dryopithecus* from France, Spain and Austria—is evidence of a wide diversity of forms across Africa and the Mediterranean basin during the relatively warm and equable climatic regimes of the Early and Middle Miocene. The youngest of the Miocene hominoids, *Oreopithecus,* is from coal beds in Italy that have been dated to 9 million years ago.

Molecular evidence indicates that the lineage of gibbons (family Hylobatidae) diverged from the line of great apes some 18–12 million years ago, and that of orangutans (subfamily Ponginae) diverged from the other great apes at about 12 million years; there are no fossils that clearly document the ancestry of gibbons, which may have originated in a so-far-unknown South East Asian hominoid population, but fossil proto-orangutans may be represented by *Sivapithecus* from India and *Griphopithecus* from Turkey, dated to around 10 million years ago.

Divergence of the human clade from other great apes

Species close to the last common ancestor of gorillas, chimpanzees and humans may be represented by *Nakalipithecus* fossils found in Kenya and *Ouranopithecus* found in Greece. Molecular evidence suggests that between 8 and 4 million years ago, first the gorillas, and then the chimpanzees (genus *Pan*) split off from the line leading to the humans. Human DNA is approximately 98.4% identical to that of chimpanzees when comparing single nucleotide polymorphisms (see human evolutionary genetics). The fossil record, however, of gorillas and chimpanzees is limited; both poor preservation—rain forest soils tend to be acidic and dissolve bone—and sampling bias probably contribute to this problem.

Other hominins probably adapted to the drier environments outside the equatorial belt; and there they encountered antelope, hyenas, dogs, pigs, elephants, horses, and others. The equatorial belt contracted after about 8 million years ago, and there is very little fossil evidence for the split—thought to have occurred around that time—of the hominin lineage from the lineages of gorillas and chimpanzees. The earliest fossils argued by some to belong to the human lineage are *Sahelanthropus tchadensis* (7 Ma) and *Orrorin tugenensis* (6 Ma), followed by *Ardipithecus* (5.5–4.4 Ma), with species *Ar. kadabba* and *Ar. ramidus.*

Figure 264: *A reconstruction of "Lucy", a female Australopithecus afarensis on exhibit in the National Museum of Natural History, Washington, D.C., US.*

It has been argued in a study of the life history of *Ar. ramidus* that the species provides evidence for a suite of anatomical and behavioral adaptations in very early hominins unlike any species of extant great ape. This study demonstrated affinities between the skull morphology of *Ar. ramidus* and that of infant and juvenile chimpanzees, suggesting the species evolved a juvenalised or paedo-morphic craniofacial morphology via heterochronic dissociation of growth tra-jectories. It was also argued that the species provides support for the notion that very early hominins, akin to bonobos (*Pan paniscus*) the less aggressive species of chimpanzee, may have evolved via the process of self-domestication. Consequently, arguing against the so-called "chimpanzee referential model" the authors suggest it is no longer tenable to use common chimpanzee (*Pan troglodytes*) social and mating behaviors in models of early hominin social evolution. When commenting on the absence of aggressive canine morphol-ogy in *Ar. ramidus* and the implications this has for the evolution of hominin social psychology, they wrote:

> *Of course Ar. ramidus differs significantly from bonobos, bonobos having retained a functional canine honing complex. However, the fact that Ar. ramidus shares with bonobos reduced sexual dimorphism, and a more paedomorphic form relative to chimpanzees, suggests that the develop-mental and social adaptations evident in bonobos may be of assistance*

*in future reconstructions of early hominin social and sexual psychology.
In fact the trend towards increased maternal care, female mate selection
and self-domestication may have been stronger and more refined in Ar.
ramidus than what we see in bonobos.:[128]*

The authors argue that many of the basic human adaptations evolved in the
ancient forest and woodland ecosystems of late Miocene and early Pliocene
Africa. Consequently, they argue that humans may not represent evolution
from a chimpanzee-like ancestor as has traditionally been supposed. This suggests many modern human adaptations represent phylogenetically deep traits
and that the behavior and morphology of chimpanzees may have evolved subsequent to the split with the common ancestor they share with humans.

Genus *Australopithecus*

The genus *Australopithecus* evolved in eastern Africa around 4 million years
ago before spreading throughout the continent and eventually becoming extinct
2 million years ago. During this time period various forms of australopiths
existed, including *Australopithecus anamensis*, *Au. afarensis*, *Au. sediba*, and
Au. africanus. There is still some debate among academics whether certain
African hominid species of this time, such as *Au. robustus* and *Au. boisei*,
constitute members of the same genus; if so, they would be considered to be
Au. robust australopiths whilst the others would be considered *Au. gracile
australopiths*. However, if these species do indeed constitute their own genus,
then they may be given their own name, the *Paranthropus*.

- *Australopithecus* (4–1.8 Ma), with species *Au. anamensis*, *Au. afarensis*,
  Au. africanus, *Au. bahrelghazali*, *Au. garhi*, and *Au. sediba*;
- *Kenyanthropus* (3–2.7 Ma), with species *K. platyops*;
- *Paranthropus* (3–1.2 Ma), with species *P. aethiopicus*, *P. boisei*, and *P.
  robustus*

A new proposed species *Australopithecus deyiremeda* is claimed to have been
discovered living at the same time period of *Au. afarensis*. There is debate
if Au. deyiremeda is a new species or is *Au. afarensis*. *Australopithecus
prometheus*, otherwise known as Little Foot has recently been dated at 3.67
million years old through a new dating technique, making the genus *Australopithecus* as old as *afarensis*.[738] Given the opposable big toe found on Little
Foot, it seems that he was a good climber, and it is thought given the night
predators of the region, he probably, like gorillas and chimpanzees, built a
nesting platform at night, in the trees.

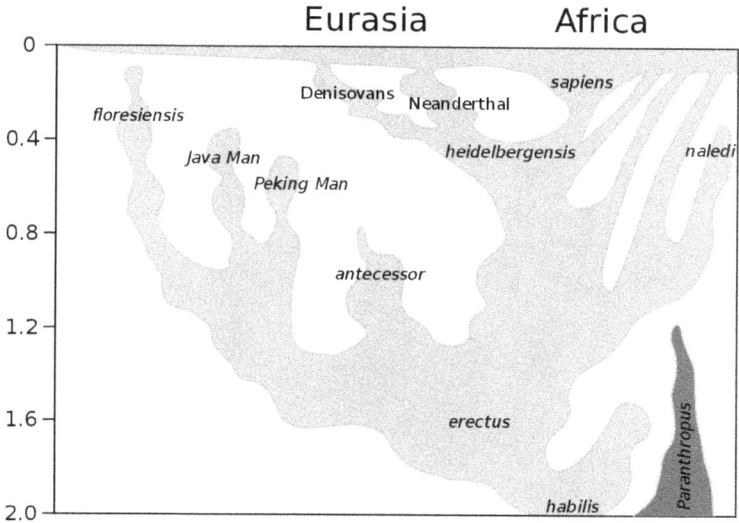

Figure 265: *A model of the evolution of the genus Homo over the last 2 million years (vertical axis). The rapid "Out of Africa" expansion of H. sapiens is indicated at the top of the diagram, with admixture indicated with Neanderthals, Denisovans, and unspecified archaic African hominins. Late survival of robust australopithecines (Paranthropus) alongside Homo until 1.2 Mya is indicated in purple.*

Evolution of genus *Homo*

The earliest documented representative of the genus *Homo* is *Homo habilis*, which evolved around 2.8[733] million years ago, and is arguably the earliest species for which there is positive evidence of the use of stone tools. The brains of these early hominins were about the same size as that of a chimpanzee, although it has been suggested that this was the time in which the human SRGAP2 gene doubled, producing a more rapid wiring of the frontal cortex. During the next million years a process of rapid encephalization occurred, and with the arrival of *Homo erectus* and *Homo ergaster* in the fossil record, cranial capacity had doubled to 850 cm^3. (Such an increase in human brain size is equivalent to each generation having 125,000 more neurons than their parents.) It is believed that *Homo erectus* and *Homo ergaster* were the first to use fire and complex tools, and were the first of the hominin line to leave Africa, spreading throughout Africa, Asia, and Europe between 1.3 to 1.8[739] million years ago.

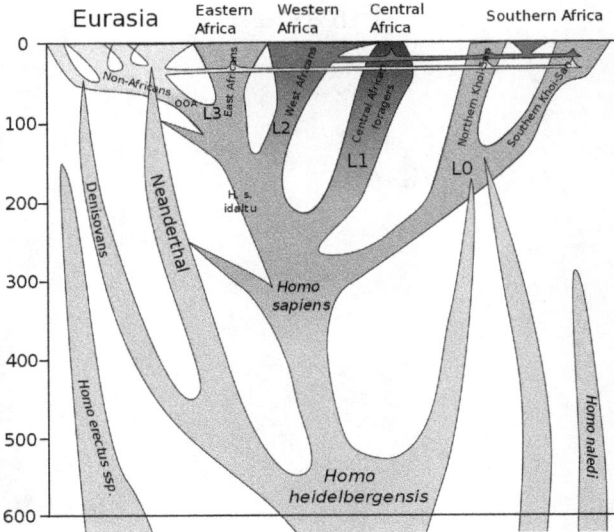

Figure 266:
A model of the phylogeny of H. sapiens during the Middle Pa-
leolithic. The horizontal axis represents geographic location;
the vertical axis represents time in thousands of years ago.[740]
Homo heidelbergensis is shown as diverging into Neanderthals, Deniso-
vans and H. sapiens. With the expansion of H. sapiens after 200 kya,
Neanderthals, Denisovans and unspecified archaic African hominins
are shown as again subsumed into the H. sapiens lineage. In addi-
tion, admixture events in modern African populations are indicated.

According to the recent African origin of modern humans theory, modern
humans evolved in Africa possibly from *Homo heidelbergensis, Homo rhode-*
siensis or *Homo antecessor* and migrated out of the continent some 50,000
to 100,000 years ago, gradually replacing local populations of *Homo erectus*,
Denisova hominins, *Homo floresiensis* and *Homo neanderthalensis*. Archaic
Homo sapiens, the forerunner of anatomically modern humans, evolved in the
Middle Paleolithic between 400,000 and 250,000 years ago. Recent DNA
evidence suggests that several haplotypes of Neanderthal origin are present
among all non-African populations, and Neanderthals and other hominins,
such as Denisovans, may have contributed up to 6% of their genome to present-
day humans, suggestive of a limited inter-breeding between these species. The
transition to behavioral modernity with the development of symbolic culture,
language, and specialized lithic technology happened around 50,000 years ago
according to some anthropologists although others point to evidence that sug-

gests that a gradual change in behavior took place over a longer time span.

Homo sapiens is the only extant species of its genus, *Homo*. While some (extinct) *Homo* species might have been ancestors of *Homo sapiens*, many, perhaps most, were likely "cousins", having speciated away from the ancestral hominin line. There is yet no consensus as to which of these groups should be considered a separate species and which should be a subspecies; this may be due to the dearth of fossils or to the slight differences used to classify species in the genus *Homo*. The Sahara pump theory (describing an occasionally passable "wet" Sahara desert) provides one possible explanation of the early variation in the genus *Homo*.

Based on archaeological and paleontological evidence, it has been possible to infer, to some extent, the ancient dietary practices of various *Homo* species and to study the role of diet in physical and behavioral evolution within *Homo*.

Some anthropologists and archaeologists subscribe to the Toba catastrophe theory, which posits that the supereruption of Lake Toba on Sumatran island in Indonesia some 70,000 years ago caused global consequences, killing the majority of humans and creating a population bottleneck that affected the genetic inheritance of all humans today.

H. habilis and *H. gautengensis*

Homo habilis lived from about 2.8 to 1.4 Ma. The species evolved in South and East Africa in the Late Pliocene or Early Pleistocene, 2.5–2 Ma, when it diverged from the australopithecines. *Homo habilis* had smaller molars and larger brains than the australopithecines, and made tools from stone and perhaps animal bones. One of the first known hominins was nicknamed 'handy man' by discoverer Louis Leakey due to its association with stone tools. Some scientists have proposed moving this species out of *Homo* and into *Australopithecus* due to the morphology of its skeleton being more adapted to living on trees rather than to moving on two legs like *Homo sapiens*.

In May 2010, a new species, *Homo gautengensis*, was discovered in South Africa.

H. rudolfensis and *H. georgicus*

These are proposed species names for fossils from about 1.9–1.6 Ma, whose relation to *Homo habilis* is not yet clear.

- *Homo rudolfensis* refers to a single, incomplete skull from Kenya. Scientists have suggested that this was another *Homo habilis*, but this has not been confirmed.
- *Homo georgicus*, from Georgia, may be an intermediate form between *Homo habilis* and *Homo erectus*, or a sub-species of *Homo erectus*.

H. ergaster and *H. erectus*

The first fossils of *Homo erectus* were discovered by Dutch physician Eugene Dubois in 1891 on the Indonesian island of Java. He originally named the material *Anthropopithecus erectus* (1892-1893, considered a this point as a chimpanzee-like fossil primate) and *Pithecanthropus erectus* (1893-1894, changing his mind as of based on its morphology, which he considered to be intermediate between that of humans and apes). Years later, in the 20th century, the German physician and paleoanthropologist Franz Weidenreich (1873-1948) compared in detail the characters of Dubois' Java Man, then named *Pithecanthropus erectus*, with the characters of the Peking Man, then named *Sinanthropus pekinensis*. Weidenreich concluded in 1940 that because of their anatomical similarity with modern humans it was necessary to gather all these specimens of Java and China in a single species of the genus *Homo*, the species *Homo erectus*.Wikipedia:Citation needed *Homo erectus* lived from about 1.8 Ma to about 70,000 years ago—which would indicate that they were probably wiped out by the Toba catastrophe; however, nearby *Homo floresiensis* survived it. The early phase of *Homo erectus*, from 1.8 to 1.25 Ma, is considered by some to be a separate species, *Homo ergaster*, or as *Homo erectus ergaster*, a subspecies of *Homo erectus*.

In Africa in the Early Pleistocene, 1.5–1 Ma, some populations of *Homo habilis* are thought to have evolved larger brains and to have made more elaborate stone tools; these differences and others are sufficient for anthropologists to classify them as a new species, *Homo erectus*—in Africa. The evolution of locking knees and the movement of the foramen magnum are thought to be likely drivers of the larger population changes. This species also may have used fire to cook meat. Richard Wrangham suggests that the fact that Homo seems to have been ground dwelling, with reduced intestinal length, smaller dentition, "and swelled our brains to their current, horrendously fuel-inefficient size", suggest that control of fire and releasing increased nutritional value through cooking was the key adaptation that separated Homo from tree-sleeping Australopithecines.[741]

A famous example of *Homo erectus* is Peking Man; others were found in Asia (notably in Indonesia), Africa, and Europe. Many paleoanthropologists now use the term *Homo ergaster* for the non-Asian forms of this group, and reserve *Homo erectus* only for those fossils that are found in Asia and meet certain skeletal and dental requirements which differ slightly from *H. ergaster*.

H. cepranensis and *H. antecessor*

These are proposed as species that may be intermediate between *H. erectus* and *H. heidelbergensis*.

- *H. antecessor* is known from fossils from Spain and England that are dated 1.2 Ma–500 ka.
- *H. cepranensis* refers to a single skull cap from Italy, estimated to be about 800,000 years old.

H. heidelbergensis

H. heidelbergensis ("Heidelberg Man") lived from about 800,000 to about 300,000 years ago. Also proposed as *Homo sapiens heidelbergensis* or *Homo sapiens paleohungaricus*.

H. rhodesiensis, and the Gawis cranium

- *H. rhodesiensis*, estimated to be 300,000–125,000 years old. Most current researchers place Rhodesian Man within the group of *Homo heidelbergensis*, though other designations such as archaic *Homo sapiens* and *Homo sapiens rhodesiensis* have been proposed.
- In February 2006 a fossil, the Gawis cranium, was found which might possibly be a species intermediate between *H. erectus* and *H. sapiens* or one of many evolutionary dead ends. The skull from Gawis, Ethiopia, is believed to be 500,000–250,000 years old. Only summary details are known, and the finders have not yet released a peer-reviewed study. Gawis man's facial features suggest its being either an intermediate species or an example of a "Bodo man" female.

Neanderthal and Denisovan

Homo neanderthalensis, alternatively designated as *Homo sapiens neanderthalensis*, lived in Europe and Asia from 400,000 to about 28,000 years ago. There are a number of clear anatomical differences between anatomically modern humans (AMH) and Neanderthal populations. Many of these relate to the superior adaptation to cold environments possessed by the Neanderthal populations. Their surface to volume ratio is an extreme version of that found amongst Inuit populations, indicating that they were less inclined to lose body heat than were AMH. From brain Endocasts, Neanderthals also had significantly larger brains. This would seem to indicate that the intellectual superiority of AMH populations may be questionable. More recent research by Eiluned Pearce, Chris Stringer, R. I. M. Dunbar, however, have shown important differences in Brain architecture. For example, in both the orbital chamber size and in the size of the occipital lobe, the larger size suggests that

Figure 267: *Reconstruction of Homo heidelbergensis which may be the direct ancestor of both Homo neanderthalensis and Homo sapiens.*

the Neanderthal had a better visual acuity than modern humans. This would give a superior vision in the inferior light conditions found in Glacial Europe. It also seems that the higher body mass of Neanderthals had a correspondingly larger brain mass required for body care and control.

The Neanderthal populations seem to have been physically superior to AMH populations. These differences may have been sufficient to give Neanderthal populations an environmental superiority to AMH populations from 75,000 to 45,000 years BP. With these differences, Neanderthal brains show a smaller area was available for social functioning. Plotting group size possible from endocrainial volume, suggests that AMH populations (minus occipital lobe size), had a Dunbars number of 144 possible relationships. Neanderthal populations seem to have been limited to about 120 individuals. This would show up in a larger number of possible mates for AMH humans, with increased risks of inbreeding amongst Neanderthal populations. It also suggests that humans had larger trade catchment areas than Neanderthals (confirmed in the distribution of stone tools). With larger populations, social and technological innovations were easier to fix in human populations, which may have all contributed to the fact that modern Homo sapiens replaced the Neanderthal populations by 28,000 BP.

Figure 268: *Dermoplastic reconstruction of a Neanderthal*

Earlier evidence from sequencing mitochondrial DNA suggested that no significant gene flow occurred between *H. neanderthalensis* and *H. sapiens*, and that the two were separate species that shared a common ancestor about 660,000 years ago. However, a sequencing of the Neanderthal genome in 2010 indicated that Neanderthals did indeed interbreed with anatomically modern humans *circa* 45,000 to 80,000 years ago (at the approximate time that modern humans migrated out from Africa, but before they dispersed into Europe, Asia and elsewhere). The genetic sequencing of a 40,000 year old human skeleton from Romania showed that 11% of its genome was Neanderthal, and it was estimated that the individual had a Neanderthal ancestor 4-6 generations previously, in addition to a contribution from earlier interbreeding in the Middle East. Though this interbred Romanian population seems not to have been ancestral to modern humans, the finding indicates that interbreeding happened repeatedly.

Nearly all modern non-African humans have 1% to 4% of their DNA derived from Neanderthal DNA, and this finding is consistent with recent studies indicating that the divergence of some human alleles dates to one Ma, although the interpretation of these studies has been questioned. Neanderthals and *Homo sapiens* could have co-existed in Europe for as long as 10,000 years, during which human populations exploded vastly outnumbering Neanderthals, possibly outcompeting them by sheer numerical strength.

In 2008, archaeologists working at the site of Denisova Cave in the Altai Mountains of Siberia uncovered a small bone fragment from the fifth finger of a juvenile member of Denisovans. Artifacts, including a bracelet, excavated in the cave at the same level were carbon dated to around 40,000 BP. As DNA had survived in the fossil fragment due to the cool climate of the Denisova Cave, both mtDNA and nuclear DNA were sequenced.

While the divergence point of the mtDNA was unexpectedly deep in time, the full genomic sequence suggested the Denisovans belonged to the same lineage as Neanderthals, with the two diverging shortly after their line split from the lineage that gave rise to modern humans. Modern humans are known to have overlapped with Neanderthals in Europe and the Near East for possibly more than 40,000 years,[742] and the discovery raises the possibility that Neanderthals, Denisovans, and modern humans may have co-existed and interbred. The existence of this distant branch creates a much more complex picture of humankind during the Late Pleistocene than previously thought. Evidence has also been found that as much as 6% of the DNA of some modern Melanesians derive from Denisovans, indicating limited interbreeding in Southeast Asia.

Alleles thought to have originated in Neanderthals and Denisovans have been identified at several genetic loci in the genomes of modern humans outside of Africa. HLA haplotypes from Denisovans and Neanderthal represent more than half the HLA alleles of modern Eurasians, indicating strong positive selection for these introgressed alleles. Corinne Simoneti at Vanderbilt University, in Nashville and her team have found from medical records of 28,000 people of European descent that the presence of Neanderthal DNA segments may be associated with a likelihood to suffer depression more frequently.[743]

The flow of genes from Neanderthal populations to modern human was not all one way. Sergi Castellano of the Max Planck Institute for Evolutionary Anthropology in Leipzig, Germany, has in 2016 reported that while Denisovan and Neanderthal genomes are more related to each other than they are to us, Siberian Neanderthal genomes show similarity to the modern human gene pool, more so than to European Neanderthal populations. The evidence suggests that the Neanderthal populations interbred with modern humans possibly 100,000 years ago, probably somewhere in the Near East.

Studies of a Neanderthal child at Gibraltar show from brain development and teeth eruption that Neanderthal children may have matured more rapidly than is the case for Homo sapiens.[744]

Figure 269: *Models representing Homo sapiens, Homo neanderthalensis and Homo floresiensis*

H. floresiensis

H. floresiensis, which lived from approximately 190,000 to 50,000 years before present, has been nicknamed *hobbit* for its small size, possibly a result of insular dwarfism. *H. floresiensis* is intriguing both for its size and its age, being an example of a recent species of the genus *Homo* that exhibits derived traits not shared with modern humans. In other words, *H. floresiensis* shares a common ancestor with modern humans, but split from the modern human lineage and followed a distinct evolutionary path. The main find was a skeleton believed to be a woman of about 30 years of age. Found in 2003, it has been dated to approximately 18,000 years old. The living woman was estimated to be one meter in height, with a brain volume of just 380 cm^3 (considered small for a chimpanzee and less than a third of the *H. sapiens* average of 1400 cm^3). Wikipedia:Citation needed

However, there is an ongoing debate over whether *H. floresiensis* is indeed a separate species. Some scientists hold that *H. floresiensis* was a modern *H. sapiens* with pathological dwarfism. This hypothesis is supported in part, because some modern humans who live on Flores, the Indonesian island where the skeleton was found, are pygmies. This, coupled with pathological dwarfism, could have resulted in a significantly diminutive human. The other major attack on *H. floresiensis* as a separate species is that it was found with tools only associated with *H. sapiens*.

The hypothesis of pathological dwarfism, however, fails to explain additional anatomical features that are unlike those of modern humans (diseased or not) but much like those of ancient members of our genus. Aside from cranial features, these features include the form of bones in the wrist, forearm, shoulder, knees, and feet. Additionally, this hypothesis fails to explain the find of multiple examples of individuals with these same characteristics, indicating they were common to a large population, and not limited to one individual. Wikipedia:Citation needed

H. sapiens

H. sapiens (the adjective *sapiens* is Latin for "wise" or "intelligent") emerged around 300,000 years ago, likely derived from *Homo heidelbergensis*. Between 400,000 years ago and the second interglacial period in the Middle Pleistocene, around 250,000 years ago, the trend in intra-cranial volume expansion and the elaboration of stone tool technologies developed, providing evidence for a transition from *H. erectus* to *H. sapiens*. The direct evidence suggests there was a migration of *H. erectus* out of Africa, then a further speciation of *H. sapiens* from *H. erectus* in Africa. A subsequent migration (both within and out of Africa) eventually replaced the earlier dispersed *H. erectus*. This migration and origin theory is usually referred to as the "recent single-origin hypothesis" or "out of Africa" theory. *H. sapiens* interbred with archaic humans both in Africa and in Eurasia, in Eurasia notably with Neanderthals and Denisovans.

The Toba catastrophe theory, which postulates a population bottleneck for *H. sapiens* about 70,000 years ago, was controversial from its first proposal in the 1990s and by the 2010s had very little support.

Distinctive human genetic variability has arisen as the result of the founder effect, by archaic admixture and by recent evolutionary pressures.

Use of tools

The use of tools has been interpreted as a sign of intelligence, and it has been theorized that tool use may have stimulated certain aspects of human evolution, especially the continued expansion of the human brain. Paleontology has yet to explain the expansion of this organ over millions of years despite being extremely demanding in terms of energy consumption. The brain of a modern human consumes about 13 watts (260 kilocalories per day), a fifth of the body's resting power consumption. Increased tool use would allow hunting for energy-rich meat products, and would enable processing more energy-rich

Figure 270: *"A sharp rock", an Oldowan pebble tool, the most basic of human stone tools.*

Figure 271: *The harnessing of fire was a pivotal milestone in human history.*

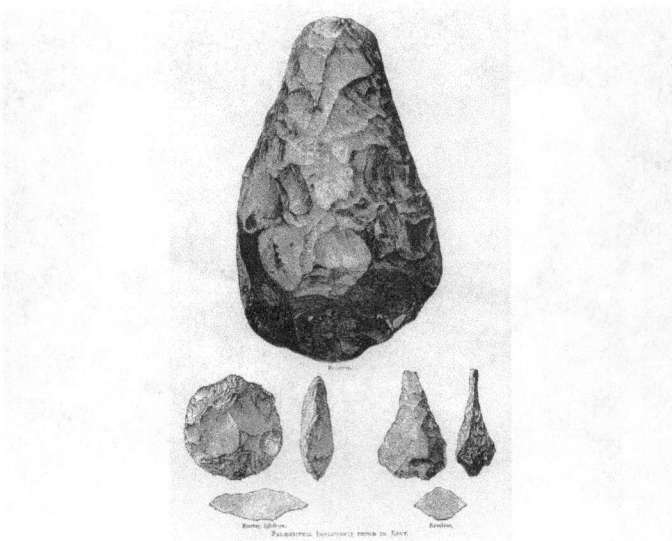

Figure 272: *Acheulean hand-axes from Kent. Homo erectus flint work. The types shown are (clockwise from top) cordate, ficron and ovate.*

Figure 273: *Venus of Willendorf, an example of Paleolithic art, dated 24–26,000 years ago.*

plant products. Researchers have suggested that early hominins were thus under evolutionary pressure to increase their capacity to create and use tools.

Precisely when early humans started to use tools is difficult to determine, because the more primitive these tools are (for example, sharp-edged stones) the more difficult it is to decide whether they are natural objects or human artifacts. There is some evidence that the australopithecines (4 Ma) may have used broken bones as tools, but this is debated.

It should be noted that many species make and use tools, but it is the human genus that dominates the areas of making and using more complex tools. The oldest known tools are the Oldowan stone tools from Ethiopia, 2.5–2.6 million years old. A *Homo* fossil was found near some Oldowan tools, and its age was noted at 2.3 million years old, suggesting that maybe the *Homo* species did indeed create and use these tools. It is a possibility but does not yet represent solid evidence. The third metacarpal styloid process enables the hand bone to lock into the wrist bones, allowing for greater amounts of pressure to be applied to the wrist and hand from a grasping thumb and fingers. It allows humans the dexterity and strength to make and use complex tools. This unique anatomical feature separates humans from apes and other nonhuman primates, and is not seen in human fossils older than 1.8 million years.

Bernard Wood noted that *Paranthropus* co-existed with the early *Homo* species in the area of the "Oldowan Industrial Complex" over roughly the same span of time. Although there is no direct evidence which identifies *Paranthropus* as the tool makers, their anatomy lends to indirect evidence of their capabilities in this area. Most paleoanthropologists agree that the early *Homo* species were indeed responsible for most of the Oldowan tools found. They argue that when most of the Oldowan tools were found in association with human fossils, *Homo* was always present, but *Paranthropus* was not.

In 1994, Randall Susman used the anatomy of opposable thumbs as the basis for his argument that both the *Homo* and *Paranthropus* species were toolmakers. He compared bones and muscles of human and chimpanzee thumbs, finding that humans have 3 muscles which are lacking in chimpanzees. Humans also have thicker metacarpals with broader heads, allowing more precise grasping than the chimpanzee hand can perform. Susman posited that modern anatomy of the human opposable thumb is an evolutionary response to the requirements associated with making and handling tools and that both species were indeed toolmakers.

Stone tools

Stone tools are first attested around 2.6 Million years ago, when *H. habilis* in Eastern Africa used so-called pebble tools, choppers made out of round pebbles that had been split by simple strikes. This marks the beginning of the Paleolithic, or Old Stone Age; its end is taken to be the end of the last Ice Age, around 10,000 years ago. The Paleolithic is subdivided into the Lower Paleolithic (Early Stone Age), ending around 350,000–300,000 years ago, the Middle Paleolithic (Middle Stone Age), until 50,000–30,000 years ago, and the Upper Paleolithic, (Late Stone Age), 50,000-10,000 years ago.

Archaeologists working in the Great Rift Valley in Kenya claim to have discovered the oldest known stone tools in the world. Dated to around 3.3 million years ago, the implements are some 700,000 years older than stone tools from Ethiopia that previously held this distinction.

The period from 700,000–300,000 years ago is also known as the Acheulean, when *H. ergaster* (or *erectus*) made large stone hand axes out of flint and quartzite, at first quite rough (Early Acheulian), later "retouched" by additional, more-subtle strikes at the sides of the flakes. After 350,000 BP the more refined so-called Levallois technique was developed, a series of consecutive strikes, by which scrapers, slicers ("racloirs"), needles, and flattened needles were made. Finally, after about 50,000 BP, ever more refined and specialized flint tools were made by the Neanderthals and the immigrant Cro-Magnons (knives, blades, skimmers). In this period they also started to make tools out of bone.

Transition to behavioral modernity

Until about 50,000–40,000 years ago, the use of stone tools seems to have progressed stepwise. Each phase (*H. habilis*, *H. ergaster*, *H. neanderthalensis*) started at a higher level than the previous one, but after each phase started, further development was slow. Currently paleoanthropologists are debating whether these *Homo* species possessed some or many of the cultural and behavioral traits associated with modern humans such as language, complex symbolic thinking, technological creativity etc. It seems that they were culturally conservative maintaining simple technologies and foraging patterns over very long periods.

Around 50,000 BP, modern human culture started to evolve more rapidly. The transition to behavioral modernity has been characterized by most as a Eurasian "Great Leap Forward", or as the "Upper Palaeolithic Revolution", due to the sudden appearance of distinctive signs of modern behavior and big game hunting in the archaeological record. Some other scholars consider the transition to have been more gradual, noting that some features had already appeared among archaic African *Homo sapiens* since 200,000 years ago. Recent evidence suggests that the Australian Aboriginal population separated from the African population 75,000 years ago, and that they made a sea journey of up to 160 km 60,000 years ago, which may diminish the evidence of the Upper Paleolithic Revolution.

Modern humans started burying their dead, using animal hides to make clothing, hunting with more sophisticated techniques (such as using trapping pits or driving animals off cliffs), and engaging in cave painting. As human culture advanced, different populations of humans introduced novelty to existing technologies: artifacts such as fish hooks, buttons, and bone needles show signs of variation among different populations of humans, something that had not been seen in human cultures prior to 50,000 BP. Typically, *H. neanderthalensis* populations do not vary in their technologies, although the Chatelperronian assemblages have been found to be Neanderthal innovations produced as a result of exposure to the Homo sapiens Aurignacian technologies.

Among concrete examples of modern human behavior, anthropologists include specialization of tools, use of jewellery and images (such as cave drawings), organization of living space, rituals (for example, burials with grave gifts), specialized hunting techniques, exploration of less hospitable geographical areas, and barter trade networks. Debate continues as to whether a "revolution" led to modern humans ("the big bang of human consciousness"), or whether the evolution was more "gradual".

Recent and current human evolution

Evolution has continued in anatomically modern human populations, which are affected by both natural selection and genetic drift. Although selection pressure on some traits, such as resistance to smallpox, has decreased in modern human life, humans are still undergoing natural selection for many other traits. Some of these are due to specific environmental pressures, while others are related to lifestyle changes since the development of agriculture (10,000 years ago), urban civilization (5,000), and industrialization (250 years ago). It has been argued that human evolution has accelerated since the development of agriculture 10,000 years ago and civilization some 5,000 years ago, resulting, it is claimed, in substantial genetic differences between different current human populations.

Particularly conspicuous is variation in superficial characteristics, such as Afro-textured hair, or the recent evolution of light skin and blond hair in some populations, which are attributed to differences in climate. Particularly strong selective pressures have resulted in high-altitude adaptation in humans, with different ones in different isolated populations. Studies of the genetic basis show that some developed very recently, with Tibetans evolving over 3,000 years to have high proportions of an allele of EPAS1 that is adaptive to high altitudes.

Other evolution is related to endemic diseases: the presence of malaria selected for sickle cell trait (the heterozygote form of sickle cell gene), while the absence of malaria and the health effects of sickle-cell anemia select against this trait. For example, the population at risk of the severe debilitating disease kuru has significant over-representation of an immune variant of the prion protein gene G127V versus non-immune alleles. The frequency of this genetic variant is due to the survival of immune persons.

Recent human evolution related to agriculture includes genetic resistance to infectious disease that has appeared in human populations by crossing the species barrier from domesticated animals, as well as changes in metabolism due to changes in diet, such as lactase persistence.

In contemporary times, since industrialization, some trends have been observed: for instance, menopause is evolving to occur later. Other reported trends appear to include lengthening of the human reproductive period and reduction in cholesterol levels, blood glucose and blood pressure in some populations.

Species list

See also: List of Homo species, Human evolution/Species chart

This list is in chronological order across the table by *genus*. Some species/subspecies names are well-established, and some are less established – especially in genus *Homo*. Please see articles for more information.

Sahelanthropus	*Homo* (human)
S. tchadensis	*H. gautengensis*
Orrorin	*H. habilis*
O. tugenensis	*H. rudolfensis*
Ardipithecus	*H. floresiensis*
A. kadabba	*H. ergaster*
A. ramidus	*H. erectus*
Australopithecus	• *H. e. georgicus*
A. anamensis	*H. cepranensis*
A. afarensis	*H. antecessor*

A. bahrelghazali	H. heidelbergensis
A. africanus	H. rhodesiensis
A. garhi	H. naledi
A. sediba	H. helmei
Kenyanthropus	H. neanderthalensis
K. platyops	H. sapiens
Paranthropus	• H. s. idaltu
P. aethiopicus	• H. s. sapiens (early)
P. boisei	• H. s. sapiens (modern)
P. robustus	

Bibliography

<templatestyles src="Template:Refbegin/styles.css" />

• Aiello, Leslie; Dean, Christopher (1990). *An Introduction to Human Evolutionary Anatomy*. London; San Diego: Elsevier Academic Press. ISBN 0-12-045591-9. LCCN 95185095[745]. OCLC 33408268[746].

• American Heritage Dictionaries (editors) (2006). *More Word Histories and Mysteries: From Aardvark to Zombie*. Boston, MA: Houghton Mifflin. ISBN 978-0-618-71681-4. LCCN 2006020835[747]. OCLC 70199867[748].

• Bogin, Barry (1997). "The Evolution of Human Nutrition". In Romanucci-Ross, Lola; Moerman, Daniel E.; Tancredi, Laurence R. *The Anthropology of Medicine: From Culture to Method* (3rd ed.). Westport, CT: Bergin & Garvey. ISBN 0-89789-516-9. LCCN 96053993[749]. OCLC 36165190[750].

• Bown, Thomas M.; Rose, Kenneth D. (1987). *Patterns of Dental Evolution in Early Eocene Anaptomorphine Primates (Omomyidae) From the Bighorn Basin, Wyoming*[751] (PDF). Memoir (Paleontological Society). **23**. Tulsa, OK: Paleontological Society. OCLC 16997265[752].

• Boyd, Robert; Silk, Joan B. (2003). *How Humans Evolved* (3rd ed.). New York: Norton. ISBN 0-393-97854-0. LCCN 2002075336[753]. OCLC 49959461[754].

• Brues, Alice M.; Snow, Clyde C. (1965). "Physical Anthropology". In Siegel, Bernard J. *Biennial Review of Anthropology 1965*. *Biennial Review of Anthropology*. **4**. Stanford, CA: Stanford University Press. ISBN 0-8047-1746-X. ISSN 0067-8503[755]. LCCN 59012726[756]. OCLC 01532912[757].

• Bryson, Bill (2004) [Originally published 2003]. "The Mysterious Biped". *A Short History of Nearly Everything.* Toronto, Canada: Anchor Canada. ISBN 0-385-66004-9. LCCN 2003046006[758]. OCLC 55016591[759].

• Cameron, David W. (2004). *Hominid Adaptations and Extinctions.* Sydney, NSW: UNSW Press. ISBN 0-86840-716-X. LCCN 2004353026[760]. OCLC 57077633[761].

• Cochran, Gregory; Harpending, Henry (2009). *The 10,000 Year Explosion: How Civilization Accelerated Human Evolution.* New York: Basic Books. ISBN 978-0-465-00221-4. LCCN 2008036672[762]. OCLC 191926088[763].

• Curry, James R. (2008). *Children of God: Children of Earth.* Bloomington, IN: AuthorHouse. ISBN 978-1-4389-1846-4. OCLC 421466369[764].

• Darwin, Charles (1981) [Originally published 1871; London: John Murray]. *The Descent of Man, and Selection in Relation to Sex.* Introduction by John Tyler Bonner and Robert M. May (Reprint ed.). Princeton, NJ: Princeton University Press. ISBN 0-691-02369-7. LCCN 80008679[765]. OCLC 7197127[766].

• Dawkins, Richard (2004). *The Ancestor's Tale: A Pilgrimage to the Dawn of Evolution.* Boston, MA: Houghton Mifflin. ISBN 0-618-00583-8. LCCN 2004059864[767]. OCLC 56617123[768].

• DeSalle, Rob; Tattersall, Ian (2008). *Human Origins: What Bones and Genomes Tell Us About Ourselves.* Texas A&M University Anthropology Series. **13** (1st ed.). College Station, TX: Texas A&M University Press. ISBN 978-1-58544-567-7. OCLC 144520427[769].

• Diamond, Jared (1999). *Guns, Germs, and Steel: The Fates of Human Societies.* New York: W. W. Norton & Company. ISBN 0-393-31755-2. LCCN 2005284124[770]. OCLC 35792200[771].

• Freeman, Scott; Herron, Jon C. (2007). *Evolutionary Analysis* (4th ed.). Upper Saddle River, NJ: Pearson Prentice Hall. ISBN 978-0-13-227584-2. LCCN 2006034384[772]. OCLC 73502978[773].

• Galinon-Melenec, Béatrice (2015). "From "TRACES" and "HUMAN TRACE" to "HUMAN-TRACE PARADIGM"". In Parrend, Pierre; Bourgine, Paul; Collet, Pierre. *First Complex systems Digital Campus World E-Conference.* Tempe Arizona USA: Springer.

• Johanson, Donald; Edey, Maitland (1981). *Lucy, the Beginnings of Humankind.* St Albans: Granada. ISBN 0-586-08437-1.

• Montgomery, William M. (1988) [Originally published 1974]. "Germany". In Glick, Thomas F. *The Comparative Reception of Darwinism.* Chicago, Illinois: University of Chicago Press. ISBN 0-226-29977-5. LCCN 87035814[774]. OCLC 17328115[775]. "The Conference on the Comparative Reception of Darwinism was held in Austin, Texas, on April 22 and 23, 1972, under the joint sponsorship of the American Council of

Learned Societies and the University of Texas at Austin"
- Kondo, Shiro, ed. (1985). *Primate Morphophysiology, Locomotor Analyses, and Human Bipedalism*. Tokyo: University of Tokyo Press. doi: 10.1002/ajpa.1330700214[776]. ISBN 4-13-066093-4. LCCN 85173489[777]. OCLC 12352830[778].
- Leakey, Richard E. (1994). *The Origin of Humankind*. Science Masters Series. New York: Basic Books. ISBN 0-465-03135-8. LCCN 94003617[779]. OCLC 30739453[780].
- M'charek, Amade (2005). *The Human Genome Diversity Project: An Ethnography of Scientific Practice*. Cambridge Studies in Society and the Life Sciences. Cambridge; New York: Cambridge University Press. ISBN 0-521-83222-5. LCCN 2004052648[781]. OCLC 55600894[782].
- Martin, Robert D. (2001). "Primates, Evolution of". In Smelser, Neil J.; Baltes, Paul B. *International Encyclopedia of the Social & Behavioral Sciences*. *International Encyclopedia of the Social & Behavioral Sciences* (1st ed.). Amsterdam; New York: Elsevier. p. 12032. doi: 10.1016/B0-08-043076-7/03083-7[783]. ISBN 978-0-08-043076-8. LCCN 2001044791[784]. OCLC 47869490[785].
- Maxwell, Mary (1984). *Human Evolution: A Philosophical Anthropology*. New York: Columbia University Press. ISBN 0-231-05946-9. LCCN 83024005[786]. OCLC 10163036[787].
- McHenry, Henry M. (2009). "Human Evolution". In Ruse, Michael; Travis, Joseph. *Evolution: The First Four Billion Years*. Foreword by Edward O. Wilson. Cambridge, Massachusetts: Belknap Press of Harvard University Press. ISBN 978-0-674-03175-3. LCCN 2008030270[788]. OCLC 225874308[789].
- Ramachandran, Sohini; Hua Tang; Gutenkunst, Ryan N.; Bustamante, Carlos D. (2010). "Genetics and Genomics of Human Population Structure". In Speicher, Michael R.; Antonarakis, Stylianos E.; Motulsky, Arno G. *Vogel and Motulsky's Human Genetics: Problems and Approaches* (4th completely rev. ed.). Heidelberg; London; New York: Springer. doi: 10.1007/978-3-540-37654-5[790]. ISBN 978-3-540-37653-8. LCCN 2009931325[791]. OCLC 549541244[792].
- Robinson, J. T. (2008) [Originally published 1963; Chicago, Illinois: Aldine Transaction]. "Adaptive Radiation in the Australopithecines and the Origin of Man". In Howell, F. Clark; Bourlière, François. *African Ecology and Human Evolution*. New Brunswick, NJ: Transaction Publishers. ISBN 978-0-202-36136-9. LCCN 2007024716[793]. OCLC 144770218[794].
- Srivastava, R. P. (2009). *Morphology of the Primates and Human Evolution*. New Delhi: PHI Learning Private Limited. ISBN 978-81-203-3656-8. OCLC 423293609[795].

- Stanford, Craig; Allen, John S.; Antón, Susan C. (2009). *Biological Anthropology: The Natural History of Humankind* (2nd ed.). Upper Saddle River, NJ: Pearson Prentice Hall. ISBN 978-0-13-601160-6. LCCN 2007052429[796]. OCLC 187548835[797].
- Strickberger, Monroe W. (2000). *Evolution* (3rd ed.). Sudbury, MA: Jones and Bartlett Publishers. ISBN 0-7637-1066-0. LCCN 99032072[798]. OCLC 41431683[799].
- Stringer, Chris B. (1994) [First published 1992]. "Evolution of Early Humans". In Jones, Steve; Martin, Robert D.; Pilbeam, David. *The Cambridge Encyclopedia of Human Evolution*. Foreword by Richard Dawkins (1st paperback ed.). Cambridge, UK: Cambridge University Press. ISBN 0-521-32370-3. LCCN 92018037[800]. OCLC 444512451[801].
- Swisher, Carl C., III; Curtis, Garniss H.; Lewin, Roger (2001) [Originally published 2000]. *Java Man: How Two Geologists Changed Our Understanding of Human Evolution*. Chicago, Illinois: University of Chicago Press. ISBN 0-226-78734-6. LCCN 2001037337[802]. OCLC 48066180[803].
- Tishkoff, S.A.; Reed, F.A.; et al. (2009). "The Genetic Structure and History of Africans and African Americans"[804]. *Science* (published 22 May 2009). **324** (5930): 1035–1044. Bibcode: 2009Sci...324.1035T[805]. doi: 10.1126/science.1172257[806]. PMC 2947357[804] ◌. PMID 19407144[807].
- Trent, Ronald J. (2005). *Molecular Medicine: An Introductory Text* (3rd ed.). Burlington, MA: Elsevier Academic Press. ISBN 0-12-699057-3. LCCN 2004028087[808]. OCLC 162577235[809].
- Trevathan, Wenda R. (2011) [Originally published 1987; New York: Aldine De Gruyter]. *Human Birth: An Evolutionary Perspective*. New Brunswick, NJ: Transaction Publishers. ISBN 978-1-4128-1502-4. LCCN 2010038249[810]. OCLC 669122326[811].
- Ungar, Peter S.; Teaford, Mark F., eds. (2002). *Human Diet: Its Origin and Evolution*. Westport, CT: Bergin & Garvey. ISBN 0-89789-736-6. LCCN 2001043790[812]. OCLC 537239907[813].
- Walker, Alan (2007). "Early Hominin Diets: Overview and Historical Perspectives". In Ungar, Peter. *Evolution of the Human Diet: The Known, the Unknown, and the Unknowable*. Human Evolution Series. Oxford; New York: Oxford University Press. ISBN 978-0-19-518346-7. LCCN 2005036120[814]. OCLC 132816551[815].
- Wallace, David Rains (2004). *Beasts of Eden: Walking Whales, Dawn Horses, and Other Enigmas of Mammal Evolution*. Berkeley, CA: University of California Press. ISBN 0-520-24684-5. LCCN 2003022857[816]. OCLC 53254011[817].
- Webster, Donovan (2010). *Meeting the Family: One Man's Journey Through His Human Ancestry*. Foreword by Spencer Wells. Washington,

D.C.: National Geographic Society. ISBN 978-1-4262-0573-6. LCCN 2009050471[818]. OCLC 429022321[819].

- Wood, Bernard A. (2009). "Where Does the Genus *Homo* Begin, and How Would We Know?". In Grine, Frederick E.; Fleagle, John G.; Leakey, Richard E. *The First Humans: Origin and Early Evolution of the Genus Homo*. Vertebrate Paleobiology and Paleoanthropology. Dordrecht, the Netherlands: Springer Netherlands. doi: 10.1007/978-1-4020-9980-9_3[820]. ISBN 978-1-4020-9979-3. ISSN 1877-9077[821]. LCCN 2009927083[822]. OCLC 310400980[823]. "Contributions from the Third Stony Brook Human Evolution Symposium and Workshop October 3–7, 2006."

Further reading

<templatestyles src="Template:Refbegin/styles.css" />

- Alexander, Richard D. (1990). *How Did Humans Evolve? Reflections on the Uniquely Unique Species*[824] (PDF). Special Publication. Ann Arbor, MI: Museum of Zoology, University of Michigan. pp. 1–38. LCCN 90623893[825]. OCLC 22860997[826].
- Barton, Nicholas H.; Briggs, Derek E. G.; Eisen, Jonathan A.; et al. (2007). *Evolution*. Cold Spring Harbor, NY: Cold Spring Harbor Laboratory Press. ISBN 978-0-87969-684-9. LCCN 2007010767[827]. OCLC 86090399[828].
- Enard, Wolfgang; Przeworski, Molly; Fisher, Simon E.; et al. (August 22, 2002). "Molecular evolution of *FOXP2*, a gene involved in speech and language". *Nature*. London: Nature Publishing Group. **418** (6900): 869–872. Bibcode: 2002Natur.418..869E[829]. doi: 10.1038/nature01025[830]. ISSN 0028-0836[831]. PMID 12192408[832].
- Flinn, Mark V.; Geary, David C.; Ward, Carol V. (January 2005). "Ecological dominance, social competition, and coalitionary arms races: Why humans evolved extraordinary intelligence"[833] (PDF). *Evolution and Human Behavior*. Amsterdam, the Netherlands: Elsevier. **26** (1): 10–46. doi: 10.1016/j.evolhumbehav.2004.08.005[834]. ISSN 1090-5138[835]. Retrieved 2015-05-05.
- Gibbons, Ann (2006). *The First Human: The Race to Discover our Earliest Ancestors* (1st ed.). New York: Doubleday. ISBN 978-0-385-51226-8. LCCN 2005053780[836]. OCLC 61652817[837].
- Hartwig, Walter C., ed. (2002). *The Primate Fossil Record*. Cambridge Studies in Biological and Evolutionary Anthropology. **33**. Cambridge; New York: Cambridge University Press. ISBN 0-521-66315-6. LCCN 2001037847[838]. OCLC 47254191[839].

- Heizmann, Elmar P. J.; Begun, David R. (November 2001). "The oldest Eurasian hominoid". *Journal of Human Evolution*. Amsterdam, the Netherlands: Elsevier. **41** (5): 463–481. doi: 10.1006/jhev.2001.0495[840]. ISSN 0047-2484[841]. PMID 11681862[842].
- Hill, Andrew; Ward, Steven (1988). "Origin of the hominidae: The record of African large hominoid evolution between 14 my and 4 my". *American Journal of Physical Anthropology*. Hoboken, NJ: John Wiley & Sons for the American Association of Physical Anthropologists. **31** (59): 49–83. doi: 10.1002/ajpa.1330310505[843]. ISSN 0002-9483[844].
- Hoagland, Hudson (1964). "Science and the New Humanism". *Science*. **143** (3602): 111–114. Bibcode: 1964Sci...143..111H[845]. doi: 10.1126/science.143.3602.111[846].
- Ijdo, Jacob W.; Baldini, Antonio; Ward, David C.; et al. (October 15, 1991). "Origin of human chromosome 2: An ancestral telomere-telomere fusion"[847] (PDF). *Proc. Natl. Acad. Sci. U.S.A.* Washington, D.C.: National Academy of Sciences. **88** (20): 9051–9055. Bibcode: 1991PNAS...88.9051I[848]. doi: 10.1073/pnas.88.20.9051[849]. ISSN 0027-8424[850]. PMC 52649[851] ∂ . PMID 1924367[852]. Retrieved 2015-05-05. — two ancestral ape chromosomes fused to give rise to human chromosome 2
- Johanson, Donald; Wong, Kate (2010). *Lucy's Legacy: The Quest for Human Origins*. New York: Three Rivers Press. ISBN 978-0-307-39640-2. LCCN 2010483830[853]. OCLC 419801728[854].
- Jones, Steve; Martin, Robert D.; Pilbeam, David, eds. (1994) [First published 1992]. *The Cambridge Encyclopedia of Human Evolution*. Foreword by Richard Dawkins (1st paperback ed.). Cambridge, UK: Cambridge University Press. ISBN 0-521-32370-3. LCCN 92018037[800]. OCLC 444512451[801]. (Note: this book contains very useful, information dense chapters on primate evolution in general, and human evolution in particular, including fossil history).
- Leakey, Richard E.; Lewin, Roger (1992). *Origins Reconsidered: In Search of What Makes us Human*. New York: Doubleday. ISBN 0-385-41264-9. LCCN 92006661[855]. OCLC 25373161[856].
- Lewin, Roger (1997). *Bones of Contention: Controversies in the Search for Human Origins* (2nd ed.). Chicago, Illinois: University of Chicago Press. ISBN 0-226-47651-0. LCCN 97000972[857]. OCLC 36181117[858].
- Morwood, Mike; van Oosterzee, Penny (2007). *A New Human: The Startling Discovery and Strange Story of the 'Hobbits' of Flores, Indonesia* (1st Smithsonian Books ed.). New York: Smithsonian Books/HarperCollins. ISBN 978-0-06-089908-0. LCCN 2006052267[859]. OCLC 76481584[860].
- Oppenheimer, Stephen (2003). *Out of Eden: The Peopling of the*

World. london: Constable & Robinson. ISBN 1-84119-697-5. LCCN 2005482222[861]. OCLC 52195607[862].

- Ovchinnikov, Igor V.; Götherström, Anders; Romanova, Galina P.; et al. (March 30, 2000). "Molecular analysis of Neanderthal DNA from the Northern Caucasus". *Nature*. London: Nature Publishing Group. **404** (6777): 490–493. doi: 10.1038/35006625[863]. ISSN 0028-0836[831]. PMID 10761915[864].

- Roberts, Alice M. (2009). *The Incredible Human Journey: The Story of How We Colonised the Planet*. London: Bloomsbury Publishing. ISBN 978-0-7475-9839-8. OCLC 310156315[865].

- Shreeve, James (1995). *The Neandertal Enigma: Solving the Mystery of Modern Human Origins*. New York: Morrow. ISBN 0-688-09407-4. LCCN 95006337[866]. OCLC 32088673[867].

- Stringer, Chris B. (2011). *The Origin of Our Species*. London: Allen Lane. ISBN 978-1-84614-140-9. LCCN 2011489742[868]. OCLC 689522193[869].

- Stringer, Chris B.; Andrews, Peter (2005). *The Complete World of Human Evolution*. London; New York: Thames & Hudson. ISBN 0-500-05132-1. LCCN 2004110563[870]. OCLC 224377190[871].

- Stringer, Christopher; McKie, Robin (1997). *African Exodus: The Origins of Modern Humanity* (1st American ed.). New York: Henry Holt and Company. ISBN 0-8050-2759-9. LCCN 96037718[872]. OCLC 36001167[873].

- Tattersall, Ian (2008). *The Fossil Trail: How We Know What We Think We Know About Human Evolution* (2nd ed.). New York: Oxford University. ISBN 978-0195367669. LCCN 2008013654[874]. OCLC 218188644[875].

- van Oosterzee, Penny (1999). *Dragon Bones: The Story of Peking Man*. St Leonards, New South Wales: Allen & Unwin. ISBN 1-86508-123-X. LCCN 00300421[876]. OCLC 45853997[877].

- Wade, Nicholas (2006). *Before the Dawn: Recovering the Lost History of Our Ancestors*. New York: Penguin Press. ISBN 1-59420-079-3. LCCN 2005055293[878]. OCLC 62282400[879].

- Walker, Alan; Shipman, Pat (1996). *The Wisdom of the Bones: In Search of Human Origins*. London: Weidenfeld & Nicolson. ISBN 0-297-81670-5. OCLC 35202130[880].

- Weiss, Mark L.; Mann, Alan E. (1985). *Human Biology and Behavior: An Anthropological Perspective* (4th ed.). Boston: Little Brown. ISBN 0-316-92894-1. LCCN 85000158[881]. OCLC 11726796[882]. (Note: this book contains very accessible descriptions of human and non-human primates, their evolution, and fossil history).

- Wells, Spencer (2003) [Originally published 2002; Princeton, NJ: Princeton University Press]. *The Journey of Man: A Genetic Odyssey* (Random House trade paperback ed.). New York: Random House Trade Paperbacks. ISBN 0-8129-7146-9. LCCN 2003066679[883]. OCLC 53287806[884].

External links

> Wikimedia Commons has media related to *Human evolution*.

- "BBC - Science & Nature - The evolution of man"[885]. BBC. Retrieved 2015-05-06.
- "Becoming Human"[886]. Arizona State University's Institute of Human Origins. Retrieved 2015-05-06.
- "Bones, Stones and Genes: The Origin of Modern Humans"[887] (Video lecture series). Howard Hughes Medical Institute. Retrieved 2015-05-06.
- "*Evolution* Figures: Chapter 25"[888]. Cold Spring Harbor Laboratory Press. Retrieved 2015-05-06. — Illustrations from the book *Evolution* (2007)
- "Human Evolution"[889]. Smithsonian Institution's Human Origins Program. Retrieved 2013-06-24.
- "Human Evolution Timeline"[890]. ArchaeologyInfo.com. Retrieved 2013-06-24.
- "Human Trace" video 2015 [[Normandy University[891]]] UNIHAVRE, CNRS, IDEES, E.Laboratory on Human Trace Unitwin Complex System Digital Campus UNESCO.
- Lambert, Tim (Producer) (June 24, 2015). *First Peoples*[892]. London: Wall to Wall Television. OCLC 910115743[893]. Retrieved 2015-07-18.
- Shaping Humanity Video[894] 2013 Yale University
- Human Timeline (Interactive)[895] – Smithsonian, National Museum of Natural History (August 2016).
- Human Evolution[896], BBC Radio 4 discussion with Steve Jones, Fred Spoor & Margaret Clegg (*In Our Time*, Feb. 16, 2006)

Mass extinctions

Extinction event

The blue graph shows the apparent *percentage* (not the absolute number) of marine animal genera becoming extinct during any given time interval. It does not represent all marine species, just those that are readily fossilized. The labels of the traditional "Big Five" extinction events and the more recently recognised End-Capitanian extinction event are clickable hyperlinks. *(source and image info)*

An **extinction event** (also known as a **mass extinction** or **biotic crisis**) is a widespread and rapid decrease in the biodiversity on Earth. Such an event is identified by a sharp change in the diversity and abundance of multicellular organisms. It occurs when the rate of extinction increases with respect to the rate of speciation. Because most diversity and biomass on Earth is microbial, and thus difficult to measure, recorded extinction events affect the easily observed, biologically complex component of the biosphere rather than the total diversity and abundance of life.

Extinction occurs at an uneven rate. Based on the fossil record, the background rate of extinctions on Earth is about two to five taxonomic families of marine animals every million years. Marine fossils are mostly used to measure extinction rates because of their superior fossil record and stratigraphic range compared to land animals.

The Great Oxygenation Event was probably the first major extinction event.Wikipedia:Citation needed Since the Cambrian explosion five further major mass extinctions have significantly exceeded the background extinction rate. The most recent and arguably best-known, the Cretaceous–Paleogene extinction event, which occurred approximately 66 million years ago (Ma), was a large-scale mass extinction of animal and plant species in a geologically short period of time. In addition to the five major mass extinctions, there are numerous minor ones as well, and the ongoing mass extinction caused by human

Figure 274: *Badlands near Drumheller, Alberta, where erosion has exposed the K–Pg boundary*

activity is sometimes called the sixth extinction.[897] Mass extinctions seem to be a mainly Phanerozoic phenomenon, with extinction rates low before large complex organisms arose.

Estimates of the number of major mass extinctions in the last 540 million years range from as few as five to more than twenty. These differences stem from the threshold chosen for describing an extinction event as "major", and the data chosen to measure past diversity.

Major extinction events

In a landmark paper published in 1982, Jack Sepkoski and David M. Raup identified five mass extinctions. They were originally identified as outliers to a general trend of decreasing extinction rates during the Phanerozoic, but as more stringent statistical tests have been applied to the accumulating data, it has been established that multicellular animal life has experienced five major and many minor mass extinctions. The "Big Five" cannot be so clearly defined, but rather appear to represent the largest (or some of the largest) of a relatively smooth continuum of extinction events.

Figure 275: *Trilobites were highly successful marine animals until the Permian–Triassic extinction event wiped them all out*

1. **Ordovician–Silurian extinction events** (End Ordovician or O–S): 450–440 Ma (million years ago) at the Ordovician–Silurian transition. Two events occurred that killed off 27% of all families, 57% of all genera and 60% to 70% of all species. Together they are ranked by many scientists as the second largest of the five major extinctions in Earth's history in terms of percentage of genera that became extinct.

2. **Late Devonian extinction**: 375–360 Ma near the Devonian–Carboniferous transition. At the end of the Frasnian Age in the later part(s) of the Devonian Period, a prolonged series of extinctions eliminated about 19% of all families, 50% of all genera and at least 70% of all species. This extinction event lasted perhaps as long as 20 million years, and there is evidence for a series of extinction pulses within this period.

3. **Permian–Triassic extinction event** (End Permian): 252 Ma at the Permian–Triassic transition. Earth's largest extinction killed 57% of all families, 83% of all genera and 90% to 96% of all species (53% of marine families, 84% of marine genera, about 96% of all marine species and an estimated 70% of land species, including insects). The highly successful marine arthropod, the trilobite, became extinct. The evidence regarding plants is less clear, but new taxa became dominant after the extinction.

The "Great Dying" had enormous evolutionary significance: on land, it ended the primacy of mammal-like reptiles. The recovery of vertebrates took 30 million years, but the vacant niches created the opportunity for archosaurs to become ascendant. In the seas, the percentage of animals that were sessile dropped from 67% to 50%. The whole late Permian was a difficult time for at least marine life, even before the "Great Dying".

4. **Triassic–Jurassic extinction event** (End Triassic): 201.3 Ma at the Triassic–Jurassic transition. About 23% of all families, 48% of all genera (20% of marine families and 55% of marine genera) and 70% to 75% of all species became extinct. Most non-dinosaurian archosaurs, most therapsids, and most of the large amphibians were eliminated, leaving dinosaurs with little terrestrial competition. Non-dinosaurian archosaurs continued to dominate aquatic environments, while non-archosaurian diapsids continued to dominate marine environments. The Temnospondyl lineage of large amphibians also survived until the Cretaceous in Australia (e.g., *Koolasuchus*).

5. **Cretaceous–Paleogene extinction event** (End Cretaceous, K–Pg extinction, or formerly K–T extinction): 66 Ma at the Cretaceous (Maastrichtian) – Paleogene (Danian) transition interval. The event formerly called the Cretaceous-Tertiary or K–T extinction or K–T boundary is now officially named the Cretaceous–Paleogene (or K–Pg) extinction event. About 17% of all families, 50% of all genera and 75% of all species became extinct. In the seas all the ammonites, plesiosaurs and mosasaurs disappeared and the percentage of sessile animals (those unable to move about) was reduced to about 33%. All non-avian dinosaurs became extinct during that time. The boundary event was severe with a significant amount of variability in the rate of extinction between and among different clades. Mammals and birds, the latter descended from theropod dinosaurs, emerged as dominant large land animals.

Despite the popularization of these five events, there is no definite line separating them from other extinction events; using different methods of calculating an extinction's impact can lead to other events featuring in the top five.

Older fossil records are more difficult to interpret. This is because:

- Older fossils are harder to find as they are usually buried at a considerable depth.
- Dating older fossils is more difficult.
- Productive fossil beds are researched more than unproductive ones, therefore leaving certain periods unresearched.
- Prehistoric environmental events can disturb the deposition process.
- The preservation of fossils varies on land, but marine fossils tend to be better preserved than their sought after land-based counterparts.[898]

It has been suggested that the apparent variations in marine biodiversity may actually be an artifact, with abundance estimates directly related to quantity of rock available for sampling from different time periods. However, statistical analysis shows that this can only account for 50% of the observed pattern,Wikipedia:Citation needed and other evidence (such as fungal spikes)Wikipedia:Please clarify provides reassurance that most widely accepted extinction events are real. A quantification of the rock exposure of Western Europe indicates that many of the minor events for which a biological explanation has been sought are most readily explained by sampling bias.

Research completed after the seminal 1982 paper has concluded that a sixth mass extinction event is ongoing:

6. **Holocene extinction**: Currently ongoing. Extinctions have occurred at over 1000 times the background extinction rate since 1900. The mass extinction is considered a result of human activity.

More recent research has indicated that the **End-Capitanian extinction event** likely constitutes a separate extinction event from the Permian–Triassic extinction event; if so, it would be larger than many of the "Big Five" extinction events.

List of extinction events

This is a list of extinction events:[899]

Period or supereon	Extinction	Date	Possible causes
Quaternary	**Holocene extinction**	c. 10,000 BCE — Ongoing	Humans
	Quaternary extinction event	640,000, 74,000, and 13,000 years ago	Unknown; may include climate changes, massive volcanic eruptions and human overhunting
Neogene	Pliocene–Pleistocene boundary extinction	2 Ma	Supernova? Eltanin impact?
	Middle Miocene disruption	14.5 Ma	– climate change due to change of ocean circulation patterns and perhaps related to the Milankovitch cycles?.
Paleogene	Eocene–Oligocene extinction event	33.9 Ma	Popigai impactor?

Creta-ceous	**Cretaceous–Paleogene extinction event**	66 Ma	Chicxulub impactor; Deccan Traps?
	Cenomanian-Turonian boundary event	94 Ma	Caribbean large igneous province
	Aptian extinction	117 Ma	
Jurassic	End-Jurassic (Tithonian) extinction	145 Ma	
	Toarcian turnover	183 Ma	Karoo-Ferrar Provinces
Triassic	**Triassic–Jurassic extinction event**	201 Ma	Central Atlantic magmatic province; impactor
	Carnian Pluvial Event	230 Ma	Wrangellia flood basalts
Permian	**Permian–Triassic extinction event**	252 Ma	Siberian Traps; Wilkes Land Crater;Anoxic event
	End-Capitanian extinction event	260 Ma	Emeishan Traps?
	Olson's Extinction	270 Ma	
Carbonif-erous	Carboniferous rainforest collapse	305 Ma	
Devonian	**Late Devonian extinction**	375–360 Ma	Viluy Traps
Silurian	Lau event	420 Ma	Changes in sea level and chemistry?
	Mulde event	424 Ma	Global drop in sea level?
	Ireviken event	428 Ma	Deep-ocean anoxia; Milankovitch cycles?
Ordovi-cian	**Ordovician–Silurian extinction events**	450–440 Ma	Global cooling and sea level drop; Gamma-ray burst?
Cambrian	Cambrian–Ordovician extinction event	488 Ma	
	Dresbachian extinction event	502 Ma	
	End-Botomian extinction event	517 Ma	
Precam-brian	End-Ediacaran extinction	542 Ma	
	Great Oxygenation Event	2400 Ma	Rising oxygen levels in the atmosphere due to the development of photosynthesis

Evolutionary importance

Life timeline

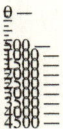

θ —
500
1000
1500
2000
2500
3000
3500
4000
4500

Axis scale: million years

🖑

Also see: *Human timeline* and *Nature timeline*

Mass extinctions have sometimes accelerated the evolution of life on Earth. When dominance of particular ecological niches passes from one group of organisms to another, it is rarely because the new dominant group is "superior" to the old and usually because an extinction event eliminates the old dominant group and makes way for the new one.

For example, mammaliformes ("almost mammals") and then mammals existed throughout the reign of the dinosaurs, but could not compete for the large terrestrial vertebrate niches which dinosaurs monopolized. The end-Cretaceous mass extinction removed the non-avian dinosaurs and made it possible for mammals to expand into the large terrestrial vertebrate niches. Ironically, the dinosaurs themselves had been beneficiaries of a previous mass extinction, the end-Triassic, which eliminated most of their chief rivals, the crurotarsans.

Another point of view put forward in the Escalation hypothesis predicts that species in ecological niches with more organism-to-organism conflict will be less likely to survive extinctions. This is because the very traits that keep a species numerous and viable under fairly static conditions become a burden once population levels fall among competing organisms during the dynamics of an extinction event.

Furthermore, many groups which survive mass extinctions do not recover in numbers or diversity, and many of these go into long-term decline, and these are often referred to as "Dead Clades Walking".

Darwin was firmly of the opinion that biotic interactions, such as competition for food and space—the 'struggle for existence'—were of considerably greater importance in promoting evolution and extinction than changes in the physical environment. He expressed this in *The Origin of Species*: "Species are produced and exterminated by slowly acting causes...and the most import of all causes of organic change is one which is almost independent of altered...physical conditions, namely the mutual relation of organism to organism-the improvement of one organism entailing the improvement or extermination of others".[900]

Patterns in frequency

It has been suggested variously that extinction events occurred periodically, every 26 to 30 million years, or that diversity fluctuates episodically every ~62 million years.[901] Various ideas attempt to explain the supposed pattern, including the presence of a hypothetical companion star to the sun, oscillations in the galactic plane, or passage through the Milky Way's spiral arms. However, other authors have concluded the data on marine mass extinctions do not fit with the idea that mass extinctions are periodic, or that ecosystems gradually build up to a point at which a mass extinction is inevitable. Many of the proposed correlations have been argued to be spurious. Others have argued that there is strong evidence supporting periodicity in a variety of records, and additional evidence in the form of coincident periodic variation in nonbiological geochemical variables.

File:Phanerozoic biodiversity blank 01.png

Phanerozoic biodiversity as shown by the fossil record

Mass extinctions are thought to result when a long-term stress is compounded by a short term shock. Over the course of the Phanerozoic, individual taxa appear to be less likely to become extinct at any time, which may reflect more robust food webs as well as less extinction-prone species and other factors such as continental distribution. However, even after accounting for sampling bias, there does appear to be a gradual decrease in extinction and origination rates during the Phanerozoic. This may represent the fact that groups with higher turnover rates are more likely to become extinct by chance; or it may be an artefact of taxonomy: families tend to become more speciose, therefore less prone to extinction, over time; and larger taxonomic groups (by definition) appear earlier in geological time.

It has also been suggested that the oceans have gradually become more hospitable to life over the last 500 million years, and thus less vulnerable to mass extinctions,[902] but susceptibility to extinction at a taxonomic level does not appear to make mass extinctions more or less probable.

Causes

There is still debate about the causes of all mass extinctions. In general, large extinctions may result when a biosphere under long-term stress undergoes a short-term shock. An underlying mechanism appears to be present in the correlation of extinction and origination rates to diversity. High diversity leads to a persistent increase in extinction rate; low diversity to a persistent increase in origination rate. These presumably ecologically controlled relationships likely amplify smaller perturbations (asteroid impacts, etc.) to produce the global effects observed.

Identifying causes of particular mass extinctions

A good theory for a particular mass extinction should: (i) explain all of the losses, not just focus on a few groups (such as dinosaurs); (ii) explain why particular groups of organisms died out and why others survived; (iii) provide mechanisms which are strong enough to cause a mass extinction but not a total extinction; (iv) be based on events or processes that can be shown to have happened, not just inferred from the extinction.

It may be necessary to consider combinations of causes. For example, the marine aspect of the end-Cretaceous extinction appears to have been caused by several processes which partially overlapped in time and may have had different levels of significance in different parts of the world.

Arens and West (2006) proposed a "press / pulse" model in which mass extinctions generally require two types of cause: long-term pressure on the ecosystem ("press") and a sudden catastrophe ("pulse") towards the end of the period of pressure.[903] Their statistical analysis of marine extinction rates throughout the Phanerozoic suggested that neither long-term pressure alone nor a catastrophe alone was sufficient to cause a significant increase in the extinction rate.

Most widely supported explanations

Macleod (2001) summarized the relationship between mass extinctions and events which are most often cited as causes of mass extinctions, using data from Courtillot et al. (1996),[904] Hallam (1992) and Grieve et al. (1996):

- Flood basalt events: 11 occurrences, all associated with significant extinctions[905,906] But Wignall (2001) concluded that only five of the major extinctions coincided with flood basalt eruptions and that the main phase of extinctions started before the eruptions.
- Sea-level falls: 12, of which seven were associated with significant extinctions.
- Asteroid impacts: one large impact is associated with a mass extinction, i.e. the Cretaceous–Paleogene extinction event; there have been many smaller impacts but they are not associated with significant extinctions.

The most commonly suggested causes of mass extinctions are listed below.

Flood basalt events

The formation of large igneous provinces by flood basalt events could have:

* produced dust and particulate aerosols which inhibited photosynthesis and thus caused food chains to collapse both on land and at sea[907]
* emitted sulfur oxides which were precipitated as acid rain and poisoned many organisms, contributing further to the collapse of food chains
* emitted carbon dioxide and thus possibly causing sustained global warming once the dust and particulate aerosols dissipated.

Flood basalt events occur as pulses of activity punctuated by dormant periods. As a result, they are likely to cause the climate to oscillate between cooling and warming, but with an overall trend towards warming as the carbon dioxide they emit can stay in the atmosphere for hundreds of years.

It is speculated that massive volcanism caused or contributed to the End-Permian, End-Triassic and End-Cretaceous extinctions. The correlation between gigantic volcanic events expressed in the large igneous provinces and mass extinctions was shown for the last 260 Myr.[908] Recently such possible correlation was extended for the whole Phanerozoic Eon.

Sea-level falls

These are often clearly marked by worldwide sequences of contemporaneous sediments which show all or part of a transition from sea-bed to tidal zone to beach to dry land – and where there is no evidence that the rocks in the relevant areas were raised by geological processes such as orogeny. Sea-level falls could reduce the continental shelf area (the most productive part of the oceans) sufficiently to cause a marine mass extinction, and could disrupt weather patterns enough to cause extinctions on land. But sea-level falls are very probably the result of other events, such as sustained global cooling or the sinking of the mid-ocean ridges.

Sea-level falls are associated with most of the mass extinctions, including all of the "Big Five"—End-Ordovician, Late Devonian, End-Permian, End-Triassic, and End-Cretaceous.

A study, published in the journal Nature (online June 15, 2008) established a relationship between the speed of mass extinction events and changes in sea level and sediment. The study suggests changes in ocean environments related to sea level exert a driving influence on rates of extinction, and generally determine the composition of life in the oceans.[909]

Impact events

The impact of a sufficiently large asteroid or comet could have caused food chains to collapse both on land and at sea by producing dust and particulate aerosols and thus inhibiting photosynthesis. Impacts on sulfur-rich rocks could have emitted sulfur oxides precipitating as poisonous acid rain, contributing further to the collapse of food chains. Such impacts could also have caused megatsunamis and/or global forest fires.

Most paleontologists now agree that an asteroid did hit the Earth about 66 Ma ago, but there is an ongoing dispute whether the impact was the sole cause of the Cretaceous–Paleogene extinction event.

Global cooling

Sustained and significant global cooling could kill many polar and temperate species and force others to migrate towards the equator; reduce the area available for tropical species; often make the Earth's climate more arid on average, mainly by locking up more of the planet's water in ice and snow. The glaciation cycles of the current ice age are believed to have had only a very mild impact on biodiversity, so the mere existence of a significant cooling is not sufficient on its own to explain a mass extinction.

It has been suggested that global cooling caused or contributed to the End-Ordovician, Permian–Triassic, Late Devonian extinctions, and possibly others. Sustained global cooling is distinguished from the temporary climatic effects of flood basalt events or impacts.

Global warming

This would have the opposite effects: expand the area available for tropical species; kill temperate species or force them to migrate towards the poles; possibly cause severe extinctions of polar species; often make the Earth's climate wetter on average, mainly by melting ice and snow and thus increasing the volume of the water cycle. It might also cause anoxic events in the oceans (see below).

Global warming as a cause of mass extinction is supported by several recent studies.

The most dramatic example of sustained warming is the Paleocene–Eocene Thermal Maximum, which was associated with one of the smaller mass extinctions. It has also been suggested to have caused the Triassic–Jurassic extinction event, during which 20% of all marine families became extinct. Furthermore, the Permian–Triassic extinction event has been suggested to have been caused by warming.

Clathrate gun hypothesis

Clathrates are composites in which a lattice of one substance forms a cage around another. Methane clathrates (in which water molecules are the cage) form on continental shelves. These clathrates are likely to break up rapidly and release the methane if the temperature rises quickly or the pressure on them drops quickly—for example in response to sudden global warming or a sudden drop in sea level or even earthquakes. Methane is a much more powerful greenhouse gas than carbon dioxide, so a methane eruption ("clathrate gun") could cause rapid global warming or make it much more severe if the eruption was itself caused by global warming.

The most likely signature of such a methane eruption would be a sudden decrease in the ratio of carbon-13 to carbon-12 in sediments, since methane clathrates are low in carbon-13; but the change would have to be very large, as other events can also reduce the percentage of carbon-13.

It has been suggested that "clathrate gun" methane eruptions were involved in the end-Permian extinction ("the Great Dying") and in the Paleocene–Eocene Thermal Maximum, which was associated with one of the smaller mass extinctions.

Anoxic events

Anoxic events are situations in which the middle and even the upper layers of the ocean become deficient or totally lacking in oxygen. Their causes are complex and controversial, but all known instances are associated with severe and sustained global warming, mostly caused by sustained massive volcanism.

It has been suggested that anoxic events caused or contributed to the Ordovician–Silurian, late Devonian, Permian–Triassic and Triassic–Jurassic extinctions, as well as a number of lesser extinctions (such as the Ireviken, Mulde, Lau, Toarcian and Cenomanian–Turonian events). On the other hand, there are widespread black shale beds from the mid-Cretaceous which indicate anoxic events but are not associated with mass extinctions.

The bio-availability of essential trace elements (in particular selenium) to potentially lethal lows has been shown to coincide with, and likely have contributed to, at least three mass extinction events in the oceans, i.e. at the end of the Ordovician, during the Middle and Late Devonian, and at the end of the Triassic. During periods of low oxygen concentrations very soluble selenate (Se^{6+}) is converted into much less soluble selenide (Se^{2+}), elemental Se and organo-selenium complexes. Bio-availability of selenium during these extinction events dropped to about 1% of the current oceanic concentration, a level that has been proven lethal to many extant organisms.

Hydrogen sulfide emissions from the seas

Kump, Pavlov and Arthur (2005) have proposed that during the Permian–Triassic extinction event the warming also upset the oceanic balance between photosynthesising plankton and deep-water sulfate-reducing bacteria, causing massive emissions of hydrogen sulfide which poisoned life on both land and sea and severely weakened the ozone layer, exposing much of the life that still remained to fatal levels of UV radiation.[910,911]

Oceanic overturn

Oceanic overturn is a disruption of thermo-haline circulation which lets surface water (which is more saline than deep water because of evaporation) sink straight down, bringing anoxic deep water to the surface and therefore killing most of the oxygen-breathing organisms which inhabit the surface and middle depths. It may occur either at the beginning or the end of a glaciation, although an overturn at the start of a glaciation is more dangerous because the preceding warm period will have created a larger volume of anoxic water.

Unlike other oceanic catastrophes such as regressions (sea-level falls) and anoxic events, overturns do not leave easily identified "signatures" in rocks and are theoretical consequences of researchers' conclusions about other climatic and marine events.

It has been suggested that oceanic overturn caused or contributed to the late Devonian and Permian–Triassic extinctions.

A nearby nova, supernova or gamma ray burst

A nearby gamma-ray burst (less than 6000 light-years away) would be powerful enough to destroy the Earth's ozone layer, leaving organisms vulnerable to ultraviolet radiation from the Sun. Gamma ray bursts are fairly rare, occurring only a few times in a given galaxy per million years. It has been suggested that a supernova or gamma ray burst caused the End-Ordovician extinction.

Geomagnetic reversal

One theory is that periods of increased geomagnetic reversals will weaken Earth's magnetic field long enough to expose the atmosphere to the solar winds, causing oxygen ions to escape the atmosphere in a rate increased by 3–4 orders, resulting in a disastrous decrease in oxygen.

Plate tectonics

Movement of the continents into some configurations can cause or contribute to extinctions in several ways: by initiating or ending ice ages; by changing ocean and wind currents and thus altering climate; by opening seaways or land bridges which expose previously isolated species to competition for which they are poorly adapted (for example, the extinction of most of South America's native ungulates and all of its large metatherians after the creation of a land bridge between North and South America). Occasionally continental drift creates a super-continent which includes the vast majority of Earth's land area, which in addition to the effects listed above is likely to reduce the total area of continental shelf (the most species-rich part of the ocean) and produce a vast, arid continental interior which may have extreme seasonal variations.

Another theory is that the creation of the super-continent Pangaea contributed to the End-Permian mass extinction. Pangaea was almost fully formed at the transition from mid-Permian to late-Permian, and the "Marine genus diversity" diagram at the top of this article shows a level of extinction starting at that time which might have qualified for inclusion in the "Big Five" if it were not overshadowed by the "Great Dying" at the end of the Permian.

Other hypotheses

Many other hypotheses have been proposed, such as the spread of a new disease, or simple out-competition following an especially successful biological innovation. But all have been rejected, usually for one of the following reasons: they require events or processes for which there is no evidence; they assume mechanisms which are contrary to the available evidence; they are based on other theories which have been rejected or superseded.

Scientists have been concerned that human activities could cause more plants and animals to become extinct than any point in the past. Along with human-made changes in climate (see above), some of these extinctions could be caused by overhunting, overfishing, invasive species, or habitat loss. A study published in May 2017 in *Proceedings of the National Academy of Sciences* argued that a "biological annihilation" akin to a sixth mass extinction event is underway as a result of anthropogenic causes, such as over-population and over-consumption. The study suggested that as much as 50% of the number of animal individuals that once lived on Earth were already extinct, threatening the basis for human existence too.

Future biosphere extinction/sterilization

The eventual warming and expanding of the Sun, combined with the eventual decline of atmospheric carbon dioxide could actually cause an even greater mass extinction, having the potential to wipe out even microbes (in other words, the Earth is completely sterilized), where rising global temperatures caused by the expanding Sun will gradually increase the rate of weathering, which in turn removes more and more carbon dioxide from the atmosphere. When carbon dioxide levels get too low (perhaps at 50 ppm), all plant life will die out, although simpler plants like grasses and mosses can survive much longer, until CO_2 levels drop to 10 ppm.

With all photosynthetic organisms gone, atmospheric oxygen can no longer be replenished, and is eventually removed by chemical reactions in the atmosphere, perhaps from volcanic eruptions. Eventually the loss of oxygen will cause all remaining aerobic life to die out via asphyxiation, leaving behind only simple anaerobic prokaryotes. When the Sun becomes 10% brighter in about a billion years, Earth will suffer a moist greenhouse effect resulting in its oceans boiling away, while the Earth's liquid outer core cools due to the inner core's expansion and causes the Earth's magnetic field to shut down. In the absence of a magnetic field, charged particles from the Sun will deplete the atmosphere and further increase the Earth's temperature to an average of $\sim$420 K (147 °C, 296 °F) in 2.8 billion years, causing the last remaining life on Earth to die out. This is the most extreme instance of a climate-caused extinction event. Since this will only happen late in the Sun's life, such will cause the final mass extinction in Earth's history (albeit a very long extinction event).

Effects and recovery

The impact of mass extinction events varied widely. After a major extinction event, usually only weedy species survive due to their ability to live in diverse habitats. Later, species diversify and occupy empty niches. Generally, biodiversity recovers 5 to 10 million years after the extinction event. In the most severe mass extinctions it may take 15 to 30 million years.

The worst event, the Permian–Triassic extinction, devastated life on earth, killing over 90% of species. Life seemed to recover quickly after the P-T extinction, but this was mostly in the form of disaster taxa, such as the hardy *Lystrosaurus*. The most recent research indicates that the specialized animals that formed complex ecosystems, with high biodiversity, complex food webs and a variety of niches, took much longer to recover. It is thought that this long recovery was due to successive waves of extinction which inhibited recovery, as well as prolonged environmental stress which continued into the

Early Triassic. Recent research indicates that recovery did not begin until the start of the mid-Triassic, 4M to 6M years after the extinction; and some writers estimate that the recovery was not complete until 30M years after the P-T extinction, i.e. in the late Triassic. Subsequent to the P-T extinction, there was an increase in provincialization, with species occupying smaller ranges – perhaps removing incumbents from niches and setting the stage for an eventual rediversification.

The effects of mass extinctions on plants are somewhat harder to quantify, given the biases inherent in the plant fossil record. Some mass extinctions (such as the end-Permian) were equally catastrophic for plants, whereas others, such as the end-Devonian, did not affect the flora.

External links

- Calculate the effects of an Impact[912]
- Species Alliance (nonprofit organization producing a documentary about Mass Extinction titled "Call of Life: Facing the Mass Extinction)[913]
- Interstellar Dust Cloud-induced Extinction Theory[914]
- Sepkoski's Global Genus Database of Marine Animals[915] – Calculate extinction rates for yourself!

Appendix

References

[1] "International Stratigraphic Chart". International Commission on Stratigraphy
[2] Early edition, published online before print.
[3] Charles Frankel, 1996, *Volcanoes of the Solar System*, Cambridge University Press, pp. 7–8,
[4] Pluto's satellite Charon is relatively larger,<ref>
[5] The Sun's evolution http://faculty.wcas.northwestern.edu/~infocom/The%20Website/evolution.html
[6] http://sp.lyellcollection.org/content/190/1/205.abstract
[7] http://adsabs.harvard.edu/abs/2001GSLSP.190..205D
[8] //doi.org/10.1144/GSL.SP.2001.190.01.14
[9] https://web.archive.org/web/20121028022719/http://www.nysm.nysed.gov/nysgs/resources/images/geologicaltimescale.pdf
[10] http://www.nysm.nysed.gov/nysgs/resources/images/geologicaltimescale.pdf
[11] http://adsabs.harvard.edu/abs/1991Sci...253..535W
[12] //doi.org/10.1126/science.253.5019.535
[13] //www.ncbi.nlm.nih.gov/pubmed/17745185
[14] https://www.theguardian.com/technology/2005/dec/20/comment.science
[15] http://www.johnkyrk.com/evolution.html
[16] http://www.bbc.com/earth/bespoke/story/20150123-earths-25-biggest-turning-points/
[17] http://historystack.com/30_Major_Events_in_History_of_the_Earth
[18] http://www.bbc.co.uk/programmes/p00547hl
[19] http://www.bbc.co.uk/programmes/p005493g
[20] https://web.archive.org/web/20140715055239/http://taxonomicon.taxonomy.nl/TaxonTree.aspx?id=1&src=0
[21] https://web.archive.org/web/20051222163318/http://sn2000.taxonomy.nl/Main/Classification/1.htm
[22] https://www.biolib.cz/en/taxon/id14772
[23] https://www.google.com/search?q=earliest+known+life+forms&source=lnms&tbm=isch&sa=X&ved=0ahUKEwjR9YXu36HbAhXyuFkKHXUFAmkQ_AUICigB&biw=1800&bih=961
[24] //en.wikipedia.org/w/index.php?title=Template:Evolutionary_biology&action=edit
[25] Land, M.F. and Nilsson, D.-E., *Animal Eyes*, Oxford University Press, Oxford (2002) .
[26] The picture labeled "Human Chromosome 2 and its analogs in the apes" in the article Comparison of the Human and Great Ape Chromosomes as Evidence for Common Ancestry http://www.gate.net/~rwms/hum_ape_chrom.html is literally a picture of a link in humans that links two separate chromosomes in the nonhuman apes creating a single chromosome in humans. Also, while the term originally referred to fossil evidence, this too is a trace from the past corresponding to some living beings that, when alive, physically embodied this link.
[27] The New York Times report *Still Evolving, Human Genes Tell New Story* https://www.nytimes.com/2006/03/07/science/07evolve.html, based on *A Map of Recent Positive Selection in the Human Genome* http://biology.plosjournals.org/perlserv/?request=get-document&doi=10.1371/journal.pbio.0040072, states the International HapMap Project is "providing the strongest evidence yet that humans are still evolving" and details some of that evidence.
[28] "Converging Evidence for Evolution." http://phylointelligence.org/combined.html Phylointelligence: Evolution for Everyone. 26 November 2010.
[29] MacAndrew, Alec. Human Chromosome 2 is a fusion of two ancestral chromosomes http://www.evolutionpages.com/chromosome_2.htm. Accessed 18 May 2006.
[30] Evidence of Common Ancestry: Human Chromosome 2 https://www.youtube.com/watch?v=x-WAHpC0Ah0 (video) 2007
[31] Human and Ape Chromosomes http://www.gate.net/~rwms/hum_ape_chrom.html ; accessed 8 September 2007.

[32] Amino acid sequences in cytochrome c proteins from different species http://www.indiana.edu/~ensiweb/lessons/molb.ws.pdf, adapted from Strahler, Arthur; Science and Earth History, 1997. page 348.

[33] Lambert, Katie. (2007-10-29) HowStuffWorks "How Atavisms Work" http://animals.howstuffworks.com/animal-facts/atavism.htm. Animals.howstuffworks.com. Retrieved on 2011-12-06.

[34] 29+ Evidences for Macroevolution: Part 1 http://www.talkorigins.org/faqs/comdesc/section1.html#nested_hierarchy. Talkorigins.org. Retrieved on 2011-12-06.

[35] "Obviously vertebrates must have had ancestors living in the Cambrian, but they were assumed to be invertebrate forerunners of the true vertebrates — protochordates. Pikaia has been heavily promoted as the oldest fossil protochordate." Richard Dawkins 2004 The Ancestor's Tale Page 289,

[36] Academy of Natural Sciences - Joseph Leidy - American Horses http://www.ansp.org/museum/leidy/paleo/equus.php

[37] Continental Drift and Evolution http://biology.clc.uc.edu/courses/bio303/contdrift.htm . Biology.clc.uc.edu (2001-03-25). Retrieved on 2011-12-06.

[38] Davis, Paul and Kenrick, Paul. 2004. Fossil Plants. Smithsonian Books (in association with the Natural History Museum of London), Washington, D.C.

[39] Mills, William James. Exploring Polar Frontiers: A Historical Encyclopedia, ABC-CLIO, 2003.

[40] Evolution and Information: The Nylon Bug http://www.nmsr.org/nylon.htm. Nmsr.org. Retrieved on 2011-12-06.

[41] Why scientists dismiss 'intelligent design' http://www.msnbc.msn.com/id/9452500/page/2/, Ker Than, MSNBC, Sept. 23, 2005

[42] Miller, Kenneth R. Only a Theory: Evolution and the Battle for America's Soul (2008) pp. 80–82

[43] Science News, Dark Power: Pigment seems to put radiation to good use http://www.sciencenews.org/articles/20070526/fob5.asp, Week of May 26, 2007; Vol. 171, No. 21, p. 325 by Davide Castelvecchi

[44] Soy and Lactose Intolerance https://web.archive.org/web/20071215230655/http://www.soynutrition.com/SoyHealth/SoyLactoseIntolerance.aspx Wayback: Soy Nutrition

[45] Liebherr, James K.; McHugh, Joseph V. in Resh, V. H.; Cardé, R. T. (Editors) 2003. Encyclopedia of Insects. Academic Press.

[46] Supporting Online Material http://www.sciencemag.org/cgi/content/full/sci;310/5747/502/DC1

[47] Simulated Evolution Gets Complex http://www.trnmag.com/Stories/2003/052103/Simulated_evolution_gets_complex_052103.html. Trnmag.com (2003-05-08). Retrieved on 2011-12-06.

[48] Molecular evolution https://web.archive.org/web/20080430031245/http://bio.kaist.ac.kr/~jsrhee/research03.html. kaist.ac.kr

[49] In Vitro Molecular Evolution http://www.isgec.org/gecco-2005/free-tutorials.html#ivme. Isgec.org (1975-08-04). Retrieved on 2011-12-06.

[50] http://citeseerx.ist.psu.edu/viewdoc/download?rep=rep1&type=pdf&doi=10.1.1.217.1490

[51] https://pdfs.semanticscholar.org/ceb2/49234e102e35db835a8846b9b5064c28e455.pdf

[52] //doi.org/10.1111/j.1096-0031.1999.tb00279.x

[53] http://max2.ese.u-psud.fr/epc/conservation/Publi/abstracta/AE_TREE2000.pdf

[54] //doi.org/10.1016/s0169-5347%2899%2901780-2

[55] http://nationalacademies.org/evolution/

[56] http://www.talkorigins.org/faqs/comdesc/

[57] http://www.talkorigins.org/faqs/faq-transitional.html

[58] https://web.archive.org/web/20100610231107/http://evolution.berkeley.edu/evosite/evo101/index.shtml

[59] http://www.nap.edu/books/0309063647/html/

[60] https://www.pbs.org/wgbh/evolution/index.html

[61] http://www.genomenewsnetwork.org/categories/index/genome/evolution.php

[62] https://web.archive.org/web/20081201133548/http://www.chainsofreason.org/wiki/Chain_3

[63] http://science.howstuffworks.com/evolution.htm/printable

[64] http://www.nature.com/nature/newspdf/evolutiongems.pdf

[65] //en.wikipedia.org/w/index.php?title=Template:Evolutionary_biology&action=edit

[66] Now called homology.

[67] : "Despite all the variety among these forms, they seem to have been produced according to a common archetype, and this analogy among them reinforces our suspicion that they are actually akin, produced by a common original mother."

[68] Krogh, David. (2005). *Biology: A Guide to the Natural World*. Pearson/Prentice Hall. p. 323. "Descent with modification was accepted by most scientists not long after publication of Darwin's *On the Origin of Species by Means of Natural Selection* in 1859. Scientists accepted it because it explained so many facets of the living world."

[69] Kellogg, Vernon L. (1907). *Darwinism To-Day* https://archive.org/stream/darwinismtodaydi00kell#page/3/mode/2up. Henry Holt and Company. p. 3

[70] Gregory, T. Ryan. (2008). *Evolution as Fact, Theory, and Path* https://link.springer.com/article/10.1007/s12052-007-0001-z. *Evolution: Education and Outreach* 1 (1): 46–52.

[71] //lccn.loc.gov/57014935

[72] //www.worldcat.org/oclc/7588392

[73] //lccn.loc.gov/06017473

[74] //www.worldcat.org/oclc/741260650

[75] http://darwin-online.org.uk/content/frameset?pageseq=1&itemID=F373&viewtype=side

[76] https://archive.org/details/2551016RX1.nlm.nih.gov

[77] //lccn.loc.gov/81002555

[78] //www.worldcat.org/oclc/7278190

[79] //lccn.loc.gov/86014852

[80] //www.worldcat.org/oclc/13796153

[81] //lccn.loc.gov/85000255

[82] //www.worldcat.org/oclc/11623262

[83] http://www.talkorigins.org/faqs/comdesc/

[84] http://tolweb.org/tree/phylogeny.html

[85] The alternative terms "homogeny" and "homogenous" were also used in the late 1800s and early 1900s. However, these terms are now archaic in biology, and the term "homogenous" is now generally found as a misspelling of the term "homogeneous" which refers to the uniformity of a mixture.<ref>"homogeneous, adj.". OED Online. March 2016. Oxford University Press. http://www.oed.com/view/Entry/88045? (accessed April 09, 2016).

[86] "homogenous, adj.". OED Online. March 2016. Oxford University Press. http://www.oed.com/view/Entry/88055? (accessed April 09, 2016).

[87] If the two pairs of wings are considered as interchangeable, homologous structures, this may be described as a parallel reduction in the number of wings, but otherwise the two changes are each divergent changes in one pair of wings.

[88] These are coloured in the lead image: humerus brown, radius pale buff, ulna red.

[89] Cf. Butler, A. B.: *Homology and Homoplasty*. In: Squire, Larry R. (Ed.): *Encyclopedia of Neuroscience*, Academic Press, 2009, pp. 1195–1199.

[90] Brower, A. V. Z. and V. Schawaroch. 1996. Three steps of homology assessment. *Cladistics* **12**:265-272.

[91] Brower, A. V. Z. and M. C. C. de Pinna. (2012). "Homology and errors". *Cladistics* **28**:529-538 doi/510.1111/j.1096-0031.2012.00398.x

[92] Patterson, C. 1982. Morphological characters and homology. Pp. 21-74 in K. A. Joysey, and A. E. Friday, eds. Problems of Phylogenetic Reconstruction. Academic Press, London and New York.

[93] Haas, O. and G. G. Simpson. 1946. Analysis of some phylogenetic terms, with attempts at redefinition. *Proc. Amer. Phil. Soc.* **90**:319-349.

[94] Larson 2004, p. 112.

[95] http://embryo.asu.edu/handle/10776/1754

[96] https://www.worldcat.org/search?fq=x0:jrnl&q=n2:1940-5030

[97] //doi.org/10.1111/j.1096-0031.1991.tb00045.x

[98] //doi.org/10.1093/hmg/ddl056

[99] //www.ncbi.nlm.nih.gov/pubmed/16651369

[100] //doi.org/10.1016/S0168-9525%2800%2902005-9

[101] //www.ncbi.nlm.nih.gov/pubmed/10782117

[102] https://web.archive.org/web/20100627024721/http://euplotes.biology.uiowa.edu/web/IBS593/week4/Homologyevolving.pdf

[103] //doi.org/10.1016/S0169-5347%2801%2902206-6

[104] http://euplotes.biology.uiowa.edu/web/IBS593/week4/Homologyevolving.pdf

[105] //doi.org/10.1016/j.tig.2008.08.009

[106] //www.ncbi.nlm.nih.gov/pubmed/18819722

[107] //en.wikipedia.org/w/index.php?title=Template:Evolutionary_biology&action=edit

[108] There is a common misconception that definitions of LUCA and progenote are the same; however, progenote is defined as an organism "still in the process of evolving the relationship between genotype and phenotype", and it is only hypothesed that LUCA is a progenote.

[109] Darwin, Charles. *On the Origin of Species*. London: John Murray, Albermarle Street. 1859. pp. 484, 490.

[110] Brown, J. R., and W. F. Doolittle. 1995. "Root of the Universal Tree of Life Based on Ancient Aminoacyl-tRNA Synthetase Gene Duplications." Proc Natl Acad Sci USA 92 (7): 2441–45. PMID 7708661 https://www.ncbi.nlm.nih.gov/pubmed/?term=7708661

[111] Gogarten, J. P., H. Kibak, P. Dittrich, L. Taiz, E. J. Bowman, B. J. Bowman, M. F. Manolson, et al. 1989. "Evolution of the Vacuolar H+-ATPase: Implications for the Origin of Eukaryotes." Proc Natl Acad Sci USA 86 (17): 6661–65. PMID 2528146 https://www.ncbi.nlm.nih.gov/pubmed/?term=2528146

[112] Gribaldo, S, and P Cammarano. 1998. "The Root of the Universal Tree of Life Inferred from Anciently Duplicated Genes Encoding Components of the Protein-Targeting Machinery." Journal of Molecular Evolution 47 (5): 508–16. PMID: 9797401 https://www.ncbi.nlm.nih.gov/pubmed/?term=9797401

[113] Iwabe, Naoyuki, Kei-Ichi Kuma, Masami Hasegawa, Syozo Osawa, Takashi Miyata Source, Masami Hasegawat, Syozo Osawat, and Takashi Miyata. 1989. "Evolutionary Relationship of Archaebacteria, Eubacteria, and Eukaryotes Inferred from Phylogenetic Trees of Duplicated Genes." Proc Natl Acad Sci USA 86 (86): 9355–59. PMID 2531898 https://www.ncbi.nlm.nih.gov/pubmed/?term=2531898

[114]

[115] Nick Lane: The Vital Question – Energy, Evolution, and the Origins of Complex Life https://books.google.de, WW Norton, 2015,

[116] Joseph F. Sutherland: on The Origin Of Tha Bacteria And The Archaea http://prehistoricict.blogspot.de/2014/08/on-origin-of-bacteria-and-archaea.html, auf B.C vom 16. August 2014

[117] https://www.nytimes.com/2016/07/26/science/last-universal-ancestor.html

[118] (requires nonfree AAAS member subscription)

[119] Russell, Michael (Ed), (2010), "Origins, Abiogenesis and the Search for Life in the Universe" (Cosmology Science Publications)

[120] Forward planetary contamination like *Tersicoccus phoenicis*, that has shown resistance to methods usually used in spacecraft assembly clean rooms:

[121] A variation of the panspermia hypothesis is **necropanspermia** which astronomer Paul Wesson describes as follows: "The vast majority of organisms reach a new home in the Milky Way in a technically dead state ... Resurrection may, however, be possible."

[122] Hoyle, F. and Wickramasinghe, N.C. (1981). *Evolution from Space*. Simon & Schuster Inc., NY, and J.M. Dent and Son, London (1981), ch3 pp. 35–49.

[123] Wickramasinghe, J., Wickramasinghe, C. and Napier, W. (2010). *Comets and the Origin of Life* http://alpha.sinp.msu.ru/~panov/LibBooks/LIFE/10972__legalreads.com.pdf. World Scientific, Singapore. ch. 6 pp. 137–154.

[124] Margaret O'Leary (2008) Anaxagoras and the Origin of Panspermia Theory, iUniverse publishing Group,

[125] Arrhenius, S. (1908) *Worlds in the Making: The Evolution of the Universe*. New York, Harper & Row.

[126] Early edition, published online before print.

[127] Studies Focus On Spacecraft Sterilization https://web.archive.org/web/20060502194219/http://www.aero.org/news/newsitems/sterilization073001.html. aero.org (July 30, 2000)

[128] Dry heat sterilisation process to high temperatures https://web.archive.org/web/20120201224127/http://www.esa.int/esaMI/Aurora/SEMBJG9ATME_0.html. European Space Agency (22 May 2006)

[129] Slow-moving rocks better odds that life crashed to Earth from space http://www.princeton.edu/main/news/archive/S34/82/42M30/. News at Princeton, September 24, 2012.

[130] "Die Verbreitung des Lebens im Weltenraum" (the "Distribution of Life in Space"). Published in *Die Umschau*. 1903.

[131] Gold, T. "Cosmic Garbage", Air Force and Space Digest, 65 (May 1960).

[132] " Anticipating an RNA world. Some past speculations on the origin of life: where are they today? http://www.fasebj.org/cgi/reprint/7/1/238.pdf" by L. E. Orgel and F. H. C. Crick in *FASEB J*. (1993) Volume 7 pages 238–239.

[133] Finley, Dave (February 28, 2013) Discoveries Suggest Icy Cosmic Start for Amino Acids and DNA Ingredients https://www.nrao.edu/pr/2013/newchem/. *The National Radio Astronomy Observatory*

[134] http://www.astrochem.org/pahdb/

[135] http://sci.esa.int/rosetta/57863-altwegg-et-al-2016/

[136] Discovery of New Microorganisms in the Stratosphere http://www.physorg.com/news156626262.html. Physorg (March 18, 2009)

[137] *A.A. Imshenetsky, S.V. Lysenko, G.A. Kazakov, "Upper boundary of the biosphere," Appl Environ Microbiol, vol. 35, pp. 1-5, 1978. • A.A. Imshenetsky, S.V. Lysenko, G.A. Kazakov, N.V. Ramkova, "On micro-organisms of the stratosphere," Life Sci Space Res, vol. 14, pp. 359–362, 1976. • Y. Yang, T. Itoh, S. Yokobori, et al., "Deinococcus aetherius sp. nov., isolated from the stratosphere," Int J Syst Evol Microbiol, vol. 60, pp. 776–779, 2010. • S. Shivaji, S. Ara, S.K. Singh, et al., "Draft genome sequence of Bacillus isronensis strain B3W22, isolated from the upper atmosphere," J Bacteriol, vol. 194, pp. 6624–6625, 2012.

[138]

[139] Surviving the Final Frontier http://www.astrobio.net/exclusive/318/surviving-the-final-frontier. astrobio.net (25 November 2002).

[140] Bacterium revived from 25 million year sleep http://commtechlab.msu.edu/sites/dlc-me/news/ns595ap1.html Digital Center for Microbial Ecology

[141] Yokobori, Shin-ichi et al (2010) Microbe space exposure experiment at International Space Station (ISS) proposed in "Tanpopo" mission https://www.researchgate.net/publication/241270775_Microbe_space_exposure_experiment_at_International_Space_Station_(ISS)_proposed_in_Tanpopo_mission. Research Gate.

[142] Yano, H. et al. (2014) " Tanpopo Experiment for Wastrobiology Exposure and Micrometeoroid Capture Onboard the ISS-JEM Exposed Facility http://www.hou.usra.edu/meetings/lpsc2014/pdf/2934.pdf." 45th Lunar and Planetary Science Conference.

[143] Tanpopo mission to search space for origins of life http://the-japan-news.com/news/article/0002066967. *The Japan News*, April 16, 2015.

[144] http://web.snauka.ru/en/issues/2013/12/30018

[145] http://profiles.nlm.nih.gov/SC/B/B/Y/P/_/scbbyp.pdf

[146] //doi.org/10.1038/news040216-20

[147] http://www.scientificamerican.com/article.cfm?id=did-life-come-from-anothe

[148] http://adsabs.harvard.edu/abs/2005SciAm.293e..64W

[149] //doi.org/10.1038/scientificamerican1105-64

[150] http://tools.wmflabs.org/timescale/?Ma=3,500

[151] Lucas J. Stal: Physiological ecology of cyanobacteria in microbial mats and other communities, New Phytologist (1995), 131, 1–32

[152] Garcia-Pichel F., Mechling M., Castenholz R.W., Diel Migrations of Microorganisms within a Benthic, Hypersaline Mat Community https://www.ncbi.nlm.nih.gov/pmc/articles/PMC201509/pdf/aem00022-0118.pdf, Appl. and Env. Microbiology, May 1994, pp. 1500–1511

[153] Bebout B.M., Garcia-Pichel F., UV B-Induced Vertical Migrations of Cyanobacteria in a Microbial Mat http://aem.asm.org/cgi/content/abstract/61/12/4215, Appl. Environ. Microbiol., Dec 1995, 4215–4222, Vol 61, No. 12

[154] http://tools.wmflabs.org/timescale/?Ma=3,480

[155] http://tools.wmflabs.org/timescale/?Ma=4,000
[156] – abstract with link to free full content (PDF)
[157] http://tools.wmflabs.org/timescale/?Ma=3,000
[158] http://tools.wmflabs.org/timescale/?Ma=1,600
[159] http://tools.wmflabs.org/timescale/?Ma=2,100
[160] http://tools.wmflabs.org/timescale/?Ma=2,700
[161] http://tools.wmflabs.org/timescale/?Ma=1,200
[162] ; ; cites U.S. Patents 7351005 and 7374670
[163] https//books.google.com
[164] http://www.uta.edu/paleomap/homepage/Schieberweb/microbial_mat_page.htm
[165] Holland, Heinrich D. "The oxygenation of the atmosphere and oceans" http://rstb.royalsocietypublishing.org/content/361/1470/903.full.pdf. *Philosophical Transactions of the Royal Society: Biological Sciences*. Vol. 361. 2006. pp. 903–915.
[166] //en.wikipedia.org/w/index.php?title=Great_Oxygenation_Event&action=edit
[167] Oxygen oasis in Antarctic lake reflects Earth in distant past https://www.sciencedaily.com/releases/2015/09/150901140759.htm
[168] http://tools.wmflabs.org/timescale/?Ma=3,400
[169] First breath: Earth's billion-year struggle for oxygen https://www.newscientist.com/article/mg20527461.100-first-breath-billionyear-struggle-for-oxygen.html *New Scientist*, #2746, 5 February 2010 by Nick Lane.
[170] "Evolution of Minerals" http://www.scientificamerican.com/article.cfm?id=evolution-of-minerals, *Scientific American*, March 2010
[171] Bernstein H, Bernstein C. Sexual communication in archaea, the precursor to meiosis. pp. 103-117 in Biocommunication of Archaea (Guenther Witzany, ed.) 2017. Springer International Publishing DOI 10.1007/978-3-319-65536-9
[172] Bernstein, H., Bernstein, C. Evolutionary origin and adaptive function of meiosis. In "Meiosis", Intech Publ (Carol Bernstein and Harris Bernstein editors), Chapter 3: 41-75 (2013).
[173] https://www.newscientist.com/article/mg20527461.100-first-breath-earths-billionyear-struggle-for-oxygen/
[174] https://web.archive.org/web/20110106141826/http://ptc-cam.blogspot.com/2010/02/first-breath-earths-billion-year.html
[175] To date, only one eukaryote, *Monocercomonoides*, is known to have completely lost its mitochondria.<ref name="Karn">
[176] Lynn Margulis, Heather I. McKhann & Lorraine Olendzenski (ed.), *Illustrated Glossary of Protoctista*, Jones and Bartlett Publishers, Boston, 1993, p.xviii.
[177] //en.wikipedia.org/w/index.php?title=Eukaryote&action=edit
[178] http://tools.wmflabs.org/timescale/?Ma=800
[179] https://www.ncbi.nlm.nih.gov/About/primer/index.html
[180] http://www.tolweb.org/Eukaryotes/3
[181] http://www.eol.org/pages/2908256
[182] Kolattukudy, P.E. (1996) "Biosynthetic pathways of cutin and waxes, and their sensitivity to environmental stresses", pp. 83-108 in: *Plant Cuticles*. G. Kerstiens (ed.), BIOS Scientific publishers Ltd., Oxford
[183] https://web.archive.org/web/20050614082020/http://www.plantphys.net/article.php?ch=e&id=122
[184] //doi.org/10.1111/j.1365-313X.2005.02482.x
[185] //www.ncbi.nlm.nih.gov/pubmed/16167891
[186] http//arjournals.annualreviews.org
[187] //doi.org/10.1146/annurev.genet.35.102401.090231
[188] //www.ncbi.nlm.nih.gov/pubmed/11700280
[189] http://www.hos.ufl.edu/ctdcweb/Birky01.pdf
[190] http://www.nature.com/scitable/topicpage/the-origin-of-plastids-14125758
[191] http://rstb.royalsocietypublishing.org/content/365/1541/729.long
[192] http://www.coextra.eu/projects/project199.html
[193] http://tolweb.org/Eukaryotes/3
[194] //en.wikipedia.org/w/index.php?title=Template:Evolutionary_biology&action=edit

[195] John Maynard Smith *The Evolution of Sex* 1978.

[196] Rolf Hoekstra 1987 *The Evolution of Sex and its Consequences* 1988 Birkhauser.

[197] Beukeboom, L. & Perrin, N. (2014). *The Evolution of Sex Determination*. Oxford University Press, p. 5-6 https://books.google.com/books?id=d4cLBAAAQBAJ&lpg=PP1&hl=pt-BR&pg=PA5#v=onepage&q&f=false. Online resources, http://www.oup.co.uk/companion/beukeboom.

[198] Crow J.F. (1994). Advantages of Sexual Reproduction, Dev. Gen., vol.15, pp. 205-213.

[199] Matt Ridley 1995 *The Red Queen: Sex and the Evolution of Human Nature* 1995 Penguin.

[200] MacIntyre, Ross J.; Clegg, Michael, T (Eds.), Springer. Hardcover , Softcover .

[201] Bernstein H, Bernstein C, Michod RE (2011). "Meiosis as an evolutionary adaptation for DNA repair." In "DNA Repair", Intech Publ (Inna Kruman, editor), Chapter 19: 357-382 Available online from: http://www.intechopen.com/books/dna-repair/meiosis-as-an-evolutionary-adaptation-for-dna-repair

[202] Darwin CR (1876). The effects of cross and self fertilisation in the vegetable kingdom. London: John Murray. http://darwin-online.org.uk/converted/published/1881_Worms_F1357/1876_CrossandSelfFertilisation_F1249/1876_CrossandSelfFertilisation_F1249.html see page 462

[203] Griffiths *et al.* 1999. Gene mutations, p197-234, *in* Modern Genetic Analysis, New York, W.H. Freeman and Company.

[204] Ridley M (2004) *Evolution*, 3rd edition. Blackwell Publishing.

[205] Charlesworth B, Charlesworth D (2010) *Elements of Evolutionary Genetics*. Roberts and Company Publishers.

[206] Nicholas J. Butterfield, "Bangiomorpha pubescens n. gen., n. sp.: implications for the evolution of sex, multicellularity, and the Mesoproterozoic/Neoproterozoic radiation of eukaryotes" http://paleobiol.geoscienceworld.org/cgi/content/abstract/26/3/386

[207] Abstract in English available online: http://www.tlu.ee/~toenu/ingl/euk_rakk.htm

[208]

[209] Bernstein H, Bernstein C. Sexual communication in archaea, the precursor to meiosis. pp. 103-117 in Biocommunication of Archaea (Guenther Witzany, ed.) 2017. Springer International Publishing DOI 10.1007/978-3-319-65536-9

[210]

[211] T. Togashi, P. Cox (Eds.) *The Evolution of Anisogamy*. Cambridge University Press, Cambridge; 2011, p. 22-29.

[212] Beukeboom, L. & Perrin, N. (2014). *The Evolution of Sex Determination*. Oxford University Press, p. 25 https://books.google.com/books?id=d4cLBAAAQBAJ&lpg=PP1&hl=pt-BR&pg=PA25#v=onepage&q&f=false. Online resources, http://www.oup.co.uk/companion/beukeboom.

[213] //doi.org/10.1016/0169-5347%2896%2981041-X

[214] //www.ncbi.nlm.nih.gov/pubmed/21237760

[215] http://www.msnbc.msn.com/id/27927661/

[216] http://www.livescience.com/health/050330_sex_good.html

[217] http://www.philippwesche.org/old1/es.html

[218] John Maynard Smith *The Evolution of Sex* 1978.

[219] Ridley M (2004) Evolution, 3rd edition. Blackwell Publishing, p. 314.

[220] Dimijian, G. G. (2005). Evolution of sexuality: biology and behavior. Proceedings (Baylor University. Medical Center), 18, 244–258.

[221] Research conducted by Patricia Adair Gowaty. Reported by

[222] BONY FISHES - Reproduction http://www.seaworld.org/animal-info/info-books/bony-fish/reproduction.htm

[223] Bernstein H, Bernstein C, Michod RE. (2012) " DNA Repair as the Primary Adaptive Function of Sex in Bacteria and Eukaryotes https://www.novapublishers.com/catalog/product_info.php?products_id=31918". Chapter 1, pp. 1–50, in *DNA Repair: New Research*, Editors S. Kimura and Shimizu S. Nova Sci. Publ., Hauppauge, New York. Open access for reading only.

[224] http://www.biolreprod.org/

[225] https://www.sciencedaily.com/releases/2003/02/030203071703.htm

[226] https://www.youtube.com/watch?v=kaSIjIzAtYA

[227] http://public.wsu.edu/~lange-m/Documnets/Teaching2011/Popper2011.pdf

[228] Margulis, L. & Chapman, M.J. (2009). Kingdoms and Domains: An Illustrated Guide to the Phyla of Life on Earth ([4th ed.]. ed.). Amsterdam: Academic Press/Elsevier. p. 116.

[229] Seravin L. N. (2001) The principle of counter-directional morphological evolution and its significance for constructing the megasystem of protists and other eukaryotes. *Protistology* 2: 6–14, http//cyberleninka.ru.

[230] Parfrey, L.W. & Lahr, D.J.G. (2013), p. 344.

[231] Seckbach, Joseph, Chapman, David J. [eds.]. (2010). *Red algae in the genomic age*. New York, NY, U.S.A.: Springer, p. 252, https://books.google.com/books?id=fegCa9G-c90C&lpg=PP1&hl=pt-BR&pg=PA252#v=onepage&q&f=false.

[232] Richter, Daniel Joseph: The gene content of diverse choanoflagellates illuminates animal origins http://escholarship.org/uc/item/7xc2p94p, 2013.

[233] Richter, D. J. (2013), p. 11.

[234] Lauckner, G. (1980). Diseases of protozoa. In: *Diseases of Marine Animals*. Kinne, O. (ed.). Vol. 1, p. 84 http://www.int-res.com/archive/doma_books/DOMA_Vol_I_(general_aspects,_protozoa_to%20gastropoda).pdf, John Wiley & Sons, Chichester, UK.

[235] Ridley M (2004) Evolution, 3rd edition. Blackwell Publishing, p. 295-297.

[236] Niklas, K. J. (2014) The evolutionary-developmental origins of multicellularity http://www.amjbot.org/content/101/1/6.long.

[237] In a Single-Cell Predator, Clues to the Animal Kingdom's Birth https//www.nytimes.com

[238] A H Knoll, 2003. *Life on a Young Planet*. Princeton University Press. (hardcover), (paperback). An excellent book on the early history of life, very accessible to the non-specialist; includes extensive discussions of early signatures, fossilization, and organization of life.

[239] AlgaeBase. Volvox Linnaeus, 1758: 820. http://www.algaebase.org/search/genus/detail/?genus_id=43497

[240] Mikhailov K. V., Konstantinova A. V., Nikitin M. A., Troshin P. V., Rusin L., Lyubetsky V., Panchin Y., Mylnikov A. P., Moroz L. L., Kumar S. & Aleoshin V. V. (2009). The origin of Metazoa: a transition from temporal to spatial cell differentiation. *Bioessays*, 31(7), 758–768, http://www.kumarlab.net/pdf_new/MikhailovAleoshin09.pdf.

[241] Jamin, M, H Raveh-Barak, B Podbilewicz, FA Rey (2014) "Structural basis of eukaryotic cell-cell fusion" (Cell, Volume 157, Issue 2, 10 April 2014), Pages 407–419

[242] Slezak, Michael (2016), "No Viruses? No skin or bones either" (New Scientist, No. 2958, 1 March 2014) p.16

[243] http://tolweb.org/Eukaryotes/3

[244] //en.wikipedia.org/w/index.php?title=Template:CEXNAV&action=edit

[245] Simple multicellular organisms such as red algae evolved at least . The status of the Francevillian biota of is unclear, but they may represent earlier multicellular forms of a more complex nature.

[246] http://tools.wmflabs.org/timescale/?Ma=575

[247] Two Explosive Evolutionary Events Shaped Early History Of Multicellular Life https://www.sciencedaily.com/releases/2008/01/080103144451.htm

[248] http://tools.wmflabs.org/timescale/?Ma=600

[249] http://tools.wmflabs.org/timescale/?Ma=542

[250] Reprint, 2004 original available here http://www.stratigraphy.org/bak/ediacaran/Knoll_et_al_2004a.pdf (PDF).

[251] MacGhabhann, 2014, Geosciences Frontiers, 5: 53–62

[252] For example, ,

[253] e.g. , summarised by

[254] Xiao *et al.*.'s response to Bailey *et al.*.'s original paper : And Bailey *et al.*.'s reply:

[255] http://tools.wmflabs.org/timescale/?Ma=632.5

[256] (a) The only current description, far from universal acceptance, appears as:

[257] For a reinterpretation, see

[258] http://tools.wmflabs.org/timescale/?Ma=1,100

[259] A. Yu. Ivantsov. (2008). "Feeding traces of the Ediacaran animals" http://www.cprm.gov.br/33IGC/1323085.html. HPF-17 Trace fossils : ichnological concepts and methods. International Geological Congress – Oslo 2008.

[260] According to
For a more cynical perspective see

[261] Discussed at length in

[262] Williams, G.C. 1997. Preliminary assessment of the phylogenetics of pennatulacean octocorals, with a reevaluation of Ediacaran frond-like fossils, and a synthesis of the history of evolutionary thought regarding the sea pens. Proceedings of the Sixth International Conference of Coelenterate Biology: 497–509.

[263] Gough, Myles (7 June 2016). "Origin of mystery deep-sea mushroom revealed". BBC News. Retrieved 7 June 2016.

[264] http://tools.wmflabs.org/timescale/?Ma=3,460

[265] http://tools.wmflabs.org/timescale/?Ma=2,700

[266] http://tools.wmflabs.org/timescale/?Ma=1,200

[267] http://tools.wmflabs.org/timescale/?Ma=610

[268] http://tools.wmflabs.org/timescale/?Ma=770

[269] http://tools.wmflabs.org/timescale/?Ma=565

[270] http://tools.wmflabs.org/timescale/?Ma=560

[271] http://tools.wmflabs.org/timescale/?Ma=550

[272]

[273] //www.worldcat.org/oclc/3758852

[274] //www.worldcat.org/oclc/43945263

[275] http://www.bbc.co.uk/programmes/b00lh2s3

[276] https://web.archive.org/web/20070528175308/http://geol.queensu.ca/museum/exhibits/ediac/drook/

[277] https://web.archive.org/web/20070405091608/http://geol.queensu.ca/museum/exhibits/ediac/ediac.html

[278] https://web.archive.org/web/20101010035019/http://www.peripatus.gen.nz/paleontology/Ediacara.html

[279] https://web.archive.org/web/20110725191154/http://www.complex-life.org/database

[280] http://www.abc.net.au/news/2013-08-03/sa-pastoral-property-nilpena-holds-animal-fossils/4862432

[281] http://www.abc.net.au/landline/content/2013/s3817544.htm

[282] //en.wikipedia.org/w/index.php?title=Template:CEXNAV&action=edit

[283] http://tools.wmflabs.org/timescale/?Ma=541

[284] This included at least animals, phytoplankton and calcimicrobes. UNIQ-ref-0-b019e64a2f5b3fa1-QINU

[285] At 610 million years ago, *Aspidella* disks appeared, but it is not clear that these represented complex life forms.

[286] http://tools.wmflabs.org/timescale/?Ma=3,850

[287] http://tools.wmflabs.org/timescale/?Ma=1,400

[288] http://tools.wmflabs.org/timescale/?Ma=580–543

[289] Whittington, H. B. (1979). Early arthropods, their appendages and relationships. In M. R. House (Ed.), The origin of major invertebrate groups (pp. 253–268). The Systematics Association Special Volume, 12. London: Academic Press.

[290] e.g.

[291]

Non-technical summary http://palaeo.gly.bris.ac.uk/cladestrat/news.html

[292] e.g.

[293] "Classifications of organisms in hierarchical systems were in use by the 17th and 18th centuries. Usually, organisms were grouped according to their morphological similarities as perceived by those early workers, and those groups were then grouped according to their similarities, and so on, to form a hierarchy."

[294] "Evolutionary biologists often make sense of the conflicting diversity of form – not always does a relationship between internal and external parts. Early in the history of the subject, it became obvious that internal organisations were generally more important to the higher classification of animals than are external shapes. The internal organisation puts general restrictions on how an animal can exchange gases, obtain nutrients, and reproduce."

[295] summarised in

[296] http://tools.wmflabs.org/timescale/?Ma=2,700

[297] http://tools.wmflabs.org/timescale/?Ma=1,250
[298] http://tools.wmflabs.org/timescale/?Ma=2,000
[299] http://tools.wmflabs.org/timescale/?Ma=1,000
[300] http://tools.wmflabs.org/timescale/?Ma=580
[301] http://tools.wmflabs.org/timescale/?Ma=565
[302] Older marks found in billion-year-old rocks<ref name="Seilacher1998">
[303] http://tools.wmflabs.org/timescale/?Ma=542
[304] Early Global Warming Was Unexpectedly Caused by a Burst of Tiny Life Forms - Inverse https://www.inverse.com/article/46647-global-warming-animals-cambrian-carbon-dioxide
[305] "Scientific American" April 2014 http://www.scientificamerican.com/article/death-valleys-first-life-came-in-by-land-not-by-sea/
[306] http://tools.wmflabs.org/timescale/?Ma=549–542
[307] http://tools.wmflabs.org/timescale/?Ma=510–515
[308] http://tools.wmflabs.org/timescale/?Ma=550–536
[309] http://tools.wmflabs.org/timescale/?Ma=536–521
[310] http://tools.wmflabs.org/timescale/?Ma=550
[311] As defined in terms of the extinction and origination rate of species.
[312] The analysis considered the bioprovinciality of trilobite lineages, as well as their evolutionary rate.<ref name="Lieberman2003">
[313]
[314] Novel Evolutionary Theory For The Explosion Of Life http://www.ibecbarcelona.eu/novel-evolutionary-theory-for-the-explosion-of-life/
[315] http://tools.wmflabs.org/timescale/?Ma=450
[316] http://tools.wmflabs.org/timescale/?Ma=400
[317] //doi.org/10.1111/j.1469-185X.1999.tb00046.x
[318] //www.ncbi.nlm.nih.gov/pubmed/10881389
[319] http://www.ucmp.berkeley.edu/phyla/metazoafr.html
[320] //www.ncbi.nlm.nih.gov/pmc/articles/PMC1578734
[321] //doi.org/10.1098/rstb.2006.1846
[322] //www.worldcat.org/issn/0962-8436
[323] //www.ncbi.nlm.nih.gov/pubmed/16754615
[324] //doi.org/10.1126/science.311.5766.1341c
[325] //doi.org/10.1126/science.284.5423.2129
[326] //www.ncbi.nlm.nih.gov/pubmed/10381872
[327] https://www.sciencedirect.com/science/article/pii/S1871174X07000030
[328] //doi.org/10.1016/j.palwor.2007.01.002
[329] //doi.org/10.1007/bf03004567
[330] //www.ncbi.nlm.nih.gov/pmc/articles/PMC1689654
[331] //doi.org/10.1098/rspb.1999.0617
[332] //www.worldcat.org/issn/0962-8452
[333] //www.ncbi.nlm.nih.gov/pubmed/10097391
[334] http://adsabs.harvard.edu/abs/1998Natur.391..553X
[335] //doi.org/10.1038/35318
[336] //www.worldcat.org/issn/0090-9556
[337] http://adsabs.harvard.edu/abs/2000Sci...288..841M
[338] //doi.org/10.1126/science.288.5467.841
[339] //www.ncbi.nlm.nih.gov/pubmed/10797002
[340] http://www.ijdb.ehu.es/web/contents.php?vol=47&issue=7–8&doi=14756326
[341] https://web.archive.org/web/20080407005007/http://genome6.cu-genome.org/andrey/GouldComment.pdf
[342] http://www.pnas.org/content/97/9/4426.full
[343] http://adsabs.harvard.edu/abs/2000PNAS...97.4426C
[344] //doi.org/10.1073/pnas.97.9.4426
[345] //www.ncbi.nlm.nih.gov/pmc/articles/PMC34314
[346] //www.ncbi.nlm.nih.gov/pubmed/10781036
[347] http://www.bbc.co.uk/programmes/p003k9bg

[348] http://www.bbc.co.uk/programmes/b006qykl

[349] http://burgess-shale.rom.on.ca/en

[350] http://www.kumip.ku.edu/cambrianlife/

[351] http://paleobiology.si.edu/geotime/main/htmlversion/cambrian2.html

[352] //en.wikipedia.org/w/index.php?title=Template:Burgess_Shale&action=edit

[353] Palaeontology's hidden agenda https//www.newscientist.com

[354] e.g.

[355] https://pubs.er.usgs.gov/#search:advance/report_number=81-743:0

[356] Haeckel, E. (1874). *Anthropogenie oder Entwicklungsgeschichte des Menschen.* Leipzig: Engelmann.

[357] "Classifications of organisms in hierarchical systems were in use by the seventeenth and eighteenth centuries. Usually, organisms were grouped according to their morphological similarities as perceived by those early workers, and those groups were then grouped according to **their** similarities, and so on, to form a hierarchy".

[358] R.C.Brusca, G.J.Brusca. *Invertebrates.* Sinauer Associates, Sunderland Mass 2003 (2nd ed.), p. 47, .

[359] Oxford English Dictionary, Third Edition, January 2009: Urochordata

[360] http://tools.wmflabs.org/timescale/?Ma=555

[361] http://tools.wmflabs.org/timescale/?Ma=558

[362] http://tools.wmflabs.org/timescale/?Ma=541

[363] http://tools.wmflabs.org/timescale/?Ma=549–543

[364] *Ernettia* is from the Kuibis formation, approximate date given by

[365] http://tools.wmflabs.org/timescale/?Ma=518

[366] http://tools.wmflabs.org/timescale/?Ma=542

[367] http://tools.wmflabs.org/timescale/?Ma=900

[368] http://tools.wmflabs.org/timescale/?Ma=896

[369] Benton, M.J. (2004). *Vertebrate Palaeontology*, Third Edition. Blackwell Publishing. The classification scheme is available online http://palaeo.gly.bris.ac.uk/benton/vertclass.html

[370] http://www.bbc.co.uk/nature/21458115

[371] https://www.amnh.org/about-the-museum/press-center/new-study-doubles-the-estimate-of-bird-species-in-the-world

[372] http://www.eol.org/pages/694

[373] https://web.archive.org/web/20110425151037/http://www.globaltwitcher.com/taxa_class.asp?phylaid=1

[374] http://tolweb.org/Chordata/2499

[375] https://www.ncbi.nlm.nih.gov/Taxonomy/Browser/wwwtax.cgi?id=7711

[376] Lecointre & Le Guyader 2007

[377] Benton, M. J. (2005) *Vertebrate Palaeontology* https//books.google.com John Wiley, 3rd edition, page 14.

[378] Romer 1970.

[379] Dawkins 2004, p. 357.

[380] Lancelet (amphioxus) genome and the origin of vertebrates https://arstechnica.com/science/2008/06/lancelet-amphioxus-genome-and-the-origin-of-vertebrates/ *Ars Technica*, 19 June 2008.

[381] http://tools.wmflabs.org/timescale/?Ma=518

[382] Haines & Chambers 2005.

[383] Encyclopædia Britannica 1954, p. 107.

[384] Berg 2004, p. 599.

[385] SOED

[386] Janvier, P. 2010. "MicroRNAs revive old views about jawless vertebrate divergence and evolution." Proceedings of the National Academy of Sciences (USA) 107:19137-19138. http://www.pnas.org/content/107/45/19137.full.pdf+html *Although I was among the early supporters of vertebrate paraphyly, I am impressed by the evidence provided by Heimberg et al. and prepared to admit that cyclostomes are, in fact, monophyletic. The consequence is that they may tell us little, if anything, about the dawn of vertebrate evolution, except that the intuitions of 19th century zoologists were correct in assuming that these odd vertebrates (notably, hagfishes) are strongly degenerate and have lost many characters over time.*

387 Benton, M. J. (2005) Vertebrate Palaeontology, Blackwell, 3rd edition, Fig 3.25 on page 73.

388 Janvier, Philippe (1997) Vertebrata. Animals with backbones http://tolweb.org/Vertebrata/ 14829/1997.01.01. Version 01 January 1997 in The Tree of Life Web Project http://tolweb.org/

389 Benton 1999, p. 44.

390 Vertebrate jaw design locked down early http://www.cosmosmagazine.com/news/4492/ vertebrate-jaw-design-locked-down-early

391 Where We Split From Sharks http://sciencelife.uchospitals.edu/2012/06/13/where-we-split-from-sharks/

392 Benton, M. J. (2005) Vertebrate Palaeontology, Blackwell, 3rd edition, Fig 7.13 on page 185.

393 https://www.youtube.com/watch?v=k-5oQlnXSTM

394 Helfman & others 2009, p. 198.

395 Clack, J. A. (2002) Gaining Ground. Indiana University

396 Nelson 2006.

397 Forey 1998.

398 Introduction to the Actinopterygii http://www.ucmp.berkeley.edu/vertebrates/actinopterygii/ actinintro.html Museum of Palaeontology, University of California.

399 http://burgess-shale.rom.on.ca/en/fossil-gallery/view-species.php?id=101

400 Dawkins 2004, p. 289: "Obviously vertebrates must have had ancestors living in the Cambrian, but they were assumed to be invertebrate forerunners of the true vertebrates — protochordates. Pikaia has been heavily promoted as the oldest fossil protochordate."

401 Palmer 2000, p. 66-67.

402 Shu2003.

403 Sarjeant & Halstead.

404 Donoghue 2000, p. 206.

405 Turner1999, p. 42–78.

406 Ahlberg 2001, p. 188.

407 Patterson 1987, p. 142.

408 Hall & Hanken 1993, p. 131.

409 Janvier 2003.

410 Benton 2005, p. 65.

411 The Marshall Illustrated Encyclopedia of Dinosaurs and Prehistoric Animals 1999, p. 43.

412 The Marshall Illustrated Encyclopedia of Dinosaurs and Prehistoric Animals 1999, p. 45.

413 The Marshall Illustrated Encyclopedia of Dinosaurs and Prehistoric Animals 1999, p. 26.

414 The Marshall Illustrated Encyclopedia of Dinosaurs and Prehistoric Animals 1999, p. 32.

415 http://animal.discovery.com/tv-shows/other/videos/animal-armageddon-bothriopelis

416 Long 1996.

417 http://animal.discovery.com/tv-shows/other/videos/animal-armageddon-dunkleosteus

418 The Marshall Illustrated Encyclopedia of Dinosaurs and Prehistoric Animals 1999, p. 33.

419 http://www.nature.com/nature/videoarchive/themotherfish/

420 http://tools.wmflabs.org/timescale/?Ma=363

421 http://animal.discovery.com/tv-shows/other/videos/animal-armageddon-eusthenopteron

422 http://dinosaurs.about.com/od/tetrapodsandamphibians/p/gogonasus.htm

423 Nature: The pelvic fin and girdle of Panderichthys and the origin of tetrapod locomotion http://www.nature.com/nature/journal/v438/n7071/edsumm/e051222-13.html

424 Jennifer A. Clack, Scientific American, Getting a Leg Up on Land http://www.scientificamerican.com/article.cfm?id=getting-a-leg-up-on-land Nov. 21, 2005.

425 John Noble Wilford, The New York Times, Scientists Call Fish Fossil the Missing Link https://www.nytimes.com/2006/04/05/science/05cnd-fossil.html?hp&ex=1144296000&en=fe3427d67e965e46&ei=5094&partner=homepage, Apr. 5, 2006.

426 " Acanthostega gunneri http://www.devoniantimes.org/Order/re-acanthostega.html," Devonian Times.

427 http://animal.discovery.com/tv-shows/other/videos/animal-armageddon-ichthyostega.htm

428 https://www.youtube.com/watch?v=qY6cHGVG8-c

429 https://www.youtube.com/watch?v=8g0_vCd71bw

430 https://www.youtube.com/watch?v=0IPmHJ_3nug

431 https://www.youtube.com/watch?v=lzEzAC5DsUc

[432] https://www.youtube.com/watch?v=TWst4N70Wr4

[433] http://www.scientificamerican.com/article.cfm?id=fossil-illuminates-evolut

[434] Fossil Fish of Bear Gulch 2005 by Richard Lund and Eileen Grogan Accessed 2009-01-14

[435] Article on Acanthodes as ancestor of Man, http://www.sci-news.com/paleontology/article00396.html, accessed 15 June 2012

[436] Journal article on Acanthodes, http://www.nature.com/nature/journal/v486/n7402/full/nature11080.html, accessed 15 June 2012

[437] http://geology.about.com/od/extinction/a/aa_permotrias.htm

[438] http://www.kgs.ku.edu/Extension/fossils/massExtinct.html

[439] Bony fishes http://www.seaworld.org/animal-info/info-books/bony-fish/scientific-classification.htm *SeaWorld*. Retrieved 2 February 2013.

[440] The Marshall Illustrated Encyclopedia of Dinosaurs and Prehistoric Animals 1999, p. 38–39.

[441] Motani, R. (2000), Rulers of the Jurassic Seas, Scientific American vol.283, no. 6

[442] Liston, Jeff, M. Newbrey, T. Challands and C. Adams (2013) "Growth, age and size of the Jurassic pachycormid *Leedsichthys problematicus* (Osteichthyes: Actinopterygii) http://eprints.gla.ac.uk/81797/1/81797.pdf." in G. Arratia, H. P. Schultze and M. V. H. Wilson (Eds) *Mesozoic Fishes* 5 – Global Diversity and Evolution, : Pages 145-175, F. Pfeil.

[443] "Biggest Fish Ever Found" Unearthed in U.K. http://news.nationalgeographic.com/news/2003/10/1001_031001_biggestfish.html *National Geographic News*, 1 October 2003.

[444] B. G. Gardiner (1984) Sturgeons as living fossils. Pp. 148–152 in N. Eldredge and S.M. Stanley, eds. Living fossils. Springer-Verlag, New York.

[445] https://www.youtube.com/watch?v=Jla2g1iM8tE

[446] Rafferty, John P (2010) *The Mesozoic Era: Age of Dinosaurs* https//books.google.com Page 219, Rosen Publishing Group.

[447] Fossils (Smithsonian Handbooks) by David Ward (Page 200)

[448] The paleobioloy Database Ptychodus entry http://www.paleodb.org/cgi-bin/bridge.pl?action=checkTaxonInfo&taxon_no=34513&is_real_user=1 accessed on 8/23/09

[449] "BBC - Earth News - Giant predatory shark fossil unearthed in Kansas"

[450] https://www.youtube.com/watch?v=bIjNqDvlyeI

[451] https://www.youtube.com/watch?v=_9LAWve5Xq4

[452] https://www.youtube.com/watch?v=Spo8vkrJFRo

[453] https://www.youtube.com/watch?v=muof-wVRJZE

[454] Myxini http://www.ucmp.berkeley.edu/vertebrates/basalfish/myxini.html - University of California Museum of Paleontology

[455] http://tools.wmflabs.org/timescale/?Ma=66

[456] //en.wikipedia.org/w/index.php?title=Template:Paleontology&action=edit

[457] http://www.nhm.ac.uk/research-curation/collections/our-collections/fossil-vertebrate-collections/fishes/index.html

[458] https://web.archive.org/web/20121111161458/http://www.synthesys.info/de_taf_mfn.htm

[459] http://fieldmuseum.org/explore/department/geology/fossil-fishes

[460] https://books.google.com/books?id=zeyRZNZl-74C&pg=PA188

[461] https//books.google.com

[462] //doi.org/10.1111/j.1469-185X.1999.tb00045.x

[463] //www.ncbi.nlm.nih.gov/pubmed/10881388

[464] https://books.google.com/books?id=fB-hO7i50esC&pg=PA131

[465] http://www.blackwellpublishing.com/helfman/

[466] https://books.google.com/books?id=DL_KQPX3AmIC&pg=PA142

[467] https://books.google.com/?id=B3leE3TTeuIC&pg=PA146

[468] https://www.youtube.com/watch?v=wJunXtPFK-0

[469] https://www.youtube.com/watch?v=ypYesuV3PoI

[470] http://palaeo.gly.bris.ac.uk/Essays/vertfr/default.html

[471] http://www.fishlarvae.com/common/sitemedia/Cloutier%202010%20Sem%20Cell%20Dev%20Biol%20The%20fossil%20record%20of%20fish%20ontogenies.pdf

[472] //doi.org/10.1016/j.semcdb.2009.11.004

[473] https://books.google.com/books?id=dEP_kQAACAAJ&dq=%22The+rise+of+fishes%22&hl=en&sa=X&ei=8WgLUdOMGo2IkgWZzYGwBw&ved=0CC8Q6AEwAA

677

[474] https//books.google.com

[475] https://books.google.com/books?id=gAiAPwAACAAJ&dq=editions:_y8O0MW7spIC&hl= en&sa=X&ei=UmgLUZGFNMemkQXjzoGoDA&ved=0CC8Q6AEwAA"Discovering

[476] //www.ncbi.nlm.nih.gov/pmc/articles/PMC3732986

[477] //doi.org/10.1073/pnas.1304661110

[478] //www.ncbi.nlm.nih.gov/pubmed/23858462

[479] https://books.google.com/books?id=c008kdNwR1cC&dq=%22Your+Inner+Fish%22&hl= en&sa=X&ei=SePbUKKxHIavkgXllIHAAQ&ved=0CDQQ6AEwAA

[480] http://www.ucmp.berkeley.edu/vertebrates/vertintro.html

[481] http://hoopermuseum.earthsci.carleton.ca/12.html

[482] http://www.ukwetlandhabitats.co.uk/fishbiologyandbehavior.html

[483] https://www.youtube.com/watch?v=gZpsVSVRsZk

[484] http://www.ibiology.org/ibioseminars/evolution-ecology/marc-w-kirschner-part-1.html

[485] http://blogs.scientificamerican.com/artful-amoeba/2013/08/29/150-million-years-of-fish-evolution-in-one-handy-figure/

[486] T. Cavalier Smith 2007, Evolution and relationships of algae major branches of the tree of life. https://books.google.com/books?id=-YEYFhgUBsQC&pg=PA21 from: Unravelling the algae, by Brodie & Lewis. CRC Press

[487] Theodor Cole & Hartmut Hilger 2013 Bryophyte Phylogeny http://www2.biologie.fu-berlin.de/sysbot/poster/BPP-E.pdf

[488] Theodor Cole & Hartmut Hilger 2013 Trachaeophyte Phylogeny http://www2.biologie.fu-berlin.de/sysbot/poster/TPP-E.pdf

[489] Theodor Cole & Hartmut Hilger 2015 Angiosperm Phylogeny, Flowering Plant Systematics. http://www2.biologie.fu-berlin.de/sysbot/poster/poster1.pdf Freie Universität Berlin

[490] http://tools.wmflabs.org/timescale/?Ma=850

[491] http://tools.wmflabs.org/timescale/?Ma=470

[492] http://tools.wmflabs.org/timescale/?Ma=390

[493] http://tools.wmflabs.org/timescale/?Ma=370

[494] http://tools.wmflabs.org/timescale/?Ma=200

[495] http://tools.wmflabs.org/timescale/?Ma=40

[496] http://tools.wmflabs.org/timescale/?Ma=10

[497] http://tools.wmflabs.org/timescale/?Ma=1,200

[498] http://tools.wmflabs.org/timescale/?Ma=1,000

[499] Why Christmas trees are not extinct http://www.physorg.com/news9298.html

[500] Hagemann, W. 1976. Sind Farne Kormophyten? Eine Alternative zur Telomtheorie. Plant Systematics and Evolution 124: 251–277.

[501] Sattler, R. 1998. On the origin of symmetry, branching and phyllotaxis in land plants. In: R.V. Jean and D. Barabé (eds) Symmetry in Plants. World Scientific, Singapore, pp. 775-793.

[502] A perspective on the CO_2 theory of early leaf evolution http://www.corante.com/loom/archives/004766.html

[503] Bernstein C, Bernstein H. (1991) Aging, Sex, and DNA Repair. Academic Press, San Diego.

[504] http://tools.wmflabs.org/timescale/?Ma=300

[505] http://tools.wmflabs.org/timescale/?Ma=130

[506] In fact, *Archaefructus* probably didn't bear true flowers: see

[507] http://tools.wmflabs.org/timescale/?Ma=125

[508] http://tools.wmflabs.org/timescale/?Ma=66

[509] //en.wikipedia.org/w/index.php?title=Evolutionary_history_of_plants&action=edit

[510] Translational Biology: From Arabidopsis Flowers to Grass Inflorescence Architecture. Beth E. Thompson* and Sarah Hake, 2009 http://www.plantphysiol.org/content/149/1/38.full, Plant Physiology 149:38–45.

[511] http://tools.wmflabs.org/timescale/?Ma=25–32

[512] http://tools.wmflabs.org/timescale/?Ma=6–7

[513] Above 35% atmospheric oxygen, the spread of fire is unstoppable. Many models have predicted higher values and had to be revised, because there was not a total extinction of plant life.

[514] http://tools.wmflabs.org/timescale/?Ma=14–9

[515] http://tools.wmflabs.org/timescale/?Ma=9–7

[516] as PDF http://usf.usfca.edu/fac_staff/dever/tetrapod_review.pdf

[517] The World Conservation Union. 2014. *IUCN Red List of Threatened Species*, 2014.3. Summary Statistics for Globally Threatened Species. Table 1: Numbers of threatened species by major groups of organisms (1996–2014) http://cmsdocs.s3.amazonaws.com/summarystats/2014_3_Summary_Stats_Page_Documents/2014_3_RL_Stats_Table_1.pdf.

[518] http://tools.wmflabs.org/timescale/?Ma=390

[519] Lucas, Spencer G. (2015) Thinopus and a critical review of Devonian tetrapod footprints. Ichnos, 22:3-4, 136-154. doi: 10.1080/10420940.2015.1063491

[520] Stossel, I. (1995) The discovery of a new Devonian tetrapod trackway in SW Ireland. Journal of the Geological Society, London, 152, 407-413.

[521] Stossel, I., Williams, E.A. & Higgs, K.T. (2016) Ichnology and depositional environment of the Middle Devonian Valentia Island tetrapod trackways, south-west Ireland. Palaeogeography, Palaeoclimatology, Palaeoecology, 462, 16-40.

[522] Research project: The Mid-Palaeozoic biotic crisis: Setting the trajectory of Tetrapod evolution http://www.southampton.ac.uk/oes/research/projects/the_mid_palaeozoic_biotic_crisis.page#overview

[523] (2nd ed. 1955; 3rd ed. 1962; 4th ed. 1970)

[524] Latreielle, P.A. (1804): Nouveau Dictionnaire à Histoire Naturelle, xxiv., cited in Latreille's *Familles naturelles du règne animal, exposés succinctement et dans un ordre analytique*, 1825

[525] Smith, C.H. (2005): Romer, Alfred Sherwood (United States 1894-1973) http://www.wku.edu/~smithch/chronob/ROME1894.htm, homepage from Western Kentucky University

[526] Benton, M. J. (1998) The quality of the fossil record of vertebrates. Pp. 269-303, in Donovan, S. K. and Paul, C. R. C. (eds), The adequacy of the fossil record, Fig. 2. Wiley, New York, 312 pp.

[527] Neill, J.D. (ed.) (2006): Knobil and Neill's Physiology of Reproduction, Vol 2, *Academic Press*, 3rd edition (p. 2177)

[528] Better than fish on land? Hearing across metamorphosis in salamanders http://rspb.royalsocietypublishing.org/content/282/1802/20141943

[529] https://books.google.com/books?id=VThUUUtM8A4C&pg=PA1

[530] https://books.google.com/books?id=6Ztrhm8uLQ0C&pg=PA1

[531] https://books.google.com/books?id=fa6gOvRdl9sC&pg=PA163

[532] https://books.google.com/books?id=wFqrAgAAQBAJ&pg=PA92

[533] https://books.google.com/books?id=9cjTW7FVBXgC&pg=PA59

[534] http://adsabs.harvard.edu/abs/2009AREPS..37..163C

[535] //doi.org/10.1146/annurev.earth.36.031207.124146

[536] https://books.google.com/books?id=Z0YWn5F9sWkC

[537] http://adsabs.harvard.edu/abs/2006Natur.444..199L

[538] //doi.org/10.1038/nature05243

[539] //www.ncbi.nlm.nih.gov/pubmed/17051154

[540] //en.wikipedia.org/w/index.php?title=Template:Paleontology&action=edit

[541] as PDF http://usf.usfca.edu/fac_staff/dever/tetrapod_review.pdf

[542] (2nd ed. 1955; 3rd ed. 1962; 4th ed. 1970)

[543] Monash University. " West Australian Fossil Find Rewrites Land Mammal Evolution https://www.sciencedaily.com/releases/2006/10/061019093718.htm." ScienceDaily 19 October 2006. Accessed 11 March 2009

[544] 375 million-year-old Fish Fossil Sheds Light on Evolution From Fins to Limbs http://www.natureworldnews.com/articles/5632/20140114/ancient-fish-began-developing-legs-before-it-moved-to-land.htm

[545] Vertebrate Land Invasions—Past, Present, and Future: An Introduction to the Symposium http://icb.oxfordjournals.org/content/early/2013/05/09/icb.ict048.full

[546] A Small Step for Lungfish, a Big Step for the Evolution of Walking https://www.sciencedaily.com/releases/2011/12/111212153117.htm

[547] When the Invasion of Land Failed: The Legacy of the Devonian Extinctions https://books.google.com/books?id=wFqrAgAAQBAJ&pg=PA263

[548] Research project: The Mid-Palaeozoic biotic crisis: Setting the trajectory of Tetrapod evolution http://www.southampton.ac.uk/oes/research/projects/the_mid_palaeozoic_biotic_crisis.page# overview

[549] Weishampel, Dodson & Osmolska, 2004, The Dinosauria

[550] Senter, P. (2007). "A new look at the phylogeny of Coelurosauria (Dinosauria: Theropoda)." *Journal of Systematic Palaeontology*, ()

[551] PC Sereno (1997) "The origin and evolution of dinosaurs" Annu. Rev. Earth Planet. Sci. 25:435-489

[552] Richard J. Butler, Paul Upchurch and David B. Norman (2008). The phylogeny of the ornithischian dinosaurs. Journal of Systematic Palaeontology, 6, pp 1-40 doi:10.1017/S1477201907002271

[553] link https://archive.org/details/proceedingsofaca19acad

[554] For example in 1923, three years before Heilmans's book, Roy Chapman Andrews found a good *Oviraptor* fossil in *Mongolia*, but Henry Fairfield Osborn, who analyzed the fossil in 1924, misidentified the furcula as an interclavicle; described in

[555] In an *Oviraptor*: See the summary and pictures at

[556] - full text currently online at This lists a large number of theropods in which furculae have been found, as well as describing those of *Suchomimus Tenerensis* and *Tyrannosaurus rex*.

[557] University of Maryland department of geology home page, "Theropoda I" on *Avetheropoda* http://www.geol.umd.edu/~tholtz/G104/10422ther.htm, 14 July 2006.

[558] **Scienceblogs:** *Limusaurus* is awesome http://scienceblogs.com/tetrapodzoology/2009/06/limusaurus_is_awesome.php .

[559] Developmental Biology 8e Online. Chapter 16: Did Birds Evolve From the Dinosaurs? http://8e.devbio.com/article.php?ch=16&id=161

[560] Vargas AO, Wagner GP and Gauthier, JA. 2009. Limusaurus and bird digit identity. Available from Nature Precedings http://hdl.handle.net/10101/npre.2009.3828.1

[561] Feduccia, A. (2012). *Riddle of the Feathered Dragons: Hidden Birds of China*. Yale University Press, ,

[562] Foth, C. (2012). "On the identification of feather structures in stem-line representatives of birds: evidence from fossils and actuopalaeontology." *Paläontologische Zeitschrift*,

[563] Chinsamy, Anusuya; and Hillenius, Willem J. (2004). "Physiology of nonavian dinosaurs". *The Dinosauria*, 2nd. 643–659.

[564] See commentary on the article https://www.theguardian.com/life/news/story/0,12976,1326559,00.html

[565] Also covers the Reproduction Biology paragraph in the Feathered dinosaurs and the bird connection section.

[566] Organ CL, Schweitzer MH, Zheng W, Freimark LM, Cantley LC, Asara JM, 2008, "Molecular phylogenetics of mastodon and *Tyrannosaurus rex*", *Science* **320**: 499

[567] Kaye TG, Gaugler G, Sawlowicz Z, 2008, "Dinosaurian soft tissues interpreted as bacterial biofilms", *PLoS ONE* **3**, e2808

[568] Peterson JE, Lenczewski ME, Scherer RP, 2010, "Influence of microbial biofilms on the preservation of primary soft tissue in fossil and extant archosaurs", *PLoS ONE* **5**, e13334

[569] Bern M, Phinney BS, Goldberg D, 2009, "Reanalysis of *Tyrannosaurus rex* mass spectra", *Journal of Proteome Research*, **8**: 4328–4332

[570] Cleland TP et al., 2015, "Mass spectrometry and antibody-based characterization of blood vessels from *Brachylophosaurus canadensis, Journal of Proteome Research* **14**: 5252–5262

[571] Michael Buckley, Stacey Warwood, Bart van Dongen, Andrew C. Kitchener, Phillip L. Manning, 2017, "A fossil protein chimera; difficulties in discriminating dinosaur peptide sequences from modern cross-contamination", *Proceedings of the Royal Society B* **284**: 20170544

[572] See also

[573] Feduccia, A. (1993).

[574] Cretaceous tracks of a bird with a similar lifestyle have been found -

[575] Videler, J.J. 2005: Avian Flight. Oxford University. Press, Oxford.

[576] Summarized at

[577] There is a video clip of a very young chick doing this at

[578] Summarized in

[579] Chatterjee, Sankar, Templin, R.J. (2004) "Feathered coelurosaurs from China: new light on the arboreal origin of avian flight" pp. 251-281. In Feathered Dragons: Studies on the Transition from Dinosaurs to Birds (P. J. Currie, E. B. Koppelhus, M. A. Shugar, and J. L. Wright (eds.). Indiana University Press, Bloomington.

[580] Samuel F. Tarsitano, Anthony P. Russell, Francis Horne1, Christopher Plummer and Karen Millerchip (2000) On the Evolution of Feathers from an Aerodynamic and Constructional View Point" *American Zoologist* 2000 40(4):676-686;

[581] Hopson, James A. "Ecomorphology of avian and nonavian theropod phalangeal proportions:Implications for the arboreal versus terrestrial origin of bird flight" (2001) From New Perspectives on the Origin and Early Evolution of Birds: Proceedings of the International Symposium in Honor of John H. Ostrom. J. Gauthier and L. F. Gall, eds. New Haven: Peabody Mus. Nat. Hist., Yale Univ. © 2001 Peabody Museum of Natural History, Yale University. All rights reserved.

[582] Paul, G.S. (2002). "Dinosaurs of the Air: The Evolution and Loss of Flight in Dinosaurs and Birds." Baltimore: Johns Hopkins University Press. page 257

[583] Parsons, William L.; Parsons, Kristen M. (2015). "Morphological Variations within the Ontogeny of Deinonychus antirrhopus (Theropoda, Dromaeosauridae)". PLoS ONE 10 (4). doi:10.1371/journal.pone.0121476. e0121476.

[584] Joel D. Hutson & Kelda N. Hutson, 2018, "Retention of the flight-adapted avian finger-joint complex in the Ostrich helps identify when wings began evolving in dinosaurs", *Ostrich: Journal of African Ornithology* DOI: 10.2989/00306525.2017.1422566

[585] https://www.webcitation.org/5QU8U23Dq?url=http://ravenel.si.edu/paleo/paleoglot/files/Barsbold_83b.pdf

[586] http://apnews.excite.com/article/20140731/us-sci-shrinking-dinosaurs-a5c053f221.html

[587] https://doi.org/10.1016/S0748-3007(03)00069-0

[588] http://www.cumv.cornell.edu/pdf/Bostwick_2003.pdf

[589] http://www.dinosauria.com

[590] https://web.archive.org/web/20070209063731/http://dinosauria.com/jdp/archie/archie.htm

[591] https://web.archive.org/web/20060812210343/http://www.dinosauria.com/jdp/archie/dinoarch.htm

[592] https://doi.org/10.1126/science.1120331

[593] http://www.ucmp.berkeley.edu/diapsids/avians.html

[594] http://www.talkorigins.org/faqs/archaeopteryx.html

[595] http://www.amnh.org/exhibitions/dinosaurs-among-us

[596] http://news.nationalgeographic.com/2016/04/160405-dinosaurs-feathers-birds-museum-new-york-science/

[597] //en.wikipedia.org/w/index.php?title=Template:Evolutionary_biology&action=edit

[598] http://animaldiversity.org/accounts/Prototheria/

[599] Mammalia: Overview – Palaeos http://www.palaeos.com/Vertebrates/Units/430Mammalia/430.000.html

[600] Carroll R.L. (1991): The origin of reptiles. In: Schultze H.-P., Trueb L., (ed) *Origins of the higher groups of tetrapods — controversy and consensus*. Ithaca: Cornell University Press, pp 331-353.

[601] Darren Naish, Episode 38: A Not Too Shabby Podcarts http://tetzoo.com/podcast/2015/1/22/episode-38-a-not-too-shabby-podcarts

[602] Kielan-Jaworowska et al. (2004), p.5

[603] http://palaeos.com/vertebrates/cynodontia/dendrogram.html

[604] See also the news item at

[605] http://www.scotese.com/cretaceo.htm

[606] Kielan-Jaworowska et al. (2004), p. 299

[607] http://www.palaeos.com/Vertebrates/Units/430Mammalia/430.500.html#Theria

[608] Paleontology and Geology of the Upper Jurassic Morrison Formation: Bulletin 36

[609] Meng Chen, Gregory Philip Wilson, A multivariate approach to infer locomotor modes in Mesozoic mammals, Article in Paleobiology 41(02) · February 2015

[610] abstract, in English http://dml.cmnh.org/2001May/msg00969.html

[611] Michael J. Benton,Mikhail A. Shishkin,David M. Unwin, The Age of Dinosaurs in Russia and Mongolia

[612] CHRISTIAN DE MUIZON and BRIGITTE LANGE-BADRÉ, Carnivorous dental adaptations in tribosphenic mammals and phylogenetic reconstruction, Article first published online: 29 MAR 2007

[613] https://web.archive.org/web/20101220204717/http://palaeos.com/Vertebrates/Lists/Cladograms/360Mammalia.html

[614] (pdf version http://biology.plosjournals.org/perlserv/?request=get-pdf&file=10.1371_journal.pbio.0040091-S.pdf)

[615] http://tools.wmflabs.org/timescale/?Ma=7.6

[616] Historical perspective (the Dynamic Earth, USGS) http://pubs.usgs.gov/gip/dynamic/historical.html

[617] Cretaceous map http://www.scotese.com/cretaceo.htm

[618] Insectivora Overview – Palaeos http://www.palaeos.com/Vertebrates/Units/460Insectivora/460.000.html

[619] http://tools.wmflabs.org/timescale/?Ma=85

[620] — a similar paper by these authors is free online at New light on the dates of primate origins and divergence http://www-hto.usc.edu/people/stavare/STpapers-pdf/SWTMM03.pdf

[621] For other opinions see "Technical comments" linked from same Web page

[622] Lactating on Eggs http://nationalzoo.si.edu/ConservationAndScience/SpotlightOnScience/oftedalolav20030714.cfm

[623] Mammals of the Mesozoic: The least mammal-like mammals https://www.webcitation.org/query?url=http://www.geocities.com/trevor_dykes/mammalsbasal.htm&date=2009-10-25+23:17:19

[624] Elsa Panciroli; Roger B. J. Benson; Stig Walsh (2017). "The dentary of Wareolestes rex (Megazostrodontidae): a new specimen from Scotland and implications for morganucodontan tooth replacement". Papers in Palaeontology. in press. doi:10.1002/spp2.1079.

[625] P Bajdek, Microbiota and food residues including possible evidence of pre-mammalian hair in Upper Permian coprolites from Russia, G. 2015,

[626] Olsen, Paul E. (2012): Cynodontipus: A procolophonid burrow – not a hairy cynodont track (Middle-Late Triassic: Europe, Morocco, Eastern North America) http://gsa.confex.com/gsa/2012NE/finalprogram/abstract_200184.htm. Program for Northeastern Section – 47th Annual Meeting (18–20 March 2012) Hartford, Connecticut, the Geological society of America

[627] Bakker 1975

[628] JENNIFER BOTHA-BRINK and KENNETH D. ANGIELCZYK, Do extraordinarily high growth rates in Permo-Triassic dicynodonts (Therapsida, Anomodontia) explain their success before and after the end-Permian extinction?, Version of Record online: 26 JUL 2010

[629] Microbiota and food residues including possible evidence of pre-mammalian hair in Upper Permian coprolites from Russia Piotr Bajdek1, Martin Qvarnström2, Krzysztof Owocki3, Tomasz Sulej3, Andrey G. Sennikov4,5, Valeriy K. Golubev4,5 andGrzegorz Niedźwiedzki2 Article first published online: 25 NOV 2015

[630] Markus Lambertz et al, A caseian point for the evolution of a diaphragm homologue among the earliest synapsids, Annals of the New York Academy of Sciences (2016).

[631] https://web.archive.org/web/20121124020009/http://carnegiemnh.net/assets/science/vp/Luo%202007%20%28Mesozoic%20mammal%20review%29%5B1%5D.pdf

[632] http://adsabs.harvard.edu/abs/2007Natur.450.1011L

[633] //doi.org/10.1038/nature06277

[634] //www.ncbi.nlm.nih.gov/pubmed/18075580

[635] http://carnegiemnh.net/assets/science/vp/Luo%202007%20%28Mesozoic%20mammal%20review%29%5B1%5D.pdf

[636] http://www.nasmus.co.za/PALAEO/jbotha/the_cynodontia.htm

[637] https://www.bbc.co.uk/programmes/p003k9ds

[638] APG 2016.

[639] Cronquist 1960.

[640] Takhtajan 1964.

[641] Takhtajan 1980.

[642] The major exception to the dominance of terrestrial ecosystems by flowering plants is the coniferous forest.

[643] Brown R., Character and description of *Kingia*, a new genus of plants found on the southwest coast of New Holland: with observations on the structure of its unimpregnated ovulum; and on the female flower of Cycadeae and Coniferae, in: King P.P. (Ed.) *Narrative of a Survey of the Intertropical and western coasts of Australia, performed between years 1818 and 1822*. John Murray, London, 1827, vol. 2., pp. 534–565, https://books.google.com/books?id= RjdCAAAAIAAJ&hl=&pg=PA534#v=onepage&q&f=false.

[644] APG 2003.

[645] APG 2009.

[646] Chase & Reveal 2009.

[647] , Figure 2 http://www.amjbot.org/cgi/content/full/91/10/1437/F2

[648] Zeng et al 2014.

[649] Bell et al 2010.

[650] http://tools.wmflabs.org/timescale/?Ma=319

[651] http://tools.wmflabs.org/timescale/?Ma=192

[652] http://tools.wmflabs.org/timescale/?Ma=160

[653] Oily Fossils Provide Clues To The Evolution Of Flowers https://www.sciencedaily.com/releases/2001/04/010403071438.htm — ScienceDaily (April 5, 2001)

[654] NOVA — Transcripts — First Flower https://www.pbs.org/wgbh/nova/transcripts/3405_flower.html — PBS Airdate: April 17, 2007

[655] South Pacific plant may be missing link in evolution of flowering plants http://www.eurekalert.org/pub_releases/2006-05/uoca-spp051506.php — Public release date: 17 May 2006

[656] Age-Old Question On Evolution Of Flowers Answered http://unisci.com/stories/20012/0615015.htm — 15-Jun-2001

[657] Human Affection Altered Evolution of Flowers http://www.livescience.com/295-human-affection-altered-evolution-flowers.html — By Robert Roy Britt, LiveScience Senior Writer (posted: 26 May 2005 06:53 am ET)

[658] http://tools.wmflabs.org/timescale/?Ma=178

[659] http://tools.wmflabs.org/timescale/?Ma=198

[660] Foster CSP, Ho SYW (2017) Strategies for partitioning clock models in phylogenomic dating: Application to the angiosperm evolutionary timescale. Genome Biol Evol

[661] Raven, Peter H., Ray F. Evert, & Susan E. Eichhorn, 2005. *Biology of Plants*, 7th edition. (New York: W. H. Freeman and Company).

[662] Snustad DP, Simmons MJ (2008). Principles of Genetics (5th ed.). Wiley.

[663] Dilcher et al 2016.

[664] http://www.blackwell-synergy.com/links/doi/10.1046/j.1095-8339.2003.t01-1-0158.x/full/

[665] //doi.org/10.1046/j.1095-8339.2003.t01-1-00158.x

[666] http://www3.interscience.wiley.com/journal/122630309/abstract

[667] //doi.org/10.1111/j.1095-8339.2009.00996.x

[668] http://onlinelibrary.wiley.com/doi/10.1111/boj.12385/abstract

[669] //doi.org/10.1111/boj.12385

[670] //doi.org/10.2307/1218032

[671] https://books.google.com/books?id=SM3khPHXhKEC

[672] https://archive.org/details/plantform00adri

[673] //doi.org/10.3732/ajb.0900346

[674] //www.ncbi.nlm.nih.gov/pubmed/21616882

[675] //doi.org/10.1111/j.1095-8339.2009.01002.x

[676] https://web.archive.org/web/20030714120417/http://www.news.harvard.edu/gazette/1999/12.16/angiosperms.html

[677] //doi.org/10.1007/BF02940572

[678] //doi.org/10.1111/j.1095-8339.1980.tb01661.x

[679] //doi.org/10.1111/j.1756-1051.1983.tb01448.x

[680] //www.ncbi.nlm.nih.gov/pmc/articles/PMC34380

[681] http://adsabs.harvard.edu/abs/2000PNAS...97.7030D

[682] //doi.org/10.1073/pnas.97.13.7030

[683] https://www.britannica.com/plant/angiosperm

[684] https://www.researchgate.net/publication/275374707_Evolutionary_Cladistics_and_the_ origin_of_Angiosperms

[685] https://books.google.com/books?id=2YbwF7tH6dUC

[686] https://books.google.com/books?id=3pZxuyUllJIC

[687] https://books.google.com/books?id=dj8KRImgyf4C

[688] //doi.org/10.1016/j.pbi.2016.03.015

[689] //doi.org/10.2307/1216134

[690] //www.jstor.org/stable/10.2307/1216134

[691] //doi.org/10.1007/bf02861558

[692] //www.jstor.org/stable/10.2307/4353970

[693] http://adsabs.harvard.edu/abs/2014NatCo...5E4956Z

[694] //doi.org/10.1038/ncomms5956

[695] http://www.biologie.fu-berlin.de/sysbot/poster/poster1.pdf

[696] https://web.archive.org/web/20140802080838/http://www.biologie.uni-hamburg.de/b-online/ delta/angio

[697] http://www.biologie.uni-hamburg.de/b-online/delta/angio

[698] http://www.eol.org/pages/282

[699] Hilton, Jason, and Richard M. Bateman. 2006. Pteridosperms are the backbone of seed-plant phylogeny. *Journal of the Torrey Botanical Society* 133: 119-168 (abstract http: //www.bioone.org/perlserv/?request=get-abstract&doi=10.3159%2F1095-5674(2006)133% 5B119%3APATBOS%5D2.0.CO%3B2&ct=1)

[700] Campbell and Reece; Biology, Eighth edition

[701] http://tools.wmflabs.org/timescale/?Ma=319

[702] Jiao Y, Wickett NJ, Ayyampalayam S, Chanderbali AS, Landherr L, Ralph PE, Tomsho LP, Hu Y, Liang H, Soltis PS, Soltis DE, Clifton SW, Schlarbaum SE, Schuster SC, Ma H, Leebens-Mack J, Depamphilis CW (2011) Ancestral polyploidy in seed plants and angiosperms. Nature

[703] Catalogue of Life: 2007 Annual checklist - Conifer database http://www.catalogueoflife.org/ show_database_details.php?database_name=Conifer+Database

[704] Campbell, Reece, "Phylum Coniferophyta."Biology. 7th. 2005. Print. P.595

[705] http://www.conifers.org/

[706] https://web.archive.org/web/20080409044237/http://www.huh.harvard.edu/research/ mathews-lab/atolHtmlSite/

[707] Conway, John R. "The Biology of Honey Ants." *The American Biology Teacher*, Vol. 48, No. 6 (Sep., 1986), pp. 335–343.

[708] Costa-Leonardo AM, Haifig I. (2014). Termite Communication During Different Behavioral Activities. In: Biocommunication of Animals. Dortrecht, Springer, 161-190.

[709] Darwin, Charles. *On the Origin of Species*, 1859. Chapter 8

[710] http://www.iussi.org/

[711] http://academic.reed.edu/biology/professors/srenn/pages/teaching/web_2007/molerats_lb_jg/ index.html

[712] http://tools.wmflabs.org/timescale/?Ma=85

[713] http://tools.wmflabs.org/timescale/?Ma=55

[714] http://tools.wmflabs.org/timescale/?Ma=14

[715] Zuk, Marlene (2014), "Paleofantasy: What Evolution Really Tells Us About Sex, Diet, and How We Live" (W. W. Norton & Company)

[716] Hrdy, Sarah Blaffer (2011), "Mothers and Others: The Evolutionary Origins of Mutual Under-standing" (Harvard Uni Press)

[717] //en.wikipedia.org/w/index.php?title=Human_evolution&action=edit

[718] http://tools.wmflabs.org/timescale/?Ma=4.2

[719] http://tools.wmflabs.org/timescale/?Ma=7.2

[720] • Tishkoff and Reed (2009)

[721] Peter B. deMenocal, (2016) "Climate Shocks" (Scientific American Vol 25, No 4)

[722] Barras, Colin (2016), "Stone Tools hint humans reached Asia much earlier" (New Scientist 6 February 2016)

[723] The results were published in the online edition of the journal *Science*.

[724] Huertha Sanchez, Emilia et al (2014), "Altitude adaptation in Tibetans caused by introgression of Denisovan-like DNA" (Nature Vol 512, 14 August 2014)

[725] Oppenheimer, Stephen (2012), "Out of Eden: The Peopling of the World" (Robinson; New Ed edition (March 1, 2012))

[726] http://tools.wmflabs.org/timescale/?Ma=4–8

[727] http://tools.wmflabs.org/timescale/?Ma=7–13

[728] http://tools.wmflabs.org/timescale/?Ma=5–6

[729] http://tools.wmflabs.org/timescale/?Ma=7

[730] http://tools.wmflabs.org/timescale/?Ma=5.7

[731] http://tools.wmflabs.org/timescale/?Ma=5.6

[732] http://tools.wmflabs.org/timescale/?Ma=4

[733] http://tools.wmflabs.org/timescale/?Ma=2.8

[734] http://tools.wmflabs.org/timescale/?Ma=1.9

[735] http://tools.wmflabs.org/timescale/?Ma=1.8–1.3

[736] http://tools.wmflabs.org/timescale/?Ma=0.09

[737] http://tools.wmflabs.org/timescale/?Ma=2

[738] Gardner., Elizabeth K.; Purdue University (April 1, 2015). "New instrument dates old skeleton before 'Lucy'; 'Little Foot' 3.67 million years old". Science Daily. Retrieved April 3, 2015.

[739] http://tools.wmflabs.org/timescale/?Ma=1.3–1.8

[740] based on Schlebusch et al., "Southern African ancient genomes estimate modern human divergence to 350,000 to 260,000 years ago" Science, 28 Sep 2017, DOI: 10.1126/science.aao6266 http://science.sciencemag.org/content/early/2017/09/27/science.aao6266.full, Fig. 3 https://d2ufo47lrtsv5s.cloudfront.net/content/sci/early/2017/09/27/science.aao6266/F3.large.jpg (H. sapiens divergence times) and (archaic admixture).

[741] Wrangham, Richard (2011), "Catching Fire: How cooking made us human"

[742] "Kaufman, Danial (2002), "Comparisons and the Case for Interaction among Neanderthals and Early Modern Humans in the Levant" (Oxford Journal of Anthropology)

[743] Science, doi.org/bch3

[744] Dean, MC, Stringer, CB et al, (1986) "Age at death of the Neanderthal child from Devil's Tower, Gibraltar and the implications for studies of general growth and development in Neanderthals" (American Journal of Physical Anthropology, Vol 70 Issue 3, July 1986)

[745] //lccn.loc.gov/95185095

[746] //www.worldcat.org/oclc/33408268

[747] //lccn.loc.gov/2006020835

[748] //www.worldcat.org/oclc/70199867

[749] //lccn.loc.gov/96053993

[750] //www.worldcat.org/oclc/36165190

[751] http://digitalcommons.unl.edu/cgi/viewcontent.cgi?article=1217&context=usgsstaffpub

[752] //www.worldcat.org/oclc/16997265

[753] //lccn.loc.gov/2002075336

[754] //www.worldcat.org/oclc/49959461

[755] //www.worldcat.org/issn/0067-8503

[756] //lccn.loc.gov/59012726

[757] //www.worldcat.org/oclc/01532912

[758] //lccn.loc.gov/2003046006

[759] //www.worldcat.org/oclc/55016591

[760] //lccn.loc.gov/2004353026

[761] //www.worldcat.org/oclc/57077633

[762] //lccn.loc.gov/2008036672

[763] //www.worldcat.org/oclc/191926088

[764] //www.worldcat.org/oclc/421466369

[765] //lccn.loc.gov/80008679

[766] //www.worldcat.org/oclc/7197127

[767] //lccn.loc.gov/2004059864

[768] //www.worldcat.org/oclc/56617123

[769] //www.worldcat.org/oclc/144520427

[770] //lccn.loc.gov/2005284124
[771] //www.worldcat.org/oclc/35792200
[772] //lccn.loc.gov/2006034384
[773] //www.worldcat.org/oclc/73502978
[774] //lccn.loc.gov/87035814
[775] //www.worldcat.org/oclc/17328115
[776] //doi.org/10.1002/ajpa.1330700214
[777] //lccn.loc.gov/85173489
[778] //www.worldcat.org/oclc/12352830
[779] //lccn.loc.gov/94003617
[780] //www.worldcat.org/oclc/30739453
[781] //lccn.loc.gov/2004052648
[782] //www.worldcat.org/oclc/55600894
[783] //doi.org/10.1016/B0-08-043076-7/03083-7
[784] //lccn.loc.gov/2001044791
[785] //www.worldcat.org/oclc/47869490
[786] //lccn.loc.gov/83024005
[787] //www.worldcat.org/oclc/10163036
[788] //lccn.loc.gov/2008030270
[789] //www.worldcat.org/oclc/225874308
[790] //doi.org/10.1007/978-3-540-37654-5
[791] //lccn.loc.gov/2009931325
[792] //www.worldcat.org/oclc/549541244
[793] //lccn.loc.gov/2007024716
[794] //www.worldcat.org/oclc/144770218
[795] //www.worldcat.org/oclc/423293609
[796] //lccn.loc.gov/2007052429
[797] //www.worldcat.org/oclc/187548835
[798] //lccn.loc.gov/99032072
[799] //www.worldcat.org/oclc/41431683
[800] //lccn.loc.gov/92018037
[801] //www.worldcat.org/oclc/444512451
[802] //lccn.loc.gov/2001037337
[803] //www.worldcat.org/oclc/48066180
[804] //www.ncbi.nlm.nih.gov/pmc/articles/PMC2947357
[805] http://adsabs.harvard.edu/abs/2009Sci...324.1035T
[806] //doi.org/10.1126/science.1172257
[807] //www.ncbi.nlm.nih.gov/pubmed/19407144
[808] //lccn.loc.gov/2004028087
[809] //www.worldcat.org/oclc/162577235
[810] //lccn.loc.gov/2010038249
[811] //www.worldcat.org/oclc/669122326
[812] //lccn.loc.gov/2001043790
[813] //www.worldcat.org/oclc/537239907
[814] //lccn.loc.gov/2005036120
[815] //www.worldcat.org/oclc/132816551
[816] //lccn.loc.gov/2003022857
[817] //www.worldcat.org/oclc/53254011
[818] //lccn.loc.gov/2009050471
[819] //www.worldcat.org/oclc/429022321
[820] //doi.org/10.1007/978-1-4020-9980-9_3
[821] //www.worldcat.org/issn/1877-9077
[822] //lccn.loc.gov/2009927083
[823] //www.worldcat.org/oclc/310400980
[824] http://qcpages.qc.edu/Biology/LahtiSites/RDAlexander/Pubs/Alexander90.pdf
[825] //lccn.loc.gov/90623893

[826] //www.worldcat.org/oclc/22860997
[827] //lccn.loc.gov/2007010767
[828] //www.worldcat.org/oclc/86090399
[829] http://adsabs.harvard.edu/abs/2002Natur.418..869E
[830] //doi.org/10.1038/nature01025
[831] //www.worldcat.org/issn/0028-0836
[832] //www.ncbi.nlm.nih.gov/pubmed/12192408
[833] http://web.missouri.edu/~gearyd/Flinnetal2005.pdf
[834] //doi.org/10.1016/j.evolhumbehav.2004.08.005
[835] //www.worldcat.org/issn/1090-5138
[836] //lccn.loc.gov/2005053780
[837] //www.worldcat.org/oclc/61652817
[838] //lccn.loc.gov/2001037847
[839] //www.worldcat.org/oclc/47254191
[840] //doi.org/10.1006/jhev.2001.0495
[841] //www.worldcat.org/issn/0047-2484
[842] //www.ncbi.nlm.nih.gov/pubmed/11681862
[843] //doi.org/10.1002/ajpa.1330310505
[844] //www.worldcat.org/issn/0002-9483
[845] http://adsabs.harvard.edu/abs/1964Sci...143..111H
[846] //doi.org/10.1126/science.143.3602.111
[847] http://www.pnas.org/content/88/20/9051.full.pdf
[848] http://adsabs.harvard.edu/abs/1991PNAS...88.9051I
[849] //doi.org/10.1073/pnas.88.20.9051
[850] //www.worldcat.org/issn/0027-8424
[851] //www.ncbi.nlm.nih.gov/pmc/articles/PMC52649
[852] //www.ncbi.nlm.nih.gov/pubmed/1924367
[853] //lccn.loc.gov/2010483830
[854] //www.worldcat.org/oclc/419801728
[855] //lccn.loc.gov/92006661
[856] //www.worldcat.org/oclc/25373161
[857] //lccn.loc.gov/97000972
[858] //www.worldcat.org/oclc/36181117
[859] //lccn.loc.gov/2006052267
[860] //www.worldcat.org/oclc/76481584
[861] //lccn.loc.gov/2005482222
[862] //www.worldcat.org/oclc/52195607
[863] //doi.org/10.1038/35006625
[864] //www.ncbi.nlm.nih.gov/pubmed/10761915
[865] //www.worldcat.org/oclc/310156315
[866] //lccn.loc.gov/95006337
[867] //www.worldcat.org/oclc/32088673
[868] //lccn.loc.gov/2011489742
[869] //www.worldcat.org/oclc/689522193
[870] //lccn.loc.gov/2004110563
[871] //www.worldcat.org/oclc/224377190
[872] //lccn.loc.gov/96037718
[873] //www.worldcat.org/oclc/36001167
[874] //lccn.loc.gov/2008013654
[875] //www.worldcat.org/oclc/218188644
[876] //lccn.loc.gov/00300421
[877] //www.worldcat.org/oclc/45853997
[878] //lccn.loc.gov/2005055293
[879] //www.worldcat.org/oclc/62282400
[880] //www.worldcat.org/oclc/35202130
[881] //lccn.loc.gov/85000158

[882] //www.worldcat.org/oclc/11726796
[883] //lccn.loc.gov/2003066679
[884] //www.worldcat.org/oclc/53287806
[885] http://www.bbc.co.uk/sn/prehistoric_life/human/human_evolution/index.shtml
[886] http://www.becominghuman.org/
[887] http://www.hhmi.org/biointeractive/bones-stones-and-genes-origin-modern-humans-0
[888] http://www.evolution-textbook.org/content/free/figures/ch25.html
[889] http://humanorigins.si.edu/
[890] http://archaeologyinfo.com/human-evolution-timeline/
[891] https//www.canal-u.tv
[892] https://www.pbs.org/first-peoples/home/
[893] //www.worldcat.org/oclc/910115743
[894] http://yalebooks.com/yupbooks/book.asp?isbn=9780300182026
[895] http://humanorigins.si.edu/evidence/human-evolution-timeline-interactive
[896] https://www.bbc.co.uk/programmes/p003hyfl
[897] • (PBS Digital Studios, November 17, 2014)
[898] Sole, R.V., and Newman, M., 2002. "Extinctions and Biodiversity in the Fossil Record – Volume Two, *The Earth system: biological and ecological dimensions of global environment change*" pp. 297–391, *Encyclopedia of Global Environmental Change* John Wiley & Sons.
[899] Partial list from Image:Extinction Intensity.png
[900] Hallam, Anthony, & Wignall, P. B. (2002). Mass Extinctions and Their Aftermath. New York: Oxford University Press Inc.
[901] Different cycle lengths have been proposed; e.g. by
[902] Dissolved oxygen became more widespread and penetrated to greater depths; the development of life on land reduced the run-off of nutrients and hence the risk of eutrophication and anoxic events; and marine ecosystems became more diversified so that food chains were less likely to be disrupted.
[903] Arens, N.C. and West, I.D. (2006). "Press/Pulse: A General Theory of Mass Extinction?" 'GSA Conference paper' Abstract http://gsa.confex.com/gsa/2006AM/finalprogram/abstract_111772.htm
[904] Courtillot, V., Jaeger, J-J., Yang, Z., Féraud, G., Hofmann, C. (1996). "The influence of continental flood basalts on mass extinctions: where do we stand?" in Ryder, G., Fastovsky, D., and Gartner, S, eds. "The Cretaceous-Tertiary event and other catastrophes in earth history". *The Geological Society of America*, Special Paper 307, 513–525.
[905] The earliest known flood basalt event is the one which produced the Siberian Traps and is associated with the end-Permian extinction.
[906] Some of the extinctions associated with flood basalts and sea-level falls were significantly smaller than the "major" extinctions, but still much greater than the background extinction level.
[907] http://www.nature.com/scientificamerican/journal/v263/n4/pdf/scientificamerican1090-85.pdf
[908] Courtillot, V. E., Renne, P. R., 2003. On the ages of flood basalt events. Comptes Rendus Geosciences 335 (1), 113–140.
[909] Newswise: Ebb and Flow of the Sea Drives World's Big Extinction Events http://newswise.com/articles/view/541743/ Retrieved on June 15, 2008.
[910] Berner, R. A., and Ward, P. D. (2004). " Positive Reinforcement, H_2S, and the Permo-Triassic Extinction: Comment and Reply http://www.gsajournals.org/perlserv/?request=get-static&name=i0091-7613-31-6-e100" describes possible positive feedback loops in the catastrophic release of hydrogen sulfide proposed by Kump, Pavlov and Arthur (2005).
[911] Summarised by Ward (2006).
[912] http://www.lpl.arizona.edu/impacteffects/
[913] http://www.speciesalliance.com
[914] http://www.space.com/scienceastronomy/050503_mass_extinctions.html
[915] http://strata.geology.wisc.edu/jack

Article Sources and Contributors

The sources listed for each article provide more detailed licensing information including the copyright status, the copyright owner, and the license conditions.

History of Earth *Source:* https://en.wikipedia.org/w/index.php?oldid=853019601 *License:* Creative Commons Attribution-Share Alike 3.0 *Contributors:* A Great Catholic Person, Alumnum, AntiVan, Apokryltaros, Aquarius-1, Arnav mahani, Bear-rings, Beland, Bender235, Bgwhite, Bobbie73, Byteflush, CLCStudent, CV9933, Caehlla, ClueBot NG, CommonsDelinker, Cryptic, Crystallizedcarbon, Dane, Dawnseeker2000, Dcirovic, Discospinster, Djwiofeoi, Dolotta, Donner60, DrStrauss, Dragonpurl, Drbogdan, Dudley Miles, DuncanHill, Dunkleosteus77, Dw122339, EarthOcean, Edward, Filursiax, G0mx, Gap9551, Gareth Griffith-Jones, GeneralizationsAreBad, GeoWriter, Gilliam, God's Godzilla, Gorthian, GrapefruitSculpin, HMSLavender, Hafeez Depar, Harizotoh9, Headbomb, HiLo48, Hut 8.5, Iacobus, Ilyakor0676, IronGargoyle, Isambard Kingdom, IyrandrarSarhana, IznoRepeat, JWNoctis, JamesBWatson, Jarble, Jdaloner, Jerod Franko, Jim1138, Jon Kolbert, Julietdeltalima, Karanblood2525, Khirurg, Kku, KylieTastic, Lappspira, Magyar25, Mikee butt, Minituremeow, My Chemistry romantic, Newone, Nyttend, Osh33m, Oshwah, PaleoNeonate, Pauli133, Peterye2005, PlyrStar93, Quibilia, Rcsprinter123, Red Planet X (Hercolubus), Renamed user 943a06d1c3, Rhinopias, Rjwilmsi, RockMagnetist, Roxy the dog, Ryangosh, SA 13 Bro, Sanfranman59, Seaweed, Semmendinger, Serafart, Serols, Slightsmile, Soul2251974, Specane111, Spike Wilbury, Tajotep, Theroadislong, Thrif, Titiawolflover, UnsungKing123, Vanamonde93, Vsmith, Wgolf, Yamaguchi先生 , Ynoss, Z0, Zboy Muer, 164 anonymous edits 1

Earliest known life forms *Source:* https://en.wikipedia.org/w/index.php?oldid=851496935 *License:* Creative Commons Attribution-Share Alike 3.0 *Contributors:* Animalparty, Aquarius-1, Aspening, Ayuta Tonomura, CIAVermont, Chiswick Chap, Darsie42, Davemck, Dpleibovitz, Drbogdan, Graeme Bartlett, Habil zare, Harizotoh9, Holothurion, John Abbe, Kintetsubuffalo, Nwbeeson, OAnick, Robshort, Spike Wilbury, TheSandDoctor, Timpo, Turtlesandfoxes, Ugly Ketchup, Volcanoguy, Whoop whoop pull up, Ysangkok, 8 anonymous edits 39

Evidence of common descent *Source:* https://en.wikipedia.org/w/index.php?oldid=853778287 *License:* Creative Commons Attribution-Share Alike 3.0 *Contributors:* Alan G. Archer, Apokryltaros, Arjayay, Azcolvin429, BD2412, Bender235, Bob9944, Brandon Vun, Bueller 007, BurritoBazooka, CanadianLinuxUser, Chiswick Chap, Chris the speller, ClueBot NG, D.M. from Ukraine, Dcirovic, Dewritech, Diannaa, Donner60, Drbogdan, Editor2020, Finnusertop, Frietjes, Fyrius, Gaius Cornelius, Gap9551, Headbomb, Iridescent, Jarble, Jess, Jessicapierce, Jim1138, Jmertel23, Johnuniq, Jonesey95, Just plain Bill, Laurusnobilis, LilHelpa, Meeyola, Micromesistius, Mild Bill Hiccup, Mindmatrix, Mogism, Myasuda, Narky Blert, Niceguyedc, Nicrodemo, Nihiltres, One15969, Oshwah, PaleoNeonate, Plantdrew, Quivico, R'n'B, RZuljani, Rcronk, Rcsprinter123, Retimuko, Rjwilmsi, Sanu N, SchreiberBike, Sizeofint, Tassedethe, Theroadislong, Titodutta, Tom.Reding, Trappist the monk, Vsmith, WolfmanSF, Wyrm127, 47 anonymous edits 47

Common descent *Source:* https://en.wikipedia.org/w/index.php?oldid=852974781 *License:* Creative Commons Attribution-Share Alike 3.0 *Contributors:* 84user, Agathman, Alan G. Archer, Alan Liefting, Alaney2k, Alexs, Anypodetos, Archaeopteryx, Armchair info guy, Art LaPella, Aunt Entropy, Auntof6, Azcolvin429, Ben Moore, Bender235, Bibliophile20, Billyshiverstick, Brandmeister, Cadiomals, Capitalismojo, Categorer, Catgut, Catslash, Cement123, Cheddarbob23, Chiswick Chap, Christian Skeptic, Citation bot 1, ClueBot NG, CommonsDelinker, CsDix, Curtis Clark, Cwbm (commons), DRosenbach, DadaNeem, David Eppstein, Dawn Bard, Deeplogic, Digitwoman, Doc Tropics, Dominus Vobisdu, Donarreiskoffer, Drbogdan, Drchris65, Dtheobald, Dufusrex, Ec5618, EoGuy, Epipelagic, Equilibrium Allure, Ernsts, Finell, Firsfron, Flyer22 Reborn, ForestDim, GSS, GSlicer, Gabbe, Georgeanewman, Gilo1969, Gpvos, Guettarda, Haggiaomega, Harburg, Headbomb, Horselover Frost, Ilikelifesciences, Infraredeclipse, J.delanoy, JHunterJ, JPaps1977, Jandalhandler, Jarhed, Joefromrandb, Jogers, Johnuniq, Join Tile, Jondw, Jonesey95, Jorge 2701, JoshuaZ, JustBerry, Justeditingtoday, Keegscee, Kjkolb, Klilidiplomus, Littlealien182, Livingrm, Logicus, Magioladitis, Mama meta modal, Martial75, Mausy5043, Mccajor, Medeis, Mild Bill Hiccup, Mindmatrix, Mr. Wheely Guy, Nihiltres, Northfox, Nwbeeson, Octavian history, Orangemarlin, PaleoNeonate, Pbarnes, Persian Poet Gal, Petr Matas, Petri Krohn, Pinethicket, Pmanderson, PoolPartay, Princeofexcess, RHaworth, Rafikgl, Rich Farmbrough, Richard001, RichardWeiss, Rjwilmsi, Robert Stevens, Rogue-pilot, Rumping, Sardanaphalus, SaudiPseudonym, Seaphoto, SheffieldSteel, Skeptic from Britain, Skizzik, SpuriousQ, Stephenb, Stesmo, Stiepan Pietrov, Strictscrutiny, Sunrise, TAnthony, The Anome, Thomas Arelatensis, TomS TDotO, Tomtheman5, Trappist the monk, Traxs7, Valich, Vanished user, Vsmith, Wallace Kneeland, Webridge, Winterschlaefer, Zappernapper, 138 anonymous edits 107

Homology (biology) *Source:* https://en.wikipedia.org/w/index.php?oldid=849021762 *License:* Creative Commons Attribution-Share Alike 3.0 *Contributors:* Abrower, Ahsan cima, Ahsanur-wiki, Alexschmidt711, Allens, AlphaEta, Bazuz, Ben Moore, Bender235, Bgwhite, BillShurts, CFCF, Chiswick Chap, Chris Capoccia, ChrisGualtieri, Citation bot 1, ClueBot NG, Cryptic, Crystallizedcarbon, DadaNeem, DangerousJXD, David Eppstein, Dcirovic, DennisPietras, Desireecooley, Dewritech, Discospinster, DocWatson42, Donner60, Dopeytaylor, Doremo, Drewmutt, Drphilharmonic, Dsmoore4, Epipelagic, Evolution and evolvability, Excirial, FamAD123, Finnusertop, Fixer88, Flyer22 Reborn, Googlablenickname, Graeme Bartlett, Grafen, GünniX, HCA, Huhnra, Illia Connell, Iwilsonp, JSquish, Jamesx12345, Jasonanaggie, Jim1138, Jimmycleveland, Johnuniq, JonRichfield, Jonesey95, KH-1, Kku, Klbrain, Knowledgeable Raven, Krenair, Kvng, Kyle J Martin, Kylesalt, Makecat, Manudouz, Mark Arsten, Materialscientist, Matt Fitzpatrick, Melonkelon, Mgiganteus1, Michaelban, MotherShabubu, Mr Stephen, Mradkins96, NanoNight, NawlinWiki, Neuroschizl, NunoAgostinho, Omnipaedista, Orangemike, Peter M. Brown, Peteruetz, Plindenbaum, Protein Chemist, Puddik at, Quercus solaris, RDBury, Reuqr, Rich Farmbrough, Richard3120, RichardWeiss, Rjwilmsi, Rlcamp1, Ryk72, Sarah1x, Smith609, Sowlos, Srnec, Stemonitis, The Anome, ThePlatypusofDoom, Theroadislong, ToddDeLuca, Vycl1994, Wavelength, WilliamJamesHerath, Wyrm127, Zatelmae, Zeeyanwiki, Zppix, הסרפד, 101 anonymous edits 114

Last universal common ancestor *Source:* https://en.wikipedia.org/w/index.php?oldid=850227956 *License:* Creative Commons Attribution-Share Alike 3.0 *Contributors:* A bit iffy, A2soup, A876, Adi, Andrewrobinson01070111GroupD ext2015, Annhilator mk3, Apokryltaros, Article editor, Axl, Azcolvin429, BatteryIncluded, Bejnar, Bender235, Booklaunch, Byelf2007, Cheater no1, Chiswick Chap, Chris Capoccia, ClueBot NG, CommonsDelinker, CuriousMind01, Davemck, David Eppstein, David Gerard, Daviddwd, Dawn Bard, Dcirovic, Dharam66, Diza, Dmarquard, Dodi 8238, Donner60, Double sharp, Drbogdan, Drbug, Driftwoodzebulin, Ecthelion83, Eleassar, Epipelagic, Ernsts, EtymAesthete, Ewen, FT2, FamAD123, Fastily, Felida97, Finnusertop, Geo-Science-International, Geogene, Glenn, Gluons12, GoEThe, Greaber, GrendelGreyfur, GünniX, Hamiltondaniel, Hanif Al Husaini, Hannahquack, Harizotoh9, Headbomb, HiLo48, Hohum, Huw Powell, I dream of horses, Iamtonymayse, InsufficientData, Iridescence, Jewlrzeye, Jimw338, Jinx69, John D. Croft, Johnuniq, JorisvS, Julesd, K kisses, Kintetsubuffalo, Kopiersperre, Larrymcp, Legobot II, Liamzebedee, Linda 444, LizardJr8, Loodog, Lorricotton, Lottamiata, Macdonald-ross, Macofe, Magioladitis, Magyar25, Mama meta modal, Mathrick, Maximusthaler, MfortyoneA, Mgiganteus1, Mimihitam, Mindmatrix, Moonraker, Mythslayer7, New User, Newone, Niceguyedc, Non-dropframe, Nwbeeson, OAnick, Odysnes, PatrickFisher, Paul A, Petter Bøckman, Pgan002, Pittsburghmuggle, Poimenlaon, Prinsgezinde, Ptbotgourou, Punkminkis, Redgolpe, Reinyday, Retimuko, Richard Keatinge, RichardWeiss, Rigsofrods, Rjwilmsi, Rorro, Rtrust, Saturn comes back around, Sjö, SkepticalRaptor, Smartse, Smyth, Solomonfromfinland, Squidonius, Srich32977, StAnselm, Subversive.sound, Sunrise, Tdadamemd, Tesseract2, The Mysterious El Willstro, The Voidwalker, Thecheesykid, Thibbs, TiMike, Tom.Reding, Trappist the monk, Tryptofish, Tweenk, Unfree, Valich, Vanished user fijtji34toksdcknqrjn54yoimascj, Videsh Ramsahai, Vsmith, Was a bee, WeijiBaikeBianji, Wikishelt, WolfmanSF, Woudloper, WurmWoode, Xanzzibar, Yolol25, Zedshort, 127 anonymous edits 131

Iron–sulfur world hypothesis *Source:* https://en.wikipedia.org/w/index.php?oldid=847266014 *License:* Creative Commons Attribution-Share Alike 3.0 *Contributors:* 84user, AC+79 3888, AManWithNoPlan, Andrew c, AndrewHowse, Art LaPella, AxelBoldt, Barbara Shack, BatteryIncluded, Bender235, Cacycle, Citation bot 1, Citation bot 4, ClassicSC, ClueBot NG, Daniel.Cardenas, Dauto, Donarreiskoffer, Dr.Buttons, Drbogdan, Dthomsen8, EagerToddler39, EvenGreenerFish, Extremophile, Farid320, Favonian, Feline1, FghIJklm, Geekdiva, General Epitaph, GregorB, Grock2, Gustavocarra, Headbomb, Heurgh, IceKarma, Insorak, Iridescent, Jasonanaggie, John D. Croft, Julianonions, Koavf, Lexor, Linakieper, Lumos3, MarkGT, MegaHasher, NotWith, OMCV, Open2universe, Paleorthid, Parutakupiu, Peak, Petri Krohn, Pgan002, Pro crast in a tor, Rgas, Rich Farmbrough, Richwil, Rifleman 82, Rjwilmsi, RockMagnetist (DCO visiting scholar), Rominandreu, Rtc, Scwlong, Sid-Vicious, Siddunitw, Smokefoot, TexasAndroid, Timwi, Tom.Reding, Tornado00, Trappist the monk, Ttiotsw, Twas Now, Ugajin, V8rik, Vanished user dfvkjmet9jweflkmdkcn234, Viriditas, VoABot II, Vsmith, Yashgaroth, Ziusudra, مصعب, 55 anonymous edits 137

Panspermia *Source:* https://en.wikipedia.org/w/index.php?oldid=852581641 *License:* Creative Commons Attribution-Share Alike 3.0 *Contributors:* Ahruman, Alaney2k, Alexis Gervais, Alpha3031, AndyShepp, Antiqueight, Auric, AusLondonder, BD2412, BSmith821, Banedon, BatteryIncluded, Bender235, Bfinn, Bgwhite, BicelPhD, Broshbush, Chiswick Chap, ChrisGualtieri, ClueBot NG, ComicsAreJustAllRight, Cspoleta, DaftRose, David J Johnson, Dbachmann, Dcirovic, Dejitarob, Denny123123, Dmcq, Dom Kaos, Donkyhotay, Download, Drbogdan, Dsmith125, Dutral, Edgar181, Erik Kennedy, Eumolpo, Fmadd, Formuse, Frap, Geoceramist, Headbomb, HiBlueSky, Jnav7, Joefromrandb, John of Reading, Jonesey95, Koavf, L d allan, LeadSongDog, MaeseLeon, Mallorne, Mapsfly, Materialscientist, Memen.Forster, Modest Genius, Nagualdesign, Nairemlap, NeilN, Niceguyedc, Nøkkenbuer, Ohsin, Orestesgaolin, Ost316, Philip Trueman, PiNerd3, Plantdrew, Quebec99, Rhinopias, Rjwilmsi, Rothorpe, Rowan Forest, Rp2006, SMcCandlish, Sir Cumference, Sobreira, Sophivorus, Spicemix, Starless∼enwiki, Stickyhammer, Sunrise, Taylordw, Terretu, Thomasmeeks, Tigercompanion25, Tom.Reding, Trappist the monk, Vsmith, Wanderer57, Wavelength, WereSpielChequers, 101 anonymous edits 141

Microbial mat *Source:* https://en.wikipedia.org/w/index.php?oldid=847419184 *License:* Creative Commons Attribution-Share Alike 3.0 *Contributors:* 564dude, Abyssal, Alan Liefting, Aldis90, Alnagov, Anrnusna, Auntof6, BD2412, Bcgill, BloodDoll, CasualObserver'48, Chowbok, Chris the speller, Citation bot 1, Crusoe8181, Dawnseeker2000, Dcirovic, Dimadick, Dinkytown, Drbogdan, EagleFan, Epipelagic, Fangorn-Y, Fibbles986, Graeme Bartlett, Headbomb, HieronymousCrowley, Iztwoz, Jeff G., Kazvorpal, Koppas, LilHelpa, Me, Myself, and I are Here, NuclearWarfare, Paleorthid, Philcha, RDBrown, Renamed user sdfkjlskdfreu8r98, Rjwilmsi, Smith609, Trappist the monk, Twinsday, Twirligig, Vsmith, Wilson44691, 16 anonymous edits 163

Great Oxygenation Event *Source:* https://en.wikipedia.org/w/index.php?oldid=853227868 *License:* Creative Commons Attribution-Share Alike 3.0 *Contributors:* 0xF8E8, A2soup, Abductive, Adrian J. Hunter, Alivingcreature, Amerias, Anderson, Anomalocaris, Anonimnystefy, Antandrus, Anthony Appleyard, Artman40, Atropos235, Axl, Bettymnz4, Bkell, BrightR, Brudersohn∼enwiki, Bryan Derksen, Bunburyd, Cantseetheforest, Cesiumfrog, ChadCloman, Chaya5260, ChrisGualtieri, ClueBot NG, Count Iblis, Dan East, Dawnseeker2000, Dcirovic, Dfeldmann, Dgies, DinosaursLoveExistence, Dom

Burgess Shale type fauna *Source:* https://en.wikipedia.org/w/index.php?oldid=847110109 *License:* Creative Commons Attribution-Share Alike 3.0 *Contributors:* Albert Non, Animalparty, Axl, Blaylockjam10, Caftaric, Chiswick Chap, CommonsDelinker, Complainer, Dawnseeker2000, Dcirovic, Diucón, DuncanHill, Fleebo, Fryed-peach, Huttson, Khajidha, Lemming42, Longisquama, Malcolma, Mogism, Philcha, Psuedomorph, Rjwilmsi, Rror, Smith609, Trappist the monk, UnitedStatesian, Washiucho, Willial17, Wilson44691, Wotnow, 2 anonymous edits 315

Chordate *Source:* https://en.wikipedia.org/w/index.php?oldid=852634300 *License:* Creative Commons Attribution-Share Alike 3.0 *Contributors:* 16GD, A8v, Aa77zz, Abyssal, Adam9007, AddWittyNameHere, Aliagarh, Alifshinobi, Alumnum, Anaxial, Animalparty, Anonymoso, Apokryltaros, AsianMathemacation, Attilios, Autodidact1, Axl, Barclod, Begoon, Bender235, Bgwhite, Biblioworm, Biexx, Bri, Bubblesorg, CLCStudent, Caftaric, CanadianLinuxUser, CaptainPiggles, Chiswick Chap, ClueBot NG, CodeBadger, ComplementC3b, Connormah, Cross leg chicken face123, Curtis Clark, Cyclopia, Dcirovic, DemocraticLuntz, Demong, Dfhhhyfghjg, Dinofahad, Dinoguy2, Discospinster, Donner60, Dr.K., Drbogdan, Edgar181, EpicUltimateHacker, Epipelagic, Ericlitman, Eweidenbener, Excirial, Eyreland, Falconfly, Flyer22 Reborn, Frietjes, FrozenMan, Fugitron, GSS, Giraffedata, Glevum, Graham87, Graphwrist, Grover cleveland, HapHaxion, Howicus, Howpper, I am One of Many, I dream of horses, Ian Dalziel, Inferior Olive, IntoThinAir, Invesabelle, Iridescent, JSquish, Jabberjaw, Jabmorris, Jmv2009, Jonesey95, Jprg1966, Jts1882, KH-1, Keithbob, Khajidha, KoriganStone, Lavalizard101, LeadSongDog, Little Mountain 5, Look2See1, Luqmanh, Lythronaxargestes, MCTales, MartinezMD, Materialscientist, McGeddon, Medeis, Mevagiss, Minnelinusa, Mrjulesd, NFD9001, NeonTetraploid, NickVeys, Ninja1430, Noellembrooks, NottNott, Nousername1234567890, Nwbeeson, Obsidian Soul, Odalcet, Oddbodz, Ortizdd, Oshwah, Overmunch, Paine Ellsworth, Petrb, Petter Bøckman, Pinethicket, Plantdrew, Porcellio, Prahlad balaji, Quasar G., Quinton Feldberg, RA0808, RepBird1, Rhinopias, Riboldipj, Rich Farmbrough, RichardWeiss, Rjwilmsi, Runawayryan, Rwflammang, Samf4u, Sampadaupadhyaya, Samsara, Serols, Sfan00 IMG, SkyGazer 512, Soren27, Speight, Stemonitis, Stevey7788, Super48paul, Teoroks, Thenewguy34, Thor Dockweiler, Tideflat, Titsgolore, Titus III, Tmaslyk, Tom.Reding, Werieth, Wikipelli, Wikisuckzz, Xxxrrr23, 160 anonymous edits 321

Evolution of fish *Source:* https://en.wikipedia.org/w/index.php?oldid=851469004 *License:* Creative Commons Attribution-Share Alike 3.0 *Contributors:* 117jose, A2soup, Abyssal, Andoof2000, Animalparty, Apokryltaros, BD2412, Bender235, Bgwhite, Caftaric, Chaoyangopterus, Chris the speller, ClueBot NG, CommonsDelinker, CompAnatProf, Cookie0v, Cyclopia, DavidLeighEllis, Dcirovic, Epipelagic, Falconfly, Farside268, Flyer22 Reborn, Forgerz, Gaius Cornelius, Geoffrey.landis, Gilliam, Giraffedata, GoingBatty, Gomphos, GreenMeansGo, Gruekiller, Hamish59, Hammersoft, Headbomb, Here2help, Hike395, Imakweenlitnoot, Izno, JamesAM, Jmv2009, Joeinwiki, John of Reading, Jon Kolbert, Jonesey95, Joycesalu, Jytdog, K6ka, Kazvorpal, Kostas20142, LeQuantum, Lollipop, Luis Goslin, Magioladitis, Mariomassone, Me, Myself, and I are Here, Medeis, Mikenorton, Miracle Pen, Mrjulesd, Narky Blert, NatigKrolik, Nihiltres, North Shoreman, NotWith, Oshwah, PaleoNeonate, Pandas1997, Plantdrew, PohranicniStraze, Primefac, Qzd, Rcsprinter123, Rjwilmsi, Rzuwig, Scarlettail, Sceptic view, Sfgiants1995, Sha-256, StarryGrandma, Stesmo, Tassedethe, Tavilis, Thingg, Titus III, Tom.Reding, Trappist the monk, VeniVidiVicipedia, Wavelength, White Arabian Filly, Wilson44691, Winner 42, WolfmanSF, Zenab33881, 70 anonymous edits 336

Evolutionary history of plants *Source:* https://en.wikipedia.org/w/index.php?oldid=853784480 *License:* Creative Commons Attribution-Share Alike 3.0 *Contributors:* 564dude, Abductive, AdamoulasA, Aeioun, Anandaaa, Anaxial, Animalparty, Anon york, Anrnusna, Aremisasling, Arthur Rubin, Azcolvin429, BD2412, Bender235, Bernstein0275, Bgwhite, Blahvscarrots, Chaya5260, Chiswick Chap, Chris the speller, Christian75, ClueBot NG, Cyclopia, Dawnseeker2000, Dcirovic, Doncurzio, Drbogdan, DuncanHill, Earthdirt, Ehaaland, EncycloPetey, Ernsts, FamAD123, Fangorn-Y, Fraggle81, Funkendub, Geopersona, Giraldusfaber, Glevum, GünniX, Headbomb, Imp336, IronChris, JMilty, JamesP, JaneStillman, Joeinwiki, John of Reading, Jonesey95, Josve05a, Karlfonza, Katieh5584, Kevmin, Kolyvansky, Lappspira, Laurenprue216, Leonardo Ferreira Fontenelle, LilHelpa, Marvellous Spider-Man, Materialscientist, Maulucioni, Menchi, Mikeblas, Mikenorton, Mild Bill Hiccup, Myasuda, NotWith, Oddbodz, Pauli133, Peter coxhead, Phvantienderen, Plantsurfer, R. S. Shaw, Reo On, Rjwilmsi, Ryan Shakiba, S. Neuman, Sakura Cartelet, Smash1gordon, Sminthopsis84, Smith609, Squids and Chips, Srleffler, Stephenoff.2, Sunrise, Søren, Tavilis, The PIPE, Tom.Reding, Trappist the monk, Usernamekiran (AWB), Vieque, Vsmith, Wavelength, Webclient101, WereSpielChequers, Wesley6677, Widr, Worldbruce, Wowfactor900, Wundalous, Максим Підліснюк, 100 anonymous edits 381

Tetrapod *Source:* https://en.wikipedia.org/w/index.php?oldid=850635651 *License:* Creative Commons Attribution-Share Alike 3.0 *Contributors:* Abyssal, Al-Andalus, Apokryltaros, ArchPope Sextus VI, BD2412, Bender235, Bgwhite, BlackBeast, Chermundy, Chiswick Chap, Chris the speller, Circulationsys, ClueBot NG, CodeBadger, Conty～enwiki, Cyclopia, DVdm, Darylgolden, Dcirovic, Dinnodal, Dinoguy2, Epipelagic, ErikHaugen, Escocialingua, Falconfly, FamAD123, Frietjes, Glevum, Grover cleveland, Gustav Rammelsberg, Hallothere!, Harizotoh9, Howpper, Jenks24, Jhorthos, Jmv2009, JoJan, Joe Decker (alt), Jonathan Tweet, Jts1882, KatalogHTE, KylieTastic, Lavateraguy, LilHelpa, Line 8 the Pink, Maczkopeti, ManeatingLemur, Mariomassone, Martino3, Matthias Thalmann, Mike Rosoft, Moonraker, Mrjulesd, NicoScribe, NightXenon999, Nihiltres, Omnipaedista, Ost316, Paul 012, Pauli133, Peter M. Brown, Pgan002, Plantdrew, Prahlad balaji, Rcsprinter123, Rhinopias, RichardWeiss, Rjwilmsi, Rohan s pandey, SDC, Sanajeh, SandKitty256, SeminulaOolite, Serendipodous, Shrimp tetrapod, Sietecolores, SkyGazer 512, Stranger forever, Tom.Reding, Topbanana, Trappist the monk, Turtlepanther566, TuxLibNit, Y7trewu7y6t5r4e, Yamaguchi先生 , Zedshort, Zyxwv99, 大作曲家 , 65 anonymous edits 425

Evolution of tetrapods *Source:* https://en.wikipedia.org/w/index.php?oldid=848150044 *License:* Creative Commons Attribution-Share Alike 3.0 *Contributors:* Animalparty, BD2412, Bender235, Bgpaulus, Blaylockjam10, Carece, ClueBot NG, Conty～enwiki, Dcirovic, DuncanHill, EAR47, Epipelagic, HCA, Harizotoh9, Headbomb, Inatan, Ira Leviton, Keith D, Klbrain, LilHelpa, Macdonald-ross, Mann jess, Nihiltres, NotWith, Petersond5798, Petter Bøckman, Rjwilmsi, SJ Defender, Saturn comes back around, Sceptic view, Staylor71, Tavilis, Trappist the monk, TuxLibNit, Zyxwv99, 46 anonymous edits 455

Evolution of dinosaurs *Source:* https://en.wikipedia.org/w/index.php?oldid=851053335 *License:* Creative Commons Attribution-Share Alike 3.0 *Contributors:* AS, Abyssal, Apokryltaros, Atethnekos, Aurous One, Azcolvin429, Bongwarrior, Bustertank, Cadiomals, ClueBot NG, CuriousMind01, CyanoTex, David Gerard, DinoRhinoMammel, Drbogdan, Evangelos Giakoumatos, Fatapatate, Firsfron, Hoseumou, I dream of horses, J. Spencer, Jim1138, Jmv2009, John of Reading, Jss367, Kangfreud, Kevmin, Kintaro, Kjoonlee, Koavf, Kumioko (renamed), Leptictidium, Lithoderm, Lusotitan, Macdonald-ross, Maharishi yogi, Mangostar, Marcraymond, Mario modesto, Mgiganteus1, Mikenorton, Mollwollfumble, Mscuthbert, Myasuda, NatureA16, No Swan So Fine, NotWith, Paul H., Peter M. Brown, Robin S, Rrburke, Rynosaur, SaberToothedWhale, Serols, Serpinium, Slightsmile, Smith609, Spotty11222, Sun Creator, Switchfootvio, Tintero, Tkjtkj, TwoTwoHello, Wavelength, Wetman, Woudloper, Zach Winkler, Zureks, 60 anonymous edits 475

Origin of birds *Source:* https://en.wikipedia.org/w/index.php?oldid=853014734 *License:* Creative Commons Attribution-Share Alike 3.0 *Contributors:* Abce2, Abductive, Alansohn, Albertonykus, Alexander Vargas, Alphathon, Andrewman327, Animalparty, AnonMoos, Anrnusna, Anthony Appleyard, Attilios, B mazuki, BD2412, Battlekow, Bender235, Bkonrad, Bueller 007, CHW100, Caballero1967, Captain Occam, Chris the speller, Citation bot 1, ClueBot NG, Collieuk, CommonsDelinker, Conty～enwiki, CuriousMind01, CyberCorn Entropic, David Gerard, Dbachmann, Dcirovic, Demize, Dinoguy2, Diucón, Dmh～enwiki, Dobermanji, Dratman, Drbogdan, Drutt, Epicgenius, Epipelagic, Ettrig, Ewen, Falconfly, Ferahgo the Assassin, Firsfron, Firstbirdbeak, Floydian, FlyingAce, Focus, FunkMonk, Følelse, G.Kiruthikan, GermanJoe, GirasoleDE, Glacialfox, Grafen, HCA, Harizotoh9, Herostratus, J 1982, J. Spencer, JCSantos, Jbrougham, Jmv2009, Johnuniq, Jonesey95, Khazar2, Kintaro, Kumioko (renamed), Lambiam, Limulus, Ljosa, Look2See1, Lythronaxargestes, MWAK, Maias, Manul, Markunator, Meganesia, Mgiganteus1, Mike s, Mike.BRZ, Mindme, Mortee, Niceguyedc, Nick Number, Non-dropframe, NotWith, Omegaman99, Pzoot, Quebec99, Rainbow Shifter, Randy Kryn, Rich Farmbrough, Rjwilmsi, Robt Owen, Rufous-crowned Sparrow, Saintrain, SamX, SchreiberBike, Secondbirdbeak, SheriffIsInTown, Shirik, Shyamal, Simplexity22, Snowmanradio, Spicemix, Stephan Schulz, Sunrise, Tejendraji, The PIPE, The Thing That Should Not Be, Tktktk, TomS TDotO, Topilsky, Trappist the monk, Ucucha, Uwaga budowa, Vanished user 19794758563875, Vsmith, Writtenonsand, 86 anonymous edits 487

Evolution of mammals *Source:* https://en.wikipedia.org/w/index.php?oldid=848327612 *License:* Creative Commons Attribution-Share Alike 3.0 *Contributors:* Aegon Stargaryen, Aliafroz1901, Allcoast14, Andrew Lancaster, Animalparty, Attilios, Awesome 210, Azcolvin429, Bender235, Bgwhite, Biolprof, Bri, Chaoyangopterus, Charredbarn, Cheddartoucan, Chris Capoccia, Chrisyoung1974, ClueBot NG, Commodore Guff, CommonsDelinker, Cyclopia, Dcirovic, Deselliers, Dicklyon, Dr. Günter Bechly, Drbogdan, Dusty relic, Egsan Bacon, Epipelagic, Fafnir1, Falconfly, FamAD123, Ferrierd, Fig wright, Florian Blaschke, Fluffernutter, Gareth Griffith-Jones, GeorgeBarnick, Gilliam, Glevum, Hairy Dude, Hamish59, Harizotoh9, Headbomb, Hibernian, IWillBuildTheRoads, Iacobus, InverseHypercube, Jan1nad, Jarble, Jim.henderson, JimJJewett, JimVC3, John of Reading, JohnOwens, Johnuniq, Jonesey95, JorisvS, Jschnur, Khazar2, Kintaro, Lemonade100, Leonardo Ferreira Fontenelle, Looie496, Lou2shi1, Magioladitis, Materialscientist, Metron, Mevagiss, NTox, Nihiltres, NotWith, Oleg Alexandrov, Orenburg1, Paul H., Pauli133, Peter M. Brown, Petter Bøckman, Phileklund, PlotKrot, Prof. Squirrel, Rjwilmsi, SchreiberBike, Septrillion, Serols, ShelleyAdams, Shellwood, Shuvuuia, Steko, Stenen Bijl, Tavilis, The Quixotic Potato, TomS TDotO, Trappist the monk, TwoTwoHello, Vanamonde93, Wavelength, Wetman, WolfmanSF, 78 anonymous edits 509

Flowering plant *Source:* https://en.wikipedia.org/w/index.php?oldid=853126616 *License:* Creative Commons Attribution-Share Alike 3.0 *Contributors:* 00getbuckets00, 78.26, AJ2265, Abroddy, Abyssal, Aishasaleemkhan, Aishuparab, Anaofgreengables, Anaxial, Anomalocaris, Anthony Appleyard, Apokryltaros, Ayzmo, BattleBorn89, Bender235, Bermicourt, Bgwhite, Bobandvageneplz, CLCStudent, Chaya5260, Chiswick Chap, Choess, Closeapple, ClueBot NG, CommonsDelinker, Corinne, Cottonshirt, Dcirovic, Dndhdhsjdu, Drbogdan, DuncanHill, El C, EncycloPetey, Entranced98, Eurodyne, Falconfly, FamAD123, Finnusertop, Foxdimi, Frietjes, Fuckyouoennon, Gdgzfallen, Gilliam, Gustav Rammelsberg, GünniX, HapHaxion, IdreamofJeanie, In ictu oculi, J D, J. N. Squire, Jackfork, James111555, Jasmine Rhyas, Jaworskicoline, Jbeans, Jim1138, JoJan, Jts1882, Julesd, K6ka, Karlfonza, Kevmin, Kolyvansky, KylieTastic, Lavateraguy, Lennonwoof, Looksie Brooksie, Martin of Sheffield, Materialscientist, MathKnight, Matterington, Matthew small, Me, Myself, and I are Here, Mean as custard, Medeis, MfortyoneA, Michael Goodyear, MisterEcotone425, Mrjulesd, Myasuda, NICHOLAS NEEDLEHAM, Neelix, Nicolsontrueman, NsTaGaTr, Oshwah, PaleCloudedWhite, Peter coxhead, Pinethicket, Plantdrew, Prachi Teli, Qaz1122, QuaeriSolet, Quenhitran, Quisqualis, Renamed user 943a06d1c3, Rjwilmsi, Rkitko, Ronz, RossiElisabeth36, Samsara, Sarr Cat, Scott McNay, Selenamarie, Ser Amantio di Nicolao, Serols, Shavill, Shearflyer, Silver Samurai, Simuliid, SmokierClover, Stupid girl, Sweepy, The Mysterious El Willstro, This lousy T-shirt, Thor Dockweiler, Thạch Kim Tuấn, Titodutta, Tom.Reding, Trappist the monk, Twistthespine, UY Scuti, User-duck, VaibhavChaurasia, Videsh Ramsahai, Virion123, Vsmith, Whitakerhoskins, Wiae, Widr, WolfmanSF, Workmail60, Xxspecial1victoryxx, Zorahia, Zygimantus, Максим Підліснюк, 空间的拓荒者 , 144 anonymous edits 547

Gymnosperm *Source:* https://en.wikipedia.org/w/index.php?oldid=849309254 *License:* Creative Commons Attribution-Share Alike 3.0 *Contributors:* 512bits, Aboctok, Abyssal, Agent Fury, Anaxial, Arjay 007, Ayush12gupta, BUBBAFRIENDOFALL, Bear-rings, Bender235, Bentogoa, Bgwhite, Bob Burkhardt, Boomer Vial, Boud, Brian Crawford, CAPTAIN RAJU, Cbdorsett, Chiswick Chap, Choess, Chrissymad, Citation bot 1, Citron, Clarince63,

Image Sources, Licenses and Contributors

The sources listed for each image provide more detailed licensing information including the copyright status, the copyright owner, and the license conditions.

Image *Source:* https://en.wikipedia.org/w/index.php?title=File:Cynognathus.JPG *License:* Creative Commons Attribution-Sharealike 3.0 *Contributors:* Ghedoghedo ... 67
Image *Source:* https://en.wikipedia.org/w/index.php?title=File:Glyptodon-1.jpg *License:* GNU Free Documentation License *Contributors:* Arent derivative work: WolfmanSF (talk) ... 68
Image *Source:* https://en.wikipedia.org/w/index.php?title=File:Dasypus_novemcinctus.jpg *License:* Public Domain *Contributors:* Boivie, Common Good, Eduardo P, Nordelch, Pfctdayelise, Ruff tuff cream puff, 3 anonymous edits ... 68
Figure 33 *Source:* https://en.wikipedia.org/w/index.php?title=File:Horseevolution.png *License:* GNU Free Documentation License *Contributors:* Alex brollo, Angusmclellan, AnonMoos, AnselmiJuan, Benlisquare, Cwbm (commons), Eventer∼commonswiki, Flappiefh, FunkMonk, HenkvD, Kersti Nebelsiek, MGA73bot2, OgreBot 2, Trijnstel, Túrelio, WolfmanSF, 5 anonymous edits ... 72
Figure 34 *Source:* https://en.wikipedia.org/w/index.php?title=File:Darwin's_finches.jpeg *License:* Public Domain *Contributors:* John Gould (14.Sep.1804 - 3.Feb.1881) ... 75
Figure 35 *Source:* https://en.wikipedia.org/w/index.php?title=File:Ring_Species_(gene_flow_around_a_barrier).png *Contributors:* User:Azcolvin429 ... 77
Image *Source:* https://en.wikipedia.org/w/index.php?title=File:Pangaea_Glossopteris.jpg *License:* Public Domain *Contributors:* Petter Bøckman 78
Image *Source:* https://en.wikipedia.org/w/index.php?title=File:Phylogenetic_tree_of_marsupials_derived_from_retroposon_data_-_journal.pbio.1000436.g002.png *License:* Creative Commons Attribution 2.5 *Contributors:* Daniel Mietchen, Was a bee ... 78
Image *Source:* https://en.wikipedia.org/w/index.php?title=File:Global_Camelid_Distribution_and_Migration.png *Contributors:* User:Azcolvin429 78
Figure 36 *Source:* https://en.wikipedia.org/w/index.php?title=File:Big_and_little_dog_1.jpg *License:* GNU Free Documentation License *Contributors:* Ellen Levy Finch / en:User:Elf (uploaded by TBjornstad 14:51, 17 August 2006 (UTC)) ... 81
Figure 37 *Source:* https://en.wikipedia.org/w/index.php?title=File:Guppy_CS_pair_4_20130113.jpg *License:* Creative Commons Attribution-Sharealike 3.0 *Contributors:* User:Emilio17 ... 85
Figure 38 *Source:* https://en.wikipedia.org/w/index.php?title=File:WhiteSandsBleachedEarlessLizard.jpg *License:* Creative Commons Attribution-ShareAlike 3.0 Unported *Contributors:* Kevinp2 ... 88
Figure 39 *Source:* https://en.wikipedia.org/w/index.php?title=File:Globorotalia_Speciation_and_Phylogeny.png *Contributors:* User:Azcolvin429 91
Figure 40 *Source:* https://en.wikipedia.org/w/index.php?title=File:Drosophila_melanogaster_-_side_(aka).jpg *License:* Creative Commons Attribution-Sharealike 2.5 *Contributors:* André Karwath aka Aka ... 93
Figure 41 *Source:* https://en.wikipedia.org/w/index.php?title=File:Arabidopsis_thaliana.jpg *License:* GNU Free Documentation License *Contributors:* Original uploader was Brona at en.wikipedia. User:Roepers at nl.wikipedia ... 99
Figure 42 *Source:* https://en.wikipedia.org/w/index.php?title=File:Tragopogon_porrifolius_flower.jpg *License:* GNU Free Documentation License *Contributors:* Stephen Lea ... 100
Figure 43 *Source:* https://en.wikipedia.org/w/index.php?title=File:Rock_Ptarmigan_(Lagopus_Muta).jpg *License:* Creative Commons Attribution-Sharealike 3.0 *Contributors:* Jan Frode Haugseth ... 103
Figure 44 *Source:* https://en.wikipedia.org/w/index.php?title=File:Phylogenetic_tree.svg *Contributors:* This vector version: Eric Gaba (Sting - fr:Sting) ... 112
Figure 45 *Source:* https://en.wikipedia.org/w/index.php?title=File:Tree_Of_Life_(with_horizontal_gene_transfer).svg *Contributors:* User:Azcolvin429 ... 113
Image *Source:* https://en.wikipedia.org/w/index.php?title=File:Symbol_support_vote.svg *License:* Public Domain *Contributors:* Anomie, Fastily, Jo-Jo Eumerus ... 114
Figure 46 *Source:* https://en.wikipedia.org/w/index.php?title=File:Homology_vertebrates-en.svg *Contributors:* Волков Владислав Петрович 115
Figure 47 *Source:* https://en.wikipedia.org/w/index.php?title=File:BelonBirdSkel.jpg *License:* Public Domain *Contributors:* L C Miall ... 116
Figure 48 *Source:* https://en.wikipedia.org/w/index.php?title=File:Eupatorus_gracilicornis_Vol.jpg *Contributors:* Didier Descouens ... 117
Figure 49 *Source:* https://en.wikipedia.org/w/index.php?title=File:Nephrotoma_guestfalica.jpg *License:* Creative Commons Attribution 3.0 *Contributors:* Lymantria, Steinsplitter ... 118
Figure 50 *Source:* https://en.wikipedia.org/w/index.php?title=File:Acer_pseudoplatanus_MHNT.BOT.2004.0.461.jpg *Contributors:* Didier Descouens ... 119
Figure 51 *Source:* https://en.wikipedia.org/w/index.php?title=File:PAX6_Phenotypes_Washington_etal_PLoSBiol_e1000247.png *License:* Creative Commons Attribution 2.5 *Contributors:* Washington NL, Haendel MA, Mungall CJ, Ashburner M, Westerfield M, Lewis SE. ... 120
Figure 52 *Source:* https://en.wikipedia.org/w/index.php?title=File:Arthropod_segment_Hox_gene_expression.svg *Contributors:* User:Chiswick Chap ... 121
Image *Source:* https://en.wikipedia.org/w/index.php?title=File:Acadoparadoxides_sp_4343.JPG *Contributors:* User:Ghedoghedo ... 121
Image *Source:* https://en.wikipedia.org/w/index.php?title=File:Araneus_quadratus_MHNT.jpg *Contributors:* Didier Descouens ... 121
Image *Source:* https://en.wikipedia.org/w/index.php?title=File:Scolopendridae_-_Scolopendra_cingulata.jpg *License:* Creative Commons Attribution-Sharealike 3.0 *Contributors:* Hectonichus ... 121
Image *Source:* https://en.wikipedia.org/w/index.php?title=File:Cerf-volant_MHNT_Dos.jpg *Contributors:* Didier Descouens ... 121
Image *Source:* https://en.wikipedia.org/w/index.php?title=File:GarneleCrystalRed20.jpg *License:* GNU Free Documentation License *Contributors:* Ricks ... 121
Figure 53 *Source:* https://en.wikipedia.org/w/index.php?title=File:EurAshLeaf.jpg *License:* GNU Free Documentation License *Contributors:* AnRo0002, Farbenfreude, MGA73bot2, Quadell ... 123
Figure 54 *Source:* https://en.wikipedia.org/w/index.php?title=File:Detail_on_a_palm_frond_(8297623365).jpg *License:* Creative Commons Attribution 2.0 *Contributors:* Frank Kovalchek from Anchorage, Alaska, USA ... 123
Figure 55 *Source:* https://en.wikipedia.org/w/index.php?title=File:Ocotillothron02262006.JPG *License:* Creative Commons Attribution-ShareAlike 3.0 Unported *Contributors:* User:Miskatonic ... 124
Figure 56 *Source:* https://en.wikipedia.org/w/index.php?title=File:Musa_acuminata_Gran_Canaria_2.JPG *License:* Creative Commons Attribution-Sharealike 3.0,2.5,2.0,1.0 *Contributors:* Toffel ... 124
Figure 57 *Source:* https://en.wikipedia.org/w/index.php?title=File:Split_Aloe.jpg *License:* GNU Free Documentation License *Contributors:* user:Raul654 ... 124
Figure 58 *Source:* https://en.wikipedia.org/w/index.php?title=File:Venus_Flytrap_showing_trigger_hairs.jpg *License:* Creative Commons Attribution-Sharealike 2.5 *Contributors:* Aroche, BRUTE, ComputerHotline, Denis Barthel, Graphium, Hystrix, NoahElhardt, Pirttroy, Thiotrix, 4 anonymous edits ... 125
Figure 59 *Source:* https://en.wikipedia.org/w/index.php?title=File:Nepenthes_muluensis.jpg *License:* Public domain *Contributors:* -Jeremiah- 125
Figure 60 *Source:* https://en.wikipedia.org/w/index.php?title=File:Onions_002.jpg *License:* Creative Commons Zero *Contributors:* User:Ocdp 126
Figure 61 *Source:* https://en.wikipedia.org/w/index.php?title=File:ABC_flower_developement.jpg *Contributors:* Amada44, Chiswick Chap, VTDK, Тинкер ... 127
Figure 62 *Source:* https://en.wikipedia.org/w/index.php?title=File:Pachyrhachis_problematicus_45.JPG *Contributors:* User:Ghedoghedo ... 128
Figure 63 *Source:* https://en.wikipedia.org/w/index.php?title=File:Histone_Alignment.png *Contributors:* User:Evolution and evolvability ... 128
Figure 64 *Source:* https://en.wikipedia.org/w/index.php?title=File:Weeper_Capuchin_01_(cropped).JPG *License:* Public Domain *Contributors:* Pacman ... 129
Image *Source:* https://en.wikipedia.org/w/index.php?title=File:Commons-logo.svg *License:* logo *Contributors:* Anomie, Callanecc, CambridgeBayWeather, Jo-Jo Eumerus, RHaworth ... 131
Figure 65 *Source:* https://en.wikipedia.org/w/index.php?title=File:Phylogenetic_tree.svg *Contributors:* This vector version: Eric Gaba (Sting - fr:Sting) ... 134
Figure 66 *Source:* https://en.wikipedia.org/w/index.php?title=File:Tree_Of_Life_(with_horizontal_gene_transfer).svg *Contributors:* User:Azcolvin429 ... 135
Figure 67 *Source:* https://en.wikipedia.org/w/index.php?title=File:Reduktiver_Acetyl-CoA-Weg.png *License:* Creative Commons Attribution-Sharealike 3.0 *Contributors:* Yikrazuul ... 136
Figure 68 *Source:* https://en.wikipedia.org/w/index.php?title=File:Panspermie.svg *License:* Creative Commons Attribution-Sharealike 3.0,2.5,2.0,1.0 *Contributors:* Silver Spoon Sokpop ... 142
Figure 69 *Source:* https://en.wikipedia.org/w/index.php?title=File:Blacksmoker_in_Atlantic_Ocean.jpg *License:* Public Domain *Contributors:* P. Rona ... 154
Figure 70 *Source:* https://en.wikipedia.org/w/index.php?title=File:STS-46_EURECA_deployment.jpg *License:* Public Domain *Contributors:* NASA ... 156
Figure 71 *Source:* https://en.wikipedia.org/w/index.php?title=File:1990_s32_LDEF_stow.jpg *License:* Public Domain *Contributors:* NASA/ exploitcorporations ... 158
Figure 72 *Source:* https://en.wikipedia.org/w/index.php?title=File:EXPOSE_location_on_the_ISS.jpg *License:* Public Domain *Contributors:* BatteryIncluded, Huntster, Morio, Ras67 ... 159
Figure 73 *Source:* https://en.wikipedia.org/w/index.php?title=File:Stardust_Dust_Collector_with_aerogel.jpg *License:* Public Domain *Contributors:* Bricktop, Gerrit41, Li-sung, Ojan, 1 anonymous edits ... 160

696

Image *Source:* https://en.wikipedia.org/w/index.php?title=File:Lepidodendron_aculeatum2.jpg *License:* GNU Free Documentation License *Contributors:* Abyssal, BotBln, DanielCD~commonswiki, EncycloPetey, Hameryko, Kevmin, MGA73bot2, Pharaoh han, WayneRay, Wieralee, 2 anonymous edits ... 399
Image *Source:* https://en.wikipedia.org/w/index.php?title=File:Lepido_root_top.jpg *License:* Creative Commons Attribution-Sharealike 3.0 *Contributors:* Verisimilus T ... 399
Figure 170 *Source:* https://en.wikipedia.org/w/index.php?title=File:Psaronius_double_section.JPG *License:* Creative Commons Attribution 3.0 *Contributors:* Verisimilus ... 401
Figure 171 *Source:* https://en.wikipedia.org/w/index.php?title=File:LepidodendronOhio.jpg *License:* Public Domain *Contributors:* Wilson44691 402
Figure 172 *Source:* https://en.wikipedia.org/w/index.php?title=File:Trigonocarpus.jpg *License:* Creative Commons Attribution 3.0 *Contributors:* Verisimilus ... 404
Figure 173 *Source:* https://en.wikipedia.org/w/index.php?title=File:Runcaria_megasporangium_and_cupule_drawing.jpg *License:* Creative Commons Attribution-Sharealike 3.0 *Contributors:* User:Chiswick Chap ... 404
Figure 174 *Source:* https://en.wikipedia.org/w/index.php?title=File:Crossotheca_nodule.JPG *License:* Creative Commons Attribution 3.0 *Contributors:* Verisimilus ... 407
Figure 175 *Source:* https://en.wikipedia.org/w/index.php?title=File:Syncarp_evolution.svg *License:* Creative Commons Attribution 3.0 *Contributors:* Verisimilus ... 408
Figure 176 *Source:* https://en.wikipedia.org/w/index.php?title=File:Bennettitales-cycadeoidaceae.jpg *License:* Creative Commons Attribution 3.0 *Contributors:* Smith609 ... 409
Figure 177 *Source:* https://en.wikipedia.org/w/index.php?title=File:Amborella.jpg *License:* Creative Commons Attribution 3.0 *Contributors:* Gauravm1312 ... 411
Image *Source:* https://en.wikipedia.org/w/index.php?title=File:Wikiversity-logo.svg *License:* Creative Commons Attribution-Sharealike 3.0 *Contributors:* Snorky (optimized and cleaned up by verdy_p) ... 412
Figure 178 *Source:* https://en.wikipedia.org/w/index.php?title=File:Pink_rose_albury_botanical_gardens.jpg *Contributors:* Basvb, BotMultichill, Fir0002, MILEPRI, Stickpen, 1 anonymous edits ... 413
Figure 179 *Source:* https://en.wikipedia.org/w/index.php?title=File:Microrna_secondary_structure.png *License:* GNU Free Documentation License *Contributors:* Opabinia regalis ... 415
Figure 180 *Source:* https://en.wikipedia.org/w/index.php?title=File:Maize-teosinte.jpg *License:* Attribution *Contributors:* John Doebley ... 416
Figure 181 *Source:* https://en.wikipedia.org/w/index.php?title=File:Cauliflower.JPG *License:* Public Domain *Contributors:* User Anthony DiPierro on en.wikipedia ... 416
Figure 182 *Source:* https://en.wikipedia.org/w/index.php?title=File:HatchSlackpathway2.svg *License:* Creative Commons Attribution-Sharealike 2.5 *Contributors:* HatchSlackpathway.svg: *HatchSlackpathway.png: Adenosine derivative work: Jamouse derivative work: Adenosine (talk) ... 418
Figure 183 *Source:* https://en.wikipedia.org/w/index.php?title=File:Azadirachtin.png *License:* Public Domain *Contributors:* Edgar181 ... 421
Image *Source:* https://en.wikipedia.org/w/index.php?title=File:Extant_tetrapoda.jpg *License:* Public Domain *Contributors:* Petter Bøckman .. 425
Image *Source:* https://en.wikipedia.org/w/index.php?title=File:Lithobates_pipiens.jpg *License:* Creative Commons Attribution 2.0 *Contributors:* Animalparty, FlickreviewR, Fraf, Jacopo Werther ... 428
Image *Source:* https://en.wikipedia.org/w/index.php?title=File:Florida_Box_Turtle_Digon3.jpg *Contributors:* "Jonathan Zander (Digon3)" ... 428
Image *Source:* https://en.wikipedia.org/w/index.php?title=File:Squirrel_(PSF).png *Contributors:* BotMultichill, Oksmith, Ruff tuff cream puff . 428
Figure 184 *Source:* https://en.wikipedia.org/w/index.php?title=File:Devonianfishes_ntm_1905_smit_1929.gif *License:* Public Domain *Contributors:* by Joseph Smit (1836-1929), from Nebula to Man, 1905 England ... 429
Figure 185 *Source:* https://en.wikipedia.org/w/index.php?title=File:Eusthenopteron_BW.jpg *License:* Creative Commons Attribution 2.5 *Contributors:* Nobu Tamura (http://spinops.blogspot.com) ... 429
Figure 186 *Source:* https://en.wikipedia.org/w/index.php?title=File:Tiktaalik_BW.jpg *License:* Creative Commons Attribution 2.5 *Contributors:* Nobu Tamura (http://spinops.blogspot.com) ... 429
Figure 187 *Source:* https://en.wikipedia.org/w/index.php?title=File:Acanthostega_BW.jpg *License:* Creative Commons Attribution 2.5 *Contributors:* Nobu Tamura (http://spinops.blogspot.com) ... 430
Image *Source:* https://en.wikipedia.org/w/index.php?title=File:Barramunda_coloured.jpg *License:* Public Domain *Contributors:* W H Flower 432
Image *Source:* https://en.wikipedia.org/w/index.php?title=File:Tiktaalik_BW_white_background.jpg *License:* Creative Commons Attribution 2.5 *Contributors:* Nobu Tamura (http://spinops.blogspot.com) ... 432
Figure 188 *Source:* https://en.wikipedia.org/w/index.php?title=File:Ichthyostega_BW.jpg *License:* Creative Commons Attribution 2.5 *Contributors:* Nobu Tamura (http://spinops.blogspot.com) ... 434
Figure 189 *Source:* https://en.wikipedia.org/w/index.php?title=File:Edops_craigi12DB.jpg *License:* GNU Free Documentation License *Contributors:* DiBgd, Haplochromis, Kevmin, Putnik, 1 anonymous edits ... 435
Figure 190 *Source:* https://en.wikipedia.org/w/index.php?title=File:Diadectes1DB.jpg *License:* GNU Free Documentation License *Contributors:* DiBgd, Haplochromis, Kevmin, Putnik, 1 anonymous edits ... 435
Figure 191 *Source:* https://en.wikipedia.org/w/index.php?title=File:Linnaeus_-_Regnum_Animale_(1735).png *License:* Public Domain *Contributors:* User:Fastfission ... 437
Image *Source:* https://en.wikipedia.org/w/index.php?title=File:Protopterus_dolloi_Boulenger2.jpg *License:* Public Domain *Contributors:* MM. P. J. Smit & J. Green ... 439
Image *Source:* https://en.wikipedia.org/w/index.php?title=File:Osteolepis_BW.jpg *License:* Creative Commons Attribution 2.5 *Contributors:* Nobu Tamura (http://spinops.blogspot.com) ... 440
Image *Source:* https://en.wikipedia.org/w/index.php?title=File:Greererpeton_BW.jpg *License:* Creative Commons Attribution 2.5 *Contributors:* Nobu Tamura (http://spinops.blogspot.com) ... 441
Image *Source:* https://en.wikipedia.org/w/index.php?title=File:Crassigyrinus_BW.jpg *License:* Creative Commons Attribution 2.5 *Contributors:* Nobu Tamura (http://spinops.blogspot.com) ... 441
Image *Source:* https://en.wikipedia.org/w/index.php?title=File:Seymouria_BW.jpg *License:* Creative Commons Attribution 2.5 *Contributors:* Nobu Tamura (http://spinops.blogspot.com) ... 441
Image *Source:* https://en.wikipedia.org/w/index.php?title=File:The_tailless_batrachians_of_Europe_(Page_194)_(Pelobates_fuscus).jpg *License:* Creative Commons Attribution 2.0 *Contributors:* Mariomassone, Ruff tuff cream puff ... 442
Image *Source:* https://en.wikipedia.org/w/index.php?title=File:Archeria_BW_(white_background).jpg *License:* Creative Commons Attribution 2.5 *Contributors:* Nobu Tamura (http://spinops.blogspot.com) ... 443
Image *Source:* https://en.wikipedia.org/w/index.php?title=File:Gefyrostegus22DB.jpg *License:* Creative Commons Attribution-Sharealike 3.0,2.5,2.0,1.0 *Contributors:* Abyssal, DiBgd, Haplochromis, Histmole, Nicolás10~commonswiki, Putnik, 1 anonymous edits ... 443
Image *Source:* https://en.wikipedia.org/w/index.php?title=File:Karpinskiosaurus1DB.jpg *License:* Creative Commons Attribution-Sharealike 3.0,2.5,2.0,1.0 *Contributors:* Abyssal, DiBgd, Haplochromis, Mariomassone, Nicolás10~commonswiki, Putnik, 1 anonymous edits ... 443
Image *Source:* https://en.wikipedia.org/w/index.php?title=File:Diadectes1DB_(flipped).jpg *License:* GNU Free Documentation License *Contributors:* Mariomassone ... 443
Image *Source:* https://en.wikipedia.org/w/index.php?title=File:Zoology_of_Egypt_(1898)_(Varanus_griseus).png *Contributors:* Elisfkc, Mariomassone, Ruff tuff cream puff ... 443
Image *Source:* https://en.wikipedia.org/w/index.php?title=File:Hyloplesion.jpg *License:* Creative Commons Attribution-Sharealike 3.0 *Contributors:* Smokeybjb ... 443
Image *Source:* https://en.wikipedia.org/w/index.php?title=File:Lysorophus.jpg *License:* Creative Commons Attribution-Sharealike 3.0 *Contributors:* Smokeybjb ... 443
Image *Source:* https://en.wikipedia.org/w/index.php?title=File:Diplocaulus_BW_(flipped).jpg *License:* Creative Commons Attribution 2.5 *Contributors:* Nobu Tamura (http://spinops.blogspot.com) ... 443
Image *Source:* https://en.wikipedia.org/w/index.php?title=File:Acanthostega_BW_(flipped).jpg *License:* Creative Commons Attribution 2.5 *Contributors:* Nobu Tamura (http://spinops.blogspot.com) ... 443
Image *Source:* https://en.wikipedia.org/w/index.php?title=File:Ichthyostega_BW_(flipped).jpg *License:* Creative Commons Attribution 2.5 *Contributors:* Nobu Tamura (http://spinops.blogspot.com) ... 443
Image *Source:* https://en.wikipedia.org/w/index.php?title=File:Eryops1DB.jpg *License:* Creative Commons Attribution-Sharealike 3.0,2.5,2.0,1.0 *Contributors:* Abyssal, DiBgd, Haplochromis, Kevmin, Mariomassone, Nicolás10~commonswiki, Putnik, 1 anonymous edits ... 444
Image *Source:* https://en.wikipedia.org/w/index.php?title=File:Syphonops_annulatus_cropped.jpg *License:* Public Domain *Contributors:* Syphonops_annulatus.jpg: From the from the first edition of Dictionnaire D'Histoire Naturelle by Charles Orbigny. 1849 d ... 445
Figure 192 *Source:* https://en.wikipedia.org/w/index.php?title=File:Labyrinthodon_Mivart.png *Contributors:* Abyssal, FunkMonk, Keith Edkins, Kevmin, Man vyi, Preto(m) ... 448
Figure 193 *Source:* https://en.wikipedia.org/w/index.php?title=File:Fishapods.png *License:* Creative Commons Attribution-Share Alike *Contributors:* Graphic by dave souza, incorporating images by others, as description ... 456
Figure 194 *Source:* https://en.wikipedia.org/w/index.php?title=File:Devonianfishes_ntm_1905_smit_1929.gif *License:* Public Domain *Contributors:* by Joseph Smit (1836-1929), from Nebula to Man, 1905 England ... 460

700

License

Index

Coelom, 322, 324
Coelomate, 295, 297, 306
Coelophysidae, 484
Coelophysis, 480
Coelophysoidea, 480
Coelurosauria, 480, 484, 494, 502, 505
Coenocyte, 255
Coenocytic, 194
Coenzyme A, 136
Coevolution, 139, 310–312
Cofactor (biochemistry), 136, 181
Cognition, 605
Cohesion-tension theory, 390
Cold War, 35
Coleoptera, 588, 592
Colin Patterson (biologist), 377
Collagen, 308, 500
Collodictyon, 199
Colobus guereza, 61
Colonial organism, 176, 253
Colonisation of land, 313
Colonization of land, 41
Colonization of the land, 312
Colony (biology), 253, 285, 326, 334
Color vision, 472
Colosteidae, 441
Colour, 56
Colugo, 539
Colugos, 537
Columbia (supercontinent), 3
Columbia University Press, 643
Comet, 145, 151
Comets, 141–143
Commelinales, 560
Commelinids, 560
Commensal, 86
Common ancestor, 58
Common basilisk, 502
Common chimpanzee, 32, 53, 185, 623
Common descent, 47, **107**, 114, 117, 131, 223, 602
Commons:Category:Aves fossils, 508
Commons:Category:Gymnosperms, 581
Commons:Category:Human evolution, 648
Commons:Category:Last universal ancestor, 137
Commons:Category:Magnoliophyta, 575
Commons:Category:Tetrapoda, 454
Commons:Homology, 131
Communication, 420
Community (ecology), 284
Companion plant, 565
Comparative advantage, 167
Comparative anatomy, 56
Competition (biology), 258, 565

Complementation (genetics), 225, 228, 232, 387
Complexity, 549
Complex life, 3
Complex organism, 245
Complex system, 104
Complex traits, 82, 312
Compound eye, 62
Compound leaf, 123
Compound leaves, 399
Compsognathidae, 495
Compsognathus, 480, 489, 495
Computational phylogenetics, 49
Computed tomography, 497
Computer, 35
Computer science, 104
Computer simulation, 12
Concealed ovulation, 606
Concentration, 307, 311
Concretion, 265
Confuciusornis, 481, 485, 504, 506
Conifer, 191, 409, 561, 563, 578
Conifer cone, 576
Coniferous forest, 683
Conifers, 548
Conodont, 333, 336, 343, 354
Conodonta, 330
Conodonts, 341, 344
Conservative mutation, 128
Conserved sequence, 113, 128
Consilience, 49
Consolea, 60
Constable & Robinson, 647
Continent, 73, 290
Continental collision, 21
Continental crust, 9
Continental drift, 74, 284, 525
Continental shelf, 26, 660, 662
Control of fire by early humans, 603
Convergent evolution, 110, 112, 118, 256, 293, 319, 355, 422, 489, 527, 578
Convolvulaceae, 568
Convolvulus, 568
Cooksonia, 388, 391
Copenhagen University, 149
Coprolite, 291
Copulation (zoology), 227
Copyright status of work by the U.S. government, 215
Coracoid, 526, 529
Coracoid process, 529
Coral, 24, 358
Coral reef, 359
Coral snake, 427
Core angiosperms, 548
Core eudicots, 560

Lagomorph, 74
Lagomorpha, 537
Lake Baikal, 76
Lake Louise (Alberta), 292
Lake Thetis, 16
Lake Toba, 627
Lake Turkana, 609, 619
Lake Vostok, 155
Lama (genus), 80
Lamellibrachia, 400
Lamiaceae, 568
Lamiales, 560
Lamprey, 325, 336, 340, 343, 354, 356, 458
Lampreys, 317
Lancelet, 319, 322, 327, 339
Land animal, 649
Land Grove Quarry, Mitcheldean, 375
Landline (TV series), 285
Land plant, 254
Land plants, 192
Language, 601
Language acquisition, 605
Lanugo, 57
Large igneous province, 658
Larry Martin, 493
Larva, 326
Larvacea, 333
Larynx, 66
Lasioglossum aeneiventre, 587
Last Common Ancestor, 458
Last glacial period, 638
Lastovo, 86
Last universal ancestor, 15
Last Universal Common Ancestor, 39, 47, 107, **131**, 139
Late Cretaceous, 601
Late Devonian, 352, 356, 361, 425
Late Devonian extinction, 27, 28, 337, 348, 355, 356, 359, 361, 399, 433, 434, 463, 651, 654, 658–661
Late Heavy Bombardment, 6, 137
Late Jurassic, 371, 487
Late Miocene, 616
Latent heat, 396
Late Ordovician, 359
Late Pleistocene, 632
Lateral gene transfer, 15, 190, 245
Lateral line, 452
Late Silurian, 382
Latimeria, 352
Latin, 608
Latitude, 73
Lau event, 654, 660
Lauraceae, 568
Laurales, 559
Laurasia, 3, 25, 462, 465, 536

Laurasiatheria, 539
Laurentia, 21
Laurus, 568
Laurussia, 25
Laws of science, 111
Leaf, 122, 393, 394, 397, 410, 578
Leaf trace, 396
Leaf vasculature, 396
LEAFY, 411
League of Nations, 35
Leaves, 393, 395
Lechuguilla Cave, 101
Leedsichthys, 370
Legless lizard, 61
Leishmania, 196
Lemon, 572
Lemur, 77, 537, 621
Lemurs, 76
Lens (anatomy), 453
Leonardo da Vinci, 34
Leopard, 73
Lepidodendrales, 396, 403
Lepidodendron, 402
Lepidodinium viride, 61
Lepidosauria, 426, 436, 473
Lepidosauromorpha, 333
Lepospondyl, 434, 442
Lepospondyli, 445, 470
Leptocardii, 322, 329
Leslie C. Aiello, 641
Leslie Orgel, 146
Lesothosaurus, 62, 476, 484
Letter to the editor, 151
Leucine, 110
Leucoplast, 216, 217
Leucoplasts, 60
Levallois technique, 638
Levant, 615
Levels of selection, 245
Lewis Carroll, 230
Liaoconodon, 533
Liaoning Province, 529
Library, 34
Library of Congress Control Number, 113, 114, 641–648
Lichen, 157, 256, 275, 276, 400
Lichens, 262
Life, 2, 39, 41, 42, 47, 107, 131, 141, 142, 150, 154, 655
Life history, 386
Life history theory, 85
Lift (force), 457
Light, 397
Light harvesting complex, 220
Lightning, 12
Light skin, 639

Oxford Museum, 288
Oxford University Press, 644
Oxidation, 181
Oxidative stress, 182
Oxidized, 182
Oxlestes, 535
Oxygen, 1, 6, 53, 164, 277, 307
Oxygenate, 149
Oxygen catastrophe, 18
Oxygenic photosynthesis, 219
Ozone, 17
Ozone layer, 3, 6, 181, 308, 661

Pachycephalosauria, 483
Pachycormiformes, 369, 371
Pachyrhachis problematicus, 127, 128
Paddlefish, 353, 373
Paedomorphic, 623
Paenungulata, 537, 539
PAH world hypothesis, 150
Pair bond, 606
Palaeolatitude, 284
Palaeonisciformes, 368
Palaeontologist, 263
Palaeontologists, 426, 456
Palaeoproterozoic, 256
Palaeoscolecid, 316
Palaeospondylus, 322
Palaeoworld, 314
Palaeozoic, 264
Palate, 359, 446, 510
Paleoanthropology, 609, 610, 612, 628
Paleoarchean, 131, 134
Paleoatmosphere, 179
Paleocene, 23, 472, 537, 539, 541, 601
Paleocene–Eocene Thermal Maximum, 659, 660
Paleogene, 23, 321, 381, 425, 509, 547, 575, 652, 653
Paleoichthyologist, 366
Paleolithic, 636, 638
Paleomagnetism, 18
Paleontological Society, 641
Paleontologist, 359, 361
Paleontologists, 535
Paleontology, 68, 290, 354, 369, 376, 455, 469, 493, 496, 511, 565, 601
Paleopolyploidy, 561
Paleosol, 275
Paleozoic, 23, 74, 340, 343, 395, 542
Palmaris longus, 61
Palmitic acid, 215
Palm (plant), 123, 568
Pambdelurion whittingtoni, 316
Pamela Soltis, 574
Pamela S. Soltis, 573

Pan-African orogeny, 21
Pandanales, 560
Panderichthys, 363, 365, 366, 441, 456, 461, 466
Pangaea, 3, 23, 24, 367, 486, 540, 662
Panina, 601
Pannotia, 3, 21
Pan paniscus, 623
Panspermia, **141**, 161
Pantherinae, 55
Paper, 572
Paraceratherium, 30
Parachute, 501
Paracryphiales, 560
Paralog, 414
Paraneoptera, 592
Paranthropus, 32, 617, 624, 625, 641
Paranthropus aethiopicus, 624, 641
Paranthropus boisei, 624, 641
Paranthropus robustus, 619, 624, 641
Paraphyletic, 323, 330, 341, 343, 356, 383, 554, 575
Paraphyly, 337
Paraphysomonas, 60
Parapithecus, 621
Parasite, 19, 230
Parasitism, 218
Paraves, 494, 507
Parchment, 167
Parenchyma, 295, 390
Parietal eye, 543
Parsimony analysis, 119
Parsley, 568, 572
Parthenocarpic, 238
Parthenogenesis, 97, 223, 231, 248, 571
Parthenogenetic, 247
Parthenogenetically, 588
Partial melting, 9
Partial pressure, 10
Particulates, 658
Parvancorina, 271, 299
Passeriformes, 61
Patagonia, 525
Patent, 175
Patricia Adair Gowaty, 671
Paul Baltes, 643
Paul Davies, 37
Paul E. Olsen, 377
Paul Hermann (botanist), 552
Pax6, 120
PBS Digital Studios, 688
Pea, 399, 411
Pear, 572
Pearl millet, 572
Peat, 69
Pectin, 192, 194

Pectoral fin, 351, 356, 359, 369, 457
Pectoral girdle, 496, 526
Pederpes, 363, 367
Pedipalps, 121
Pedopenna, 487, 503, 507
Pedosphere, 41
Peer review, 492, 500
Pegasoferae, 539
Peking Man, 628
Pelagic zone, 456
Pelomyxa, 189
Pelvic fin, 351, 369
Pelvic girdle, 60
Pelvis, 33, 65, 448, 527
Pelycodus, 93
Pelycosaur, 514, 528
Pemphigus spyrothecae, 588
Penguin, 427
Penguin Books, 647
Penguin Press, 647
Penguins, 74, 472
Penicillium, 152
Penicillium roqueforti, 156
Penis, 66
Pennaraptora, 507
Pennsylvanian (geology), 434, 470
Penstemon centranthifolius, 60
Pentadactyl limb, 65
Peoples Republic of China, 359
Peppered moth evolution, 82, 103
Peptide, 110
Peptide nucleic acid, 13
Peptide-RNA world, 13
Peptides, 13, 500
Peptidoglycan, 194, 218
Percolozoa, 206
Peregrine falcon, 322, 330
Perfume, 579
Pericarp, 570
Period (geology), 3, 69, 266
Perissodactyla, 537, 539
Peritidal zone, 168
Peritoneum, 324
Perleidus, 369
Permafrost, 22
Permian, 79, 321, 369, 403, 408, 425, 435, 471, 510, 514, 517, 547, 562, 575, 651, 654
Permian–Triassic extinction event, 23, 313, 337, 428, 435, 471, 651, 654, 658–661, 663, 688
Permian-Triassic event, 367
Permian-Triassic extinction event, 352, 369, 516, 517, 662
Permo-Triassic extinction event, 381
Perodicticus potto, 61

Peroxide, 189
Peroxisome, 189, 200
Peroxisomes, 20
Perturbation (astronomy), 4
Peştera cu Oase, 631
Petal, 56, 127, 551
Petalichthyida, 333, 360
Petalichthyidae, 349
Petalodontiformes, 368
Petals, 412, 413
Peter F. Stevens, 573
Peter Ungar, 644
Petiole (botany), 124, 551
PETM, 397
Petrifaction, 68
Petromyzontida, 330
Petrosaviales, 560
Petunia, 412, 556
Phagocyte, 192
Phalange, 65
Phalanges of the hand, 493
Phalanx bones, 493
Phanerozoic, 3, 23, 170, 185, 253, 264, 471, 650, 656, 657
Pharyngeal arch, 58, 346
Pharyngeal slit, 322, 323
Pharyngula (blog), 148
Pharynx, 323, 325, 334
Phase transition, 181
Phenotype, 415, 418
Phenotypes, 412
Phenotypic trait, 119, 245
Phenylalanine, 110
Pheromone, 597
Pheromones, 587, 595
Philippe Janvier, 342, 377, 675
Philippines, 76
Phloem, 549
Pholidophorus, 370
Pholidota, 537, 539
Phoronida, 317
Phosphorus, 389, 498
Photodissociation, 182
Photo-oxidation of polymers, 219
Photorespiration, 417
Photosynthesis, 2, 16, 53, 146, 163, 167, 169, 171, 176, 178, 186, 190, 194, 215, 217, 219, 250, 274, 277, 307, 393, 417, 654, 661
Photosynthesis and respiration, 171
Photosynthetic, 20, 122, 393, 395
Phototaxis, 93
Phycobilisomes, 220
Phyllanthaceae, 568
Phyllanthus, 568
Phyllotaxy, 398

740

Triassic, 23, 321, 352, 369, 381, 425, 436, 470, 475, 484, 509, 511, 547, 548, 561, 575, 651, 652, 654
Triassic–Jurassic extinction event, 28, 337, 370, 652, 654, 655, 658–660
Triassic-Jurassic extinction event, 29
Tribosphenic molar, 525
Triceratops, 475, 483
Trichomonas, 189
Trichomonas vaginalis, 195
Trichromacy, 525
Tricolpates, 555
Triconodonta, 524
Trigonocarpus, 404
Trilobite, 26, 67, 70, 121, 303, 316, 358, 359, 651
Trilobites, 25, 288
Trilobitomorpha, 121
Trimerophyte, 395
Trinidad, 84
Trinil, 608
Tripeptide, 139
Triploblastic, 295, 307
Trirachodon, 517
Tristichopteridae, 347, 430, 440, 459
Tristichopterus, 440
Tritheledontidae, 521
Tritylodontidae, 518, 521, 542
TRNA, 189, 217
Trochodendrales, 560
Trondhjemite, 9
Troodontidae, 481, 494, 498, 506
Trophic level, 278
Tropical, 413
Tropics, 659
Trucidocynodon, 518
T. Ryan Gregory, 109
Tryptophan, 110
Tsukubamonas globosa, 206
T Tauri star, 4
Tuatara, 76, 333, 436, 473
Tube feet, 334
Tubercle (anatomy), 527
Tubulidentata, 537, 539
Tubulin, 191
Tucson, Arizona, 37
Tugen Hills, 612
Tulerpeton, 468
Tungsten, 137
Tunica (biology), 325
Tunicata, 322
Tunicate, 322, 325, 328, 333, 338
Tupinambis, 543
Turbidite, 268
Turborotalia, 92
Turgor, 194

Turkey, 622
Turkey (bird), 54
Turtle, 370, 426, 436, 470, 473, 511
Tympanal organ, 453
Type (biology), 609
Types of pits, 392
Tyrannosauridae, 485, 502
Tyrannosauroidea, 480, 495, 506
Tyrannosaurus, 362, 481, 500
Tyrosine, 110

UCP1, 494
Ulna, 65, 118, 366, 449, 459
Ultraviolet, 157, 181, 182, 661
Ultraviolet radiation, 26, 142, 155
Ulvophyceae, 254
Umbilical cord, 362
Uncoupling protein, 494
Undulipodia, 191
Ungulate, 79, 535, 541
Ungulates, 30, 473
Ungulatomorpha, 537
Unicellular organism, 186, 253
Unikont, 199
Unikonts, 202
United Kingdom, 491
United Nations, 35
United States, 35, 609
Universe, 2, 39, 42, 141, 143, 150
University of Arizona, 37
University of California, Berkeley, 199
University of California Press, 644
University of California, Santa Cruz, 329
University of Chicago Press, 642
University of Edinburgh, 41
University of Exeter, 101
University of Florida, 100
University of Freiburg, 101
University of Maryland, 680
University of Naples Federico II, 152
University of Texas at Austin, 643
University of Tokyo Press, 643
Unsupported attributions, 176
Upper Carboniferous, 402
Upper Paleolithic, 638
Uracil, 148, 150
Uralian orogeny, 25
Uranium, 277
Urban wildlife, 87
Ur (continent), 3
Urea, 139, 464, 518
Ureter, 66
Urethra, 66
Uric acid, 518
Uridine, 133
Urochordata, 329

Uronemus, 368
Uropygid, 64
Urticaceae, 568
Utahraptor, 481
Uterus, 529, 532
UV, 144, 158–160
UV radiation, 152, 661

Vaalbara, 3
Vacuole, 188, 192, 194
Vagus nerve, 66
Vahliales, 560
Valentia Island, 433
Valine, 110
Vancomycin, 82
Variscan Orogeny, 25
Varisulca, 204, 206
Varnish, 579
Vascular bundle, 549
Vascular cambium, 381, 402
Vascular plant, 383, 386, 397, 553
Vascular plants, 579
Vascular tissue, 389
Vas deferens, 66
V:Bloom Clock, 412
Vegetation, 72
Vegetative reproduction, 250
Velociraptor, 481, 505
Venation .28arrangement of the veins.29, 393
Vendian, 274
Ventastega, 441
Ventral, 526
Ventral aorta, 458
Venus flytrap, 122, 125
Venus of Willendorf, 33, 636
Vernalization, 413
Vernanimalcula, 271, 297, 306
Verne Grant, 96
Vernon Lyman Kellogg, 109, 667
Vertebra, 324, 469
Vertebrae, 361
Vertebral column, 26, 114, 321–323, 325, 336, 339, 370, 470
Vertebrata, 330, 340
Vertebrate, 3, 26, 70, 117, 322, 323, 325, 333, 336, 338, 340, 346, 354, 373, 426, 456, 469, 494, 513, 598
Vertebrate Palaeontology (Benton), 329, 455
Vertebrate paleontology, 438
Vertebrate Paleontology (Romer), 438
Vertebrates, 66, 114, 354, 457
Vesicle (biology), 188
Vespidae, 586, 592
Vespinae, 586
Vespula vulgaris, 596
Vessel element, 389, 414

Vestigiality, 48, 494
Vestigial structure, 54
Vetulicolia, 322, 329
Victoriapithecus, 621
Vincent Sarich, 610
Vindhya, 207
Violet (plant), 551
Viral Eukaryogenesis, 207
Viridiplantae, 204, 206
Virus, 83, 151
Visean, 470
Vitales, 560
Vitruvian Man, 34
Viviparity, 362
Viviparous, 247
Volaticotherium, 510, 533–535
Volatiles, 10
Volcanic ash, 266
Volcanic eruption, 653
Volcano, 9, 171
Volvocaceae, 257
Volvox, 255, 257
Volvox carteri, 185
Vomeronasal organ, 61
Von Baers laws, 116
Voyager program, 36

Wall to Wall (production company), 648
Walter Hartwig, 645
Walter Max Zimmermann, 396
Walter M. Fitch, 129
Wangiella dermatitidis, 84
Waptia, 280, 319
War, 35
Warm-blooded, 494, 510, 543
Warrawoona, Australia, 288
Washington, D.C., 623
Washington (U.S. state), 99
Wasp, 246, 565, 583, 586
Waste, 146
Wastebasket taxon, 269
Water, 133, 172
Water bird, 427
Water cycle, 659
Water flea, 230
Water gas shift reaction, 138
Water in the development of Earth, 39, 41
Water opossum, 526
Water vapor, 10
Water vascular system, 334
Watt, 634
Wattieza, 402
Wax, 62
W:de:Otto Kandler, 139
W. D. Hamilton, 225, 590
W.D. Hamilton, 593